COMBUSTION
MEASUREMENTS

Combustion: An International Series
Norman Chigier, *Editor*

Chigier, Combustion Measurements
Lefebvre, Atomization and Sprays
Kuznetsov and Sabel'nikov, Turbulence and Combustion

Forthcoming titles
Chigier, Stevenson, and Hirleman, Flow Velocity and Particle Size Measurement
Libby, Introduction to Turbulence

COMBUSTION MEASUREMENTS

Edited by

Norman Chigier

Department of Mechanical Engineering
Carnegie Mellon University
Pittsburgh, Pennsylvania

●HEMISPHERE PUBLISHING CORPORATION
A member of the Taylor & Francis Group
New York   Washington   Philadelphia   London

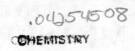
COMBUSTION MEASUREMENTS

1 2 3 4 5 6 7 8 9 0   E B E B   9 8 7 6 5 4 3 2 1

This book was set in Times Roman by Harper Graphics. The editors were Dave Weber and Carol Taylor Edwards.
Cover design by Sharon M. DePass.
Printing and binding by Edwards Brothers, Inc.

A CIP catalog record for this book is available from the British Library.

Library of Congress Cataloging-in-Publication Data

Combustion measurements / edited by Norman Chigier.
        p.   cm -- (Combustion)
    Includes bibliographical references and index.

    1. Combustion—Measurement.   I. Chigier, N. A.   II. Series:
Combustion (Hemisphere Publishing Corporation)
QD516.C6154   1991
621.402'3—dc20   89-40823
                                                              CIP

ISBN 1-56032-028-1
ISSN 1040-2756

CONTENTS

CONTRIBUTORS

W. D. Bachalo
Aerometrics, Inc.
894 Ross Drive, Suite 105
Sunnyvale, CA 94086

D. R. Ballal
Research Institute
University of Dayton
Dayton, OH 45469

F. Beretta
Istituto di Ricerche sulla Combustione
CNR, Naples
Italy

P. E. Best
Physics Department and Institute of
    Materials Science
University of Connecticut
Storrs, CT 06268

A. Breña de la Rosa
Aerometrics, Inc.
894 Ross Drive, Suite 105
Sunnyvale, CA 94086

J. Cao
Department of Mechanical and Process
    Engineering
Chemical Engineering and Fuel
    Technology
University of Sheffield
Mappin Street
Sheffield S1 3JD
England

A. Cavaliere
Dipartimento di Ingegneria Chimica
Università di Pisa
Pisa
Italy

Norman Chigier
Department of Mechanical
    Engineering
Carnegie Mellon University
Pittsburgh, PA 15213

A. D'Alessio
Dipartimento di Ingegneria Chimica
Università Degli Studi Di Napoli
Piazzale V. Tecchio
80125 Naples
Italy

Richard C. Flagan
Department of Environmental
    Engineering Science
California Institute of Technology
Pasadena, CA 91125

G. Gouesbet
Laboratoire d'Energétique des
    Systèmes et Procédés
INSA de Rouen - BP8
76131 Mont-Saint-Aignan Cedex
France

G. Gréhan
Laboratoire d'Energétique des
    Systèmes et Procédés
INSA de Rouen - BP8
76131 Mont-Saint-Aignan Cedex
France

A. A. Hamidi
Department of Mechanical and Process
    Engineering
Chemical Engineering and Fuel
    Technology
University of Sheffield
Mappin Street
Sheffield S1 3JD
England

Donald J. Holve
Insitec
2110 Omega Road, Suite F
San Ramon, CA 94583

Normand M. Laurendeau
School of Mechanical Engineering
Purdue University
West Lafayette, IN 47907

B. Maheu
Laboratoire d'Energétique des
    Systèmes et Procédés
INSA de Rouen - BP8
76131 Mont-Saint-Aignan Cedex
France

Patricia L. Meyer
Insitec
2110 Omega Road, Suite F
San Ramon, CA 94583

Ranajit Sahu
Department of Environmental
    Engineering Science
California Institute of Technology
Pasadena, CA 91125

S. V. Sankar
Aerometrics, Inc.
894 Ross Drive, Suite 105
Sunnyvale, CA 94086

P. R. Solomon
Advanced Fuel Research
P.O. Box 18343
East Hartford, CT 06118

J. Swithenbank
Department of Mechanical and Process
    Engineering
Chemical Engineering and Fuel
    Technology
University of Sheffield
Mappin Street
Sheffield S1 3JD
England

J. D. Trolinger
Metro Laser
18006 Skypark Boulevard, Suite 108
Irvine, CA 92714-6428

Measurements in combustion systems are required for purposes of analysis and control. Flow rates, temperatures, and concentration of fluids and materials introduced into the system and leaving the system need to be continuously monitored. This information is fed directly into the control system. Temperatures of the walls of the combustor and of gases flowing through the combustor also need to be monitored to prevent damage to inner walls and surfaces of the combustor.

Major advances have been made in the development of instrumentation for making measurements of velocity, temperature, species concentrations, and particle size inside flames. High-powered lasers have replaced water-cooled probes for measurements in both single-phase and multiphase turbulent reacting flow systems. Laser anemometry for measurement of velocity is well established. Velocity and size of particles can be measured simultaneously by measurement of phase and frequency. Use of laser sheet lighting and holography has improved visualization of flow fields. A number of methods have proved to be effective for temperature measurement based on light scattering, Raman spectroscopy, fluorescence, and coherent anti-Stokes Raman spectroscopy. These techniques also yield information on species concentrations. Joint measurements of temperature, species concentration, and velocity have been made in flames by combining laser anemometry and Raman spectroscopy.

The measurement of liquid and solid particle size has made rapid advances in recent years. Several diagnostic techniques have been proposed, and some of these instruments are commercially available. They are generally based upon the scattering of light by single particles. The theory and practice of using these light scattering techniques for particle sizing are presented in this book. Fourier transform infrared spectrometry is an alternative technique that is showing considerable promise for detailed analysis of gases. Instrumentation for the analysis of coal particles is discussed and, finally, the special difficulties involved in solid propellant combustion measurements. Burning rate measurements, combustion wave structure, pressure, chemical decomposition, and ignition require special instrumentation because of the harsh environment in solid propellant rockets.

Chapter 1 is an introduction to the book. The general problems of making measurements in high temperature, chemically reacting flow systems are presented. Each measuring instrument is described with the various diagnostic techniques. Measurements that have been made in furnaces, gas turbine combustors, flames, and rocket engines are discussed.

Chapter 2 covers velocity measurements by laser anemometer in single-phase and two-phase combustion systems using frequency shifting, forward scatter and backscatter collection, signal processing, and data analysis by computer. Instantaneous measurements are made of three orthogonal velocity components. Conditional sampling is used to determine form intermittency. Comparisons of mean drop sizes, number densities, and velocities under burning and nonburning conditions show the extent that flow fields are modified by combustion.

Chapter 3 shows how particle and flow field optical holography is used to record the interference patterns of scattered and unscattered waves. Holography is used under dynamic conditions to record high-resolution optical conditions over a large volume in a short time. The hologram stores all optical information, including phase and wave fronts. In holographic interferometry, two mutually coherent waves produce interference fringes that are used to determine the phase distribution. In particle field holography, dynamic fields are optically frozen to provide a three-dimensional high-resolution image. Phase shift interferometry and heterodyne interferometry make automated data reduction easier.

Chapter 4 describes temperature measurements by light-scattering methods using spontaneous vibrational and rotational Raman scattering, Rayleigh scattering, and laser-induced fluorescence. Fluorescence provides intensity and spectral selectivity in flows moderately laden with particles where there is interference due to blackbody radiation, flame emission, or laser-induced incandescence. Combinations of two-dimensional Rayleigh thermometry with planar laser-induced fluorescence measurements of C_2 radical are used to identify the location of the flame front in turbulent nonpremixed flames.

Chapter 5 shows the integration of laser Raman spectroscopy with laser Doppler anemometry. For each laser pulse, concentration of the individual major species (H_2, N_2, O_2, and H_2O) is determined from respective Stokes vibrational Raman intensities. The temperature is determined from the ratio of anti-Stokes-to-Stokes N_2 vibrational scattering intensities. Temperature is calculated from the total number density. Raman and laser anemometer beams are brought on focus in the same location. Joint probability density functions and correlations have been measured between velocity and concentration as well as autocorrelations and power spectral density functions of temperature fluctuations.

Chapter 6 examines a wide range of particle size measurement techniques including photography and holography with automatic scanning and image analysis. Particle-sizing interferometry is based on visibility measurements using a laser anemometer. Fraunhofer diffraction is used to measure particle size distribution and concentration in low particle density systems. Tomographic techniques are used to calculate particle size distribution variations by scanning across the flow field.

Chapter 7 describes the phase Doppler particle analyzer developed for detailed diagnostics of liquid fuel sprays. Using a combination of laser anemometry and phase shift, the velocity and size of individual particles passing through the measurement control volume are measured simultaneously. Number density, volume flux, angle of trajectory, and time-resolved information on particles are determined.

Chapter 8 deals with instruments specially designed for particle size measurements in large-scale combustion environments that will lead to reduction of slagging, fouling, and erosion of surfaces. Measurements of particle size, concentration, and velocity distributions inside furnaces are made using light-scattering techniques. Measurements are made of liquid, solid, or slurry particles with regular or irregular shapes and with wide ranges of refractive index and surface reflectivity.

Chapter 9 deals with the determination of the number concentration and average size of soot particles. Scattering and extinction measurements at different wavelengths reveal the structural properties of soot particles. Polarization ratio measurements allow a clear discrimination between soot and liquid fuel droplets.

Chapter 10 presents a generalized Lorenz-Mie theory for computation of the properties of light scattered by a Mie scatter center illuminated by a nonplane wave. The Maxwell equations are solved, accounting for the specific boundary conditions using the Bromwich formulation, which allows solutions satisfying the boundary conditions in special curvilinear coordinate systems.

Chapter 11 discusses Fourier transform infrared spectroscopy for measurement of gas, particle, and soot temperatures and concentration in flames. Separate temperature and concentration for individual gas species as well as for solid particles are determined by employing different regions of the infrared spectrum. Measurements of particle size can be made in densely loaded streams. The instrument has the capability of separating radioactive contributions from soot and from char particles.

Chapter 12 provides a survey of measurement techniques used in the study of coal and char combustion. Emphasis is placed on reactivity measurements, mineral matter transformations, the measurement of particle temperatures, and particle structure characterization. Thermogravimetric analyzers are used for studies of pyrolysis and char oxidation reactions. The electrodynamic balance is used to suspend and weigh single charged particles in an electric field. The electrodynamic thermogravimetric analyzer uses laser heating to study particles at temperatures at which reaction occurs.

The detailed measurement techniques described in this book cover a wide spectrum of applications in combustion systems, including gas turbine, rocket, and

Measurement techniques that were developed in laboratories are being used in high-temperature, reacting, particle-laden flow systems. Information obtained on detailed temperature, velocity, particle size, and gas concentration distribution is leading to improve understanding of the chemical combustion process and to design improvements in combustors.

Norman Chigier

MEASUREMENTS IN COMBUSTION SYSTEMS

Norman Chigier

1-1 INTRODUCTION

Before the advent of the laser, measurements in flames were made by inserting probes such as pitot tubes for velocity, suction pyrometers for temperature, and suction probes for gas concentration measurements. Particles were collected on filters inside suction probes for subsequent removal and size analysis. In the past two decades, important developments have taken place in combustion diagnostic techniques. There has been a large-scale increase in the quantity of research that has been carried out in university, government, and industrial laboratories, resulting in a much better understanding of fundamental processes in turbulent, reacting, high-temperature systems. Instead of relying on global measurements, instruments have been developed that probe into flames and combustion environments, allowing measurements to be made as a function of both space and time. The presence of particles of solid fuel, liquid drops, and soot has created special difficulties in making measurements of temperature and concentration. Interest developed in obtaining more information about these particles and led to a number of new concepts for particle size and concentration measurements. At the same time, alternative techniques were developed, allowing measurements to be made of temperature in heavily laden, radiating, particle flow fields.

Computer models to predict turbulent mixing and combustion flow fields are being developed and are becoming more sophisticated and reliable. Computational fluid dynamics uses the fundamental equations of conservation of mass, momentum, energy, and individual species, but models are required in order to obtain closure and solution to these equations. Detailed measurements of velocity, temperature, species concentration, particle size and velocity, and number density are required for formulation and verification of computer models. There is particular interest in accurate determination of initial and boundary conditions and providing information on distributions as a function of both space and time. The new developments in measurement techniques and the increasing sophistication and quantity of these measurements are resulting in improved understanding of fundamental mechanisms as well as more accurate predictions.

1-2 VELOCITY

Measurements of the distribution of gas velocity provide the information for determination of the flow field. Streamlines determined from velocity measurements show recirculation and reverse flow zones as well as regions of acceleration and deceleration. Velocity and streamline distributions are the foundation for computational schemes. Information on flow patterns can help designers identify locations for excessive erosion and particle deposition. Changes in design geometry can frequently be assessed by comparing flow patterns. Information on turbulence is obtained almost exclusively from velocity measurements. Three components of turbulence intensity, shear stresses, and kinetic energy of turbulence can be measured. Turbulent length scales and eddy sizes can be derived. Conditional sampling is being used to identify coherent structures and periodic events. Separate measurements are being made of particle and gas velocities in high-temperature flows.

Laser anemometry (see Chapter 2) is used for measurement of local instantaneous velocity of particles suspended in the flow. In the dual-beam fringe anemometer, two light beams of equal intensity are focused to intersect, producing a pattern of successive light and dark fringes. When a particle crosses these fringes, a periodic variation in intensity of the scattered light is generated as a Doppler burst. The measured frequency difference is directly proportional to the instantaneous component of velocity of the particle lying in the plane of the beams. For measurements in recirculating and high turbulence intensity flows, Bragg cells or rotating diffraction gratings are used for frequency shifting.

The measured intensity of scattered light is a complex function of scattering angle, beam angle, particle size and shape, particle material, and ratio of fringe spacing to particle diameter. Frequency counters use high-frequency clocks to measure the time durations for a number of zero crossings, and transient recorders are very fast analog-to-digital converters able to fully digitize a Doppler burst using fast Fourier transform instrumentation. Aluminum oxide particles, 0.5 mm in diameter, are used for seeding with particle densities of the order of 10^{10} particles/m^3. Data rates are of the order of 10,000/s. Power spectral and autocorrelations are obtained from continuous records of velocity versus time using high seeding levels.

Three-color laser Doppler velocimeter (LDV) systems allow simultaneous measurement of three velocity components. Independent selectable frequency shifting is provided for each component, so that three-dimensional flows with swirl, recirculation, large fluctuations, and high turbulence intensities can be readily measured. Receiving optics modules may be placed in on-axis forward scatter or in off-axis backscatter. Fiber optics are being used for measurements in small or restricted spaces and in complex geometric configurations. Fiber links separate the optics from the laser and the system electronics, which can be protected from the harsh environment.

Menu-driven software packages are used to process raw data to provide individual velocity components and their projections in three orthogonal planes with statistical properties including mean, turbulence intensity, skewness, and flatness of velocity components. Correlations, Reynolds stresses, velocity, and flow angle histograms and vectors expressed in terms of magnitude and angle are displayed or printed in tabular or graphic form. Software packages are also available to control the automatic traversing of the flow or instrument system with three axes of translation. These software packages are greatly increasing the accuracy of measurements and the ease and convenience of making LDV measurements.

Comparison of measurements made in noncombusting and combusting jet flows shows significant differences in mean velocity and kinetic energy of turbulence. Detailed measurements of axial and radial distribution in combusting and noncombusting flows are made of mean axial velocity, turbulent kinetic energy, temperature, normal turbulent intensities, shear stress, relative axial turbulent intensity, probability density functions, power spectra, and turbulent macroscale.

Detailed measurements of axial and azimuthal mean and rms velocity have been made with a two-color LDV in a swirl-stabilized model combustor, which has the important features of practical gas turbine combustors. Comparison of measurements made in reacting and nonreacting flows showed that, in the case of reaction, the recirculation zone was stronger, more compact, and radially wider than for nonreacting flows. The rms axial velocities and normalized Reynolds stresses were substantially higher as a result of combustion.

Heat release accompanying combustion results in acceleration of the flow. In shear flows with combustion, the mean shear is enhanced by the heat release. Reynolds stresses, turbulence levels, and production of turbulence kinetic energy by the mean shear are all increased.

Simultaneous measurement of velocity, temperature, and species concentration allows direct determination of correlation terms in the energy and species concentration conservation equations. Scalar measurements can be made jointly with LDV using Mie scattering, laser Rayleigh scattering, laser-induced fluorescence, spontaneous Raman scattering, and coherent anti-Stokes Raman scattering.

Measurements of velocity, drop size distribution, drop velocity distribution, drop number density, and liquid flux are made in air-assisted spray flames with the phase Doppler laser anemometer. Comparisons between burning and nonburning sprays show reduction in drop size and number density as a result of evaporation and changes in shape of the drop size distribution. The presence of the flame did not affect the shape of the mean axial velocity distribution.

1-3 HOLOGRAPHY

Holography, using pulsed lasers with a pulse duration of several nanoseconds, allows the freezing of events in three dimensions. Reconstruction of the hologram allows the recorded events to be analyzed. Measurements of particle size and shape can be readily made at a large number of planes from a single hologram. Velocities and trajectories of particles can be measured by using multiple-exposure holography. New developments in automatic scanning image analyzers are allowing more rapid data acquisition and analysis.

In-line holograms (see Chapter 3) are made by passing a plane or spherical wave front through the field and recording the interference pattern of the scattered and unscattered waves. The diffraction process can be caused by amplitude transmission variations, phase shifting, or pathlength difference on reflection. The process can occur in a single plane, in a volume, on transmission, or on reflection. Holography is essentially an information buffer memory between the experiment and data analysis. Holography is used in dynamic cases where the requirement is to record high-resolution optical information over a large volume in a short time. A hologram retains all optical information including phase. Wave front information can be stored. In experiments where time is short or extremely expensive, data reduction can be done after the experiment. Holocameras are designed to use pulsed ruby, neodymuim:yttruim/aluminum/garnet (Nd:YAG), and dye lasers for recording and a collinear HeNe CW laser for alignment and reconstruction.

In holographic interferometry, two mutually coherent waves produce interference fringes that are used to determine the phase distribution of one of the waves relative to the other. Several special cases of holographic interferometry are considered: double exposure, double reference wave, double plate, tilt in both plates during reconstruction, real time, single-exposure live reference wave, local reference wave, and glint.

In particle field holography, dynamic fields are optically frozen to provide a three-dimensional high-resolution image that can be studied microscopically. Multiple-exposure holography allows study of the dynamics of the field. By using holographic subtraction, the signal-to-noise ratio in a reconstructed image is improved. Image analyses augment data extraction by scanning the image volume while counting and sizing individual particles. The Fourier transform of a scattered light field yields instantaneous extraction of the size distribution. Image analyzers can be programed to average many data frames to improve picture quality and signal-to-noise ratio. Phase shift interferometry and heterodyne interferometry offer promise for making automated data reduction easier.

1-4 TEMPERATURE

Temperature measurements provide important information on heat release as a result of chemical reaction. Heat transfer by radiation, convection, and conduction is temperature dependent. Knowledge of temperature is required to avoid burning of and damage to surfaces. Calculations of thermodynamic processes, development of thrust, and transfer of heat require information on the variation of temperature with space and time. Computer predictions are based on mean temperature for calculation of density and the energy equation. Determination of correlation of temperature with velocity and concentration is needed to complete the turbulence information. Previous measurements by thermocouples and suction pyrometers were mainly restricted to time-averaged "point" measurements of temperature.

Temperature measurements (see Chapter 4) are now made by light-scattering methods using spontaneous vibrational and rotational Raman scattering, Rayleigh scattering, and laser-induced fluorescence. These methods provide capability for nonintrusive, in situ measurements with spatial resolution of less than 0.1 mm^3 and temporal resolution of the order of 10 ns. Precise temperatures are measured in laminar flows by averaging over many laser shots, and accurate probability density functions are obtained in turbulent flows by compiling single-shot data at one spatial location. Raman and Rayleigh scattering measurements require high species concentration and clean laboratory conditions, whereas fluorescence provides intensity and spectral selectivity in flows moderately laden with particles where there is interference due to blackbody radiation, flame emission, or laser-induced incandescence.

For laminar flows, temperatures can be measured to within 2% of thermocouple measurements. For turbulent systems the Stokes/anti-Stokes method with a pulsed laser and gated detection yields temporally resolved temperatures with a precision of 4%. For laminar systems, slow excitation or fluorescence scans can be used to determine the rotational temperature of molecular species. For excitation scans the spectral bandwidth of the detection is fixed, while the excitation wavelength is varied. For fluorescence scans the excitation frequency is fixed, and the grating of the monochromter is rotated to determine the spectral distribution of the emissive signal. In two-line fluorescence thermometry, a pair of excitation wavelengths is used to generate two broadband fluorescence signals. Two-line molecular fluorescence is employed to monitor OH flame temperatures and to measure ambient temperatures by single- and two-photon excited fluorescence from NO. The dynamic range and sensitivity of two-line fluorescence methods can be enhanced by employing two-line laser-saturated fluorescence for OH thermometry.

In thermally assisted fluorescence, the laser-induced populations of those energy levels higher than the laser-excited level are presumed to be collisionally equilibrated so that an electronic, vibrational, or rotational temperature can be extracted from the resulting fluorescent spectrum. Planar thermometry is achieved by sheet illumination with right-angle detection of scattered light using a two-dimensional array detector with a microchannel plate image intensifier. A combination of two-dimensional Rayleigh thermometry with planar laser-induced fluorescence measurements of the C_2 radical is used to identify the location of the flame front in turbulent nonpremixed flames.

1-5 JOINT MEASUREMENTS OF VELOCITY, TEMPERATURE, AND SPECIES CONCENTRATION

In turbulent reacting flows, velocity, temperature, and species concentration fluctuate. The variation of these quantities with time is related. The basic flow fields, movements of eddies, and large flow structures result in variation of each quantity, but the variations of each quantity are not necessarily the same. The local fluxes of momentum, heat, and species are directly related to the time-averaged correlation of instantaneous values of velocity, temperature, and concentration. By using a pulsed laser, it has only fairly recently become possible to make simultaneous measurements of velocity-temperature, velocity-concentration correlations and joint probability density functions (pdfs) in turbulent combustion flows.

Joint measurements of velocity, temperature, and species concentration are being made in flames (see Chapter 5) using joint laser Raman spectroscopy and anemometry. For the understanding of turbulent combustion, time-resolved measurements of velocity, temperature, and concentration are necessary. Simultaneous measurements are required of velocity-scalar correlations such as $\overline{uc}$, $\overline{vc}$, $\overline{uT}$, $\overline{vT}$, and the joint probability density function $P(u, \phi, x)$. To perform such measurements in turbulent combustion flows, both Raman spectroscopy and laser anemometry systems have to be spatially and temporally integrated.

Single-pulse vibrational Raman spectroscopy (VRS) has been used for time-resolved measurements of scalar pdfs. Concentrations of individual major species (H_2, N_2, O_2, H_2O) are determined from their respective Stokes vibrational Raman intensities. Temperature is determined from the ratio of anti-Stokes-to-Stokes N_2 vibrational Raman-scattering intensities. Due to the inherent weakness of the observed signal, the VRS does not permit measurement of temporal power spectral density functions or autocorrelation functions required to compute temperature or concentration scales in turbulent reactive flows.

In rotational Raman spectroscopy (RRS), a high-power CW laser and a multipass cell optics configuration are used to detect pure rotational Raman transitions. This combination provides signals 1000 times stronger than those observed from the N_2 Q-branch using a 1-W CW laser. Simultaneous measurement is made of Rayleigh intensity to determine either the concentration of two gases in cold flow or the temperature and concentration fluctuations of a single-gas species, such as N_2, in a flame. The laser Doppler anemometer (LDA) system used for integration with the RRS system is a two-component, real fringe system. The two LDA measurement channels are separated by polarization, and the optical train provides three parallel beams. The Raman interface unit (RIU) provides a variety of joint measurement schemes to detect CH-stretch Raman vibrational bands of the fuel species or an oxygen Raman rotational line. It also allows laser-induced fluorescence measurements and background signal measurements to be made.

Mass conservation balances carried out by integrating velocity and mass function profiles across the flow field show agreement to within 5%, and conservation of momentum is verified to within 5%. Raman measurements have been found accurate to 7% at 2000 K and 1 mole % for species mole fraction.

Combined Raman-LDA determinations of correlations, joint pdfs, and spectra in combusting flows are providing spatial and temporal resolutions as well as accuracy that are acceptable. Both Rayleigh and Raman techniques are better understood and simpler to use than the more sophisticated coherent anti-Stokes Raman scattering (CARS), inverse Raman scattering, or planar-imaging techniques. Sources of error such as Raman signal contamination due to background flame luminosity, interference from LDA seed, and soot particle Mie scattering, uncertainties due to Poisson statistics, beam steering, beam defocusing, and LDA particle bias errors have been carefully analyzed and corrected.

Measurements of joint pdfs, velocity-scalar correlations, scalar power spectra, and scalar length scales have been made in variable density nonreactive round jets, turbulent premixed flames, and turbulent diffusion flames. Measurements reveal the mixing process between jet fluid and outer airstream. Combustion measurements in turbulent premixed flames demonstrate the wrinkled, wavy nature of the flame front and the relatively high scalar dissipation rates associated with practical thick flames compared with laboratory thin flames. Measurements have been made of conserved scalar fluxes and mass fluxes in turbulent diffusion flames. Measured data demonstrate the interaction between mean gradients of the scalar and fluctuating velocity field, effects of buoyancy, countergradient diffusion at selected locations in a flame, and gradient transport at other locations.

1-6 PARTICLE SIZING

Liquid fuel is atomized during injection into combustion chambers. Droplets in the sprays penetrate into the gas flow fields, and the largest drops can have lifetimes sufficient to result in impingement on surfaces. The trajectories of droplets and their rates of evaporation determine the local air-fuel mixture ratios, which in turn determine ignition and local heat release. Atomizers are designed to provide a spectrum of drop sizes and specific spray angles. Several different techniques have been developed for measurement of particle size (see Chapter 6).

Photographic and holographic methods are used to determine particle size distributions by counting and sizing the diameter of particles from photographs and holograms. Image analysis is used for automatic scanning of negatives or video images. Resolution for drop size is approximately 5 μm. Holograms provide three-dimensional pictures, which are subsequently analyzed similar to photography. Q-switched pulse lasers with light pulse durations of several nanoseconds succeed in freezing the motion of particles. Velocities are determined by using double-pulse holography.

Particle-sizing interferometry is an extension of LDA. Using a two-color LDA for validation, visibility from Doppler signals is used for particle size measurement. Only particles that pass through the central portion of the beam, where light intensity is uniform, are measured. Alternatively, all particles crossing the measurement volume are recorded, and subsequently, corrections are made based on the distribution of particle trajectories and corresponding incident intensities, assuming that all possible trajectories through the volume are equally probable.

For measurement of particle size distribution and concentration in low particle density systems, a portable Malvern particle sizer has been developed for measurements in the range of 1–30 μm at concentrations down to 20 mg/m^3. A 20-mW solid state laser increases the detector/amplifier sensitivity. A light baffle is used to block outside light from entering the measurement volume.

The conventional Malvern particle sizer can be used for the measurement of concentration as well as the particle size distribution for obscurations between 5% and

50%. Beer-Lambert's law is used, based on measurements of light intensity measured by the central diode, light path length, mean drop diameter, extinction cross-section efficiency, and drop volume distribution.

For dense particle fields with high obscuration, a mathematical model has been constructed to simulate the effect of multiple scattering. Parametric studies show that the multiple-scattering correction factor is both a function of obscuration and the actual particle dispersion. The two-parameter size distribution correction for multiple scattering has been extended to the 15-parameter model independent size distribution model.

A new technique has been developed for on-line measurement of both particle size and velocity distribution. To obtain drop velocity information, a dual-collimated laser beam system is employed. An acousto-optic Bragg cell is used as beam splitter and a beam switch for rapid switching between the diffracted and undiffracted beams. Separation between the two parallel collimated beams is varied, and the transit time of particles crossing the two beams is measured. Velocities of drops of different size are obtained by analyzing signals received from different detector rings.

Tomographic techniques are used to measure particle size distribution and concentration in volume elements within axisymmetric sprays by scanning a cross section of the spray. Abel transformation of the series of line-of-sight measurements yields the profiles of particle size and concentration distributions. Analysis of diffraction patterns for cylindrical and rectangular particles indicates that information can be obtained from laser diffraction instruments of average dimensions of noncircular particles.

Modeling and computational fluid dynamic procedures for two-phase reacting flows are based on the FLUENT code using an interactive menu-driven interface. A finite difference method is used to solve the nonlinear differential Navier-Stokes equations using additional equations for solving turbulence models, chemical species, radiation fluxes, and enthalpy. The equations of motion and trajectories of drops and particles are solved in a Lagrangian frame of reference with the path through each finite difference cell broken into a number of time steps. Allowance is made for heating of drops and evaporation. Initial size distributions are represented by the Rosin-Rammler continuous distribution. Equations of motion and auxiliary equations are solved for ensembles of drops of differing sizes and initial conditions. Account is taken of the interactions that particles and drops can have when encountering boundary surfaces. The importance of measuring quantities to provide accurate information on spray characteristics is emphasized for computation and design of spray combustion systems.

1-7 SIMULTANEOUS PARTICLE SIZE AND VELOCITY

The trajectories of particles are governed by initial particle size, particle velocity, injection angle, and subsequent interaction with the gas flow field. Collision, coalescence, deflection, drag, acceleration, deceleration, and evaporation, all influence particle trajectories. Momentum-to-drag ratios will determine the extent that particles follow or deviate from gas flow streamlines. Asymmetries in the spray result in asymmetries in the flame, which can result in "hot spots" where flame elements can melt or burn wall surfaces. The development of the phase Doppler analyzer (see Chapter 7) that measures simultaneously and instantaneously the velocity and diameter of single particles in sprays has been a major breakthrough in instrument technology.

A phase Doppler analyzer has been developed for detailed diagnostics of liquid fuel sprays, which have important influences on air-fuel mixing, ignition, flame stability, combustion efficiency, combustor durability, and pollutant emissions. Drop spacing, spatial oxygen distribution, ambient temperature, relative velocity between drops and air, drop size, and drop interactions have been found to influence the production of pollutants. Drop number density and mass flux also contribute to the processes leading to soot formation. Drop number densities are usually high in fuel sprays, leading to problems in obtaining coincident occurrences in sample volumes, while beam extinction reduces the accuracy of measurements. Steep density gradients under burning conditions cause laser beam steering and spreading, which induce gradients in the index of refraction. The phase Doppler method developed by Bachalo and Houser (see Chapter 7) uses light-scattering interferometry to acquire simultaneously the size and velocity of spherical particles. Measurements are dependent upon the wavelength of the scattered light, which is unaffected by attenuation of the spray or intervening optics.

Bachalo derived a theory for drop sizing, utilizing the phase shift of the light transmitted through or reflected from spherical particles. The phase shift is obtained by using light-scattering interferometry produced with a standard dual-beam LDV. Drop size measurement is obtained from accurate measurement of the spatial frequency of the interference fringe pattern. The temporal frequency of the fringe pattern is the Doppler difference frequency, which is directly related to the velocity of the particle. The interference pattern is measured directly by using pairs of detectors located at known angles to the laser beam and separated by fixed spacings. Doppler burst signals are produced by each detector with a phase shift between them. The phase difference between the detectors is determined by measuring the time between zero crossings of the signals. Measurements of the phase shift are directly related to particle size.

The on-line signal processing and data management computer stores data packets, which include drop size, velocity, and time of arrival for each drop measured. Data are stored by direct memory access and processed at a continuous rate of approximately 2000 samples/s. Data are processed to form velocity pdfs for each size class. Typically, 10,000 particle measurements are acquired at each point in the flow field. Frequency shifting is used to measure the small transverse velocity components and the resolution of the directional ambiguity in recirculating flows. Information is provided on size, velocity, number density, volume flux, angle of trajectory, and time-resolved data.

To provide comparative measurements of number density and mass flux, a light extinction system was used to acquire a line-of-sight measurement of the beam attenuation on the optical path. Determination of the transmittance was made using Beer's law relating the extinction cross section to the measured drop area mean diameter. The phase Doppler instrument provided mean diameters at points along the optical path, which were integrated to estimate the transmittance, which is used as a consistency check on the number density determined by the phase Doppler instrument. Comparisons between measurements made with sprays in quiescent air and with sprays injected into the recirculation zone downstream of a bluffbody showed drastic differences in drop size/drop velocity distribution and in size-velocity correlations.

The phase Doppler particle analyzer (PDPA) keeps track of particle arrival times as they sweep through the probe volume at a particular location. New software provides a time analysis of particle arrivals at the measuring probe volume for specified drop size and drop velocity ranges. Distance between two particles is obtained from the time elapsed between particle arrival times and the instantaneous velocity of the particles. Events are timed to an accuracy of 1 μs with a mean data acquisition rate of 100,000 particles/s. Processing time is 3 μs/particle, allowing an equivalent rate of 300,000 particles/s in short-duration bursts. Results show that drops tend to form clusters that vary widely in number and concentration. Local number density within individual clusters can be determined.

1-8 PARTICLE SIZING IN LARGE FURNACES

Measurements in large furnaces are usually made with water-cooled probes that are inserted through furnace doors. Suction probes collect particles on filters, which are then withdrawn and analyzed. Attempts to make optical measurements in furnaces have encountered difficulties in preventing damage to, and deposition on, lenses and problems associated with high-temperature radiating particles in the flow. Probes, using fiber optics, have now been developed that overcome many of these problems.

Instruments have been developed for particle size measurements in large-scale combustion environments (see Chapter 8). Measurements of particle size, concentration, and velocity distributions inside furnaces provide information that leads to reduction of slagging, fouling, and erosion of surfaces. A light-scattering technique is used for making in-line real time measurements of particle-laden flows. Problems due to thermal gradients that cause changes in the refractive index and result in deflection of laser beams (beam steering) arise in large-scale combustion applications. Aerosol opacity refers to the additional problem of transmitting light signals through gas with high particle concentrations over long distances in industrial systems. It is difficult to maintain clean windows for optical access in particle-laden streams, and it is also difficult to avoid degradation of light signals through secondary scattering by other particles in the flow. Finally, the measuring instrument must be designed to cause the minimum of interference to the flow.

The Insitec probe (see Chapter 8) has been designed to overcome these problems and make measurements of liquid, solid, or slurry particles that may be regular or irregular in shape with wide ranges of refractive index and surface reflectivity.

Spatially resolved measurements are made at points in space. Count rates are up to 500,000 particles/s, and data analysis requires less than 1 s. Absorbing particles with irregular shapes such as coal or fly ash can be measured as well as nonabsorbing particles such as water drops or latex spheres. The probe is water cooled for operation up to temperatures of 1400°C and high pressures. Windows in the probe are gas purged. Fiber optic signal transmission and a computer-driven motor system allow instruments to be remotely operated. The instrument has the capability of measuring size in the range of 0.2–200 μm with particle concentrations of $10^7/cm^3$ for the submicron range and up to 10 g/m^3 for the supermicron range at particle speeds in the range 0.1–400 m/s. Continuous in situ alignment is maintained under changing thermal conditions. A large-diameter, multimode fiber optic cable is used to transfer the scattered light signals from the optical head to the signal processor for detection and analysis. Computer keyboard controls allow the fiber optic system to be translated within the optical head to ensure proper alignment of the optical system during measurements in hostile environments. A narrow band-pass interference filter centered at 632.8 nm screens out background radiation from the system. The system can be remotely operated at distances up to 200 feet ($\sim$70 m).

The principle of operation is based on single-particle counting. The peak intensity of scattered light is measured together with the width of scattered light signals produced by single particles moving through the sample volume of a single, focused laser beam. The signal processor determines particle size from the measurement of peak intensity and particle velocity from the pulse width of each scattered light signal. Since the laser light intensity varies across the measurement volume, a particle trajectory through the center of the measurement volume results in a much higher signal intensity than does a particle trajectory near the boundary. The light intensity distribution throughout the sample volume must be known in order to solve for particle size and concentration. The sample volume also varies with particle size. Larger particles experience larger sample volumes than do smaller particles. The probability of counting a larger particle is greater than that of counting small particles.

The intensity deconvolution algorithm is based on the statistical analysis of a large number of scattered light signals from single particles passing through the measurement volume. Use of the intensity deconvolution algorithm allows the absolute particle concentration and particle size distribution to be obtained directly from the experimental data. A near-forward light-scattering configuration is selected for minimum sensitivity to particle shape and refractive index. The sample volume size must be made small enough that only one particle is in the sample volume at a time. In order to cover the wide dynamic range in concentration and size, two beams with different beam diameters and different sample volume sizes are used.

The Insitec particle analyzer (see Chapter 8) has been used to make in-line measurements of particles of pulverized coal, coal water slurries, liquid fuels, limestone powders, and fly ash in laboratory and industrial environments for both ambient and high-temperature particle-laden flow streams. Measurements that have been made in large-scale combustion facilities are not amenable to measurement by conventional instrumentation.

1-9 OPTICAL PROPERTIES OF SOOT AND DROPLETS

Soot particles are in the submicron range and cannot be detected by most particle-sizing instruments. Measurements of light scattering and extinction through clouds of particles yield number concentration and average size as well as the structural properties of particles. Use of interferometry allows simultaneous measurement of velocity and size. Polarization ratio measurements allow a clear discrimination between soot and droplets. This set of optical techniques (see Chapter 9) provides information of particle size, velocity, and optical properties of submicron solid and liquid particle clouds.

Condensed phases are found in combustion systems in the form of liquid fuel drops, pulverized solid fuel, polycyclic aromatic hydrocarbons, carbon clusters, and submicron soot particles. The theories of light scattering for particles much smaller than the wavelength of the incident radiation and also for particles much larger than the wavelength of the incident radiation are summarized in Chapter 9.

Characterization of the optical properties of soot formed in flames is obtained from measurements of the angular patterns of the scattering coefficients. When the complex refractive index is known, combined measurement of the scattering and extinction coefficients yields the number concentration and average size of soot particles present in a cloud. Scattering and extinction measurements at different wavelengths reveal the structural properties of soot particles. When the average size and number concentration of particles or the optical properties at reference wavelength are known by independent measurements, the complex refractive index of soot can be quantitatively evaluated.

The experimental apparatus is based on interferometry, which allows the simultaneous determination of velocity and size from which the joint distribution can be evaluated. The same apparatus and technique are used for measuring the scattering cross sections and the polarization ratio of partially absorbing drops. Measurements of average drop size are significantly smaller than those obtained using photographic techniques. The explanation given is that the ensemble scattering method gives the lower moments of the size distribution function, which automatically takes into account the very large number of small drops that are not detectable by photography.

Fuel droplets in combustion systems undergo chemical reactions resulting in soot formation. Spectrophotometric measurements carried out on samples of liquid removed from flames show blackening of the drops. Taking into account that initially transparent drops become partially absorbing, a combined set of measurements in both forward and side scattering regions simultaneously yields the average size and the imaginary part of the refractive index of the drops.

The structure of spray flames has been investigated by measuring radial and axial distributions of the scattering coefficients and the polarization ratios. Polarization ratio measurements allow a clear discrimination between soot and droplets. The average size and volume fraction of soot particles are obtained from scattering and extinction methods.

For measurements in transient diesel sprays, quantitative imaging techniques are used. A Nd:YAG laser tuned to its second harmonic is shaped into a thin light sheet through two cylindrical lenses. The collection optics, perpendicular to the laser beam, focuses the light-scattered images using macro-objective lenses. Images are split and focused on two microchannel plate intensifiers gated on the laser pulse. Photoelectrons from the intensifier are directed through bundles of fiber optics to an interfaced CCD camera. Video signals are sent to an analog-to-digital converter and stored in a buffer memory. Selected digital images are processed by a host computer, where image enhancement, noise reduction, stray light elimination, intensity calibration, and comparison between images are performed numerically.

The simultaneous use of the two arrays allows two-dimensional frozen light-scattering measurements at two different polarization states, or different scattering angles, or different wavelengths. When the laser is operated in a double-pulse mode, instantaneous temporal gradients can be measured to determine two-dimensional velocity fields of drop and spatial correlations of other combustion quantities.

1-10 GENERALIZED LORENZ-MIE THEORY FOR LIGHT SCATTERING

For many years the Lorenz-Mie theory has been the basis for computation of the properties of light scattered by particles. Because of the limitation of the classical theory, Gouesbet and colleagues have spent several years in developing a more general theory, which is synthesized in Chapter 10 of this book. Classical Lorenz-Mie theory was found not to be appropriate for most cases where laser sources are used in combustion systems. The classical theory is based on plane wave scattering, whereas the new generalized theory describes the interaction between a laser beam and a sphere. When the particle diameter is not small enough with respect to the beam width, the classical theory can be misleading. The new theory is considered to be a breakthrough in light scattering, allowing more rigorous theories to be developed for visibility and phase Doppler instruments. The new theory is proving to be of importance in design of instruments and interpretation of measurement data.

Gouesbet, Gréhan, and Maheu (see Chapter 10) present a generalized Lorenz-Mie theory for computation of the properties of light scattered by a Mie scatter center illuminated by a nonplane wave. The Maxwell equations are solved, accounting for the specific boundary conditions using the Bromwich formulation, which allows solutions satisfying the boundary conditions in special curvilinear coordinate systems. The properties of scattered light observed at a point are computed together with associated integral quantities. Separate equations are used for external waves scattered by the particle and spherical waves inside the particle. Scattering coefficients of the external wave are determined from the tangential continuity of the electric and magnetic fields at the sphere surface. The field components of the scattered wave are obtained from the Bromwich scalar potentials. Scattered intensities are computed with the aid of the Poynting theorem. Radiative balances are carried out on spheres surrounding the scatterer. Balances are made of the incident field, and scattering cross sections are computed. The radiation pressure force exerted by the beam on the scatterer is proportional to the net momentum removed from the incident beam.

Numerical computations of the mathematical functions involved in the theory are made, and the required scattered properties are computed. Numerical results are compared for three methods: quadratures, finite series, and localized approximations.

The optical levitation technique is used to study the interaction between a laser beam and a particle. Scattered light from the levitating scatter center is collected by an optical fiber and fed to a detector. The experimental scattering diagram is compared with results from the computations. A generalized Lorenz-Mie theory has been developed by Gouesbet, Gréhan, and Maheu (see Chapter 10). Accurate predictions will be possible for the scattering of Gaussian beams by spheres and for two crossing incident Gaussian beams. Doppler signals for particles larger than the beam diameter can be computed. These new computational techniques will lead to improvement in design of instruments and data processing software packages and also to the development of new instruments for two-phase flow studies. These calculations will initiate the wave of the future in Mie-scattering measurement analysis.

1-11 FOURIER TRANSFORM INFRARED SPECTROSCOPY

Fourier transform infrared (FT-IR) spectroscopy provides measurements of temperature and concentration of individual gas species and particles in flames. Line-of-sight measurements are made over an ensemble of particles from which point measurements are derived by tomography. Separate radiation contributions can be determined for soot and char. The shape of the transmittance is used to determine particle size and composition, while the amplitude of the transmittance yields the concentration of particles.

Fourier transform infrared spectroscopy is used for measurement of gas, particle, and soot temperatures and concentrations in flames (see Chapter 11). The advantages of the FT-IR emission and transmission (E/T) technique include (1) the capability to determine separate temperature and concentration for individual gas species as well as for solid particles, including soot, by employing different regions of the infrared spectrum, (2) the capability to determine temperatures as low as 100°C and, consequently, the ability to follow particle or droplet temperatures prior to ignition, (3) the ability to make measurements in densely loaded streams to study cloud effects, (4) the capability to separate the radiative contribution from soot and from char particles, and (5) the capability to measure particle sizes.

FT-IR E/T spectroscopy makes measurements of an ensemble of particles over a line of sight. Tomographic techniques must be used to obtain spatially resolved data. Spectra acquisition times are of the order of 0.1 s, so that time-averaged properties are measured. Successful measurements have been made in coal, ethylene, hexane spray, and coal water fuel spray flames. The Fourier transform technique processes all wavelengths of a spectrum simultaneously. It can be used to measure spectral properties of particulate streams with varying flow rate. Radiation passing through the interferometer is amplitude modulated, while particle emission passes directly to the detector and does not interfere with the measurements of scattering or transmission.

Transmission tomography is used to construct three-dimensional images from two-dimensional slices of the image. For objects with axial symmetry, Abel's radial inversion (onion peeling) equations are used. In spectral regions for which Beer's law applies, the two-dimensional image reconstructed from transmittance leads to the spatial dependence of absorbances and hence concentration. A Fourier reconstruction of two-dimensional images was found to be consistent with the measured transmittance. Gas temperatures determined using an automated least squares fitting routine in a homogeneous sample gave good agreement with average thermocouple measurements with an accuracy of 10 K. Particle temperatures, obtained by fitting theoretical blackbody curves to the experimental normalized radiance, were in good agreement with calculated temperatures and temperatures measured with a thermocouple. Particle size information is obtained from extinction spectra for particles less than 80 μm in diameter based on calculations using Rayleigh theory for large particles, or Mie theory for small particles.

In a laminar diffusion flame, soot particle concentration and temperature were measured at several axial positions. Different temperatures were recorded for soot, CO_2, and HC gases that were concentrated at different radial positions in the flame. Local soot concentrations were in good agreement with measurements made by laser extinction/scattering. Excellent agreement is also found between FT-IR and CARS temperature measurements.

The shape of the transmittance provides information on the particle size and composition, while the amplitude is proportional to concentration. Experiments were performed on unignited sprays and spray flames of hexane and coal water fuels. From Mie calculations for each particle size distribution, best fit theoretical spectra were overlayed on the measured spectra. The extinction measurement was found to be a remarkably sensitive gage of particle size distribution. For spray combustion, both emission and transmission measurements were obtained to determine concentrations of pyrolysis and combustion products and their temperatures.

In coal water fuel spray flames, the following quantities were determined: gas concentrations and temperatures for CO_2, H_2O, CH_4; concentrations of particles and soot; particle temperatures; percent of particles ignited; and flame radiation intensity from individual components (particles, soot, gases).

1-12 COAL AND CHAR PARTICLES

A wide range of laboratory instruments are in use for measurements of single particles or small samples of burning coal and char particles. Measurements are made of reactivity, weight loss, particle temperature, pore volume, and particle structure. Particles are analyzed in thermogravimetric analyzers, electrodynamic balances, electrodynamic thermogravimetric analyzers, and laminar-flow drop-tube reactors. Pyrolysis and char oxidation are followed as samples are continuously weighed, and programable temperature controllers are used for heating. Single particles are held in suspension by electric fields in the electrodynamic balance, while lasers heat the particle. Particle temperatures are measured by radiant emissions from hot particles at selected wavelengths, brightness pyrometry, or by disappearing filament pyrometer. The volume of pores and pore structure are determined by gas adsorption isotherms and mercury porosimetry.

Measurement techniques used in the study of coal and char combustion are discussed by Sahu and Flagan (see Chapter 12) with emphasis on reactivity measurements, mineral matter transformations, measurement of particle temperatures, and particle structure characterization. Experimental reactors using electrical, radiant, flame, plasma, laser, and shock tubes for heating have achieved temperatures as high as those attained in pulverized coal combustors, i.e., 2000 K. Laminar flow reactors with a dilute stream of coal particles introduced on the reactor centerline are used to study rate processes in entrained flow reactors, where the reaction time is the same for all particles. Kinetic studies require that systems be large enough to permit sample extraction from the reaction zone or have optical access to facilitate noninvasive measurements of particle size, particle temperature, and gas composition. Sampling probes introduce a heat load that can alter reactor temperature profiles and possibly introduce sampling biases. Kinetic measurements examine the evolution of an ensemble of particles, e.g., bulk property measurements made on extracted samples, or optical measurements of a cloud of particles, or single-particle measurements made at a fixed point along a reactor length. Measurements of the combustion history of a single particle can be made by viewing along the axis of a drop-tube furnace and measuring temperature histories with a two-color pyrometer. To simulate coal flames in practical combustion systems, large reactors are needed with turbulence and flow recirculation.

The thermogravimetric analyzer (TGA) is used extensively in low-temperature studies of pyrolysis and char oxidation reactions. The TGA consists of a sensitive balance that weighs a sample continuously as it is heated and reacts. A programable temperature controller is used to heat the sample at a prescribed rate. A gas flow continuously supplies reactants and removes reaction products from samples ranging in mass from 1 to 100 mg. Temperatures of particles can be up to 1700 K, but the TGA is most useful for studies of reactions at temperatures below 1000 K.

The electrodynamic balance (EDB) is used to suspend and weigh a single charged particle in an electric field. Although high-temperature experiments are usually limited to particles larger than 100 μm, particles ranging in size from a fraction of a micron to several hundred microns can be held in suspension in an EDB. A charged particle is suspended in the electric field created by a potential across the top and bottom electrodes. An alternating potential on the ring electrode creates a time-varying force on the charged particle. That force increases with displacement from the center of the cell, causing a particle that is not centered to oscillate. Due to particle inertia and aerodynamic drag, the motion of the particle lags the field, leading to a time-averaged force that tends to push the particle back toward the center of the chamber. This dynamic focusing makes possible the study of particles undergoing rapid change. The EDB has a sensitivity in the range 10^{-9}–10^{-14} g.

An electrodynamic thermogravimetric analyzer (EDTGA) uses laser heating to raise a particle trapped in an electrodynamic balance to temperatures at which reaction occurs. The particle temperature is measured optically, and its mass is determined from the field required to hold it at a null point of the cell. Laser heating of a particle produces photophoretic forces that can push a particle from the veiw volume unless the forces are carefully balanced. A hot particle in a cold gas generates a buoyant flow that tends to lift the particle. At temperatures above 1400 K, the particle begins to lose charge, making mass loss measurements impossible. The EDTGA is the only technique that allows detailed characterization of a burning coal particle. Photographs are taken, and the porous microstructure of the particle is studied by gravimetric gas adsorption. A monolayer of adsorbed gas such as carbon dioxide is weighed. Reactivity and ignition characteristics of levitated char particles have been determined by EDTGA.

Measurements of reactivity at high temperatures have been made in laminar-flow drop-tube reactors with heat supplied by a flame, plasma, or electric heating. In vitiated combustion, the temperature depends on the fuel-oxidant mixture ratios. In an entrained flow reactor, an electrically heated air preheater tube is followed by a reactor tube. The pulverized sample is entrained in the reactant gas and introduced into the co-flowing hot primary stream. Optical measurements of particle size and temperature are made through transparent reactor walls. A moveable particle collection probe is inserted from the exhaust end of the reactor to collect partially burned samples or combustion products for characterization and analysis. Gas or liquid quench is used in the probes to stop reaction of the particles immediately on entering the collector. Samples can be obtained at different particle residence times. Feed rates of particulate material range from 100 mg/h to several grams per hour.

For measurements of the temperatures in a cloud of burning particles with a range of particle size, the gas environment varies greatly with position in the flame, resulting in considerable uncertainty in the temperature measurement. Particle temperature measurements are based on radiant emissions from the hot particles at selected wavelengths. In brightness pyrometry, the absolute emission intensity is measured at a single wavelength. The disappearing filament pyrometer matches the particle emission to that from a calibrated filament in the red. More recent brightness pyrometers make electronic measurement of the absolute emittance from the particle possible. Taking advantage of the wavelength dependence of the spectral emittance and assuming that the spectral emissivity does not vary with wavelength over the region of interest, two-color pyrometers compare intensity measurements at two wavelengths.

Spectrally resolved measurements are made using a monochromater or band-pass filters to select wavelength intervals viewed by separate detectors. If the difference between the two wavelengths selected for measurement is small, the variation in the emissivity is assumed not to vary. The temperature of single-burning bituminous char particles has been measured in one instance to be 1800 K over a period of 30 ms. Optical methods have been used for simultaneous measurement of particle size, velocity, and temperature in a combustion environment. Temperature measurements were made using two-color pyrometry, while particle size and velocity were measured by imaging the particle onto a coded aperture. The signal from one of the two photodetectors in the two-color pyrometers is used to determine the particle size. For a uniformly heated, spherical particle of uniform emissivity, the ratio of the peak signals determined by the photodetector is a monotonic function of the particle diameter.

Noninvasive optical measurements of particle size are made by light scattering or imaging. Typical particle concentrations are between 10^8 and 10^{11} per cm^3; hence in situ measurements of fine particles are generally based upon measurements of light scattered from a large number of particles. Conditional sampling ensures that larger particles are not present in the sample volume when measurements are made. In situ measurements are generally limited to small or dilute systems in which the region of interest is optically thin. Extractive sampling must be employed when the density of the aerosol is such that only the perimeter regions can be probed optically.

Fume particles produced by condensation of refractory vapors are generally agglomerates of small spherules. These aggregate particles have remarkably similar structures, which have been characterized as fractal structures in which particle mass scales as radius. The value of the fractal dimension conveys information about the mechanisms of particle formation and growth. The dynamics of the aggregation process depend on the aerodynamics of the particles and on particle structure. Small-angle X-ray scattering and small-angle neutron scattering are used to measure mass correlation functions for particles in the range 1–100 nm.

Extraction of samples by probes can result in sampling biases caused by probe and particle aerodynamics and deposition of particles within the probe, particularly where hot particle-laden gas is exposed to cool surfaces and changes in aerosol properties. Coagulation can be minimized by diluting and quenching the sample. The temperature gradient between the hot particle-laden gas and the cool surface of the probe can result in particle deposition on the surface due to thermophoresis. Deposition can be reduced by using a transpiration-cooled sampling probe with diluent injection. Isokinetic sampling reduces aerodynamic biases. Particles can also be lost to walls of sampling systems by convective diffusion, inertial impaction at bends in the flow, sedimentation, and electrostatic deposition. Compared with metal tubing, many plastics create localized electric fields that result in increased deposition of charged particles.

Particles larger than 0.3 μm can be classified aerodynamically in inertial impactors. Cyclone separators are used for measurements of particles larger than a few microns in diameter. Particles collected in size-classified samples are chemically analyzed. Measurement of the size distributions of particles smaller than 100 nm is most commonly performed using electrical mobility analysis, in which the particles are given an electrical charge and then classified in an electrical field. Gas adsorption isotherms are used to determine the volume of pores in char particles. Mercury porosimetry is widely used to study the pore structure properties of porous solids. Mercury, which is nonwetting with most substances, is forced under pressure to penetrate pores, openings, and voids in materials. The volume of mercury penetrated is recorded as a function of pressure. Pore surface area distributions are calculated from the pore volume.

1-13 CONCLUSIONS

Laser technology has changed the field of combustion measurements. In the past, scientists had to deduce combustion characteristics on the basis of measurements of pressure and temperature made at the periphery of the system. Global calculations were made of heat release and flow rates. The insertion of water-cooled probes into flames enabled samples of gas and particles to be withdrawn for subsequent physical and chemical analysis. These probes cause disturbances to the temperature and flow fields, and withdrawn samples are not fully representative of the flame conditions. The measurement techniques are laborious and cumbersome.

Laser anemometry has become the established technique for measuring velocity and turbulence characteristics in flames. With two- or three-component LDA systems, normal and shear stresses are measured from which kinetic energy of turbulence is determined. Distributions of these quantities provide the fundamental basis for computational fluid dynamic models and calculations.

Imaging techniques include high-speed photography, holography, and cinematography. The use of seeding and color film is revealing detailed dynamic structures within flames. Laser sheet lighting shows two-dimensional thin sections of the flow field from which dynamic movements of large eddy structures can be clearly seen and analyzed. Holography allows the freezing of events in three dimensions using pulsed lasers with pulse durations of several nanoseconds. Automatic image analysis provides more rapid data acquisition and analysis. Particle size, shape, and velocity can be readily analyzed.

Noninvasive laser techniques for simultaneous gas temperature and concentration measurement include spontaneous vibrational and rotational Raman scattering, Rayleigh scattering, coherent anti-Stokes Raman spectroscopy, and laser-induced fluorescence. The accuracy of temperature and concentration measurements is well established, and laser techniques are superior to methods using physical probes. These optical methods are providing details of the chemical processes as they are taking place in the flame, showing the wide range of chemical reactions that occur even with simple fuels. This information is leading to the formulation of complex kinetic models, which when coupled with the fluid dynamic models, are leading to more accurate predictions and analyses.

Instead of requiring a series of sequential measurements, instruments are being coupled using the same laser beams and the same measurement control volumes but with different detection systems. It has become possible to simultaneously measure velocity, gas temperature, and gas concentration. This also allows simultaneous measurement of velocity-temperature, velocity-concentration correlations, and joint pdfs in turbulent combustion flows. Eventually, we can expect incorporation of several instruments into a single hybrid instrument for making simultaneous instantaneous measurements of several quantities in flames. Earlier difficulties in making measurements in the presence of irradiating particles appear to be overcome.

Important advances have been made in the ability to measure the size of liquid and solid fuel, char, soot, and dust particles in flames. Liquid fuel sprays with atomization, break up, collision, coalescence, acceleration, deceleration, and evaporation are being analyzed. Laser diffraction with line-of-sight measurements yields average size and size distribution of particles. Phase Doppler simultaneously measures velocity and size of single spherical particles from which number density, volume flux, and size-velocity correlations are determined as well as angle of trajectory and time-resolved data. Originally, measurements based upon refraction were only made on transparent droplets. More recently, measurements based on reflection using backscatter have also been made on opaque spherical particles. For large furnaces, light scattering is used to measure velocity and concentration of liquid, solid, and slurry particles. The probes are water cooled, and fiber optics is used for signal transmission. This instrument is becoming increasingly useful for reduction of slagging, fouling, and erosion of surfaces and reduction of emission of particulates. Polarization ratio measurements are particularly useful for determining particle size and concentration in dense particle clouds such as soot or diesel sprays.

Fourier transform infrared spectroscopy provides temperature and concentration of individual gas species, particles, and soot. It is a line-of-sight measurement of an ensemble of particles. Particle size information is obtained from extinction spectra.

Interesting experiments have been performed on spray flames of hexane and coal water slurries.

Following the changes in characteristics of burning coal and char particles has become possible in electrodynamic balances and thermogravimetric analyzers, where single coal particles are held in suspension by electrodynamic fields. The particles are heated by laser. Particle temperatures are measured by radiant emissions, while pore volume and structure are determined by gas adsorption and mercury porosimetry.

Overall, major developments have taken place in combustion instruments during the past two decades. These developments are allowing more sophisticated and detailed research in a wide variety of combustion systems. Combustion environments were considered to be hostile because of the high temperatures, high gas flows, and high concentration of particles. Each one of these problems has been overcome. Laser beams that penetrate through flames without interference have made it possible to measure velocities, particle size, temperature, and concentrations of many species. These combustion instruments are also being applied to other industrial processes with high-temperature and particle-laden flows. The measurements are leading to greater understanding, greater control, and better predictions of the processes.

TWO

VELOCITY MEASUREMENTS BY LASER ANEMOMETRY

Norman Chigier

2-1 INTRODUCTION

Velocity fields in combustion systems are established by the injection of fuel and oxidant, the geometry of the combustion chamber, and the exit conditions. Reynolds numbers are usually sufficiently high for flows to be turbulent, and high shear is generated in high-velocity jet flows. Jets may impinge on surfaces, and recirculation zones are generated between jet boundaries and confining walls. Swirlers in the air and fuel flows result in swirling jets with central, pressure-induced recirculation zones with high turbulence intensity and mixing rates. Liquid fuel is sprayed into high-shear zones to promote rapid vaporization, mixing, and combustion.

Early measurements of velocity in combustion systems were made with water-cooled pitot tubes (Beer and Chigier 1972). These provided mean velocity measurements based on separate measurements of temperature to correct for gas density. Most attempts at using hot wire anemometers in flames were unsuccessful due to difficulty of hot wires withstanding the temperature and flow fields as well as the considerable difficulty in separating the influences of the velocity field from the temperature field. The nonintrusive laser anemometer (Thompson and Stevenson 1979) has proved to be a far superior, more accurate, and more comprehensive instrument for measurement of three components of mean and fluctuating velocity, normal and shear stress components, kinetic energy of turbulence, power spectra, autocorrelation coefficients, and turbulence macroscales. The possible sources of error have been carefully examined

This chapter was first published in *Combustion and Flame*, 1989 under the title "Velocity Measurements in Inhomogeneous Combustion Systems." Reprinted by permission.

and, provided that all the recommendations concerning optical components, alignment, seeding, filtering, signal processing, and calibration are followed, measurements of mean velocity within 3% and rms within 6% accuracy can be obtained. Measurement of two orthogonal components can be made instantaneously and simultaneously with little difficulty, and some simultaneous measurements have been made of three orthogonal components. Making measurements in small laboratory flames has proven to be no more difficult than making measurements in the corresponding nonburning flow. For larger flames, as in furnaces, beam steering due to refractive index gradients has caused problems with alignment and coincidence of the beams at the measurement control volume. By direct comparison of measurements made in combustion systems with the corresponding nonburning flow, important data have been collected in a variety of flows. These comparisons are providing quantitative information on changes in acceleration, expansion, recirculation zones, increase and suppression of turbulence, isotropy, and shear as a direct consequence of chemical reaction. By the use of laser-induced fluorescence, coherent anti-Stokes Raman spectroscopy, and compensated thermocouples, measurements are being made, in parallel, of instantaneous local temperature and species concentrations. Some measurements have also been reported of simultaneous temperature velocity and velocity species concentration correlations. With this information, mass, momentum, species concentration, and energy balances are being made on the basis of direct measurement of individual terms from the appropriate equations for turbulent reacting flow.

Many combustion systems involve two- or three-phase flows. For liquid fuels, droplets are present in the spray that may penetrate deep into the combustion chamber. Soot particles are formed from gaseous fuels, while solid carbon and/or mineral particles are present when liquid or solid fuels are burned. The presence of particles in the submicron range is useful for velocity measurement. If the particle size is small enough and the particle number density is within the required range, no additional seeding is required. Larger supermicron particles will not follow the higher frequency components of the gas velocity fluctuations or the accelerations and decelerations of the gas flow. At any one point in the flow field, separate measurements are required of the velocity of particles of different size. The relative velocity between particle and gas will depend upon the drag momentum ratio, which allows determination of local drag coefficient and particle Reynolds number from which local trajectories and rates of evaporation and reaction can be calculated. The phase Doppler particle analyzer is a laser anemometer for particle velocity measurement that also simultaneously measures the particle size. Three photomultipliers collect three Doppler signals from which the particle size is determined from the measurement of the phase difference. Within 1 s, thousands of measurements are made while the histograms of velocity and size and the size velocity correlation curves are generated on the computer.

2-2 LASER VELOCIMETRY

Laser velocimetry allows measurement of the local instantaneous velocity of particles suspended in the flow. If the particles are small enough for the drag to greatly exceed the momentum, the particles are in dynamic equilibrium with the flow, so that the particle velocity equals the local gas velocity. For larger particles that are not in dynamic equilibrium, there is a slip or relative velocity that is a function of the particle size, geometry, and velocity. In many combustion systems, fuel, product, or dust particles are present, which may be suitable and adequate for gas velocity measurement. Otherwise, seeding particles need to be added to the flow, so that they do not cause any physical or chemical disturbance to the flow. The time required for making individual velocity measurements is of the order of tens or hundreds of microseconds, depending upon the number density of particles and the time required for electronic data acquisition and processing. Sufficient time is allowed at each measuring point to collect several thousand individual instantaneous velocity measurements from which time mean average, rms of fluctuation, and correlations are computed. With a one-dimensional system, three components of velocity can be measured in sequence by rotating the plane of the laser beams in each orthogonal direction. In a two-dimensional system with an argon ion laser, two blue beams measure a vertical velocity component, while two green beams, focused at the same measuring point, measure a horizontal velocity component, yielding simultaneous instantaneous velocity measurements in two orthogonal planes. In a three-dimensional system, all three velocity components are measured simultaneously. From the trace of instantaneous velocity as a function of time, probability density function (pdf), conditional sampling, and frequency spectra can be determined.

Several optical arrangements such as "two scattered beam" and "reference beam" have been used, but the most frequently used arrangement is the dual-beam fringe anemometer. Two light beams of equal intensity are focused to intersect, producing a pattern of successive light and dark fringes. When a particle crosses these fringes, a periodic variation in intensity of the scattered light is generated as a Doppler burst. The measured frequency difference, ν_D, is directly proportional to the instantaneous component of velocity of the particle lying in the plane of the beams and normal to their bisector, u:

$$\nu_D = 2\frac{u \sin \theta/2}{\lambda} \tag{2-1}$$

where λ is the wavelength of the incident laser beam and θ is the beam intersection angle.

In order to make measurements in recirculating and high-turbulence intensity flows, frequency shifting is required by insertion of Bragg cells or a rotating diffraction grating in the transmitting beams. By this means, reverse flow velocities can be distinguished from forward flow velocities. When a particle crosses the control volume with the velocity component u, the scattered light has an overall Gaussian envelope, or pedestal, due to the Gaussian light intensity distribution within the control volume.

The light-collecting system transmits the scattered light from the measuring control volume to the photodetector. The intensity of scattered light is a complex function of scattering angle, beam angle, particle size and shape, particle material, and ratio of fringe spacing to particle diameter. Signal processing can be carried out by frequency tracker, frequency counter, spectrum analyzer, transient recorder, filter bank, or photon correlator, depending upon the specific flow problem and the nature of the photo-multiplier output. The instruments are linked to microcomputers for data processing. Frequency trackers provide a continuous output signal with voltage directly proportional to the input signal frequency. Frequency counters use high-frequency clocks to measure the time duration for a preset number of zero crossings. Spectrum analyzers measure the pdf of the Doppler frequency over a period of time, while transient recorders are very fast analog-to-digital converters able to fully digitize a Doppler burst using fast Fourier transform instrumentation.

A number of factors can influence measurement accuracy. Large control volumes can result in velocity gradient broadening. When particles move in both forward and reverse directions, errors will arise unless sufficient frequency shifting is applied to accommodate all reverse flow velocities. For seeding particles of MgO or TiO_2 in flames, particle diameters are required to be 1 μm for a precision of 1% in velocity up to frequencies of 10 kHz. In combustion systems, high temperatures result in density gradient effects. As flow proceeds from the unburned mixture with high particle concentration through the flame front into the hot product gases, the resultant changes in seeding concentration influence the accuracy of the measurement. It has been established that, provided the appropriate adjustments are made to the optical and signal analysis systems, accurate measurements can be made of mean and rms velocity components and correlations in combustion flows with recirculation zones and high-turbulence intensities.

Sislian, Jiang, and Cusworth (1988) made comprehensive measurements in flames using a one-dimensional laser Doppler velocimeter (LDV) operating in the dual-beam, forward scatter mode with standard DISA 55× modular optics components. Focal length of the transmitting lens was 600 mm, and the beam intersection angle was 6.5°. Spacing of the interference fringes in the probe volume was 5.72 μm, waist diameter of the focused laser beam was 226.8 μm, and the probe volume dimensions were 4.1 mm major axis and 0.227 mm minor axis. The forward off-axis collection angle was 20° with focusing onto a pinhole aperture of 0.1-mm diameter. The focused scattered light was then filtered with a DISA 55 × 38 narrow bandwidth red interference filter in order to effectively eliminate interference from flame radiation. The effective probe volume length as seen by the photomultiplier was 0.6 mm. Signals were processed by a counter-type signal processor (TSI model 1980A) and continuously monitored on an HP 1744A oscilloscope. Data reduction was performed by microcomputer.

Aluminum oxide particles, 0.5 μm in diameter, were used for seeding the cold and combusting jets. A sonic probe was used to disperse the particles in a suspension with water. The wet particles were passed through a diffusion drier and then neutralized electrostatically by a charge neutralizer. Agglomeration was thus minimized. The seeder yielded a range of particle densities of the order of 10^9–10^{10} particles/m^3. Seeding of the flow was controlled, so that only one particle was present in the probe volume at a given instant of time. Average data rates were 10,000/s at the exit central portion of the flow, 5000/s in the central portion 50 exit diameters downstream, and about 200/s at the edges of the flow. For each data point, 2000 individual realizations were averaged for the cold flow and 8000 for the combusting jet flow.

Power spectra and autocorrelations of the axial velocity component were obtained from an essentially continuous record of instantaneous axial velocity versus time by the use of high seeding levels and the analog output of the counter processor. The time-resolved velocity measurements were limited to the central lower portion of the cold and combusting jets. With high seeding levels, data rates were 60,000/s at the exit and 10,000/s in the upper part of the investigated flow region. The turbulence macroscale L was determined by multiplying the Eulerian integral time scale of turbulence times the local mean axial velocity component:

$$L = \bar{u} \int_0^{\tau_{max}} R(\tau)d\tau \tag{2-2}$$

where $R(\tau)$ is the normalized autocorrelation function of the axial fluctuating velocity component,

$$R(\tau) = \frac{\overline{u'(t)\, u'(t + \tau)}}{\overline{u'^2}} \tag{2-3}$$

τ_{max} is the value of the time lag for which $R(\tau) = 0$, and $\overline{u'^2}$ is the local turbulent intensity of the axial velocity fluctuation.

The LDV system used by Heitor and Whitelaw (1986) comprised an argon ion laser operated at a wavelength of 514.5 nm with a power of 1 W, a two-beam optical arrangement with sensitivity to the flow direction provided by light-frequency shifting from a Bragg cell at 40 MHz, a 500-mm focal length transmission lens, and forward scattered light collected by a 150-mm focal length lens at a magnification of 1.8. The half angle between the beams was 3.4°, and the calculated dimensions of the measuring volume at the e^2 intensity locations were 3.68 and 0.219 mm. The output of the photomultiplier was mixed with a signal derived from the driving frequency of the Bragg cell, and the resulting signal was processed by a frequency counter interfaced with an 8-bit microprocessor. The transfer function in the absence of frequency shift was 0.23 MHz/ms. The fuel and air flows were seeded with powdered aluminum oxide (nominal diameter between 0.6 and 1.0 μm before agglomeration) dispersed in a spark discharge particle generator.

The two velocity components and corresponding Reynolds stresses were determined from measurements with the laser beams in the axial direction and at ±45°. The number of individual velocity values used to form the averages ranged from 640 to 3840, depending on the location and flow conditions. In combusting flows the largest statistical errors were of the order of 2.5% and 5%, respectively, for the mean and rms values, considering a 95% confidence interval. Systematic errors arise when intensity levels are larger than 15%, where velocity bias effects can lead to overestimation of the mean values by up to 9% and to underestimation of the rms of the fluctuations by up to 5%. Nonturbulent Doppler broadening errors due to velocity gradient broadening can be important for the rms values of tangential velocity and can lead to an overestimation by up to 20% in the center of a combustor.

Recent Developments in LDV

The use of a three-color LDV system allows simultaneous measurement of three velocity components in applications with large flow velocities and large velocity variations. Each velocity component is measured using a separate color, imposing no inherent limits on the velocities measured. The separation of transmitted and scattered light paths provides excellent signal quality with small measurement volume and effective flare or reflection rejection from boundaries. Dispersion prisms provide clean color separation for the incident beams: 514.5 nm (green), 488 nm (blue), and 476.6 nm (violet). Independent selectable frequency shifting is provided for each component, so that three-dimensional flows with swirl recirculation, large fluctuations, and high turbulence intensities can be measured readily by selecting the appropriate shift. Use of a 3.75 beam expansion reduces the size of the measuring volume and improves signal-to-noise ratio by more than 50 times. Further improvements in signal-to-noise ratio can be obtained by using a beam expansion of $8.5 \times$. The blue and green pairs of beams are transmitted through one optical train, while the blue and green scattered light is collected through the second optical train. The violet beams are transmitted through the second optical train, and scattered light is collected through the optical train, corresponding to the transmission for blue and green wavelengths. The receiving optics modules may be placed in on-axis forward scatter or in off-axis backscatter to reduce flare effects and decrease the measuring volume in length for measurements closer to walls. The receiving optics for the blue and the green scattered light consist of a color-splitter module and separate receiving modules to image the scattered light from each color to individual photomultipliers. The three measured velocity components are processed using computer software to determine three orthogonal velocity components. Counter-type signal processors measure the Doppler frequencies that correspond to each of the velocity components.

Fiber optics are providing a new degree of flexibility that was not available with conventional LDV systems and is of particular interest in combustion and flame studies and two-phase flows. Fiber optics are being used for flow measurements in small or restricted spaces and in complex geometric configurations. Fiber optic LDV probes can be positioned inside or outside the flow and maneuvered easily. Fiber links separate the optics from the laser and the system electronics, which can be protected from the harsh environment of the flow. Special couplers are used for directing the laser beam into and out of the optical fibers. Small probes (14- and 25-mm diameter) are inserted into flows for minimum disturbance. The probe usually contains focusing and receiving lenses connected to the LDV system by fiber cable. In some cases, a beam splitter is included in the probe. The fiber link enables a user to separate the laser from the rest of the system, thus protecting the laser and providing additional flexibility during operation. Fiber links can be used with either a one-color helium neon or single-color argon laser or a two-color argon laser system.

Software packages are becoming available that provide data analysis using the output from the signal processor of LDV systems. The menu-driven format uses commands that do not require a knowledge of computer terminology. Raw data are processed to provide individual velocity components and their projections in three orthogonal planes with statistical properties including mean, turbulence intensity, skewness, and flatness of the velocity components. In addition, correlations, Reynolds stresses, velocity and flow angle histograms, and vectors, expressed in terms of magnitude and angle, are displayed or printed in tabular or graphic form. Velocity bias occurs when there is a high probability of making more velocity measurements during the period when the velocity magnitude is high than when the velocity magnitude is low. In order to avoid velocity bias, each independent velocity measurement must be weighted by a factor related to its instantaneous velocity. Transit time (particle residence time) is the commonly used weighting factor. By selecting the appropriate signal-processor settings, the software package corrects for velocity bias effects. Software packages are also available to control the automatic traversing of the table with three axes of translation. These software packages are greatly increasing the accuracy of measurements and the ease and convenience of making LDV measurements.

LDV Measurements in Flames

Measurements in turbulent diffusion flames have been reported by Bilger and Beck (1975), Ballantyne and Bray (1977), Glass and Bilger (1978), Toshimi, Hyun-Dong, and Akiza (1981), Takagi, Shin, and Ishio (1981), Lewis and Smoot (1981), You and Faeth (1982), Razdan and Stevens (1985), and Sislian, Jiang, and Cusworth (1988). Measurements in a model combustor are reported by Brum and Samuelsen (1987) and Charles et al. (1988), whereas Heitor and Whitelaw (1986) made measurements in a gas turbine combustor. Moreau et al. (1987) studied the effects of a bluffbody in a combustor. Representative measurements and conclusions have been selected from these studies to provide a general picture of velocity and turbulence fields in flames.

A detailed investigation of the turbulence structure of jet diffusion flames at a Reynolds number of 10^4 has recently been completed by Sislian, Jiang, and Cusworth (1988). They measured mean velocity components, turbulent intensities, velocity pdfs, power spectra, and autocorrelation functions of axial velocity fluctuation and spatial turbulence macroscale. Direct comparisons were made between the noncombusting case (a mixture of air and helium) and the combusting case (mixture of methane and argon). A one-dimensional laser velocimeter operating in forward scatter was used for velocity measurement. Thermocouples were used for gas temperature measurement, and Schlieren photography was used to visualize flame structure. They found that combustion suppresses turbulence in the upstream region of the jet and enhances it in the downstream region, where turbulence intensities are substantially higher than in the corresponding cold jet flow. Relative intensities were found to be smaller in the flame in the upstream region and comparable to the cold jet relative intensities in the downstream region. Turbulence macroscale is significantly smaller in the flame than in the cold jet in the upstream region and increases appreciably at downstream distances.

The measurements by Sislian, Jiang, and Cusworth (1988) of mean temperature, mean velocity components, and turbulent intensities for noncombusting and combusting flows provide detailed quantitative information on the influence of combustion and heat release on the turbulence structure of jet flows. Figure 2-1 shows the centerline distributions of mean axial velocity, turbulent kinetic energy, and mean temperature up to $X/D = 50$. The decay of mean axial velocity on the centerline is much less in the flame than in the cold jet, and the potential core in the flame is considerably longer than in the cold jet. Several investigators (Yule et al. 1981) have found similar results for jet flows with Reynolds numbers of the order of 10^4. The distribution of kinetic energy of turbulence k is quite different in the flame than in the cold jet. In the near-

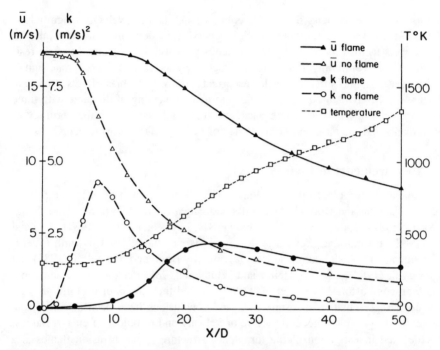

Figure 2-1 Axial distributions of mean axial velocity, turbulent kinetic energy, and temperature *(Sislian, Jiang, and Cusworth 1988).*

exit region of the flame, k is small, whereas in the corresponding cold jet flow, k reaches a maximum and then decays rapidly. This suppression of turbulence is caused partly by the increase in kinematic viscosity due to temperature rise and partly by the acceleration of the flow resulting in changes in velocity gradients. The turbulent kinetic energy in the flame increases gradually and reaches a peak at $X/D = 25$. The magnitude of this peak of k is appreciably less than the maximum value of k in the noncombusting jet. Beyond $X/D = 25$, k changes less rapidly and remains always higher in the flame than in the cold flow. Centerline distributions of normal turbulent intensities, $\overline{u'^2}$ and $\overline{v'^2}$ (Fig. 2-2) are similar to those of k. In the near-exit region of the noncombusting jet, $\overline{u'^2}$ is always larger than $\overline{v'^2}$, whereas in the same region of the flame, these normal stresses are small and almost equal. In the downstream region, turbulence becomes more anisotropic in the flame than in the noncombusting jet, with higher values of $\overline{u'^2}$ and $\overline{v'^2}$ than those in the cold flow.

Radial distributions of mean axial velocity, turbulent kinetic energy, and temperature at $X/D = 10$ are shown in Fig. 2-3. At the exit, there were no appreciable differences between the cold and hot flow quantities. The combusting jet widens very soon after the exit due to thermal expansion. The values of the turbulent kinetic energy in the shear layer are smaller in the combusting case due to the increase of kinematic viscosity with temperature. The mean position of the flame front is indicated by the radial position of the maximum mean temperature and is located in a region of low turbulent kinetic energy. In the central core region of the jet, kinetic energy levels are

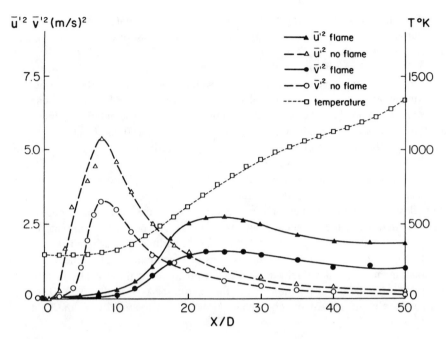

Figure 2-2 Axial distributions of normal turbulent intensities and temperature *(Sislian, Jiang, and Cusworth 1988).*

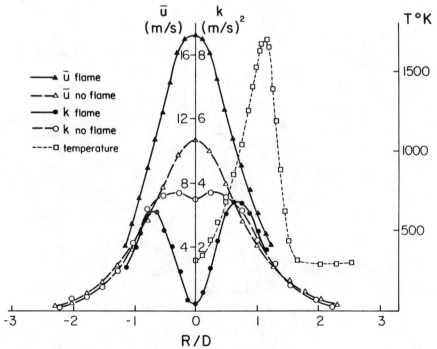

Figure 2-3 Radial distributions of mean axial velocity, turbulent kinetic energy, and temperature; $X/D =$ 10 *(Sislian, Jiang, and Cusworth 1988).*

very low in the flame, indicating that the flow is almost laminar, while the cold jet flow is turbulent with the kinetic energy reaching a maximum in the core. The mean axial velocities are consistently higher in the combusting jet. Beyond $X/D = 15$, the turbulent kinetic energy is higher almost everywhere across the flow. The growth of turbulence due to higher velocity gradients outweighs the decay of turbulence due to higher kinematic viscosities. The radial position where the turbulent kinetic energy attains its maximum value, and is a measure of the maximum turbulent mixing intensity in the flame, does not coincide with the radial position of the flame front.

Radial profiles of the turbulence normal intensities $\overline{u'^2}$ and $\overline{v'^2}$ at $X/D = 10$ are shown in Fig. 2-4. As in the case of turbulent kinetic energy, the magnitude of the normal stresses are generally smaller in the jet flame than in the cold jet. Further downstream they become appreciably larger than the corresponding cold flow values. In both flows, maximum values of $\overline{u'^2}$ and $\overline{v'^2}$ occur at the same radial positions. Radial distributions of the turbulent shear intensity $\overline{u'v'}$, in noncombusting and combusting flows are shown in Fig. 2-5. In both cases the turbulent shear intensity is zero on the jet axis and has a peak near the positions of maximum radial gradients of the corresponding mean axial velocities. The radial positions of these peaks in the hot flow tend toward the mean position of the flame front at $X/D = 40$. At $X/D = 10$ the values of $\overline{u'v'}$ are approximately equal. Further downstream, $\overline{u'v'}$ in the cold jet

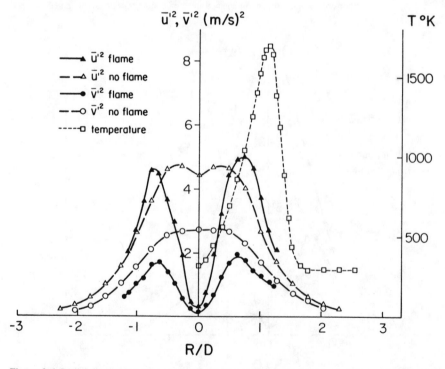

Figure 2-4 Radial distributions of normal turbulent intensities and temperature; $X/D = 10$ (Sislian, Jiang, and Cusworth 1988).

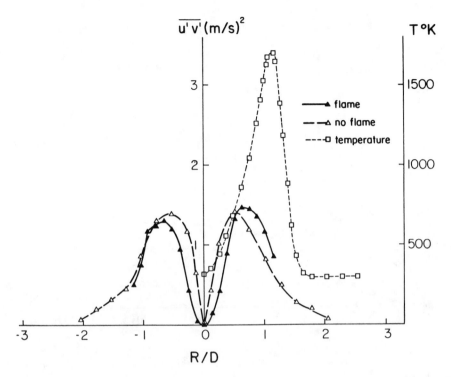

Figure 2-5 Radial distributions of shear stress and temperature; $X/D = 10$ *(Sislian, Jiang, and Cusworth 1988)*.

flow decreases rapidly, its value at $X/D = 30$ being already very small, while in the combusting jet, $\overline{u'v'}$ increases slightly up to distances $X/D = 15$ and then slowly decreases further downstream, its value being still appreciably higher than in the cold jet. In both flows, the values of $\overline{u'v'}$ are appreciably smaller than the corresponding values of normal turbulence intensities.

The ratio of axial component of fluctuating velocity and mean velocity (relative intensity) as a function of X/D is shown in Fig. 2-6. The relative axial intensity is always smaller in the flame than in the cold flow. Radial distributions showed values of relative axial turbulence intensity of 0.5 in the flame and 1.0 at the edge of the cold jet (Fig. 2-7).

Probability density functions of axial velocity fluctuations u' at $X/D = 20$, $R/D = 1.53$ are shown in Fig. 2-8. For both the combustion and noncombustion cases, deviations from the Gaussian distribution (positive skewness) occur at the edges of the jet flows, and in the combusting case, on the inner side of the flame front. The pdfs of radial velocity fluctuations v' are similar to those of axial fluctuating velocity.

The power spectrum of axial velocity fluctuation u' at various points on the jet flow axis (flame) are shown in Fig. 2-9. The ordinate represents the power spectral estimate divided by the local value of $\overline{u'^2}$. The abscissa is the wave number $k = 2\pi f \sqrt{u}$. At $X/D = 0.3$ the power spectrum for both cold and burning flows has the

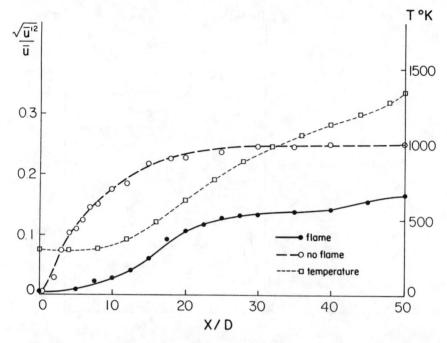

Figure 2-6 Axial distributions of relative axial turbulence intensity and temperature *(Sislian, Jiang, and Cusworth 1988).*

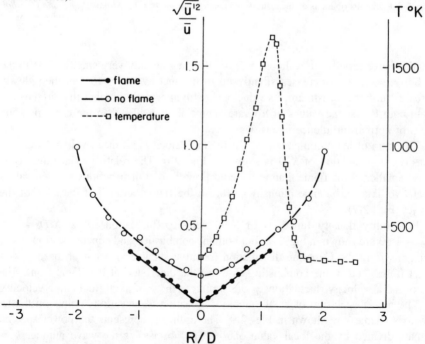

Figure 2-7 Radial distributions of relative axial turbulence instensity and temperature; $X/D = 10$ *(Sislian, Jiang, and Cusworth 1988).*

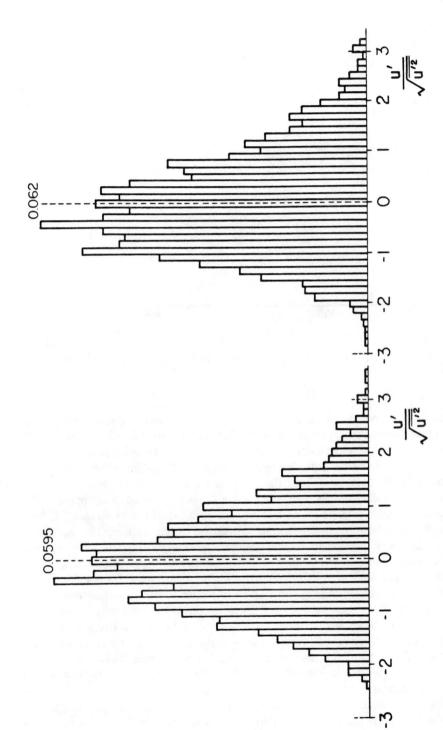

Figure 2-8 Probability density functions of axial velocity fluctuation; $X/D = 20$, $R/D = 1.53$ (*Sislian, Jiang, and Cusworth 1988*).

33

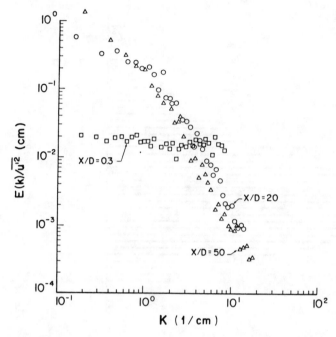

Figure 2-9 Power spectra of axial velocity fluctuation on jet flame axis *(Sislian, Jiang, and Cusworth 1988).*

typical form for homogeneous grid turbulence. The turbulence energy is mostly contained in the wave number range below 10. The energy density increases downstream, especially in the low wave number range. In the high wave number range, turbulence energy density decreases with downstream distance. Turbulent macroscale L, as a function of R/D at several X/D for flame and no flame, is shown in Fig. 2-10. At the initial section, values of L in both cold and hot flows were found to be approximately the same and equal to 0.4 mm. At $X/D = 15$, L increases rapidly with radial distance in the flame, following closely the rate of increase of temperature. The magnitudes of L are appreciably larger in the flame than in the cold jet. Corresponding increases of turbulent macroscale were observed in Schlieren photographs of the flame.

Razdan and Stevens (1985) measured velocity, temperature, and species concentration in a turbulent diffusion flame jet of CO in co-flowing air. For both laminar and turbulent flow, they have shown that unique relationships exist between thermochemical properties and the mixture fraction for the CO air reactive flow. Changes in fluid flow, and geometrical factors with spatial location in the flow, influence only the distribution of mixture fraction within the boundary layer. Once specified at a point in space, the mixture fraction determines the local species mass fractions and the temperature at that point. Comparisons made between laminar flame and chemical equilibrium calculations of thermochemical properties as functions of mixture fraction revealed no significant differences. Turbulent kinetic energy k and dissipation ϵ were calculated from the LDV measurements of velocity and turbulence intensity. Conservation balances were made at several axial locations of carbon, momentum, and total

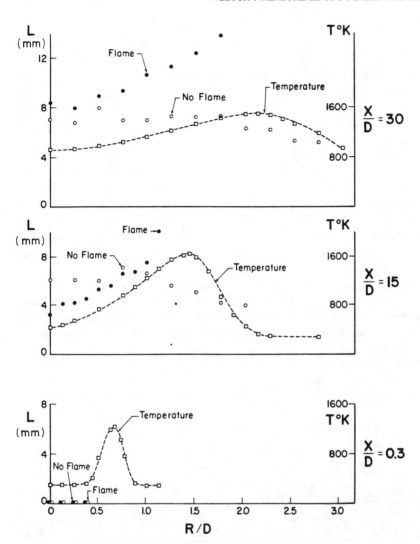

Figure 2-10 Radial distributions of turbulent macroscale and temperatures *(Sislian, Jiang, and Cusworth 1988).*

enthalpy. The study concluded that the modeling of turbulent diffusion flames based on their local laminar flame structure allows separate treatment of the chemistry and mixing problems, so that unique relationships between the thermochemical variables and the mixture fraction can be obtained. These relationships are independent of the flow geometry, initial conditions, and spatial location in the flow.

Reacting Flows in a Swirl-Stabilized Model Combustor

Detailed velocity measurements were made by Brum and Samuelsen (1987), and Charles et al. (1988) in a model combustor with the important features of practical gas turbine combustors, i.e., swirl-stabilized recirculation zone and high-turbulence intensity. A set of swirl vanes was concentrically located within a cylindrical tube for the swirl air, and dilution air was introduced through an outer annulus (Fig. 2-11). Propane was introduced through a central cone-annular gas injector to simulate the directional momentum flux of a hollow-cone liquid spray. Axial and azimuthal mean and rms velocity measurements were made using a two-color LDV system. All streams were seeded uniformly to the same level of concentration with 1-μm alumina particles. Five thousand samples were taken at all measuring points, and velocity bias was avoided by using a low seeding rate. Overall accuracy of mean velocity was ±5%, rms velocity ±10%, and shear stress ±40%. Data were presented for both reacting and nonreacting flows for two overall equivalence ratios φ = 0.1 and 0.2. Turbulence measurements showed isotropy in most of the flow field. A form intermittency with periodic fluctuations of flow field structure was detected in addition to microscale fluctuations. High-speed movies revealed that the recirculation zone experienced a ±10% length change at a rate of approximately 100 Hz, and precession of the vortex core was also detected.

Comparison of measurements made in reacting and nonreacting flows showed that the recirculation zone in the case of reaction was stronger (higher reverse flow velocities), more compact (shorter), and radially wider than for nonreacting flows. Root-mean-square axial velocity levels were approximately 50% higher for the reacting case, and reaction resulted in increased values of normalized Reynolds stress.

In a subsequent study (Charles et al. 1988) of the effect of inlet conditions on combustion performance and flow field structure, it was shown that relatively modest changes in inlet conditions can dramatically affect the flow field structure. The basic

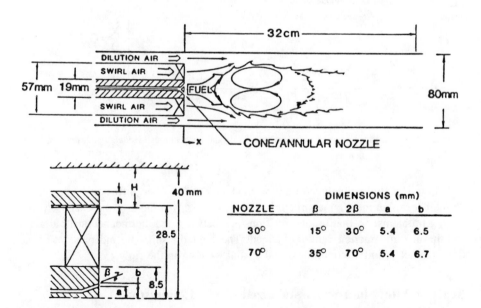

Figure 2-11 Swirl-stabilized model combustor *(Charles et al. 1988).*

aerodynamic and thermal flow structure of the reactor is shown in Fig. 2-12. Axial velocity radial profiles are shown in the top half of the upper figure, while azimuthal profiles are shown in the bottom. Radial profiles of axial velocity show a strong zone of recirculation off-axis (location 1), while the centerline profile of axial velocity reveals a small recirculation zone within the hollow cone of the fuel injector (location 2). Lines of constant stream function, obtained by radially integrating the profiles of mean axial velocity, illustrate the form of the time-averaged flow field and clearly delineate the length and radial extent of the recirculation zone (location 4). Further evidence of off-axis recirculation is given by the mean thermal field, where the peak temperatures coincide spatially with the recirculation zone (location 5). Downstream of the recirculation zone, the temperatures peak at the centerline within a spiraling core that extends to the exit of the combustor.

Bluffbody Stabilization

Moreau et al. (1987) made a detailed study of the flow behind a downstream facing step fed by premixed air methane preheated to 530 K with an equivalence ratio of 0.8. Based on LDV measurements in the flow with and without combustion, it was

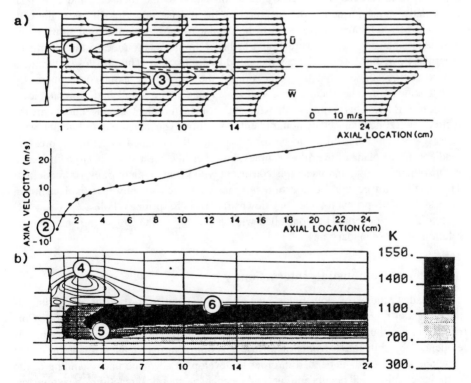

Figure 2-12 Aerodynamic and thermal structure of swirl-stabilized model combustor; 70° nozzle, 100% solidity, $h/H = 0.08$. (a) Velocity profiles and (b) streamlines and temperature *(Charles et al. 1988)*.

found that the recirculation zone behind the bluffbody in a confined flow is reduced in length by combustion due to the increase in volume flow rate of the combustion gases rather than to any turbulence modification. Turbulence within the recirculation zone is modified when combustion occurs. The anisotropy of turbulence is enhanced, the Reynolds stress is strongly increased, and the location of the maximum of turbulent kinetic energy is displaced.

Flow in a Gas Turbine Combustor

Heitor and Whitelaw (1986) measured the isothermal and combusting flow character-istics of a model can-type gas turbine combustor. Density-weighted longitudinal and tangential velocity profiles were measured by LDV with temperature by thermocouple and concentrations of O_2, UHC, H_2, CO_2, CO, and NO_x. The mean velocity results show that, relative to isothermal flow, combustion increases the strength and decreases the width of the primary vortex and, further downstream, attenuates the magnitude and strength of the swirl by the longitudinal acceleration of the flow. Maximum reverse velocity was 12 rather than 4 m/s, and velocities between the recirculation zone and the combustor walls also increased from 10 to 14 m/s with consequently steeper radial gradients across the shear layer. The double-cell structure of the isothermal flow appeared to have been destroyed by the combustion of the fuel jets, and a toroidal recirculation zone formed with a decrease in the reverse flow velocity near the cen-terline. Swirl velocities were found to be similar to those in isothermal flow. Local turbulence intensity values were found to be of similar magnitude to those with is-othermal flow, but local variances increased by up to 50%. The profiles of turbulent kinetic energy suggest that the precession of the core of the axial vortex increased with combustion. The shear stress $\overline{u''v''}$ was found to be small and comparable to isothermal flow. This suggests that the increase in turbulent kinetic energy in the recirculation zone is caused by nongradient production. The increase in turbulent kinetic energy was considered likely to be related to the self-sustained mean radial pressure gradient. In the plane of the dilution holes, however, the shear stress is large compared with isothermal flow, and its sign is consistent with mean gradient transport modeling. The results showed that combustion is more controlled by physical than chemical processes in the primary zone, but downstream of the primary holes, fuel and CO burnup rates are chemically controlled with quenching of the CO-to-CO_2 reaction by dilution of the secondary air.

Turbulence Combustion Interactions

The detailed studies of flow fields by LDV are leading toward an understanding of the effects of combustion on the structure of turbulence and turbulent transport. Cor-respondingly, the effects of turbulence on chemical reaction rates are emerging, as evidenced by thin flame fronts or diffuse reaction zones. Extensive mapping of the flow and scalar fields is needed in order to develop an adequate understanding of the complex coupling that occurs among fluid mechanics, chemical reactions, heat transfer, and other physical phenomena involved. The fundamental questions as to whether turbulence is increased or suppressed by combustion and whether flame-generated turbulence exists can be answered for specific flows that have been measured in detail.

In most combustion systems the heat release accompanying combustion results in acceleration of the flow, and hence higher velocity fluctuations are found in the higher velocity regions of the flow. This can be considered as flame-generated turbulence because turbulent kinetic energy is higher with combustion than without. In uniform density flows, turbulent kinetic energy is generated by the action of the turbulent shear stresses on the mean velocity gradients. In shear flows with combustion, the mean shear is often greatly enhanced by the heat release. Reynolds stresses are also increased in line with the increased turbulence levels, and so production of turbulent kinetic energy by the mean shear is greatly increased. The increase in turbulence can simply be explained as being due to the increase of shear without requiring the invocation of flame-generated turbulence as a separate or special turbulence-generating mechanism.

Bilger (1986) and coworkers made detailed measurements with LDV in jet diffusion flames of hydrogen that allowed the determination of a budget of turbulent kinetic energy by balancing terms in the turbulence energy equations. It was found that turbulent kinetic energy was lower when normalized by the excess velocity on the jet centerline than in nonreacting jets, as were all the terms in the turbulence energy equation. It was concluded that turbulence is shear generated and can be modeled in much the same way as isothermal flows, the lower values of the turbulent kinetic energy arising from the lower contribution of advection to the balance. The structure of the turbulence is, however, different, the normal stresses being much less isotropic, and $\overline{v''^2}/\overline{u''^2}$ being about 0.5 as against 0.9 in isothermal flows.

Joint LDV Scalar Measurements

Simultaneous measurement of velocity and temperature and/or velocity and species concentration allows direct measurement of correlation terms in the energy and species concentration conservation equations. Combined instruments use laser beams focused at a point in the flow with separate collection for the velocity, temperature, and species concentration measurements. Penner, Wang, and Bahadori (1985) reviewed the various optical methods that have been used for scalar measurements. These include Mie scattering (MS) from particles in the flow, laser Rayleigh scattering (LRS) from molecules and particles, laser-induced fluorescence (LIF), spontaneous Raman scattering (SRS), and coherent anti-Stokes Raman scattering (CARS). Bilger and coworkers (Starner and Bilger 1986; Dibble, Masri, and Bilger 1987) used joint LDV-MS to derive instantaneous density and mixture fraction from the Mie signal. Driscoll, Schefer, and Dibble (1982) reported on joint LDV-LRS from which density distributions were determined. Cheng (1984) used two-color LDV and a thin flame front assumption to allow determination of the density from the joint velocity pdf. Starner and Bilger (1986) used the joint LDV-MS technique to determine budgets for the normal stresses and Reynolds shear stresses in jet diffusion flames with and without weak swirl. In these flames the effects of the mean pressure gradient are small, and the balance of terms is much like that in isothermal flow except that advection is lower. In confined swirling flows, however, the mean pressure gradient terms are significant. Laser sheet lighting is being used to obtain instantaneous two-dimensional images from the flow, showing the evolution with time of coherent structures in flames. These images are being related to conditional averaged measurements in order to establish the mean structure of eddies and large-scale motions.

Measurements in Spray Flames

For the burning of liquid fuel in spray flames, the presence of liquid drops affects the velocity because, in general, particles of different size have different velocities. There is both an interest and necessity in knowing drop size, from which information on vaporization rate, liquid flux, and air fuel ratio can be determined. Information for individual drops of velocity and size and relative velocity provides information on local drop Reynolds numbers, drag coefficients, and trajectories. When a standard LDV is used in such two-phase flows, the local averaged velocities are based on information from drops of all sizes passing through the measurement volume, a measurement that has little value. For this reason, it became necessary to devise an instrument that could simultaneously measure drop size and velocity.

A light-scattering technique developed by Bachalo and Houser (1984) for spray analysis is used to make drop size measurements in spray flames. The technique is based upon the measurement of the interference fringe pattern produced by spherical drops passing through the intersection of two laser beams. Three detectors, separated at fixed spacing, are used to receive Doppler signals and to determine the phase shift due to the different path lengths of the laser beam. The spatial frequency of the interference fringe pattern is linearly proportional to the measured drop size. The most important capability of this instrument is that it provides simultaneous drop size and velocity measurements. Other local point information provided by this instrument includes liquid flux, number density, size velocity correlation, and spatial and temporal mean drop sizes. A detailed validation study has been performed at Carnegie Mellon University (Mao, Wang, and Chigier 1986a) to determine the limitations and accuracy of this instrument. A rotating disk and an impulsed piezoelectric drop generator were used as calibration devices. The optics were configured to provide 5.8 mm of fringe spacing. A 3.75× beam expander and a 762-mm focal length lens were used on the transmitter. The beam separation distance was 82.5 mm. The receiver was arranged at 30° off the forward axis direction with a 495-mm focal length lens.

Results from burning and nonburning sprays have been reported by Mao, Wang, and Chigier (1986a). An air-assisted research atomizer was selected to generate the mixed spray flames. Unlike turbulent diffusion flames, the structure of mixed spray flames is highly dependent upon the initial drop size and velocity distributions. On the basis of the measurements and observations of air-assisted atomizer spray flames, a physical model of the flame structure has been constructed and is shown in Fig. 2-13. Drops enter the flame front as a hollow-cone spray. The large drops are mostly distributed on the outer edge of the spray, and small drops are located in the center region. The spray boundary is wider than the flame boundary. Large drops distributed on the edge of the spray are slightly affected by the presence of the flame. Due to the slower evaporation rate of large drops, a blue lean reaction sheath is formed inside

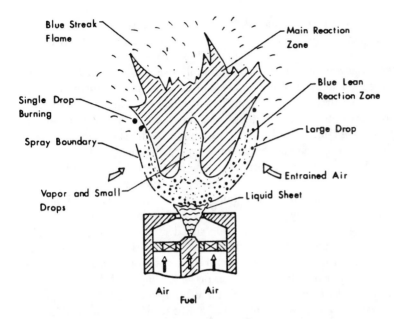

Figure 2-13 Physical model of spray flame *(Mao, Wang, and Chigier 1986a).*

the spray boundary. The hatched area indicates the main reaction zone in which the mixture of fuel vapor and air is sufficient for combustion. In the central core of the spray, the fuel vapor concentration is too rich to allow chemical reaction to take place. Above the spray flames, tiny blue streak flames can be seen flying in many directions. These tiny flames are produced by very small drops with individually attached dilute fuel vapor trails. Single-drop burning with self-sustained envelope flames is seen occasionally at the outer periphery of the spray flame. A swirling airflow is introduced inside the atomizer to atomize the liquid fuel and to stabilize the spray flame. Measurements of the mean drop size, number density, liquid flux, and drop mean axial velocity distributions were made at various radial and axial stations in the sprays. Interesting features of the spray flames are described and compared with the structure of nonburning sprays.

The measurements are concentrated in the flow field near the atomizer exit, where the light-scattering technique can be applied before drop evaporation is complete. In the flame reaction zone and far downstream of the sprays, the measurements are exceedingly difficult to make because of the high-intensity flame radiation and extremely low number density of drops. Even with the radiation line filter installed in the detector, the signal-to-noise ratio is too low to obtain accurate information. Therefore data are not reported in the flame reaction zone and far downstream regions.

Measurements of the variation of Sauter mean diameter (SMD) at several axial and radial stations in a nonburning air-assisted atomizer spray show that the small drops are located in the center region and SMD increases gradually along with the increase of radial distance. Larger drops possess higher initial momentum, which allows them to penetrate farther and to wider distances. Inside the region of 1-cm radial distance, the SMD increases with axial downstream distance. This is primarily caused by the effects of collision and spatial dispersion of drops in the spray. At axial distance $x = 7.5$ cm, the drops are distributed more uniformly over the entire cross-sectional area. It was also found that the distribution mode of the drop size spectra did not vary significantly throughout the spray. A Rosin-Rammler size distribution mode is considered to be appropriate to describe the size spectrum variation for most regions in the nonburning sprays. Figure 2-14 shows the variation of SMD at different axial and radial stations in the burning air-assisted atomizer spray. Because the flame front is located at about 1.5 cm above the atomizer exit, the liquid sheet breakup distance is shortened due to the presence of the flame. In the region before the flame front, $x = 1$ cm, the shape of the radial size distribution profile is still similar to that of the nonburning sprays. After entering the flame zone, $x = 3.5$ cm, the radial size distribution profile is changed dramatically. The drop size decreases to a minimum value at the high-temperature region, which is about 1.5 cm off-axis, the then increases toward the outer edges of the spray. A portion of the large drops manages to escape from the flame without much preheating and results in the higher SMD values on the outer edge of the spray. Beyond $x = 5$ cm, the central spray core has completely vanished, and measurements in this region could not be made.

Figure 2-15 compares the radial mean drop size distribution at axial distance $x = 3.5$ cm in the burning and nonburning sprays. In the central core region, the SMD in the burning spray is higher than that of the nonburning spray. Outside the spray core, SMD is lower in the burning spray. This result is due to the preferential evaporation of smaller drops. Because small drops are mostly concentrated in the central core region and evaporate faster than large drops, statistically, this results in an increase

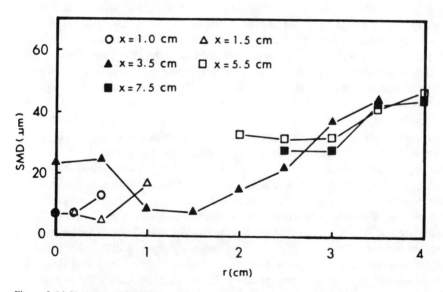

Figure 2-14 Variation of SMD in burning air-assisted atomizer spray *(Mao, Wang, and Chigier 1986a).*

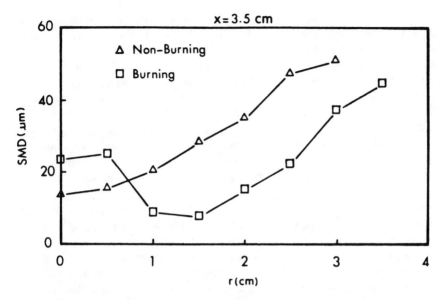

Figure 2-15 Comparison of the radial mean drop size distribution at $x = 3.5$ cm in burning and nonburning sprays *(Mao, Wang, and Chigier 1986a)*.

in average drop size due to evaporation. On the other hand, large drops are concentrated on the outside of the spray, and longer residence times are required for evaporation. In fact, the mean drop sizes approach the same value for both burning and nonburning sprays at the outer edge of the spray. This indicates that large drops at the outer edges of the spray are only slightly affected by the presence of the flame. Under burning conditions, it was found that a bimodal size distribution exists outside of the flame reaction zone, and the relative number of large drops gradually increases toward the outer edge of the spray.

Number Density Distribution

The number density is an important parameter for characterizing and understanding the collective behavior of drops in sprays. The effects of drop acceleration/deceleration and collision can also be evaluated based upon number density distribution information. From the data processing point of view, it is also crucial for obtaining accurate ensemble averages. By taking into account the different rejection rates of data validation at various stations, the total number of data samples is fixed at 2000 for each station. The radial profiles of the number density distribution are found to be rather symmetric about the centerline axis. In general, the number density distribution is uniform in the center region of the spray and then gradually increases with radial distance to reach a peak, followed by a decrease to zero at the spray boundary. The peaks of the radial profiles are located close to the spray boundary. For traverses farther downstream, the value of the maximum number density remains about the same. The radial position of the peak number density shifts outward as axial distance increases due to the effect of spray spatial dispersion. It is interesting to note that the number density increases with downstream distance in the center region of the sprays. This is due to the deceleration of individual drops and entrainment of drops from outer edges of the spray. The drop number density distribution in the burning air-assisted atomizer spray shows that the number density is about 3 orders of magnitude smaller when it is compared with the nonburning spray. Between the atomizer exit and flame front, e.g., $x = 1$ cm, the radial profile of number density is similar to that of the nonburning spray. However, the shape of the number density profiles is dramatically changed as downstream distance increases. In the central core region the number density drops quickly with increase of axial distance. Much higher number density is observed outside of the flame reaction zone, where larger drops are located. Figure 2-16 compares the radial number density distribution at $x = 3.5$ cm with and without combustion in an air-assisted atomizer spray. The effect of flame on number density seems to be insignificant near the edge of the spray. This implies that the drop sizes at the spray boundary may be reduced but the number of drops remains the same under the burning situation. In the burning spray the position of the minimum number density coincides with the highest temperature region, where the fastest evaporation rate occurs. Because the number density in the center region of the burning sprays is rather low, collision of drops could have little chance to occur. The decrease of drop number density is also related to the drop velocity acceleration while passing through the flame zone.

Liquid flux measurements for both nonburning and burning air-assisted atomizer sprays are also reported. This information provides the criteria for evaluating the

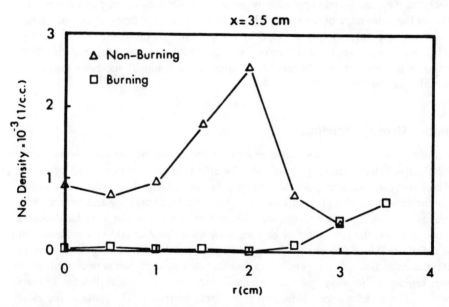

Figure 2-16 A comparison of the radial number density distribution at $x = 3.5$ cm with and without combustion *(Mao, Wang, and Chigier 1986a)*.

evaporation and mixing rate in the sprays. The local fuel-air mixture ratio determines the effectiveness of the spray combustion. The data are obtained via the validated drop size measurements in the probe volume during the entire sampling period. The variation of liquid flux is closely associated with number density change. In the nonburning sprays the liquid flux increases with the radial distance to a maximum value and then decreases toward the outer edges of the spray. The peak value of each profile decreases with the increase of axial downstream distance. The finding of the increase of liquid flux in the center region as axial distance increases is compatible with the results of the increase of number density. The relative increase in the number of large drops contributes to the increase of liquid flux. Far downstream, the liquid flux is distributed more uniformly over the whole cross section of the spray because mean drop sizes and drop velocities become more uniform.

For the burning spray the liquid flux distribution changes drastically in the radial direction. Inside the flame zone, the profile of liquid flux is similar to that of the mean drop size distribution. It first increases radially and then decreases sharply to a minimum at the high-temperature region. After passing through the flame, the liquid flux increases with radial distance. Although the number density changes very little on the edges of the spray, the liquid flux varies a great deal. Apparently, the presence of the flame only affects the size of the drops by evaporation but does not affect the total number of drops at the edges of the spray.

Further understanding of the structure of air-assisted atomizer spray flames requires the description of drop velocity distribution. The radial profiles of drop mean axial velocity and velocity fluctuation at axial distance $x = 1.5$ cm in the burning and nonburning sprays are shown in Fig. 2-17. At this axial position, it is found that the

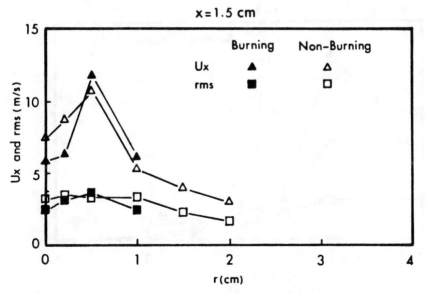

Figure 2-17 The variation of radial drop mean axial velocity and velocity fluctuation at $x = 1.5$ cm in burning and nonburning sprays *(Mao, Wang, and Chigier 1986a)*.

drop mean axial velocity increases with radial distance to reach a maximum value at about 0.5 cm off-axis and then decreases toward the outer edge of the spray. The presence of the flame does not seem to affect the shape of the drop mean axial velocity distribution, and the velocity field retains the character of a swirling jet. In the center region the drop mean axial velocity is smaller for the burning spray than the nonburning spray. To explain this observation, it is necessary to examine the pressure gradient in the region of the flame front and to describe the velocity distribution of the whole drop size spectrum. In front of the flame, the velocity of the individual drop is reduced by encountering a higher pressure gradient induced by the flame and also due to the reduction of drop size. In a previous study of nonburning air-assisted atomizer sprays (Mao, Wang, and Chigier 1986*b*), evidence was presented for an initial acceleration of all drops as they emerge from the atomizer. The large drops are accelerated at a faster rate than the smaller drops for a very short distance, followed by faster deceleration.

Farther downstream, the effect of chemical reaction plays a more important role in changing the velocity fields. Figure 2-18 shows the radial variation of drop mean axial velocity and velocity fluctuation at $x = 3.5$ cm in both burning and nonburning sprays. Both sprays consist of rapidly moving drops in the central core region and are surrounded by slow moving drops on the periphery. Due to the presence of the flame, the surviving drops have greatly increased their axial velocities. In this region, the increase of drop mean axial velocity in a burning spray could be caused by the reduction of drop drag forces, volumetric expansion of the hot gases, and considerable decrease of drop number density. In the nonburning sprays the drop mean axial velocities of different sizes are rather uniform over most of the spray area except at the edges of

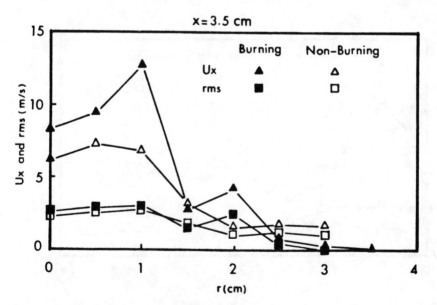

Figure 2-18 The variation of drop mean axial velocity and velocity fluctuation at $x = 3.5$ cm in burning and nonburning sprays *(Mao, Wang, and Chigier 1986a)*.

the spray. For a burning spray the small drops seem to have higher velocities than the large drops at the same measurement point. Measurements of the drop velocity fluctuations are very similar for both burning and nonburning sprays.

A comparison among the centerline variations of SMD, drop number density, and drop mean axial velocity in burning and nonburning sprays is shown in Fig. 2-19. In the nonburning air-assisted atomizer spray, the mean drop size and drop number density gradually increase with centerline downstream distance, while the drop mean axial velocity decays continuously with axial distance. For a burning spray, however, the SMD remains relatively constant until close to the end of the spray core. The sudden increase of the SMD is due to the preferential evaporation of all the small drops. Figure 2-19(*b*) shows that the position of peak number density is in front of the flame. Figure 2-19(*c*) indicates that drops start to accelerate after entering the flame reaction zone.

2-3 CONCLUSIONS

In turbulent jets the flame suppresses turbulence in the upstream region and enhances it in the downstream region, where turbulence intensities are substantially higher than in the corresponding cold jet flow.

Local ratios of turbulence intensity to mean velocity are smaller in jet diffusion flames than in cold jets and become comparable to relative turbulence intensities in downstream regions of cold jets.

Turbulence is appreciably more anisotropic in the jet diffusion flame than in the corresponding cold jet in all regions of the flow.

Turbulence macroscale is significantly smaller than in the cold jet flow in the upstream region and increases appreciably at downstream distances. The rate of this increase closely follows the rate of temperature increase.

Recirculation zones in reacting swirling confined flows are stronger, more compact, and wider than for nonreacting flows.

Turbulence within recirculation zones is modified when combustion occurs. The anisotropy of turbulence is enhanced, the Reynolds stress is greatly increased, and the location of the maximum of turbulent kinetic energy is displaced.

Combustion attenuates the magnitude and strength of swirl by the longitudinal acceleration of flow in downstream regions of swirling flow. Precession of the core of the axial vortex increases with combustion.

In shear flows with combustion, the mean shear is often greatly enhanced by heat release. Reynolds stresses are increased, and production of turbulent kinetic energy by mean shear is greatly increased.

The structure of air-assisted atomizer sprays is dramatically changed by combustion. Significant changes occur in both magnitude and shape of radial distribution profiles of each spray property.

The intermittency between burned and fresh gases contributes greatly to the turbulence intensity increase associated with combustion in premixed burner flames.

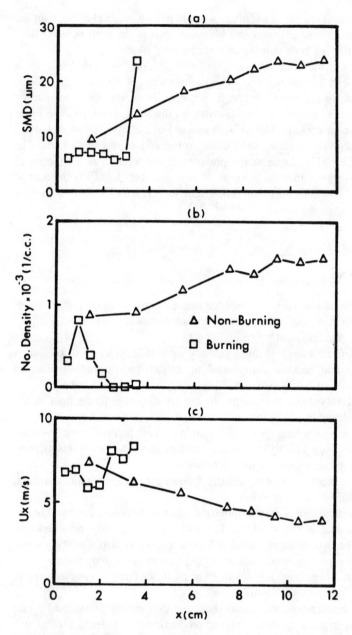

Figure 2-19 (a) SMD, (b) number density, and (c) drop mean axial velocity variation along the centerline for burning and nonburning sprays *(Mao, Wang, and Chigier 1986a).*

2-4 NOMENCLATURE

D	nozzle exit diameter
f	frequency
k	turbulent kinetic energy; wave number
L	spatial turbulence macroscale
R	radial distance
u	axial velocity component
v	radial velocity component
X	axial distance

REFERENCES

Bachalo, W. D., and Houser, M. J. 1984. *Opt. Eng.*, 23:583–90.

Ballantyne, A., and Bray, K. N. C. 1977. *Sixteenth symposium on combustion*, pp. 777–87. Pittsburgh, Pa.: The Combustion Institute.

Beer, J. M., and Chigier, N. A. 1972. *Combustion aerodynamics*. New York: Elsevier Applied Science. (Reprinted by W. Krieger, 1983.)

Bilger, R. W. 1986. *9th Australian fluid mechanics conference*, pp. 545–48.

Bilger, R. W., and Beck, R. E. 1975. *Fifteenth symposium (international) on combustion*, pp. 541–52. Pittsburgh, Pa.: The Combustion Institute.

Brum, R. D., and Samuelsen, G. S. 1987. *Exp. Fluids* 5:95–102.

Charles, R. E., Emdee, J. L., Muzio, L. J., and Samuelsen, G. S. 1988. *Twenty-second symposium (international) on combustion*. Pittsburgh, Pa.: The Combustion Institute.

Cheng, R. K. 1984. *Combustion Sci. Technol.* 41:109–42.

Dibble, R. W., Masri, A. R., and Bilger, R. W. 1987. *Combust. Flame* 67:189–206.

Driscoll, J. R., Schefer, R. W., and Dibble, R. W. 1982. *Nineteenth symposium (international) on combustion*, pp. 477–85. Pittsburgh, Pa.: The Combustion Institute.

Glass, M., and Bilger, R. W. 1978. *Combustion Sci. Technol.* 18:165–77.

Heitor, M. V., and Whitelaw, J. H. 1986. *Combust. Flame* 64:1.

Lewis, M. H., and Smoot, L. D. 1981. *Combust. Flame* 42:183–96.

Mao, C. P., Wang, G., and Chigier, N. 1986a. *Twenty-first symposium (international) on combustion*. Pittsburgh, Pa.: The Combustion Institute.

Mao, C. P., Wang, G., and Chigier, N. 1986b. *Atomisation Spray Technol.* 2:151–69.

Moreau, P., Labbe, J., Dupoirieux, F., and Borghi, R. 1987. *Turbulent shear flows*, vol. 5, pp. 337–46. New York: Springer.

Penner, S. S., Wang, C. P., and Bahadori, M. Y. 1985. *Twentieth symposium (international) on combustion*, pp. 1149–76. Pittsburgh, Pa.: The Combustion Institute.

Razdan, M. K., and Stevens, J. G. 1985. *Combust. Flame* 59:289–301.

Sislian, J. P., Jiang, L. Y., and Cusworth, R. A. 1988. *Progress in energy and combustion science*, vol. 14, pp. 99–146. New York: Pergamon.

Starner, S. H., and Bilger, R. W. 1986. *Twenty-first symposium (international) on combustion*. Pittsburgh, Pa.: The Combustion Institute.

Takagi, T., Shin, H. D., and Ishio, A. 1981. *Combust. Flame* 41:261–71.

Thompson, H. D., and Stevenson, W. H. 1979. *Laser velocimetry and particle sizing*. New York: Hemisphere.

Toshimi, T., Hyun-Dong, S., and Akiza, I. 1981. *Combust. Flame* 40:141–40.

You, H. Z., and Faeth, G. M. 1982. *Combust. Flame* 44:261–75.

Yule, A. J., Chigier, N. A., Ralph, S., Boulderstone, R., and Ventura, J. 1981. *AIAA J.* 19:752–60.

THREE

PARTICLE AND FLOW FIELD HOLOGRAPHY

J. D. Trolinger

3-1 INTRODUCTION

Background

Holography (Vest 1979; Abramson 1981; Goodman 1968; Collier, Burckhardt, and Lin 1971; Kiemle and Ross 1969; Caulfield and Lu 1970; Smith 1977) has satisfied a need that is so basic to the study of particle and flow fields that its use has continued to grow steadily since the first application. Even though the data-handling problem has not been entirely solved, the number of applications requiring the features of holography has continued to fuel development and use. With present technology, holography is the only method for accurately recording a three-dimensional image of a dynamic event. This is particularly useful, for example, in the microscopic examination of combustion and explosive events, droplet breakup and formation mechanisms, and velocity measurement of droplet fields. Holography has likewise made some types of flow visualization possible that have not been achievable by other methods. Specifically, the applications of interferometry to directly compare two flow fields and applications of interferometry in facilities with poor optical windows are outstanding.

Holography was invented in 1947 by Gabor (1948), who produced holograms of a variety of microscopic samples using partially coherent light, since lasers would not be available for another 15 years. His application exploited the magnification properties of holograms and not the three-dimensional imaging properties. Another 20 years passed before holography saw application in a field environment in the studies of fog droplets by Thompson, Ward, and Zinky (1965, 1967).

The first application to flow diagnostics was by Brooks, Heflinger, and Wuerker (1965), who used holographic interferometry to analyze the flow field around projectiles in flight. Since then, virtually every test facility in the world with a need for flow diagnostics has incorporated holography (Trolinger, Farmer, and Belz 1969a; O'Hare and Trolinger 1969; Surget, Delery, and Lacharme 1977).

The development of data-handling techniques lagged far behind what was required to compete with conventional photographic techniques. In recent years, great strides have been made in removing this obstruction. Available computer power, new codes, and detector arrays have played key roles (*Optical Engineering* 1985a, b).

Flow visualization holography has not replaced more conventional methods, but rather has become a complement. When a distinct set of conditions or requirements exists, flow visualization holography is often the best or least expensive or, quite often, the only way to obtain a desired set of data. It is otherwise usually the more difficult method, since it involves less refined, more complicated hardware.

How Holograms Work

Ideally, a hologram of a scene can be thought of as a window into the past, through which a precise image of a scene can be viewed even after the actual scene is physically removed. Having the hologram is, from a viewing standpoint, equivalent to having the actual scene behind the hologram frozen in time. The radiation field on the viewer's side of the hologram has the same information content in either case. This radiation field, which is unique to a given object field, can be extremely complex. It is synthesized by illuminating the hologram with a simple wave front, which, through the process of diffraction, is converted into the more complex one. We are referring to the coherent radiation field that originally was reflected from or passed through the object field. Radiation generated by the object field must be excluded, since it is not coherent and it simply adds a bias to the coherent radiation; it is not involved holographically.

Holograms can be made by many methods on many types of material. They can be generated by computer, drawn by hand, or produced by photographic methods. The discussion here is limited to optical holograms. A hologram can be made by recording the intensity of the sum of two or more waves, one of which must be a reproducible wave (the reference wave), and the other of which is the modulated wave carrying phase and amplitude information (the object wave).

Classifying Holograms

Holograms are classified according to the recording material, the location relative to the object and reference wave, the direction of the lighting of the object field, the type of lighting, and the use of lenses in the recording. Figure 3-1 summarizes this taxonomy. Two basic types of holography are in-line (made with the object wave almost parallel to the reference wave) and off-axis (made with a relatively large angle between the object and reference wave).

An in-line hologram can be made by passing a simple wave front (plane or

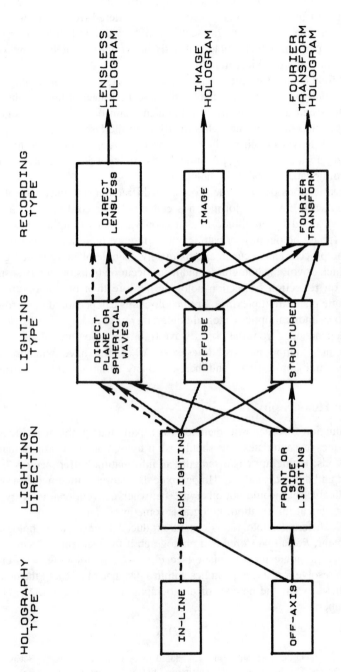

Figure 3-1 Hologram classification. Shown here are 27 different classifications of a hologram.

spherical) through the field, which must be about 80% transparent, and recording the interference pattern of the scattered (object) and unscattered (reference) waves. Off-axis holograms are made by mixing the scattered light with a mutually coherent wave that has taken a path around the field. Off-axis holography is much more versatile with respect to the type of object illumination.

An object field can be front lit or back lit, with direct, diffuse, or structured light. The most common type of structured light is the so-called sheet of light produced with a cylindrical lens and a small-diameter collimated beam of light. Direct lighting is derived by expanding or diverging light directly from the laser.

Imaging with lenses or mirrors before recording is commonly done to impart a convenient size, location, or magnification to the field of interest, or to relax the required recording capabilities of the hologram.

Other classifications describe the process of diffraction, the character of the recording material, and how the information is coded before recording. The diffraction process can be caused by amplitude transmission variations (amplitude hologram), phase shifting (phase hologram), or path length difference on reflection (surface relief hologram). The process can occur in approximately a single plane (thin hologram) or in a volume (thick or volume hologram), and it can occur on transmission (transmission hologram) or on reflection (reflection hologram). More than one hologram can be stored on a single plate (multiplexing), or the holograms can be stored in narrow strips by coding with cylindrical optics (integral holograms) or coded by filtering one of the dimensional dependences from the object wave (rainbow holography).

Clearly, a large number of possibilities exist. All of these have found use in some form in aerodynamic applications, and there are many variations yet to be tried.

When to Use Holography

When the evaluation of diagnostic methods for a particular problem includes holographic techniques as candidates, one should ask if holography offers more than other recording methods. Holography is essentially an information buffer memory between the experiment and the data analysis. Holographic data storage often makes a standard technique applicable that would not otherwise be; when conventional imagery cannot store sufficient optical information, holograms sometimes can.

How many different photographs can be produced from a single hologram? A resolution element, R, can be stored in a photograph if the element lies within a field depth $R^2/2\lambda$ near a single focused plane (see Fig. 3-2). A hologram with effective diameter D can resolve R if the element lies within a distance of RD/λ of the hologram. The ratio of these two is the number of photographs in the hologram, N_p, clearly a very large number.

$$N_p f = 2D/R \qquad (3-1)$$

EXAMPLE: A 10-cm-diameter hologram is produced to cover a field of particles having 5-μm minimum diameter. This hologram could contain over 40,000 different, useful photographs.

Holography is especially helpful in dynamic cases where the requirement is to

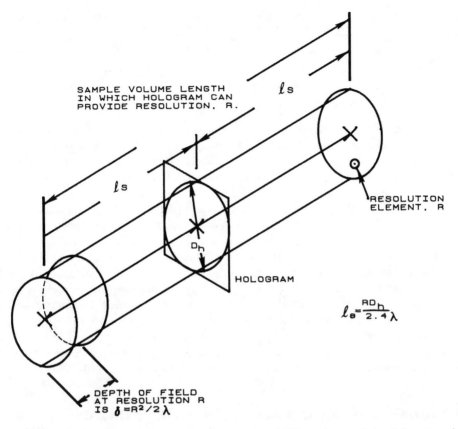

SAMPLE VOLUME LENGTH
IN WHICH HOLOGRAM CAN
PROVIDE RESOLUTION, R.

ℓ s

ℓ s

RESOLUTION
ELEMENT, R

D_h

HOLOGRAM

$$\ell_s = \frac{RD_h}{2.4\lambda}$$

DEPTH OF FIELD
AT RESOLUTION R
IS $\delta = R^2/2\lambda$

Figure 3-2 The sample volume in a hologram. The number of different photographs in a hologram can be estimated by dividing the photographic depth of field at a given resolution into the depth over which the hologram can resolve that resolution.

record high-resolution optical information over a large volume in a short time. But in stationary cases, holograms may not necessarily be superior in storing the three-dimensional (3-D) data, depending on how the data are to be processed later. When a wave front is to be analyzed for phase information, holographic storage is convenient, since the hologram retains all optical information including phase. (But holography is not the only way to store phase information.) When the wave front is extremely complex, such as that from a diffuser or from very poor optics, holography is likely to be the best if not the only candidate.

When a variety of wave front diagnostics is desired, holography provides a method to store the wave front until the diagnostics stage takes place. This is especially true when the experiment time is either short or extremely expensive, since data reduction can be done at a separate time and place at the scientist's convenience.

Finally, when ambient light levels are too great for photography, holography can be used to extend the recording range. The hologram acts like a coherent light filter.

While ambient light does fog the hologram, it plays no direct part in the reconstructed wave. The reference wave intensity can be raised to greatly exceed ambient light level at the hologram. When mixed with the object wave coherently, it provides a type of amplification.

Hologram Quality

The quality of a hologram can be defined by three parameters: (1) diffraction efficiency, (2) resolution or spatial bandwidth, and (3) signal-to-noise (S/N) power ratio, or smallest retrievable object intensity. Diffraction efficiency depends upon the holographic fringe visibility, which is maximum when the object and reference waves are of equal intensity. (This is not where S/N is largest.) In some materials, diffraction efficiencies are above 90%, resulting in extremely bright holograms. Brightness is an important parameter in holographic gratings and in display, but it is of little importance in diagnostics.

Resolution is determined by experimental geometry, hologram size, and quality of optical components. The latter requirement can be replaced by a capability to accurately reproduce the reference wave used to make the hologram, as well as to reproduce the optical geometry in which the recording was made (to subtract out aberrations during reconstruction).

The S/N is determined by the type of emulsion and by the quality of technique such as cleanliness, linear recording, use of liquid gates, and refined processing methods. Table 3-1 summarizes the techniques for improving resolution and S/N.

3-2 HOLOCAMERAS AND RECONSTRUCTION SYSTEMS

General Requirements

A holocamera is a system of lenses, mirrors, lasers, and other optics for making holograms. Choosing from the array of possibilities described above, the system must be engineered to accomplish the following tasks.

1. Properly illuminate the field of interest.
2. Place the interest field or its image in a suitable place relative to the hologram.
3. Achieve an acceptable F number for the desired resolution.
4. Deal with environmental factors.
5. Meet coherence requirements.
6. House the film or plate transport, laser, and optics.
7. Produce a hologram of acceptable quality.

The reconstruction system is the system of optics that provides for the replay of the hologram into a data reduction system. It may be part of the holocamera or it may be a completely separate system. Ideally, the hologram should be illuminated with the same wave as the original reference wave, but practical considerations often prevent

Table 3-1 Techniques for improving hologram quality

Where applied	Signal-to-noise ratio	Resolution
Design	1. Use as few optical elements as possible. Low scatter elements	1. Hologram size must be large enough
	2. Baffle stray light	2. Emulsion resolution high
	3. Spatial filter beams	3. Low object/reference beam angle
	4. Choose an emulsion with low noise, high MTR, high gamma	4. Use flat plates
	5. Have reference wave intensity at least 10 times object wave	5. Place object close to hologram
	6. Object reference angle is bisected by plate normal	6. Use direct (not diffuse light)
		7. High-quality optics (flatness, parallelism, good lenses)
		8. Image and magnify the object before recording
		9. High temporal and spatial coherence
During recording	1. Keep everything clean in the optical train	1. Keep everything stationary to $\lambda/10$
	2. Maintain uniform intensity wave fronts	2. Keep windows thin and normal to light rays
	3. Put as little as possible in the object field	
	4. Expose optimally	
During processing	1. Develop to intensity transmissivity of 0.16–0.20	Same as for high signal to noise ratio
	2. Shorten development time	
	3. Maintain cleanliness	
	4. Final distilled water wash	
During playback	1. Keep optical train free of dust	1. Orient hologram position precisely
	2. Uniform beams	2. Use real image corrected by projecting backwards through any optics originally used
	3. Use a liquid gate	3. Use correct wavelength
	4. Illuminate only the part of hologram required	4. Duplicate original reference wave
	5. Spatial filter the reference wave	5. Spatial filter the reference wave

this. So, when a hologram is made with a pulsed laser, a continuous wave (CW) laser is commonly used in the reconstruction system, and it is not always possible to match wavelengths.

Typical Holocamera

Figure 3-3 illustrates a common holocamera design, using a pulsed ruby laser as a recording laser and a collinear HeNe CW laser as an alignment laser. The CW laser can also be used for reconstruction within the holocamera. Note that the object and

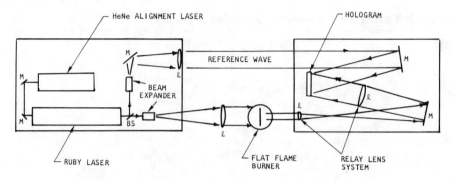

Figure 3-3 Holographic system layout.

reference pathlengths are matched. The object field in this case, a combustion flame, is imaged with magnification to a position near the hologram. In a system like this, where the same rays do not come back together at the hologram, it is important that the lasers have good spatial coherence. This type of holocamera is capable of producing 3-D images of the object field with a resolution of about 10 μm. Multiple pulsing of the laser and use of both beams make possible the recording of multiple 3-D images, with time spacings from almost zero up to about 800 μs.

A ruby laser equipped system can make holograms typically with a 20-ns exposure, which freezes the motion of most fields of interest. Other laser candidates for this type of system are yttrium/aluminum/garnet (YAG) lasers and dye lasers. Ruby lasers were historically the first used, but YAG lasers are gradually becoming more competitive. Compared with the maximum, one multiple pulse per second of commercial ruby lasers, YAG lasers can pulse at rates up to about 30/s, making them ideal for making movies.

Optimizing for Data Reduction

A principal advantage for holography is that the entire wave front passing through the interest field can be recorded and reproduced exactly. In practice, however, real effects make it important to choose a configuration that has been optimized for the highest quality and easiest image analysis. Holograms can be made without lenses; however, it is also highly advantageous to use the best properties of holograms and lenses in combination to get the best results. Even so, every additional optical element added to a system increases optical noise. For this reason, the number of lenses, windows, and mirrors should be limited.

The accuracy with which reconstruction angles are to be matched depends ultimately upon the desired resolution. Image resolution in the range 5–10 μm does not require a highly critical alignment of the hologram. Image resolution below 5 μm requires that this angle be aligned very critically. This is commonly done by making the hologram of a resolution chart and adjusting the angular position of the hologram until the resolution is optimized in the reconstructed image.

3-3 HOLOGRAPHIC INTERFEROMETRY

Interferometry is the mixing of two mutually coherent waves to produce interference fringes that can be used to determine the phase distribution of one of the waves relative to the other. In holographic interferometry (HI), one or both of the waves is derived from a hologram.

Techniques

Basically, there are about seven different types of HI (see Figs. 3-4 and 3-5). These include, from Fig. 3-4, (*a*) interferometry with one nonreconstructed wave front serving as the interferometry reference wave, (*b*) real time holographic interferometry,

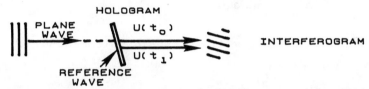

A. HOLOGRAPHIC INTERFEROMETRY WITH A SINGLE HOLOGRAM.

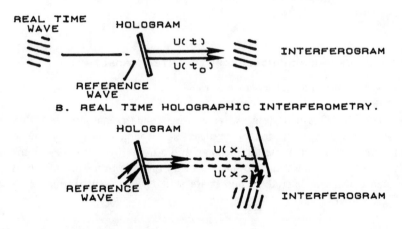

B. REAL TIME HOLOGRAPHIC INTERFEROMETRY.

C. HOLOGRAPHIC WAVE SHEARING INTERFEROMETRY.

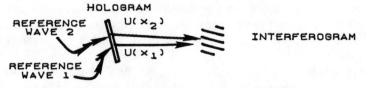

D. TWO REFERENCE WAVE SHEARING.

Figure 3-4 Holographic interferometry with singly exposed holograms.

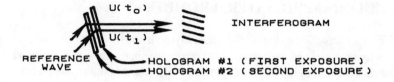

A. DOUBLE PLATE HOLOGRAPHIC INTERFEROMETRY

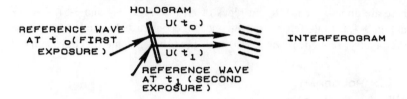

B. DOUBLE REFERENCE WAVE HOLOGRAPHIC INTERFEROMETRY

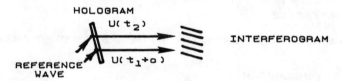

C. DOUBLE PULSED HOLOGRAPHIC INTERFEROMETRY.

Figure 3-5 Holographic interferometry with multiple recordings.

(c) "common path" interferometry (e.g., wave shearing), and (d) interferometry with two time-separated wave fronts from a single plate (e.g., double exposure or double pulsed), and from Fig. 3-5, (a) interferometry of wave fronts from two holograms (e.g., double plate or sandwich), and (b) interferometry of wave fronts reconstructed with two reference waves. Heterodyne and phase shift HI can be performed with methods in Figs. 3-4(a), 3-4(b), 3-5(a) and 3-5(b).

Equations

The following analysis considers the general case, where the interfering waves are derived from two different holograms made at different times, separated by Δz, moved laterally and tilted with respect to each other between exposures, and made with two different reference waves. The resulting general equation can then be specialized to describe all of the cases illustrated.

The general equation will describe double-plate HI. By setting Δz equal to zero, double-exposure, double-reference wave HI results. By setting the two reference waves equal, double-exposure HI results. By allowing one of the reconstructed waves to emulate a real time wave, real time HI results.

In this computation, we are interested in the effect of the relative position of the holograms on the interferogram. We assume that a plane reference wave is used with an originally plane object wave that has been modulated by a phase factor $e^{i\phi c(x)}$, and recorded first on a plate 1, then on a plate 2, which is located behind plate 1 (Fig. 3-6). Then the two holograms are illuminated by plane reference waves, simultaneously reconstructing object waves from both plates. Let us fix the z position of hologram 2 and assume that the position of hologram 1 can move after the recording, and compute a one-dimensional (x) hologram.

We ultimately require the general equation of interference of the two waves reconstructed from the two plates, respectively. The following are the object and

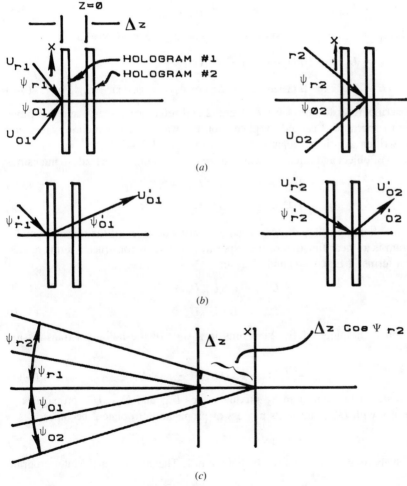

(a)

(b)

(c)

Figure 3-6 (a) Double-plate holographic interferometry; (b) double plate holography—reconstruction; (c) phase shifting between two plates.

reference wave amplitudes in the front plane of the respective holograms during recording and during reconstruction.

In general, a plane wave can be represented by

$$U = u \exp i\{k(x \sin \psi + z \cos \psi) + \phi(x) + \beta\} \tag{3-2}$$

where: U is the complex amplitude,
k is $2\pi/\lambda$,
λ is the wavelength,
ψ is the angle made by the light ray and x axis,
$\phi(x)$ is a phase variation,
β is an aberration caused by optics.

If the z position of hologram 2 is held fixed at $z = 0$, then we can write

$$U_{o1} = U_{o1} \exp i\{k(x \sin \psi_{o1}) + \phi_{o1}(x)\} \tag{3-3}$$

$$U_{o1} = U_{o2} \exp i\{k(x \sin \psi_{o2} + \Delta z \cos \psi_{o2}) + \phi_{o2}(x) + \beta'_{o2}\} \tag{3-4}$$

$$U_{r1} = U_{r1} \exp i\{k(x \sin \psi_{r1}) + \phi_{r1}(x)\} \tag{3-5}$$

$$U_{r2} = U_{r2} \exp i\{k(x \sin \psi_{r2} + \Delta z \cos \psi_{r2}) + \psi_{r2}(x) + \beta'_{r2}\} \tag{3-6}$$

where subscripts refer to object waves 1 and 2 and reference waves 1 and 2. Primes on β refer to aberrations caused by plate 1 on the wave. Reconstruction waves and angles are written as primed quantities similar to Eqs. (3-3)–(3-6).

The sum of object and reference waves on respective holograms leads to intensities

$$I_1 = U_{o1}^2 + U_{r1}^2 + U_{r1}^* U_{o1} + U_{r1} U_{o1}^* \tag{3-7}$$

$$I_2 = U_{o2}^2 + U_{r2}^2 + U_{r2}^* U_{o2} + U_{r2} U_{o2} \tag{3-8}$$

When the holograms are reilluminated with primed waves, if the transmission by the hologram is a linear function of the exposure, then the reconstructed terms related to the third terms in Eqs. (3-7) and (3-8) are

$$U'_{o1} = U'_{r1} U_{r1}^* U_{o1} \tag{3-9}$$

$$U'_{o2} = U'_{r2} U_{r2}^* U_{o2} \tag{3-10}$$

A translation of hologram 1 by Δx results in a shift of the hologram transmission function, or

$$U'_{o1} = U_{r1}^*(x - \Delta x) U_{o1}(x - \Delta x) U'_{r1} \tag{3-11}$$

We create an interferogram by mixing waves U'_{o1} and U'_{o2}. U'_{o1} propagates to plate 2 to mix with U'_{o2}, and in its process of propagation to plate 2 becomes

$$U'_{o1,2} = U'_{o1} \exp i\{k\Delta z' \cos \psi'_{o1} + \beta'_{o1}\} \tag{3-12}$$

having been aberrated an amount β'_{o1} by hologram 1. The intensity in the interferogram is

$$I = |U'_{o1,2} + U'_{o2}|^2$$

$$= U_{o1.2}'^2 + U_{o2}'^2 + 2\,\mathrm{Re}\,\{U_{o1.2}' U_{o2}'^*\} \qquad (3\text{-}13)$$

Defining

$$A = U_{r1}' U_{r1}^* U_{o1} U_{r2}'^* U_{r2} U_{o2}^*$$

and substituting Eq. (3-11) into Eq. (3-12), incorporating the result in Eq. (3-13), and applying Eqs. (3-3)–(3-6), we derive the third term of Eq. (3-13), which represents the fringes of the interferogram.

$$\begin{aligned}
I = 2A \cos\, [&kx(\sin \psi_{r1}' - \sin \psi_{r1} + \sin \psi_{o1} - \sin \psi_{o2} \\
& - \sin \psi_{r2}' + \sin \psi_{r2}) + \Delta x(\sin \psi_{o1} - \sin \psi_{r1}) \\
& + \Delta z(\cos \psi_{r2} - \cos \psi_{o2}) - \Delta z'(\cos \psi_{r2}' - \cos \psi_{o1}') \\
& + [\phi_{r1}'(x) - \phi_{r1}(x)] + [\phi_{r2}'(x) - \phi_{r2}(x)] + [\phi_{o1}(x) - \phi_{o2}(x)] \\
& - \beta_{o1}^1 + \beta_{o2}^1 + \beta_{r2}^1]
\end{aligned} \qquad (3\text{-}14)$$

Note that we have still retained two different reconstruction reference waves, so that the equation is general.

Now consider some special cases.

Case 1: Double exposure holographic interferometry.

Single plate

$$\Delta z = \Delta x = 0$$

Same reference waves

$$\psi_{r1} = \psi_{r2} \qquad \phi_{r1} = \phi_{r2}$$

Single reconstruction wave

$$\psi_{r1}' = \psi_{r2}' \qquad \phi_{r1}' = \phi_{r2}'$$

Same original object wave

$$\psi_{o1} = \psi_{o2}$$

The resulting interferogram is described by

$$I = 2A \cos\, [\Delta \phi_o(x) + \Delta \phi_r(x)] \qquad (3\text{-}15)$$

Note that all of the fixed aberrations have cancelled out of the equation. Also note that reference fringes (finite fringe interferogram) could have been added by changing conditions between the two exposures, so that $\psi_{r1} \neq \psi_{r2}$ or $\psi_{o1} \neq \psi_{o2}$ or $\Delta x \neq 0$. Equation (3-19) describes double-pulsed holographic interferometry and shows how it depicts the phase change $\Delta \phi_o(x)$ between the two exposures.

Case 2: Double reference wave holographic interferometry (DRWHI). Assume the following conditions.

Single plate

$$\Delta z = \Delta x = 0$$

Same original object wave

$$\psi_{o1} = \psi_{o2}$$

The interferogram is described by

$$I = 2A \cos [kx(\sin \psi'_{r1} - \sin \psi_{r1} = \sin \psi'_{r2} + \sin \psi_{r2})$$

$$+ \Delta\phi_o(x) + \Delta\phi_{r1}(x) + \Delta\phi_{r2}(x)] \qquad (3\text{-}16)$$

Equation (3-16) describes an extremely powerful technique. An interferogram containing the data, which is $\Delta\phi_o(x)$, is produced, and it contains a completely controllable set of reference fringes. The controls include the following.

1. Spacing adjustment of the reference fringes during reconstruction by adjusting the relative angles of the two reference waves.
2. Orientation of the reference fringes.
3. Phase shift of the reference fringes by adjusting relative phase of the reference waves.
4. Fringe contrast adjustment during reconstruction by relative intensity adjustment of the reference waves.

Note that the phrase errors in the object and reference waves carry through directly into the final interferogram. However, if the same reference waves are used for recording and reconstruction, the error cancels.

Case 3: Double-plate holographic interferometry (DPHI). Assume (1) the special case where identical reference waves are used for recording and reconstruction and (2) the hologram normal bisects the angle between the object and reference wave. The resulting interferogram is described by

$$I = 2A \cos [\Delta x(\sin \psi_o - \sin \psi_r) + \Delta(\Delta z)(\cos \psi_r - \cos \psi_o)$$

$$+ \Delta\phi_o(x) - \beta^1_{o1} + \phi^1_r(x) - \phi_r(x) + \beta^1_{r2}] \qquad (3\text{-}17)$$

where the phase error on the object beam caused by passing through the first hologram can be cancelled by that introduced onto the reconstruction reference wave for the second. Also note that the reference fringe phase can be shifted by a relative lateral adjustment between the two holograms during reconstruction.

Case 4: Double holographic interferometry with tilt in both plates during reconstruction. (1) Consider a slight tilting of the two holograms during reconstruction, and (2) assume $\psi'_r = \psi_r + \theta$, where θ is a small adjustment.

If identical reference waves are used during recording and reconstruction (see Fig. 3-7), then

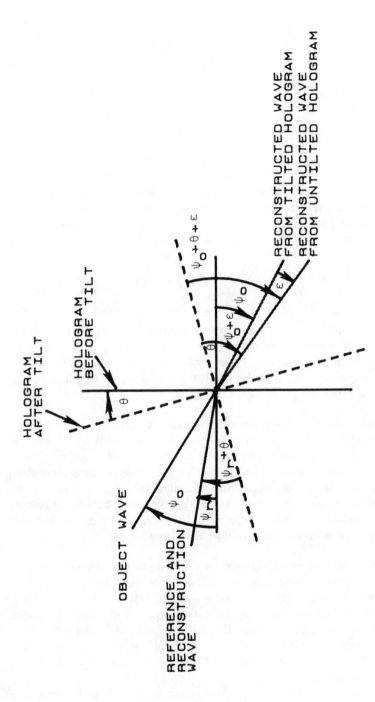

Figure 3-7 Reconstruction with a film tilt error.

$$I = 2A \cos [\Delta x(\sin \psi_{o1} - \sin \psi_r)$$

$$+ \Delta z(\cos \psi_r - \cos \psi_{o2}) - \Delta z'[\cos (\psi_r + \theta) - \cos (\psi_{o1} + \theta + \epsilon)]$$

$$+ \Delta\phi_o - \beta_{o1}^1 + \phi_r'(x) - \phi_r(x) - \beta_{r2}^1] \qquad (3\text{-}18)$$

$$\epsilon = \frac{\theta(\cos \psi_r - \cos \psi_o)}{\cos \psi_o} \qquad (3\text{-}19)$$

Note that a tilt, θ, can therefore be used to phase shift the interferogram.

Case 5: Real time holographic interferometry (RTHI). The reconstructed wave, made at time zero, is mixed with a real time wave that has passed through the region of interest, providing an interferogram that depicts the phase difference of the real time wave and the time zero wave. Equation (3-14) can be applied.

This is normally done by viewing the illuminated real time scene through the hologram. Assuming that the hologram is not moved after its original recording, Equation (3-14) reduces to

$$I = 2A \cos [kx(\sin \psi_{r1}' - \sin \psi_{r1} + \sin \psi_{o1} - \sin \psi_{o2})$$

$$+ \Delta x(\sin \psi_{o1} - \sin \psi_{r1}) + \Delta\phi_o(x) + \Delta\phi_r] \qquad (3\text{-}20)$$

Note that if the same reference wave is used, all aberrations cancel. Even the aberrations introduced on the waves by the hologram substrate are removed, since they are the same for both interfering waves. This important result allows high-quality interferometry to be done through relatively low-quality photographic film and glass plates without added error.

Case 6: Single exposure, live reference wave HI. A degenerative case of real time HI is that in which an interferometer reference wave derived from a laser is mixed with a reconstructed data wave. If the reference wave is passed through the hologram, then Eq. (3-20) applies. This method, which has been used to cancel otherwise troublesome film aberrations, has been termed the film aberration correction technique (FACT). The advantage is that single-exposure holograms can be used for holographic interferometry with considerable fringe control available during data reduction. Such control makes, for example, phase shift or heterodyne HI possible.

Case 7: Local reference wave HI (LRWHI). Transferring the reference wave around the field of interest can sometimes present impossible requirements, for example, in large wind tunnels, long-range cases, and holography between two moving platforms (such as aircraft). For such cases, the reference wave can sometimes be transmitted through a part of the field of interest that does not add phase information to it, for example, through a laminar part of the flow field. Indeed, even if phase information is added to the reference wave, it can be filtered from the wave before it is mixed with the object wave.

In LRWHI, the object wave is split into two or more waves after having traversed the field of interest, one serving as the information wave, the others serving as the reference waves. It is usually best at this point to remove from the latter that part of the wave that has traversed the strong part of the flow field by aperturing, if possible. After this operation, the wave may be a satisfactory reference wave. This can be determined by looking at the resulting interferogram, which characterizes the region from which the reference wave was extracted.

If there is no weak part of the flow field, then it is necessary to spatial filter the phase information from the wave by passing it through a pinhole spatial filter to make it a useful reference wave. Just how much error this adds to the final determination of phase can be seen by examining Eq. (3-13), which contains a term that represents the aberration of the reference wave.

Case 8: Glint holographic interferometry (GHI). A degenerative case of LRWHI occurs when some part of the object or object field contains a suitable point source (glint) that can be used as an object wave (a mirror can actually be attached to the object). Consider the holography of the surface of a fast moving object, such as a projectile that has a shiny part. Light from the region of the shiny part may serve as a reference wave.

Deriving the reference wave from the object wave may provide several benefits. Producing a hologram of the surface of the projectile may, under normal conditions, require extremely short exposures to ensure that the holographic fringes remain stationary during the recording. However, if the reference wave is derived from the object itself, both the object and reference waves are shifted equally in phase by the motion of the object, greatly relaxing the recording conditions on stability.

Of perhaps even greater importance is the impression of the same phase modulation on the object and reference waves by the media between the object and the recording. This results in the cancellation of such information in the recording as though it were not there at all. This might have application in holography through a strong flow field, or in the holography of an object at long distance through a disturbing atmospheric condition. This principle has been used in the holography of burning droplets in rocket exhausts (Trolinger, Farmer, and Belz 1969*b*).

3-4 APPLICATIONS

In the following, a selected number of applications are presented to illustrate the use and state of the art of holography for particle and flow diagnostics. Holography is now being used in hundreds of laboratories, and examples have been contributed by various individuals.

Particle Field Holography

The study of dynamic particle fields was the first real application of holography and is still one of the most important scientific applications. Holograms can optically freeze a 3-D high-resolution image of a particle field that can be studied microscopically long after the actual particle field ceases to exist. Multiple-exposure holography further allows the study of the dynamics of the field. There is currently no other way to perform such a 3-D study.

The field has been enriched by many new refinements and techniques since its first use in the early 1960s. Included in the new technologies are the following.

1. Methods for aberration corrections to provide diffraction-limited imaging with low-quality optics (Wuerker 1976).
2. Methods for subtracting out optical noise normally present in holographic images (Trolinger 1979).
3. Holographic movies (Decker 1981, 1982, 1984).
4. Methods to automatically extract the data from holograms (Hess, Trolinger, and Wilmot 1985; Ewan, Swithenbank, and Sorusbay 1984; Becker, Meier, and Wegner 1982).
5. Multiple-wavelength holography to encode exposures (Kashiwagi, Kashiwagi, and Baum 1983; Mayinger and Panknin 1976).
6. Multiple reference wave holography to provide two independent views of a phase or particle field in the same hologram (Iwata 1977).
7. Thermoplastic recording for in-place electronic processing (Umstatter, Doty, and Trolinger 1985).

Many types of particle fields have been studied by using holography, including the following.

1. Ice, snow, and water droplets in the field of meteorology (Haman 1984).
2. Pollution particulates and fibers (Brenden 1981).
3. Fiber and particle generation in, for example, the fiber insulation industry (Thompson 1980).
4. Spray nozzle development and characterization (McVey 1976).
5. Particle formation and breakup (Craig 1984).
6. Explosions (Sheffield, Hess, and Trolinger 1986).
7. Impact phenomena (Hove 1975).
8. Two-phase flow (Bakrunov 1980).
9. Combustion (Netzer 1982).
10. Crystal formation and growth (Trolinger 1974).
11. Microbiology (Briones 1980).
12. Cloud study from aircraft (Conway 1982).
13. Flow diagnostics by particle tracking (Belozerov 1974; Charwat 1977; Yano 1980).

Spray nozzle diagnostics. Figure 3-8 illustrates data taken from a doubly exposed hologram of a spray nozzle (Hess, Trolinger, and Wilmot 1985). Though the difficulty of presenting three-dimensional data in a 2-D format has not been overcome, this figure illustrates one technique of presenting such data. An overall shot of the field is recorded at relatively low resolution to provide a global look at the subject field. Then microscopic higher resolution inserts are added to show specific details. As previously computed, such a hologram contains many such photographs. The illustrated

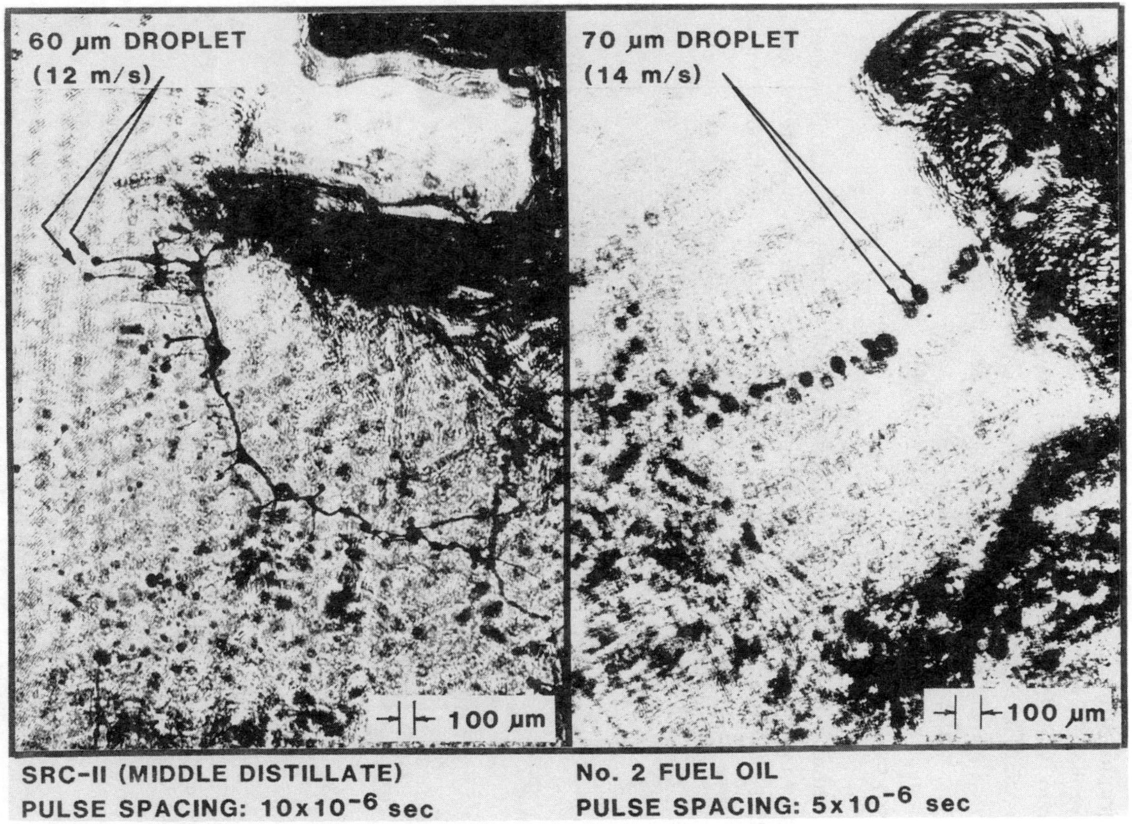
60 μm DROPLET
(12 m/s)
100 μm
70 μm DROPLET
(14 m/s)
100 μm
SRC-II (MIDDLE DISTILLATE)
PULSE SPACING: 10x10^-6 sec
No. 2 FUEL OIL
PULSE SPACING: 5x10^-6 sec

Figure 3-8 Double-pulsed hologram of droplet formation in a fuel spray. Note ligament formation and breakup.

recording is from a 10-cm-diameter hologram. So there are about 40,000 different 10-cm-diameter photographs available (with 5-μm resolution) in this hologram and many more at lower resolution.

Holographic subtraction. Figure 3-9 illustrates the use of holographic subtraction to improve the S/N ratio in a reconstructed image (Trolinger 1975). In this example, the hologram is made with two reference waves, one for the first exposure and one for the second. The reconstructed images of each exposure are illustrated in Figs. 3-9(*a*) and (*b*). In Figs. 3-9(*a*) and (*b*) the two images are reconstructed simultaneously. However, in Fig. 3-9(*c*), one of the reference waves is phase shifted by 180°, so the backgrounds subtract and the images reverse from dark to bright. The improvement in S/N is evident.

Data reduction. Particle field holography is still severely limited because of the difficulty in extracting the data from the hologram. Some image analyzers exist that augment data extraction, but as of yet, no commercial instrument can handle completely

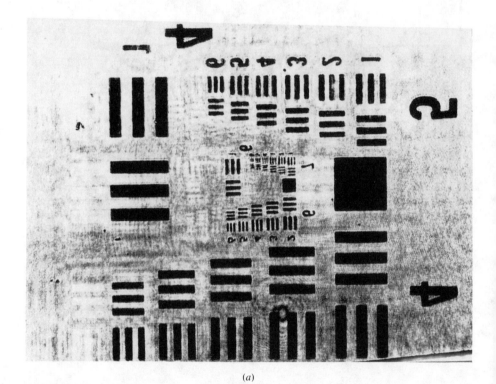

(*a*)

Figure 3-9 Reconstructions from double reference wave hologram with target present in both exposures but shifted vertically for the second exposure. (*a*) Reconstruction from reference wave 1.

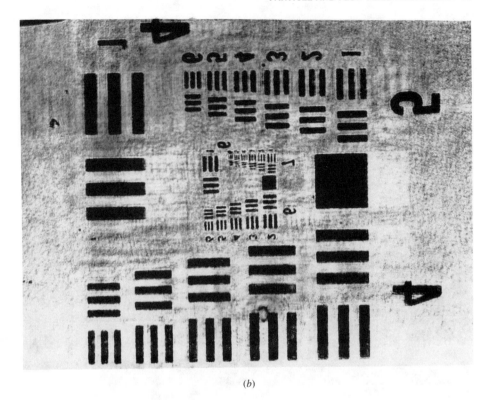

(b)

Figure 3-9 (continued) (b) Reconstruction from reference wave 2.

the problem of distinguishing in and out of focus images and the severe coherent noise commonly associated with such images.

The most basic data reduction procedure is to scan the image volume while counting and sizing individual particles. Particle field images are usually observed on a closed circuit TV monitor or in photographs. In either case, it is important to understand that the depth of field is a function of particle size, so that the sample volume must be corrected for particle size when particles are counted in a single plane.

An alternative method is to examine the scattered light field from the hologram. It can be shown that the Fourier transform of a scattered light field can yield the size distribution. Using this principle, Ewan, Swithenbank, and Sorusbay (1984) and Hess, Trolinger, and Wilmot (1985) demonstrated that the size distributions of particle fields recorded in holograms can be extracted almost instantly.

The limitations up to now on holography data reduction, and the requirement to record at relatively slow rates, is somewhat confining. Other particle sizing instruments that can provide real time data at high rates excel for some requirements, while holography is considered to be the desired method to attain global views and occasional full field examinations.

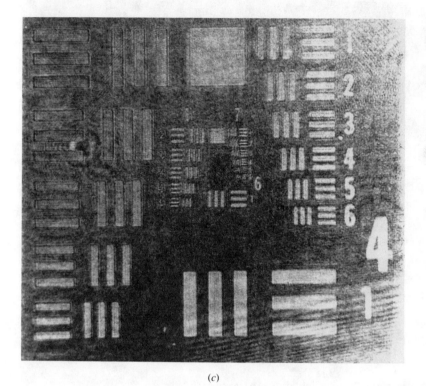

(c)

Figure 3-9 (continued) Reconstructions from double reference wave hologram with target present only during one of two exposures. (c) Two reconstructed beams subtracting.

Flow Visualization Holography

A light wave travels slower through regions of higher refractive index, so the wave fronts in such regions are phase shifted, and the emerging wave front is modulated with phase information that characterizes the flow field. The refractive index is directly related to density in gases, so the gas density is sometimes derivable from a measurement of phase of the wave front.

Shock waves are characterized by a discontinuity in refractive index, and they strongly diffract light. Therefore they can be focused in three dimensions like any object. Smaller refractive index gradients refract the light at smaller angles and cannot be observed in an ordinary image of the field; their effect on the phase of a wave front must be observed with other methods to be described.

In flow visualization, since the object is usually a phase object (for example, a distribution of refractive index), off-axis holography is almost always used. A wave front that has been modulated by the flow field can be recorded in its entirety for later analysis. The later analysis may comprise any of the conventional methods, such as

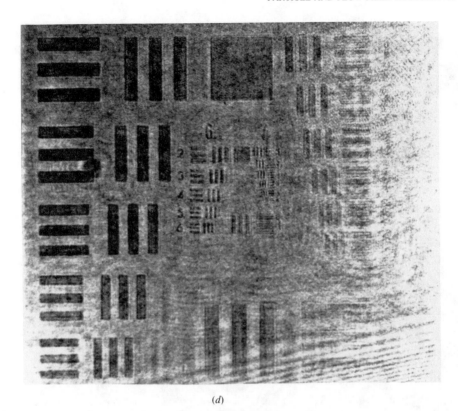

(*d*)

Figure 3-9 (continued) (*d*) Two reconstructed beams adding.

shadowgraph, Schlieren, deflectometry, Hartmann, moire, or interferometry (see Chapter 2). Choice of the method can be postponed, and indeed, all of the methods can be applied. Moreover, many variations for each method can be applied. The seven types of interferometry can be applied, the shadowgraph focus can be varied, the Schlieren knife edge can be adjusted, and the deflection sensitivity can be varied.

A unique advantage of holography is that a wave front modulated by the optical system only can be subtracted from one that has both optical system and flow field modulation, removing the effects of the optics. This can be extremely valuable in large systems, where cost of large diffraction-limited windows and mirrors can be prohibitive. This principle can be carried further to subtract effects of a steady part of an object field, leaving an unsteady part that might otherwise be overshadowed by the steady part. Figure 3-10(*a*) illustrates a case in which a weak unsteady flow is overshadowed by a strong steady flow and can be seen clearly only after subtracting off the steady flow effects. Note the vortex street in the model wake, the unsteady shock wave, and the free stream turbulence in Fig. 3-10(*b*). One can zoom in on specific parts of the flow field with very high resolution as illustrated in Fig. 3-10(*c*), while other parts of the flow field do not require such resolution for analysis.

(a)

Figure 3-10 Viewing steady and unsteady flows. (a) Potential flow field. Supersonic trailing edge flow showing the Prandtl-Meyer expansion of the flow from a Mach number of Ma = 1.3 to 1.9. Contributed by P. Bryanston-Cross.

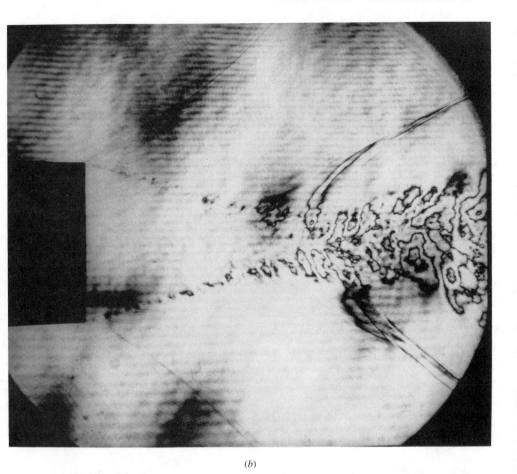

(*b*)

Figure 3-10 (continued) (*b*) Unsteady components of the stable exit flow shown in Fig. 3-12. Contributed by P. Bryanston-Cross.

State of the art systems. YAG lasers have become popular in flow diagnostics because of their repetition rate, which is higher than ruby lasers, which were the first used in this application. Other pulsed laser candidates are dye lasers and Alexandrite, which all have pulse durations in the 20-ns range. CW lasers can be used if the entire system is isolated from vibration, or if the laser can be properly shuttered to sufficiently short exposures. In general, a state of the art system includes two individually modulatable reference waves, a thermoplastic recording device, and an on-line image processor to analyze the interferograms. The methods illustrated in Figs. 3-4 and 3-5 would then be applicable.

A good example of this type of system has been developed by Craig (1982) and is in use in the NASA Ames Research Center two-foot transonic wind tunnel. The

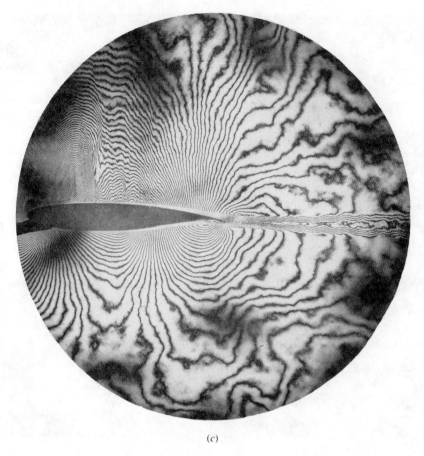

(c)

Figure 3-10 (continued) Viewing steady and unsteady flows. (c) Flow over a NASA airfoil showing high resolution in wake. Contributed by W. Bachalo.

system employs a frequency doubled YAG laser that produces 20 mJ of energy per pulse. Craig has recently demonstrated the on-line capability of the system by employing a thermoplastic device that can reconstruct data directly into an automated fringe reduction system developed by Tan and Modarress (1985).

This basic type of system has been in general use in many wind tunnels since the early 1970s (O'Hare and Trolinger 1969; Burner and Goad 1979; Burner 1973; Bachalo 1979, 1983; Surget 1976, 1982; Kittleson 1983; Ozkul 1979; Veret 1982, 1984; Havener 1983; Havener and Radley 1972; Hannah and Havener 1975) but only in recent years have hardware and technology evolved to the point of on-line data capability.

Shock wave/boundary layer interaction. The interaction of shock waves with boundary layers is an extremely important process in aerodynamics. This process is also too complex to completely describe analytically. Therefore experimental aerodynamics must provide quantitative results for the development and evaluation of theoretical models and computer codes. Holographic flow visualization has played an important role in providing such results.

Figures 3-11(*a*) from the work of Burner at NASA Langley, and 3-11(*b*) from the work of Havener (1983; Havener and Radley 1972; Hannah and Havener 1975), illustrate various capabilities offered by holographic flow visualization for shock wave and boundary layer investigations. These also provide a good comparison of two different flow visualization techniques; Schlieren and infinite fringe interferometry.

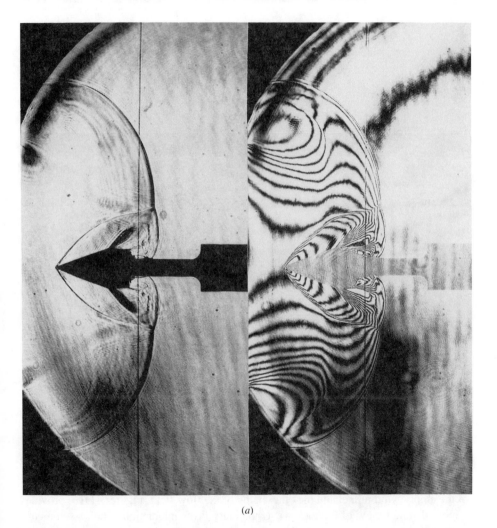

(*a*)

Figure 3-11 Holographic flow visualization providing for various phase sensitive techniques. (*a*) Schlieren on right; Infinite fringe interferometry on left. Contributed by A. Burner, NASA Langley Research Center.

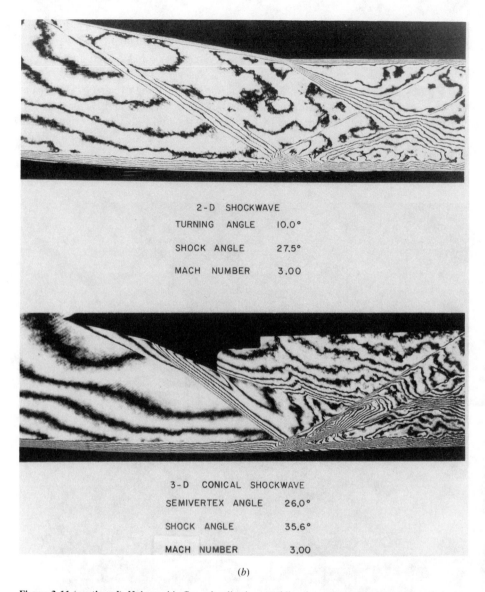

(b)

Figure 3-11 (continued) Holographic flow visualization providing for various phase sensitive techniques. (b) Interferograms showing separated flow fields produced by a shock wave impinging on a two-dimensional boundary layer. Contributed by G. Havener.

The interferograms were produced by the double-plate methods, one hologram being of the empty wind tunnel and the other being of the flow field. Figure 3-11(b) provides a more detailed view of a shock wave/boundary layer interaction. The axisymmetric parts and the two-dimensional parts of the flow field can be analyzed quantitatively from these interferograms, but the 3-D parts cannot.

Bryanston-Cross (1981, 1985; Bryanston-Cross and Dawes 1985; Bryanston-Cross and Denton 1984; Bryanston-Cross, Veretta-Piccoli, and Ott 1984; Bryanston-Cross, Edwards, and Squire 1984; Bryanston-Cross, Camus, and Richards 1983; Bryanston-Cross et al. 1981) has produced a series of comparisons of similar data with numerical solutions (Fig. 3-12).

Figure 3-13, from the works of ONERA (Office Nationale Étude Recherche Aerospatiale; Delery, Surget, and Lacharme 1977; Surget 1973; Philbert and Surget 1968; Veret et al. 1977; Surget and Chatriot 1969) is a real time holographic interferogram of a transonic channel flow, showing the density distribution in the flow at three different flow conditions. The pressure distribution in the channel and the Mach number can be determined by assuming 2-D flow and counting fringes. The Mach number, as determined from pressure gages, agrees well with the values determined from HI. The hologram is first recorded with no flow, then it is returned to the original recording position and reilluminated with the reference wave. The reconstructed wave mixes with the real time wave to produce the interferogram. This system employs a 1.5-W argon laser that is shuttered acousto-optically to provide exposures less than 1 ms. One advantage of this type of system is that it can be used to produce high-speed movies of a flow. A disadvantage is that the apparatus must be highly stable.

Internal flow. The flow in cascades (Moore et al. 1981; Parker and Jones 1985, 1986) inside combustion engines and in other confined regions presents a new range of problems, including time and optical access. Getting the light in and out is no trivial problem. The installation of windows that survive and do not disturb the flow is a science in itself. Double-pulsed HI of the flow between compressor blades in a turbomachine can make the flow visible because the position changes between exposures. This type of holography is used to locate the shock wave position in 3-D. It cannot be used to quantify the density.

Three-dimensional flow. Tomographic solutions now exist for solving 3-D flows if enough interferograms are made at the correct viewing angles through the flow. Only a few such systems are currently in operation; however, the number is likely to increase in the near future. A good example is the 60-cm-diameter system using a pulsed ruby laser in operation in the U.S. Army Aeromechanics Rotor Laboratory at Ames Research Center (Kittleson 1983; Tan and Modarress 1985). The object beam passes over the tip of a rotor blade, recording the flow in the region around in tip (Fig. 3-14).

By traversing the system laterally, the view angle through the flow is varied automatically. The system must be fired by a signal that indicates that the blade is in the proper position. Holographic interferograms are typically recorded every 5°–10°. For the views that contain large density gradients, more recordings are needed. Note the appearance of the shock wave at different angles of view. In some of the angles of view, the shock wave edge is not visible.

This type of system produces so many data that some type of automation in data reduction is imperative. Considerable efforts to this end have been successful in producing systems that can digitize and analyze data from an interferogram in a few seconds.

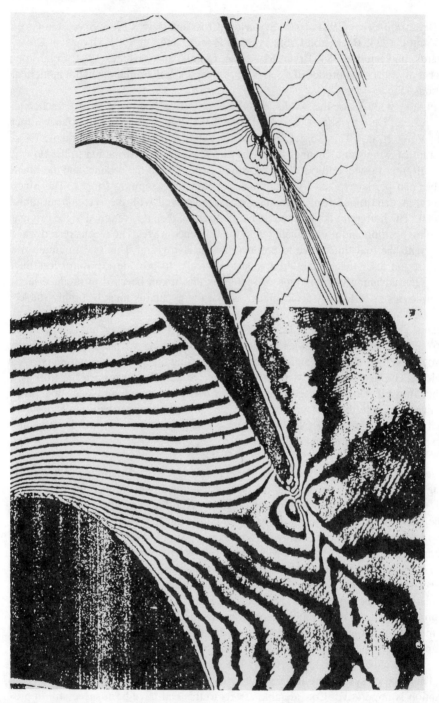

Figure 3-12 Comparison between the trailing edge flow in a linear cascade measured interferometrically and a loss predicting Navier-Stokes computed solution.

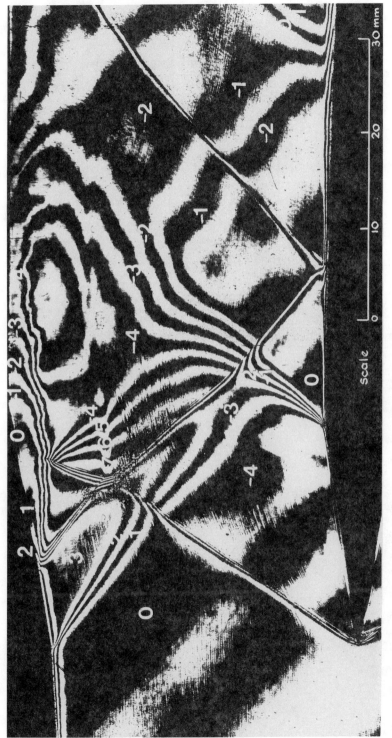

Figure 3-13 Diagnostics of the transonic flow in a channel. Contributed by ONERA.

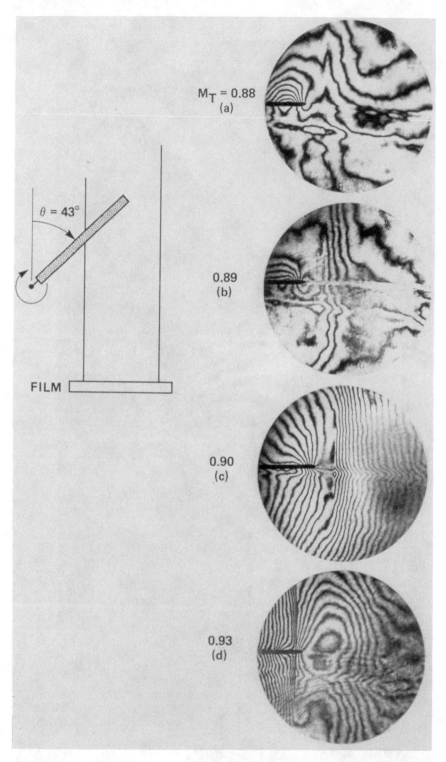

Figure 3-14 Holographic interferometry of flow over a helicopter blade tip. M_T, tip mach number.

The reconstructed density distribution (p/p_0), produced by the tomographic reconstruction code applied to 40 projections, is shown in Fig. 3-15 for the heights $z = 0.5, 1, 1.5,$ and 2 in. (chord C is 3 in., the leading edge of the blade $Y/C = -0.5$, and the aspect ratio $= 13.7$). The optimum number of iterations was 3–6 for the experimental data. The computation time on an IBM PC was approximately 3 min per iteration.

3-5 DATA REDUCTION

Figure 3-16 represents the flow of data in the holographic data reduction process. The first two steps, producing and reconstructing the holographic image, have been described above. The remaining steps are described here.

For years, publications reporting uses of holographic interferometry focused on feasibility demonstrations, and the actual data output of such research was quite limited. More recently, attention has been directed to the complete extraction of data from the holograms (Trolinger 1985; Tan, Trolinger, and Modarress 1986). Furthermore, recent laser and recording technology makes possible the recording of thousands or even tens of thousands of holograms, placing an even greater demand on automated data reduction.

The basic information contained in an interferogram is the distribution of the relative phase of two waves. This information can be extracted in essentially three different ways.

1. Analysis of a fixed interference fringe set.

Locate the coordinates of the center of interference fringes.
Identify a reference point from which to start.
Identify and number fringes integrally starting with zero at the reference point and adding (subtracting) where the phase is known to have increased (decreased). Each succeeding fringe represents an increase or decrease in the phase difference of the two waves by 2π.
The dependence of intensity in the fringe pattern is sinusoidal, and it is most practical to locate maxima or minima, although in principle, any equal intensity locus between any two maxima represents an isophase difference contour.

2. Analysis of a moving interference fringe set (heterodyne interferometry).

The fringes are swept at a known rate (by methods already described), while monitoring the intensities at discrete points over the interference region.
Locate a reference point and arbitrarily define the phase as zero. As the fringes sweep over this point, a detector will produce a time-varying sinusoidal signal.
Moving out from this point with a second detector, a sinusoid of the same frequency, but different phase, will be observed. The phase difference in the two time-varying signals is the desired phase difference between the two wave fronts. Each time

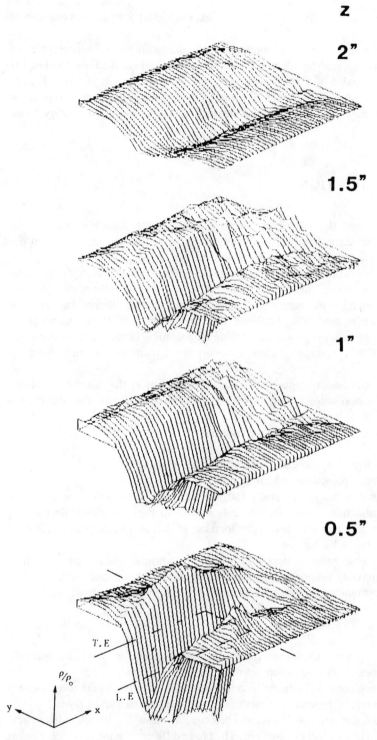

Figure 3-15 (*a*) Reconstructed density fields for different heights above chord line.

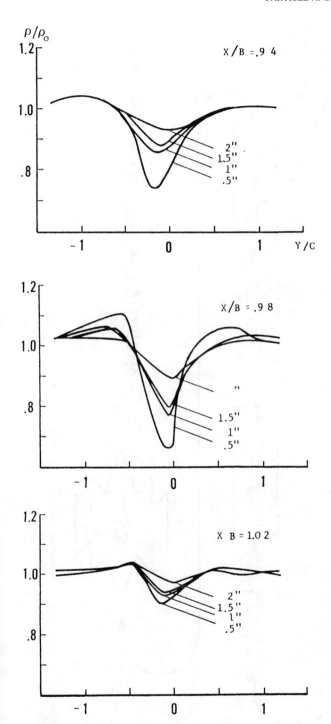

Figure 3-15 (continued) (*b*) Density ratio profiles (blade length B = 41.1 in.; chord length is 3 in.).

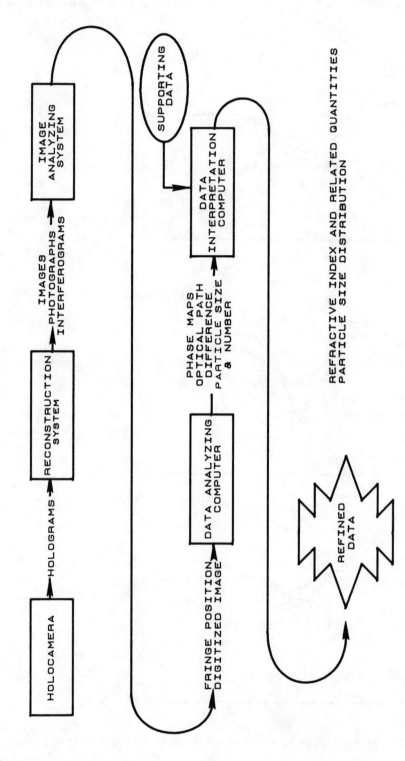

Figure 3-16 Data reduction path in automated holography reduction.

this phase difference increases through 2π, the signal appears identical to that observed at the reference point.

Numbering fringes becomes less of a problem, since the sign of phase shift is observable.

Interpolating between fringes is no longer a problem, since they sweep over the detector. The resolution of the measurement is determined by the ratio of the detector size to fringe spacing.

The entire process can be achieved with one fixed and one moving detector or, alternately, a fixed detector array.

3. Analysis of the fringe set by introducing known phase shifts while observing intensity changes at discrete points (phase shift interferometry).

A detector is positioned at a reference point where the phase is arbitrarily set at zero. Moving out from the reference point in steps of $\Delta\phi$ less than 2π, the intensity is given by Eq. (3-18).

The unknowns are I_o, I_s, and $\Delta\phi$. By introducing two known phase shifts, three equations are produced.

The three equations can then be solved for $\Delta\phi$ in terms of I_1.

A moving detector or detector array can scan the interferogram.

Unfortunately, the intensity in a real interferogram is much more complicated than that expressed in Eq. (3-13) because of optical noise and parasitic interference. A real interferogram is characterized by the following.

1. Broken, discontinuous, and split fringes.
2. Varying contrast.
3. Extraneous fringes.
4. Diffraction noise.
5. Broad cloud-like fringes.
6. Closely spaced fringes.
7. Lack of a known reference position.
8. Unknown fringe sign.
9. Uneven background.
10. Regions blocked by opaque objects.
11. Speckle noise.

Therefore considerable preprocessing is sometimes required before the task of locating fringe position or measuring phase difference is practical.

Signal-to-noise improvement. In addition to noise in the interferogram, electronic noise in the sensors also contributes to the final results. One of the most direct methods of improving S/N ratio is signal averaging. Image analyzers can be programed to average many data frames at TV rates, thus improving picture quality. Next, spatial smoothing is useful to reduce high-frequency noise due, for example, to speckle-like noise.

Thresholding and nonuniform background. Many of the techniques commonly used in pattern and line recognition fail in interferometry because the intensity is sinusoidal and the background is nonuniform. This presents a problem in thresholding. Procedures for dealing with nonuniform background include the following: logarithmic detection, normalization and referencing to a mean intensity across the interferogram, which is first established before thresholding begins, normalizing to local intensity averages, and finally, subtracting off the nonuniform background.

A floating threshold method, used by Becker and Yu (1985) requires two passes over each line across the interferogram. The first pass searches for extrema. With the found extrema, a step-like threshold function is then defined, which is the mean gray value between adjacent extrema. The second scan is referenced to this function to produce a binary fringe pattern. This produces good results on fringe fields, even with severe background intensity variation and low contrast.

The problem can also be approached optically, in double reference wave holographic interferometry, by phase shifting one of the reconstructed waves by 180° to subtract away the background entirely.

Interactive preprocessing. Until artificial intelligence is applied to this problem, there is little hope of producing a fully automated system. Image processing systems fortunately allow a direct interaction from the operator to connect broken fringes, give the correct sign to fringes, and guess where a discontinuous fringe should proceed. What can be automated is partial recognition and flagging of the occurrence of such events to allow an operator to intervene.

Coordinate extraction. After the above preprocessing, the fringe pattern is a binary pattern. Then this pattern can be reduced to a line pattern, where lines are the centers of the black or white fringes. Alternately, the edges or transition from black to white can designate the fringes. Each scan line therefore provides a y coordinate for the line and an x coordinate for the fringes. With each x coordinate, a fringe number is required. Each adjacent scan must then be referenced to the previous scan, so that like numbered fringes can be identified.

Because fringe density becomes very large in some regions, a magnification must be performed to maintain accuracy. This further complicates the numbering of fringes. A number of rules exist for fringe numbering.

1. Adjacent fringes differ in number by zero or one.
2. Fringes of different numbers cannot touch each other.
3. Fringes cannot end inside the field of view (except where they close on themselves).
4. When integrating over a closed line, the fringe number differences add up to zero.

Phase shift interferometry and heterodyne interferometry both offer promise for making automated data reduction of holograms easier yet. The output signals appear to be more amenable to filtering and noise removal. This is one of the most promising and rapidly growing areas in holographic interferometry today.

3-6 NOMENCLATURE

D	diameter
E	energy in an exposure
I	intensity of an electromagnetic wave
I_B	intensity of the bias illumination
I_1	intensity in an interferogram
N	number
R	resolution
Re	real part of a complex quantity
S/N	signal to noise ratio
T_{hi}	amplitude transmissivity of the ith hologram
U	complex amplitude of an electromagnetic wave
U^*	complex conjugate of the amplitude
U_{o1}	first object wave amplitude
U_{o2}	second object wave amplitude
U_{oi}	amplitude of the ith wave
U'_{oij}	amplitude of the ith reconstructed wave at the jth hologram
U_{r1}	first reference wave amplitude
U_{r2}	second reference wave amplitude
U_t	amplitude of a sum of waves
x, y	distance in the plane of the interferogram or hologram
x_i, y_i, z_i	reconstructed image wave position
x_o, y_o, z_o	object wave location
x_p, y_p, z_p	reconstructing reference wave position
x_r, y_r, z_r	reference wave location
z	distance normal to the interferogram or hologram
δ	depth of field
ϵ	angular error
θ	hologram tilt angle
λ	wavelength
ϕ_{oi}	phase shift angle on the ith wave
ϕ_{ri}	phase shift angle on the ith reference wave
ψ	angle measured from normal to the hologram
ψ_{ri}	angle of the ith reference wave
ψ_{oi}	angle of the ith object wave
ψ'	primes refer to angles in the reconstruction process

REFERENCES

Abramson, N. H. 1981. *The making and evaluation of holograms.* San Diego, Calif.: Academic.

Bachalo, W. D. 1979. *An investigation of transonic turbulent boundary layer separation generated on an axisymmetric flow model.* AIAA paper 79-1479.

Bachalo, W. D. 1983. An experimental investigation of supercritical and circulation control airfoils at transonic speeds using holographic interferometry. Paper AIAA-83-1793 read at symposium, 13–15 July 1983, Danvers, Massachusetts.

Bakrunov, A. O. 1980. Holographic method of determining the velocity field of a disperse phase in a two-phase flow (translation). *Fluid Dyn.* 15(1):153–55.

Becker, F., and Yu, Y. 1985. Digital fringe reduction techniques applied to the measurement of 3-D transonic flow fields. *Opt. Eng.* 24(3):429.

Becker, F., Meier, G. E. A., and Wegner, H. 1982. Automatic evaluation of interferograms. In *Applications of digital image processing IV*, SPIE vol. 359. New York: Society of Photo-Optical Instrumentation Engineers.

Belozerov, A. F. 1974. Shadow and interferometer investigations of low density gas flow by means of reconstructing the wave front from hologram. In *Proceedings of the eleventh international congress*, pp. 265–70. London: Chapman Hall.

Brenden, B. B. 1981. Miniature multiple-pulse Q-switched ruby laser holocamera for aerosol analysis. *Opt. Eng.* 20:907–11.

Briones, R. A. 1980. *Particle holography at extended distances and micron resolutions*. SPIE vol. 523, pp. 112–14. New York: Society of Photo-Optical Instrumentation Engineers.

Brooks, R. E., Heflinger, L. O., and Wuerker, R. F. 1965. Interferometry with a holographically reconstructed comparison beam. *Appl. Phys. Lett.* 7:248.

Bryanston-Cross, P. J. 1981. Three-dimensional holographic flow visualization. Paper read at the Symposium on Measuring Techniques in Transonic and Supersonic Cascade Flow, Sept. 1981, Lyon.

Bryanston-Cross, P. J. 1985. A holographic system for visualizing a vortex structure in a turbocharger. Paper read at the Meeting of Holographic Measurement, Speckle and Allied Phenomena, 16–17 April 1975, at the CEGB Headquarters, London.

Bryanston-Cross, P. J., and Dawes, W. N. 1985. Comparison of inviscid and viscous computations with an interferometrically measured transonic flow. *AIAA J.* 23(6):834.

Bryanston-Cross, P. J., and Denton, J. 1984. Comparison of a measured and predicted flow around the leading edge of an aerofoil. *AIAA J.* 22:1025–26.

Bryanston-Cross, P. J., Camus, J. J., and Richards, P. 1983. Dynamic correlation of a Schlieren image in a transonic airflow. In *Photon correlation techniques in fluid mechanics*. Series in Optical Sciences, vol. 38, pp. 270–75. New York: Springer-Verlag.

Bryanston-Cross, P. J., Edwards, J., and Squire, L. 1984. Measurements in an unsteady two-dimensional shock/boundary layer interaction. Paper read at the IUTAM Unsteady Aerodynamics Conference, September 1984, at Jesus College, Cambridge.

Bryanston-Cross, P. J., Veretta-Piccoli, F., and Ott, P. 1984. *Implementation of the ruby pulse laser holography system at the LTT/EPFL*. Ecole Polytechnique Federale de Lausanne internal report LIT-TM-16-84.

Bryanston-Cross, P. J., Lang, T., Oldfield, M. L. G., and Norton, R. J. G. 1981. Interferometric measurements in a turbine cascade using image plane holography. *Trans. ASME.* 103:124–126.

Burner, A. W. 1973. A holographic interferometer system for measuring density profiles in high-velocity flows. In *ICIASF '73 record*. New York: Institute of Electrical and Electronics Engineers.

Burner, A. W., and Goad, W. K. 1979. Holographic flow visualization at NASA Langley. Paper read at the 25th International Instrumentation Symposium, 7–10 May 1979, Anaheim, California.

Caulfield, H. J., and Lu, S. 1970. *The applications of holography*. New York: John Wiley.

Charwat, A. F. 1977. Generator of droplet tracers for holographic flow visualization in water tunnels. *Rev. Sci. Instrum.* 48:1034–36.

Collier, R. J., Burckhardt, C. B., and Lin, L. H. 1971. *Optical holography*. San Diego, Calif.: Academic.

Conway, B. J. 1982. Ground-based and airborne holography of ice and water clouds. *Atmos. Environ.* 16:1193–1207.

Craig, J. E. 1982. Nd:YAG holographic interferometer for aerodynamic research. *Opt. Eng.* 353:96–103.

Craig, J. E. 1984. Conventional and liquid metal droplet breakup in aerodynamic nozzle contractions. Paper read at 22nd AIAA Aerospace Sciences Meeting, 9–12 January 1984, Reno, Nevada.

Decker, A. J. 1981. Holographic flow visualization of time-varying shock waves. *Appl. Opt.* 20:3120–27.

Decker, A. J. 1982. Holographic cinematography of time-varying reflecting and time-varying phase objects using a Nd:YAG laser. *Opt. Lett.* 7:122–23.

Decker, A. J. 1984. Measurement of fluid properties using rapid-double-exposure and time-average holographic interferometry. Paper AIAA-84-1461 read at 20th Joint Propulsion Conference, 11–13 June, 1984, at Cincinnati, Ohio.

Delery, J. 1977. *Recherches sur la interaction onde de choc-couche limite turbulente.* Rech. Aerosp. report 1977–6.

Delery, J., Surget, J., and Lacharme, J. P. 1977. *Interferometri holographique quantitative en ecoulement bidimensionnel.* Rech. Aerosp. report 1977-2, pp. 89–101.

Ewan, B. C. R., Swithenbank, J., and Sorusbay, D. 1984. Measurement of transient spray size distributions. *Opt. Eng.* 23:620.

Gabor, D. 1948. A new microscope principle. *Nature* 161:777–78.

Goodman, J. 1968. *Introduction to Fourier optics.* New York: McGraw-Hill.

Haman, K. 1984. Preliminary results of an investigation of the spatial distribution of fog droplets by a holographic method. *R. Meteorol. Soc. Q. J.* 110:65–73. (Polska Akademia Nauk, Instytut Geofizyki, Warsaw, Poland.)

Hannah, B. W., and Havener, A. G. 1975. Application of automated holographic interferometry. In *ICIASF 1975 record*, pp. 237–46. New York: Institute of Electrical and Electronics Engineers.

Havener, G. 1983. The application of holographic interferometry to the measurement of transition in supersonic, axisymmetric boundary layers. Presented at the 16th Fluid and Plasma Dynamics Conference, *AIAA*.

Havener, G., and Radley, J. 1972. *Quantitative measurements using dual hologram interferometry.* Aerospace Research Laboratories report ARL 72-0085.

Hess, C. F., Trolinger, J. D., and Wilmot, T. M. 1985. Particle field holography data reduction by Fourier transform analysis. In *Applications of holography.* SPIE vol. 523. New York: Society of Photo-Optical Instrumentation Engineers.

Hove, D. T. 1975. Holographic analysis of particle-induced hypersonic bow-shock distortions. *AIAA J.* 13:947–49.

Iwata, K. 1977. Measurement of flow velocity distribution by means of double exposure holographic interferometry. *J. Opt. Soc. Am.* 67:1117–21.

Kashiwagi, T., Jones, W., Kashiwagi, T., and Baum, H. 1983. *Application of high-speed two-wavelength holographic interferometry to the analysis of radiative ignition.* American Society of Mechanical Engineers report 83-HT-49.

Kiemle, H., and Ross, C. 1969. *Einfuhrung in die Technik der Holographic.* Frankfurt: Akademische Versagsgesellschaft.

Kittleson, J. K. 1983. *A holographic interferometry technique for measuring transonic flow near a rotor blade.* USAAVRADCOM technical report 83-A-10.

Mayinger, F., and Panknin, W. 1976. Holographic two-wavelengths interferometry for measurement of combined heat and mass transfer. In *Combustion measurements*, ed. R. Goulard, p. 270. San Deigo, Calif.: Academic.

McVey, J. B. 1976. Diagnostic techniques for measurements in burning sprays. Paper read at Combustion Institute Fall Meeting, 18–20 October 1976, La Jolla, California.

Moore, C. J., Jones, D. G., Haxell, C. F., Bryanston-Cross, P. J., and Parker, R. J. 1981. Optical methods of flow diagnostics in turbomachinery. In *ICIASF 1981 record.* New York: Institute of Electrical and Electronics Engineers.

Netzer, D. W. 1982. An investigation of particulate behavior in solid propellant rocket motors. In *The 18th JANNAF combustion meeting*, vol. 3, pp. 93–110. Washington, D.C.: Organization of the Joint Army Navy NASA Air Force.

O'Hare, J. E., and Trolinger, J. D. 1969. Holographic color Schlieren. *Appl. Opt.* 8:204.

Optical Engineering. 1985a. Automated reduction of image and hologram data. May/June.

Optical Engineering. 1985b. Applications of holography, and holographic interferometry. Sept./Oct.

Ozkul, A. 1979. Investigation of acoustic radiation from supersonic jets by double-pulse holographic interferometry. *AIAA J.* 17(10), Article 79-4123.

Parker, R. J., and Jones, D. G. 1985. Holographic flow visualization in rotating transonic flows. Paper read at VI International Conference on Photon Correlation and Other Techniques in Fluid Mechanics.

Parker, R. J., and Jones, D. G. 1986. *Industrial holography—the Rolls Royce experience.* SPIE, vol. 699.

New York: Society of Photo-Optical Instrumentation Engineers.

Philbert, M., and Surget, J. 1968. *Application de linterferometrie holographique en soufflerie*. Rech. Aerosp. report 122, pp. 55–60.

Sheffield, S. A., Hess, C. F., and Trolinger, J. D. 1986. Holographic studies of the vapor explosion of vaporizing water-in-fuel emulsion droplets. In *Proceedings of the 2nd international colloquium on droplets and bubbles*, pp. 112–19. Pasadena, Calif.: Jet Propulsion Laboratory.

Smith, H. M. 1977. *Holographic recording materials: topics in applied physics*. New York: Springer-Verlag.

Surget, J. 1973. *Etude quantitative d'un ecoulement aerodynamique par interferometrie holographique*. Rech. Aerosp. report 1973-3, pp. 161–71.

Surget, J. 1976. Two reference beam holographic interferometry for aerodynamic flow studies. Paper read at la Conference Internationale sur les Applications de l'Holographie et le Traitement Optique des Donnees, 23–26 August 1976, Jerusalem.

Surget, J. E. 1982. *Holographic interferometry by a non-silver film process*. Rech. Aerosp. report 1982–4.

Surget, J., and Chatriot, J. 1969. *Cinematographie ultra-rapide d'interferogammes holographiques*. Rech. Aerosp. report 132, pp. 51–55.

Surget, J., Delery, J., and Lacharme, J. 1977. Holographic interferometry applied to the meteorology of gaseous flows. Paper read at 1et Congres Europeen sur l'Optique Appliquee a la Metrologie, 26–28 October 1977, at Strasbourg.

Tan, H., and Modarress, D. 1985. Algebraic reconstruction technique code for tomographic interferometry. *Opt. Eng.* 24:435–40.

Tan, H., Trolinger, J., and Modarress, D. 1986. An automated holography data reduction system. Paper read at SPIE International Symposium, August 1986, San Diego, California.

Thompson, B. J. 1980. Advances in far-field holography—theory and applications. In *Proceedings of the seminar*, vol. 523, pp. 102–11. New York: Society of Photo-Optical Instrumentation Engineers.

Thompson, B. J., Ward, J. H., and Zinky, W. 1965. Application of hologram techniques for particle size analysis. *J. Opt. Soc. Am.* 55:1566.

Thompson, B. J., Ward, J. H., and Zinky, W. 1967. Application of hologram technique for particle size analysis. *Appl. Opt.* 6:519.

Trolinger, J. D. 1974. An airborne holography system for cloud particle analysis in weather studies. In *International instrumentation automation conference proceedings*, pp. 617.1–627.8. Pittsburgh, Pa.: Instrument Society of America.

Trolinger, J. D. 1975. Particle field holography—state-of-the-art and applications. *Opt. Eng.* 14:383–392.

Trolinger, J. D. 1979. Application of generalized phase control during reconstruction to flow visualization holography. *Appl. Opt.* 18:766–74.

Trolinger, J. D. 1985. Automated holography data reduction. *Opt. Eng.* 24:840–842.

Trolinger, J. D., Farmer, W. M., and Belz, R. A. 1969a. Applications of holography in environmental science. *J. Environ. Sci.* 12:10.

Trolinger, J. D., Farmer, W. M., and Belz, R. A. 1969b. Holographic techniques for the study of dynamic particle fields. *Appl. Opt.* 8:967.

Umstatter, H. M., Doty, J. L., and Trolinger, J. D. 1985. Dual thermoplastic holography recording system. In *Applications of holography*, SPIE vol. 523. New York: Society of Photo-Optical Instrumentation Engineers.

Veret, C. 1982. Applications of flow visualization techniques in aerodynamics. Paper read at 15th International Congress on High Speed Photography and Photonics, 21–27 August 1982, San Diego, California.

Veret, C. 1984. Techniques De Visualization En Aerodynamique. Paper read at 16eme Congres International de Photographie Rapide et de Photonique, 27–31 August 1984, Strasbourg.

Veret, C., Philbert, M., Surget, J., and Fertin, G. 1977. Aerodynamic flow visualization in the ONERA facilities. Paper read at the International Symposium on Flow Visualization, October 1977, Tokyo.

Vest, C. M. 1979. *Holographic interferometry*, Highstown, NJ: McGraw-Hill.

Wuerker, R. F. 1976. Particle and flow field measurements by laser holography. In *The engineering uses of coherent optics: proceedings of the conference*, pp. 516–40. New York: Cambridge University Press.

Yano, M. 1980. Improved holographic method for the measurement of velocity in water flow. Paper A82-27108 read at Second International Symposium, 9–12 Sept. 1980, Bochum, West Germany.

TEMPERATURE MEASUREMENTS BY LIGHT-SCATTERING METHODS

Normand M. Laurendeau

4-1 INTRODUCTION

Many advances in the thermal sciences depend on accurate determination of temperature. In combustion, temperature measurements are needed to assess energy release and to study the influence of temperature on the chemical kinetics controlling fuel oxidation and pollutant formation. Temperature measurements are clearly needed in heat transfer, especially for complex convective flows below 1000 K. In gas dynamics, temperature measurements are required to understand compressible flow fields and to investigate the interaction between turbulence and chemical kinetics in chemically reacting systems.

In this chapter, I review methods for temperature measurement based on scattering of laser light. In this context, scattering is used loosely to indicate those techniques for which the signal is incoherently distributed over 4π steradians in accord with dipole radiation laws (Eckbreth, Bonczyk, and Verdieck 1979). These methods are spontaneous vibrational and rotational Raman scattering, Rayleigh scattering, and laser-

This chapter was first printed as a review paper in *Progress in Energy and Combustion Science*, vol. 14, no. 2, pp. 147–170 (1988).

I thank M. C. Drake, R. K. Hanson, M. B. Long, P. Ho, J. Keller, and R. W. Dibble for kindly providing research results for this chapter. The work on thermometry at Purdue was sponsored by the Department of Energy. This work was presented as the keynote lecture for a symposium on experimental techniques in combustion at the 24th National Heat Transfer Conference, Pittsburgh, Pennsylvania, August 1987.

induced fluorescence. The utilization of these techniques for concentration and temperature measurements has been the subject of previous reviews (Eckbreth 1981, 1987; Hanson 1986; Lapp and Penney 1977; Measures 1984). Coherent anti-Stokes Raman spectroscopy (CARS), a wave-mixing technique that has received much attention for thermometry in practical combustion systems (Eckbreth, Bonczyk, and Verdieck 1979; Eckbreth 1981, 1987), is not considered here because the light-scattering methods are generally simpler, less expensive, and thus more appropriate for fundamental studies in the broad thermal sciences.

A common feature of the laser-based scattering methods is the capability for nonintrusive, in situ measurements with high spatial (<0.1 mm^3) and temporal (<10 ns) resolution. Hence these techniques offer the unique opportunity to measure both precise temperatures in laminar flows (by averaging over many laser shots) and accurate probability distribution functions (PDFs) in turbulent flows (by compiling single-shot data at one spatial location). More recently, two-dimensional temperature measurements have become possible using Rayleigh scattering or laser-induced fluorescence. Although most of the above methods have been developed for application to combustion, an important goal of this chapter is to acquaint investigators in heat transfer and fluid mechanics with the exciting possibilities for innovative research using light-scattering thermometry.

4-2 LIGHT-SCATTERING METHODS: SIMILARITIES AND DIFFERENCES

The four light-scattering methods are compared on an energy level diagram in Fig. 4-1 (Measures 1984). Rayleigh scattering is an elastic process, so that the scattered signal occurs at the same wavelength as the excitation source. For this reason, the Rayleigh signal is subject to interference from the intense scattered light produced by particles and nearby surfaces. Raman scattering is an inelastic process in which the oscillating polarizability of the molecule modulates the scattered radiation, thereby leading to the appearance of two sideband frequencies. If a molecule gains energy from the radiative field, the resulting lower frequency scattered radiation is termed the Stokes component; if the molecule loses energy to the radiative field, the scattered radiation is referred to as the anti-Stokes component.

For a diatomic molecule possessing zero angular momentum around the internuclear axis (Σ state), the selection rules for Raman scattering are given by $\Delta v = 0$, ± 1 and $\Delta J = 0$, ± 2. Hence the pure rotational Raman spectrum ($\Delta v = 0$) is characterized predominantly by a series of Stokes and anti-Stokes rotational lines. The vibrational Raman spectrum also consists of a Stokes ($\Delta v = +1$) and an anti-Stokes ($\Delta v = -1$) component, each with three branches: S ($\Delta J = +2$), Q ($\Delta J = 0$), and O ($\Delta J = -2$). Since the rotational lines in the Q branch are stronger and not normally resolved, this branch is usually 2 orders of magnitude more intense than the S and O branches and is thus the most characteristic feature of the Stokes and anti-Stokes components of the vibrational Raman spectrum.

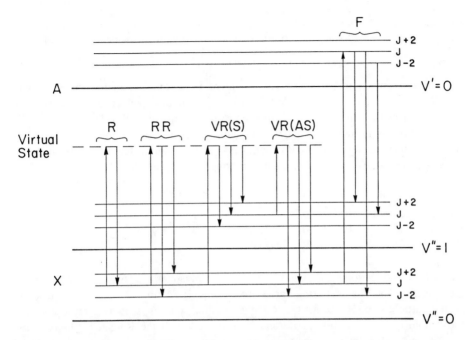

Figure 4-1 Energy-level representation of Rayleigh, rotational Raman, vibrational Raman, and fluorescence scattering methods. R, Rayleigh scattering; RR, rotational Raman; VR(S)/VR(AS), Stokes/anti-Stokes vibrational Raman; F, fluorescence.

As indicated by the virtual state in Fig. 4-1, Rayleigh or Raman scattering can occur upon excitation at any visible or ultraviolet wavelength. Fluorescence, however, requires a precise excitation frequency, since molecules in a particular rovibronic level of the ground electronic state must be excited to a specific rovibronic level in another electronic state. Spontaneous emission can then be monitored from the upper rovibronic level to a variety of such levels in the ground electronic state. Fluorescence can also occur from indirectly excited levels in the upper electronic state owing to collisional dynamics, i.e., rotational and vibrational relaxation.

The signal for the four scattering processes depicted in Fig. 4-1 can be obtained from

$$\Phi = \beta\Omega_c V_c (d\sigma/d\Omega) N_{vJ} I_L \qquad (4\text{-}1)$$

where Φ is the power of the measured signal (W), β the detection efficiency, Ω_c the solid angle of the collection optics (sr), V_c the collection volume (cm^3), $d\sigma/d\Omega$ the differential cross section (cm^2/sr), N_{vJ} the number density in the initial rovibronic level (v'', J'') of the ground electronic state (cm^{-3}), and I_L the irradiance of the laser beam (W/cm^2). At thermal equilibrium, the initial number density is related to the total number density of the scattering species N_s and the temperature T by

$$N_{vJ} = \frac{g_J(2J + 1)N_s}{Q_{rot}Q_{vib}} \exp\left[-hcE(v, J)/kT\right] \qquad (4\text{-}2)$$

where g_J is the nuclear spin degeneracy, $2J + 1$ is the rotational degeneracy, Q_{rot} and Q_{vib} are the rotational and vibrational partition functions, respectively, and

$$E(v, J) = G(v) + F_v(J) \tag{4-3}$$

is the energy (cm^{-1}) of the initial rovibronic level. Equation (4-2) assumes no electronic splitting of the rotational energy levels. The exponential dependence of N_{vJ} on temperature forms the basis for most temperature measurement methods.

Equation (4-1) indicates that for the same experimental conditions, the relative signal for each scattering method depends primarily on the differential cross section. Approximate values (cm^2/sr) for each scattering method are as follows:

vibrational Raman	10^{-30}–10^{-28}
rotational Raman	10^{-28}–10^{-27}
Rayleigh	10^{-27}–10^{-25}
fluorescence	10^{-25}–10^{-20}

where the cross section for fluorescence refers to atmospheric pressure. The differential cross section for Mie scattering varies from about 10^{-27} to 10^{-8} cm^2/sr, depending on the particle size. Hence for the Rayleigh method, particle scattering represents the most significant source of interference. Compared with the Raman methods, only fluorescence offers the intensity and spectral selectivity for thermometry in moderately dirty flows, where interferences can also arise owing to blackbody radiation, flame emission, or laser-induced incandescence. Because of their low differential cross sections, Raman and Rayleigh scattering measurements normally require high species concentrations and clean laboratory conditions.

The usual experimental setup for scattering measurements is shown schematically in Fig. 4-2. The laser can be continuous wave (CW) or pulsed, although pulsed operation with gated detection is best with respect to suppression of background noise. Normalization of the scattered signal to account for variations in laser power can be effected by monitoring the laser pulse with a photodiode. The scattered signal is usually obtained perpendicular to the direction of the laser beam; this arrangement enhances the spatial resolution of the measurement. For weak Raman processes, a mirror can be used to double the scattered signal. Spectral selection is obtained with a single or double monochromator, depending on the requirements for spectral resolution. For broadband measurements, the spectrometer can be replaced by suitable spectral filters. Signal averaging is commonly used to increase the precision of the measurement. Typical signal averagers are sampling oscilloscopes, photon-counting systems, or box-car averagers.

4-3 SPONTANEOUS RAMAN SCATTERING

Figure 4-3 schematically demonstrates the spontaneous vibrational and rotational Raman signals for air at two temperatures, 300 K and 1100 K (Lapp 1974). The central Rayleigh-scattered signal is flanked by pure rotational Raman scattering, represented

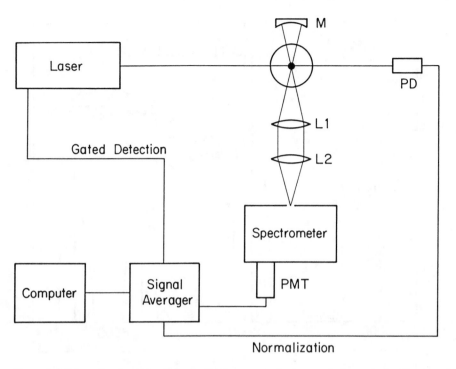

Figure 4-2 Schematic of experimental setup for thermometry by laser-scattering methods. M, mirror; L1/L2, lenses; PD, photodiode; PMT, photomultiplier tube.

here by wing envelopes for the peak rotational line intensities. The vibrational Q branches on the Stokes and anti-Stokes sides are shown for N_2 and O_2 at their characteristic Raman shifts (2331 cm^{-1} for N_2 and 1556 cm^{-1} for O_2). The strong Q branches are again surrounded by rotational wings, arising in this case from the much weaker vibrational O and S branches. Note that since the anti-Stokes vibrational Q branch signal arises from the $v'' = 1$ level, its intensity is much lower than that of the Stokes signal and increases substantially compared with the Stokes signal as temperature rises. An increase in temperature also broadens the rotational wings for both the pure rotational and the vibrational Raman signals. In the next two subsections, I will address the exploitation of these observations for Raman thermometry.

Because of the weakness of the Raman effect, the most useful Raman scatterers are major stable species such as N_2, O_2, and H_2. H_2 is especially useful for pure rotational Raman spectroscopy because its large rotational constant reduces interference from Mie scattering. Figures 4-4 and 4-5 display the pure rotational and vibrational Q branch Stokes spectra for N_2 at flame temperatures (Drake and Hastie 1981). The intensity alternation of the rotational spectrum arises from the different nuclear degeneracies for odd and even J values. Individual rotational lines are not resolved in the Q branch spectrum. However, vibrational anharmonicity gives rise to a sawtooth structure owing to Raman signals from different initial vibrational levels. This vibrational structure can obviously be used for thermometry.

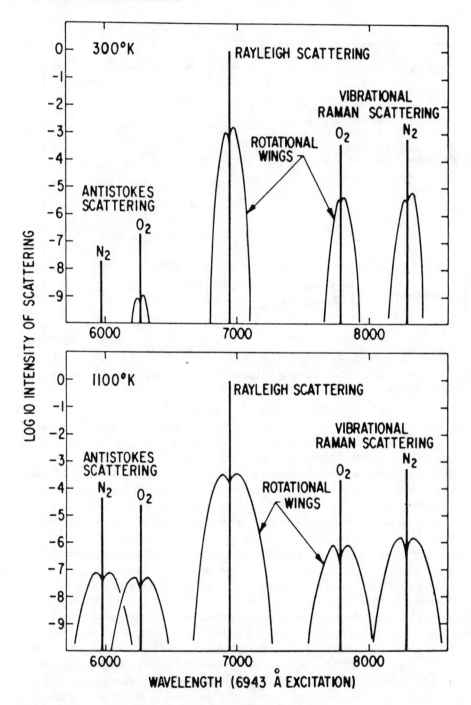

Figure 4-3 Raman scattering for air at 300 K and 1100 K *(Lapp 1974)*.

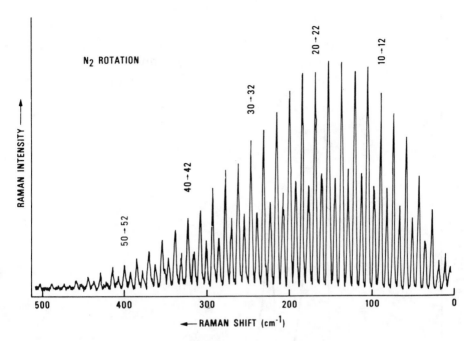

Figure 4-4 Stokes N_2 rotational Raman spectrum for an $H_2/O_2/N_2$ flame *(Drake and Hastie 1981)*. The flame temperature is 1988 ± 14 K.

The rotational and vibrational spectra in Figs. 4-4 and 4-5 are related to temperature through the latter's effect on the population distribution among the rotational and vibrational levels of N_2. Comparison between theory and experiment requires an expression for the differential cross section for spontaneous Raman scattering. A suitable relation is (Drake and Hastie 1981; Lapp, Penney, and Goldman 1972; Drake, Lapp, and Penney 1982; Drake et al. 1982)

$$(d\sigma/d\Omega) = C\gamma_v \nu_J^4 [S(J)/(2J + 1)] f(J) \qquad (4\text{-}4)$$

where C is a numerical constant that depends only on the laser polarization and collection geometry, ν_J and ν_L are the scattered and laser frequencies, respectively, $S(J)$ is the rigid-rotor line strength of the rotational transition, and $f(J)$ is a correction factor arising from rotation-vibration coupling. For N_2 and O_2 this correction is insignificant at temperatures below about 1000 K, since high J levels remain unpopulated; for H_2 the rotational constant is large enough to cause significant corrections even at room temperature (Drake, Lapp, and Penney 1982; Drake et al. 1982). For vibrational Raman spectra, $\gamma_v = v + 1$ for the Stokes Q branch and $\gamma_v = v$ for the anti-Stokes Q branch. For pure rotational Raman spectra, γ_v accounts for any corrections due to the influence of upper vibrational levels. In this case, corrections for H_2 are negligible because of the large separation between rotational transitions from the ground and

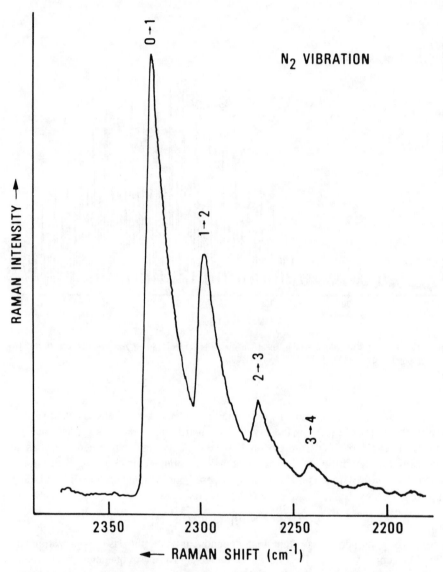

Figure 4-5 Stokes N_2 vibrational Q branch Raman spectrum for an $H_2/O_2/N_2$ flame *(Drake and Hastie 1981)*. The flame temperature is 2160 ± 50 K.

excited vibrational levels; however, corrections for N_2 and O_2 can be substantial at temperatures above 1000 K owing to overlap of spectral lines.

Thermometry by Vibrational Raman Spectroscopy

Four common methods of determining temperature from N_2 vibrational Raman scattering are summarized in Fig. 4-6 (Lapp 1974; Drake, Lapp, and Penney 1982). All

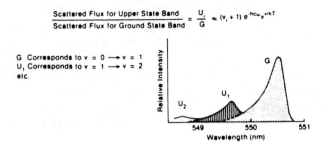

(a) Band Area Method for Temperature Measurement From Stokes Vibrational Q-Branch Contours

$$\frac{\text{Scattered Flux for Upper State Band}}{\text{Scattered Flux for Ground State Band}} = \frac{U_1}{G} \approx (v_1 + 1)\, e^{-hc\omega} e^{v/kT}$$

G Corresponds to $v = 0 \longrightarrow v = 1$
U_1 Corresponds to $v = 1 \longrightarrow v = 2$
etc.

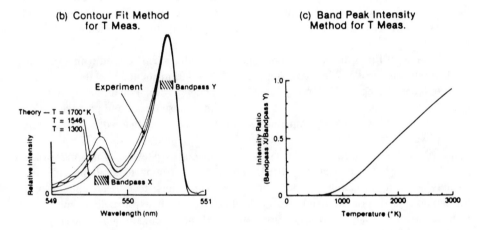

(b) Contour Fit Method for T Meas.

(c) Band Peak Intensity Method for T Meas.

(d) Stokes-AntiStokes Method for T Meas.

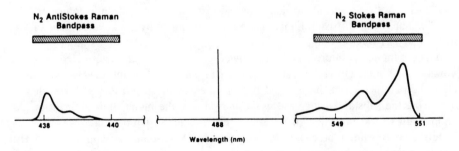

Figure 4-6 Four thermometric methods *(Drake, Lapp, and Penney 1982)* for N_2 vibrational Raman scattering: *(a)* band area method, *(b)* contour fit method, *(c)* band peak-intensity method, and *(d)* Stokes/anti-Stokes method.

of the techniques rely on the increased population in the upper N_2 vibrational levels with rising temperature. The shape of the spectrum is a convolution of the spectral power given by Eq. (4-1) and the slit function of the monochromator. In the band area method the temperature is obtained from the relative intensities of the upper and ground level vibrational bands within the Stokes Q branch. For the contour fit method the temperature is determined by theoretical replication of the relative Stokes spectral intensity profile. In the band peak intensity method the temperature is obtained by using the intensity ratio for two spectral bands at the peaks of the contour. The Stokes/anti-Stokes method is similar; however, a wider spectral band pass provides the maximum collected signal. From Eqs. (4-1)–(4-4) the Stokes/anti-Stokes ratio is related to temperature by

$$\Phi_S/\Phi_{AS} = (\nu_S/\nu_{AS})^4 \exp(-hc\Delta G/kT) \tag{4-5}$$

where Φ_S and Φ_{AS} must be corrected for any difference in the detection efficiency at the Stokes and anti-Stokes wavelengths. The energy difference ΔG between the ground and first excited vibrational levels is the Raman shift for the fundamental Q branch band.

For laminar flows the band area or contour fit methods can be used with an argon-ion laser and signal averaging to obtain temperatures within about 2% of thermocouple measurements (Lapp 1974; Drake, Lapp, and Penney 1982). For turbulent systems, the Stokes/anti-Stokes method can be applied with a pulsed laser and gated detection to obtain temporally resolved temperatures with an inherent precision of approximately 4%. However, the Stokes/anti-Stokes method for N_2 cannot be used at temperatures below about 800 K because of an insufficient number of photons in the anti-Stokes signal (Drake, Lapp, and Penney 1982). The three remaining methods provide reasonably accurate measurements down to 300 K, since these techniques employ only the Stokes signal. For the band area or band contour methods, a multichannel spectrometer would be needed for temporally resolved measurements; unfortunately, the weakness of the Raman signal precludes reliable single-shot results (Drake et al. 1982; Sochet et al. 1979; Michael-Saade et al. 1983).

Stephenson (1979) and Drake and Hastie (1981) applied the contour fit and band peak intensity methods, respectively, to determine temperature profiles using N_2 in premixed flames. Aeschliman, Cummings, and Hill (1979) employed the contour fit method for both N_2 and O_2 to determine similar time-averaged temperature profiles in a laminar H_2/air diffusion flame. The profiles obtained by Drake and Hastie (1981) in premixed $H_2/O_2/N_2$ flames are shown in Fig. 4-7. To maintain a similar precision throughout the temperature profile, the lower temperatures were determined using the vibrational Q branch of H_2. Because of its large rotational constant, the H_2 temperature could be obtained from the isolated lines in the Stokes spectrum, as shown in Fig. 4-8. Note the substantial increase in the intensity of the high-J transitions at greater temperatures. The precision of the measured H_2 and N_2 temperatures was approximately 2%. A similar precision was obtained by Aeschliman, Cummings, and Hill (1979) in their nonpremixed flame.

In recent years, Drake and coworkers (Drake et al. 1981; Drake, Bilger, and Starner 1983) have employed the N_2 Stokes/anti-Stokes method to determine temper-

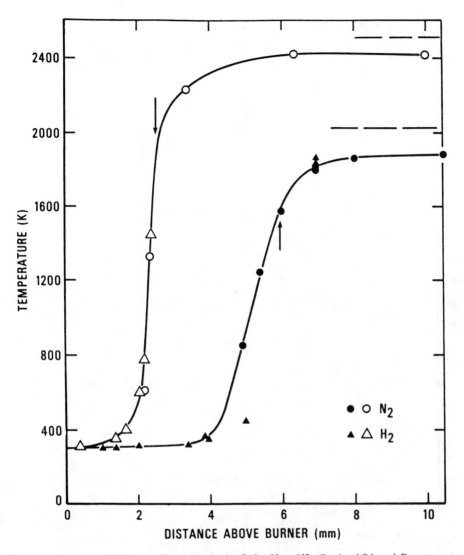

Figure 4-7 Axial temperature profiles obtained using Stokes N_2 and H_2 vibrational Q branch Raman spectra for two $H_2/O_2/N_2$ flames *(Drake and Hastie 1981)*. The horizontal dashed lines represent the adiabatic flame temperature, and the vertical arrows designate the median location of the flame front.

ature PDFs in turbulent nonpremixed $H_2/O_2/N_2$ flames. Instantaneous temperatures below 800 K were determined by a Raman measurement of total number density, in a manner similar to that employed for Rayleigh thermometry (see Section 4-4). Preliminary measurements at room temperature and in a laminar premixed flame were used for system calibration. A pulsed dye laser with a wavelength of 488 nm, a temporal bandwidth of 1–2 μs, and a repetition rate of 1 pulse per second (pps) was used for excitation of the Raman signal. The analog photomultiplier tube (PMT) signals

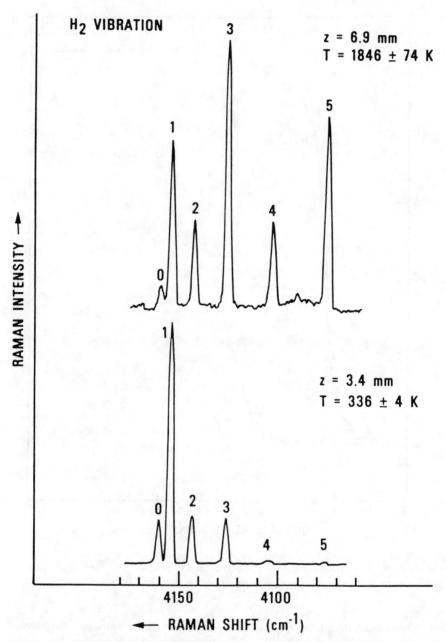

Figure 4-8 Stokes vibrational Raman scattering from H_2 in the preflame and postflame regions of an H_2/O_2/N_2 flame *(Drake and Hastie 1981)*.

were digitized, corrected for variations in laser power and background, and stored in a microcomputer. PDFs were generated from individual temperature measurements obtained by repetitively pulsing the laser at the same flame location.

Figure 4-9 shows the PDFs measured in both a laminar premixed flame and a

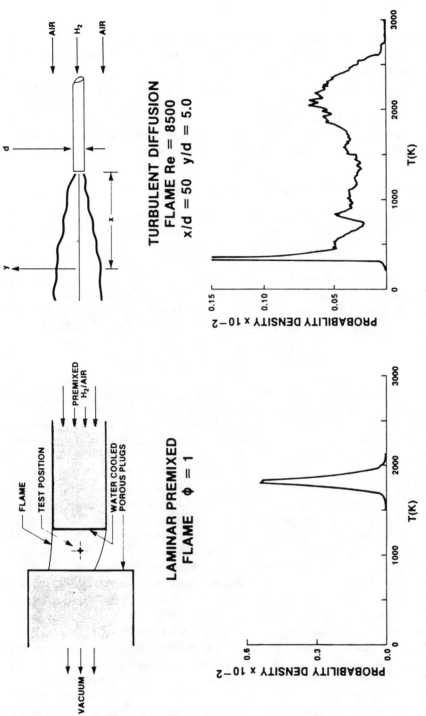

Figure 4-9 Probability density functions of temperature obtained by the Stokes/anti-Stokes method in the postflame region of a laminar premixed ($\Phi = 1$) H_2-air flame and the mixing layer of a turbulent H_2–air jet diffusion flame *(Drake, Lapp, and Penney 1982; Drake et al. 1981)*.

nonpremixed turbulent flame (Drake, Lapp, and Penney 1982; Drake et al. 1981). The width of the PDF for the laminar flame defines the precision of the Raman measurement. The PDF is nearly Gaussian, with an average temperature of 1823 K and a 4% relative standard deviation, due almost totally to photon statistics. Under these conditions, the accuracy in the average temperature is approximately ± 50 K. The PDF for the mixing layer in the turbulent diffusion flame is much wider, owing to real temperature fluctuations at this location. Similar measurements have been shown to be feasible in luminous flames but not in sooting flames.

Thermometry by Rotational Raman Spectroscopy

Thermometry by pure rotational Raman spectroscopy is ideally suited to lower temperatures because of the smaller spacing between rotational levels compared with that for vibrational levels (Drake et al. 1982). As for vibrational Raman spectroscopy, the most useful species have proved to be N_2, O_2, and H_2, although the latter usually provides less precision because of the paucity of available rotational lines compared with N_2 and O_2 (approximately six for H_2 versus 50 for N_2). Utilizing Eqs. (4-1)–(4-4) and assuming that the Raman signal has been corrected for any variations in detection efficiency and laser irradiance, the rotational temperature can be determined from

$$-hcF_0(J)/kT = \ln \left(\frac{\Phi_J}{g_J \nu_J^4 S(J) f(J)} \right) + \text{const} \qquad (4\text{-}6)$$

where Φ_J is the scattered power (W) for each rotational transition and $F_0(J)$ is the corresponding rotational energy (cm^{-1}) in the ground vibrational level.

Drake and Rosenblatt (1978) employed an argon-ion laser, a double monochromator, and photon counting to determine temperatures using rotational Raman spectroscopy in several different laminar, premixed H_2/O_2 flames. The peak heights in the rotational spectra obtained for N_2, O_2, and H_2 are plotted in Fig. 4-10 following Eq. (4-6). Both Stokes and anti-Stokes transitions are included. (The ν^3 term in the ordinate arises when photon counts are employed instead of radiative power.) The highest precision (0.5–1.0%) was obtained using the N_2 rotational temperature. The substantial increase in precision for rotational versus vibrational Raman reflects its inherent advantages: (1) higher sensitivity due to a larger differential cross section, (2) lower uncertainty because of a larger number of measurable transitions, and (3) greater dynamic range because of an increased spectral sensitivity to substantial variations in temperature. However, an important disadvantage of rotational compared with vibrational Raman spectroscopy is the propensity for overlapped spectral signatures in gaseous mixtures (Measures 1984; Drake and Rosenblatt 1978).

Rotational Raman thermometry becomes both more precise and accurate at temperatures below 1000 K because little or no correction is needed to account for rotation-vibration coupling and vibrational anharmonicity, except when H_2 is employed (Drake et al. 1982). The recent temperature measurements made in chemical vapor deposition (CVD) reactors by Breiland and Ho (1984) and Breiland, Coltrin, and Ho (1986) demonstrate this conclusion. These investigators utilized a frequency-tripled neodymium:yttrium/aluminum/garnet (Nd:YAG) laser at a wavelength of 355 nm, an energy

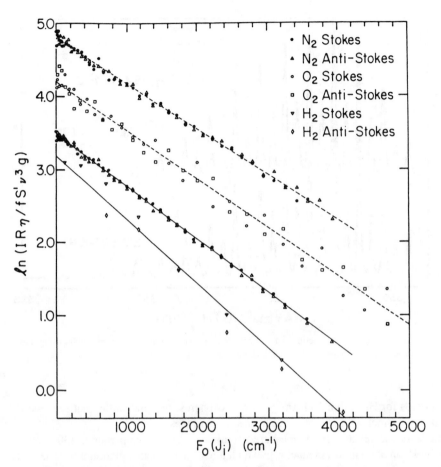

Figure 4-10 Rotational temperature determination plots for several H_2-O_2 flames *(Drake and Rosenblatt 1978)*. From top to bottom, the rotational temperatures are $T(N_2) = 2295 \pm 35$ K, $T(O_2) = 2205 \pm 105$ K, $T(N_2) = 1985 \pm 20$ K, and $T(H_2) = 1740 \pm 100$ K, where the uncertainties represent 2 standard deviations.

of 30 mJ/pulse, a temporal width of 4 ns, and a repetition rate of 10 pps. The advantages compared with an argon-ion laser include background suppression via gated detection and enhancement of the scattered signal via the fourth-power frequency dependence of the Raman cross section. The experimental and theoretical results for the N_2 spectrum are shown in Fig. 4-11; note the ± 1 K accuracy in this case.

Temperature profiles measured in a CVD reactor using H_2 are shown in Fig. 4-12 (Breiland, Coltrin, and Ho 1986). The lines represent the results from a CVD model incorporating flow over a flat plate, and the associated heat transfer, chemical kinetics, and transport processes. For H_2 the precision is $\pm 10–15$ K; this error is acceptable because a double monochromator is not needed and the CVD model contains even greater temperature uncertainties. Similar results have been obtained by Drake

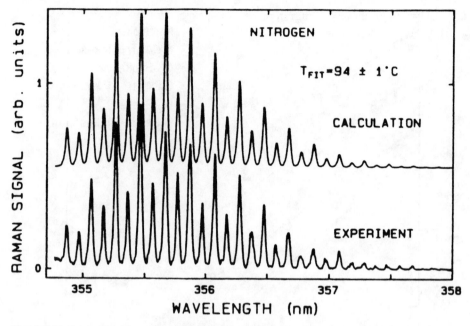

Figure 4-11 Comparison of an experimental and calculated rotational Raman spectrum for N_2 using temperature as an adjustable parameter *(Breiland and Ho 1984)*.

et al. (1979) for H_2 within 100-μm glass microspheres, thus indicating that successful Raman thermometry can be effected near solid surfaces.

Leyendecker et al. (1983) measured rotational Raman temperatures (300–1350 K) using H_2 in a model CVD reactor in which a resistance-heated platinum wire served as the substrate material. An argon-ion laser at 488 nm was employed as the excitation source, and a multichannel analyzer was applied to record an optimal pair of rotational lines specially selected to minimize the random error in the derived temperature. For $T < 600$ K the most suitable rotational lines of H_2 correspond to the $J = 1 \rightarrow 3$ and $J = 3 \rightarrow 5$ transitions; for $600 < T < 1500$ K, the appropriate pair of lines is $J = 1 \rightarrow 3$ and $J = 5 \rightarrow 3$. Figure 4-13 shows derived temperature profiles at a wire temperature of 773 K. A remarkable achievement is the reported precision of ± 4 K in the 300–1350 K range. Comparison with thermocouple measurements at 300–500 K in a homogeneously heated cell indicates an accuracy of ± 0.5 K. These data testify to the accurate and precise nature of rotational Raman thermometry at temperatures below 1000 K.

N_2 rotational Raman thermometry near 250–300 K has demonstrated both a precision and accuracy of ± 1 K (Drake et al. 1982; Cooney 1983; Arshinov et al. 1983). Such precision requires substantial signal averaging. Measurements at lower temperatures (105–135 K) have been reported by Hill et al. (1976) in an open-jet supersonic nozzle; in this case the measured temperatures agreed with theoretical temperatures to better than 2%. Temporally resolved spectra can also be obtained by employing a

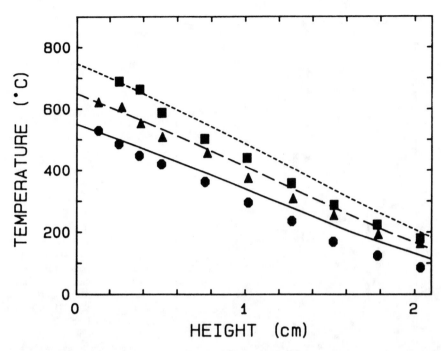

Figure 4-12 Profiles of gas temperature versus height above the CVD susceptor obtained from the rotational Raman spectrum of hydrogen at a partial pressure of 37.6 torrs *(Breiland, Coltrin, and Ho 1986)*. The susceptor temperatures are 550°C (circle), 650°C (triangle), and 750°C (square); the lines represent the results from a CVD model.

gated multichannel spectrometer (Smith and Giedt 1977; Kreutner, Stricker, and Just 1987). The accuracy and precision of the temperature are limited by the lower number of detected photons. For example, Smith and Giedt (1977) report an accuracy of ± 10 K at 300–430 K for excitation with a CW argon-ion laser and multichannel detection via a 33-ms gate. Using a frequency-doubled Nd:YAG laser (532 nm, 200 mJ/pulse) and an optical multichannel analyzer with a 50-ns gate, Kreutner, Stricker, and Just (1987) recently found that the resulting temperatures in laminar, premixed propane/ air flames for 10 successive 3000-shot measurements ranged from 1898 to 2167 K. This result is not unexpected considering the weakness of the spontaneous Raman process and the requirement for accurate determination of intensity for the individual rotational lines.

4-4 THERMOMETRY BY RAYLEIGH SCATTERING

Since Rayleigh scattering is an elastic process, the total scattered power for a single gaseous species is the summation of the contributions from the number density in each rovibronic level. Thus Eq. (4-1) becomes

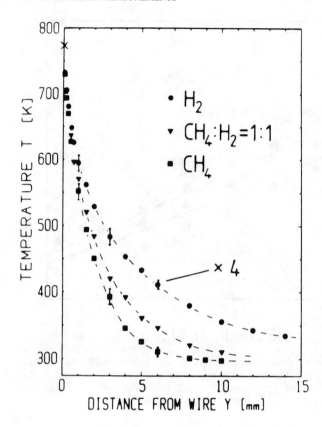

Figure 4-13 Temperature profiles in a model CVD reactor obtained by H_2 rotational Raman at a wire temperature of 773 K and a pressure of 1000 mbar *(Leyendecker et al. 1983)*. The bars indicate standard deviations calculated from the data and are expanded four times for clarity. For pure CH_4 the temperatures are obtained via vibrational Raman calibrated against H_2 rotational Raman.

$$\Phi = \beta\Omega_c V_c N_s I_L (d\sigma_R/d\Omega) \tag{4-7}$$

where the differential cross section is given by (Measures 1984; Penney 1969)

$$(d\sigma_R/d\Omega) = (3/8\pi)\sigma_R(\cos^2\phi\,\cos^2\theta + \sin^2\phi) \tag{4-8}$$

where ϕ is the polarization angle, θ is the scattering angle, and σ_R is the total Rayleigh scattering cross section, which is related to the complex refractive index η of the medium at STP conditions by

$$\sigma_R(\lambda) = \frac{8\pi}{3}\left[\frac{\pi^2(\eta^2 - 1)^2}{N_0^2\lambda^4}\right] \tag{4-9}$$

where we have neglected the small correction for the anisotropic polarization of the scattering molecules (Measures 1984) and N_0 is the number density at STP conditions (Loschmidt number). The inverse dependence of σ_R on the fourth power of the wavelength mandates the use of visible or UV lasers to obtain sufficient signal-to-noise (S/N) ratio for effective scattering measurements.

For the usual experimental configuration (see Fig. 4-1), both the polarization and scattering angles are 90°, so that for a gas mixture, we obtain from Eqs (4-7) and (4-8),

$$\Phi = (3/8\pi)\beta\Omega_c V_c N_T I_L \sigma_{eff} \qquad (4\text{-}10)$$

where N_T is the total number density of the medium and the effective Rayleigh cross section is given by (Penney 1969; Namer and Schefer 1985)

$$\sigma_{eff} = \sum x_i \sigma_{Ri} \qquad (4\text{-}11)$$

where x_i and σ_{Ri} are the mole fraction and total Rayleigh cross section of the ith species, respectively. Since the total number density of the medium $N_T = P/kT$ for an ideal gas mixture, the Rayleigh scattering method can be used for thermometry if the pressure P and the effective cross section are constant. The latter is, of course, guaranteed for a nonreactive mixture. In this case the temperature can be easily determined by

$$T/T_0 = \Phi_0/\Phi \qquad (4\text{-}12)$$

where Φ_0 is the Rayleigh scattered power at a reference temperature T_0. Calibration is usually effected at room temperature, and normalization with respect to the instantaneous laser signal is employed to reduce uncertainties caused by variations in laser power. As indicated by Eq. (4-12), an important attribute of Rayleigh thermometry is the inherently large dynamic range of the method.

Under steady laminar conditions, sufficient averaging of the Rayleigh signal can provide measured temperatures having good accuracy and precision ($\approx 1\%$). Such measurements have been made by Schefer and coworkers (Namer and Schefer 1985; Schefer, Robben, and Cheng 1980; Cattolica and Schefer 1982) as part of a study of catalyzed combustion of H_2/air mixtures in a flat plate boundary layer. Initial work in premixed flames showed that excellent agreement could be obtained with thermocouple data, after correcting for variations in the effective cross section with changing composition (Namer and Schefer 1985). In the later studies (Schefer, Robben, and Cheng 1980; Cattolica and Schefer 1982), a 1-W argon-ion laser was used; the resulting photoelectron rate was over 10^6 counts/s for air at room temperature. If the measurement error could be attributed totally to photon statistics, this count rate would give a precision better than 0.1%. Unfortunately, the measured precision is degraded by laser power fluctuations ($\approx 0.5\%$) and background noise caused by scattered radiation from dust particles and surfaces. Nevertheless, excellent measurements can be obtained, as indicated by Fig. 4-14, which shows a comparison of experimental and theoretical temperature profiles for nonreactive flow over a heated flat plate (Schefer, Robben, and Cheng 1980). Deviations between the two profiles tend to be greatest near the surface, which is reasonable considering the greater corrections for scattering at this location.

Dibble and Hollenbach (1981) and Chandran et al. (1985) demonstrated the utilization of Rayleigh thermometry in turbulent premixed and nonpremixed flames. Dibble and Hollenbach (1981) employed an argon-ion laser, but signal detection was effected by a filtered PMT assembly and an operational amplifier. The temporal resolution of the technique was controlled by the resistance-capacitance (RC) time constant of the amplifier, which ranged from 30 to 500 μs, depending on the required trade-off between resolution and precision. Calibration occurred at room temperature,

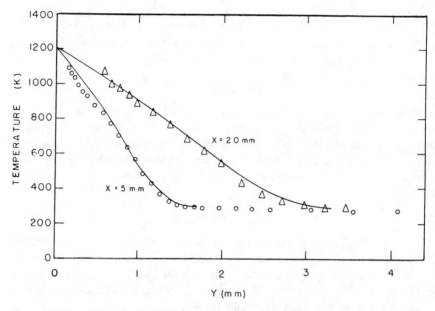

Figure 4-14 Comparison of calculated temperature profiles with those measured using Rayleigh thermometry for airflow over a heated flat plate *(Schefer, Robben, and Cheng 1980)*. Profiles are shown at two downstream locations for $T_s = 1200$ K and $U_\infty = 3.17$ m/s.

and the scattered signals were corrected for background noise due to both flame luminescence (<15%) and scattered radiation (<0.3%). Variations in the effective Rayleigh cross section were kept below 2% in both the premixed and nonpremixed flames by judicious selection of fuel composition (Namer and Schefer 1985; Dibble and Hollenbach 1981). Constructed PDFs for a turbulent jet diffusion flame are shown in Fig. 4-15. As might be expected, a bimodal distribution occurs near the mixing layer; such distributions were also observed in turbulent premixed flames because of the highly fluctuating flame front. The inherent precision of these temporally resolved measurements was about ±75 K. Mean and rms temperatures can easily be generated from the measured PDFs (Dibble and Hollenbach 1981; Chandran et al. 1985).

4-5 LASER-INDUCED FLUORESCENCE

The fundamental expressions required to understand fluorescence thermometry can be developed with the aid of the simple three-level model shown in Fig. 4-16. Levels i and j are the directly excited levels, while level k represents the summation of all atomic or molecular levels accessed by level j via spontaneous emission or collisional quenching. The rate equation for the jth level is

$$dN_j/dt = N_iW_{ij} - N_jW_{ji} - N_j(A + Q) \qquad (4\text{-}13)$$

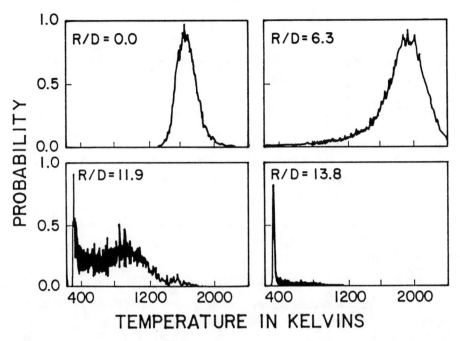

Figure 4-15 Probability distributions for a turbulent jet diffusion flame obtained by Rayleigh thermometry *(Dibble and Hollenbach 1981).* The fuel is 38% CH_4 + H_2, Re = 4400, and U_0 = 92 m/s. The axial L/D = 65, and the indicated radial R/D = 0.0, 6.3, 11.9, 13.8.

where N_j is the number density (cm^{-3}) in the jth level, W_{ij} is the rate coefficient (s^{-1}) for absorption, W_{ji} is the rate coefficient (s^{-1}) for stimulated emission, and A and Q are the total rate coefficients (s^{-1}) for spontaneous emission and collisional quenching, respectively. From Eq. (4-13), the steady state population in the jth level is then

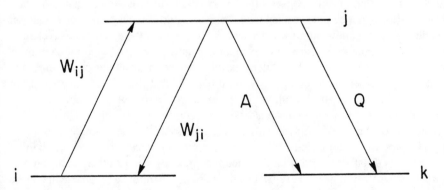

Figure 4-16 Basic three-level model for fluorescence thermometry. W_{ij} and W_{ji} are the rate coefficients (s^{-1}) for absorption and stimulated emission, respectively. A and Q are the rate coefficients (s^{-1}) for spontaneous emission and collisional quenching.

$$N_j = \frac{W_{ij}N_i}{A + Q + W_{ji}} \tag{4-14}$$

Assuming, for simplicity, a laser beam with a Lorentzian spectral profile that is much wider than the absorption linewidth, W_{ij} is given by

$$W_{ij} = \frac{2B_{ij}I_L(\nu_{ij})}{\pi c \Delta \nu} \tag{4-15}$$

where B_{ij} is the Einstein coefficient for absorption [cm^3/(J s^2)], $I_L(\nu_{ij})$ is the laser irradiance (W/cm^2), and $\Delta \nu$ is the spectral bandwidth of the laser (s^{-1}). Since spontaneous emission is isotropic, the fluorescence power can be obtained from

$$\Phi = \beta(\Omega_c/4\pi)V_c h\nu_{jk}A_{jk}N_j \tag{4-16}$$

where A_{jk} is the Einstein coefficient for spontaneous emission at the fluorescence frequency, ν_{jk}. At ordinary irradiances and pressures, $Q \gg A + W_{ji}$, so that Eqs. (4-14)–(4-16) give

$$\Phi = \beta\Omega_c V_c h\nu_{jk}(A_{jk}/Q)(B_{ij}/2\pi^2 c\Delta\nu)N_i I_L(\nu_{ij}) \tag{4-17}$$

Since $N_i = N_{vJ}$ for a molecular species, a comparison of Eqs. (4-1) and (4-17) shows that the differential cross section for fluorescence is given by

$$(d\sigma_F/d\Omega) = \frac{h\nu_{jk}(A_{jk}/Q)B_{ij}}{2\pi^2 c\Delta\nu} \tag{4-18}$$

where the usual pressure dependence arises from the influence of molecular collisions on the rate coefficient for electronic quenching.

Thermometry by Excitation/Fluorescence Spectroscopy

For laminar systems, slow excitation or fluorescence scans can be used to determine the rotational temperature of molecular species. The procedure is quite similar to ordinary absorption spectroscopy, with the advantage that the fluorescence method provides a local temperature rather than a line-of-sight temperature. For excitation scans the spectral bandwidth of the detector is fixed, while the excitation wavelength is varied. For fluorescence scans the excitation frequency is fixed, and the grating of the monochromator is rotated to determine the spectral distribution of the emissive signal. In both cases an accurate temperature requires that the fluorescence spectrum be uncontaminated by self-absorption; for the excitation case, optically thin conditions are also needed to avoid variations in laser beam absorptivity with changes in excitation wavelength (Bechtel 1979; Anderson, Decker, and Kotlar 1982). Excitation spectroscopy samples the rotational distribution in the ground electronic state, so thermal equilibrium is usually assured. Fluorescence spectroscopy, on the other hand, samples the rotational distribution in an excited electronic state. Therefore a reliable temperature necessitates sufficient rotational relaxation to ensure equilibration; this requirement constitutes the principal difficulty of the fluorescence method (Chan and Daily 1980; Zizak, Horvath, and Winefordner 1981; Furuya, Yamamoto, and Takubo 1985).

The excitation case also requires that collisional quenching not vary with excitation wavelength, while the fluorescence case mandates that collisional quenching not vary with fluorescence wavelength. These requirements can usually be achieved at high temperatures, since the collisional cross section varies rather weakly with rotational quantum number in the excited electronic state, especially for molecules with small rotational constants (Anderson, Decker, and Kotlar 1982; McKenzie and Gross 1981). For species with large rotational constants, such as OH, CH, and NH, a more substantial variation of the quenching cross section with rotational quantum number leads to additional uncertainties in the fluorescence case, since individual rovibronic transitions are utilized for the determination of temperature. For the excitation case, broadband detection is ordinarily employed, so rotational redistribution tends to ameliorate any uncertainty caused by rotational variations in electronic quenching (Anderson, Decker, and Kotlar 1982; McKenzie and Gross 1981; Smith and Crosley 1981; Crosley and Smith 1981).

Typical excitation or fluorescence scans are limited to a narrow spectral range, so that β, $I_L(v_{ij})$, and Δv should be nearly invariant. If we also assume that g_J is constant and that quenching does not vary substantially with excitation or fluorescence wavelength, then Eqs. (4-2) and (4-17) show that for the excitation case, broadband detection gives

$$\ln\left[\Phi(v_{ij})/B_{ij}(2J + 1)\right] = -hcE(v, J)/kT + \text{const} \qquad (4\text{-}19)$$

while for the fluorescence case, Eq. (4-16) leads to

$$\ln\left[\Phi(v_{jk})/v_{jk}A_{jk}(2J' + 1)\right] = -hcE(v', J')/kT + \text{const} \qquad (4\text{-}20)$$

where $E(v', J')$ refers to the rovibronic energy (cm^{-1}) in the excited electronic state. Note that Eq. (4-20) assumes thermal equilibration in the excited state; as mentioned previously, this assumption is the primary source of uncertainty for fluorescence thermometry (Chan and Daily 1980; Zizak, Horvath, and Winefordner 1981; Furuya, Yamamoto, and Takubo 1985).

Bechtel (1979), Anderson, Decker, and Kotlar (1982), and Crosley and Smith (1982) successfully used OH excitation scans with wideband detection to obtain temperature profiles in premixed flames. Figure 4-17 shows a typical Boltzmann plot (Bechtel 1979); the associated uncertainty is approximately ± 50 K (Bechtel 1979; Anderson, Decker, and Kotlar 1982; Crosley and Smith 1982). Bechtel (1979) employed excitation and detection in the strong (0, 0) band of OH, but optically thin conditions were maintained by using a specially constructed slot burner with a curved flame sheet. Anderson, Decker, and Kotlar (1982) used a standard flat-flame burner, but ensured optically thin conditions by exciting and detecting OH fluorescence in the (1, 1) band. Crosley and Smith (1982) employed the (0, 0) band, but self-absorption effects were minimized by operating near the edge of a flat-flame burner. Bechtel (1979) and Anderson, Decker, and Kotlar (1982) obtained good agreement between temperature measurements made by OH excitation and by vibrational N_2 Raman scattering ($\Delta T < 100$ K at $T = 1600\text{--}2200$ K).

Kohse-Höinghaus, Perc, and Just (1983) utilized fluorescence scans to determine OH rotational temperatures in a low-pressure C_2H_2/O_2 flame. The precision of these

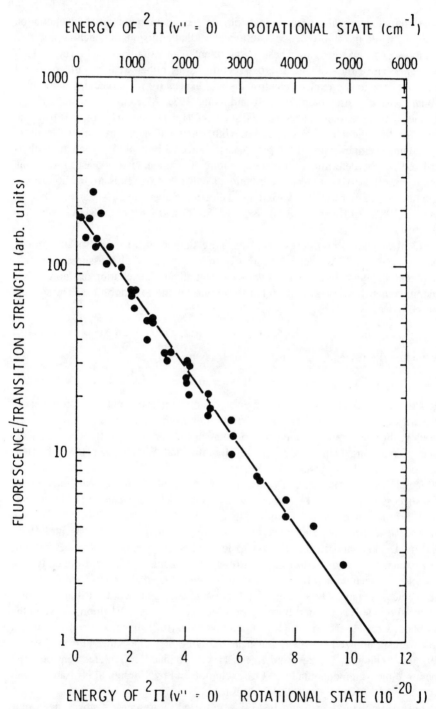

Figure 4-17 Fluorescence signal per transition strength versus rotational energy in the $^2\Pi(v'' = 0)$ electronic state of OH *(Bechtel 1979)*. The slope of the line determined by the excitation scan gives the OH rotational temperature.

measurements was only ± 100 K, which appears to confirm the sensitivity of the fluorescence method to quenching variations and residual nonthermalization. On the other hand, Zizak, Horvath, and Winefordner (1981) contend that fluorescence scans should give more precise results if only those levels having higher energies than the directly excited level are employed for thermometry. Such "thermally assisted" methods are considered in the section below on monochromatic fluorescence thermometry.

Two-Line Fluorescence Thermometry

Two-line fluorescence thermometry has evolved in response to the slow and tedious nature of excitation or fluorescence scans, which also prevents utilization of the latter in turbulent flows. In the two-line method, a pair of excitation wavelengths is used to generate two broadband fluorescence signals. Since many more spectral lines are employed in the scanning methods, the accompanying cost of the two-line technique is ordinarily a reduction in the precision of the measured temperature.

The methodology for two-line fluorescence thermometry is displayed for both atomic and molecular systems in Fig. 4-18. For an atom, an electronic temperature is measured by employing a low-lying excited electronic state. At sufficiently high pressures, the electronic temperature can be assumed to be equivalent to the translational temperature. Similarly, for a molecule, two rotational levels in the $v'' = 0$ level of the ground electronic state are used to infer the rotational temperature. In both cases, the dependence on quenching is avoided by exciting lower levels 1 and 2 to the same upper level. For atoms, level 3 is an unpopulated electronic state; for molecules, level 3 could be a single rovibronic level or, more likely, a vibrational band within an upper electronic state. For a molecule with closely spaced rotational levels, broadband fluorescence to the ground electronic state should not depend on the particular rovibronic level excited in the upper electronic state because of both fast rotational relaxation and the weak dependence of the quenching rate on rotational quantum number (Anderson, Decker, and Kotlar 1982; McKenzie and Gross 1981). This observation provides flexibility in choosing two transitions that give the highest sensitivity to a specific temperature range.

In the atomic case, detection of the fluorescence signal usually occurs by monitoring emission to that lower level not excited by the laser. Hence, from Eq. (4-17), the fluorescence ratio is given by

$$\frac{\Phi_{31}}{\Phi_{32}} = \zeta\left(\frac{I_{23}}{I_{13}}\right)\left(\frac{v_{31}\Delta v_{13}}{v_{32}\Delta v_{23}}\right)\left(\frac{A_{31}B_{23}}{A_{32}B_{13}}\right)\left(\frac{N_2}{N_1}\right) \tag{4-21}$$

where ζ accounts for the different light collection and detection efficiencies at the two fluorescence wavelengths. Employing the usual thermodynamic and radiative relations (Measures 1984),

$$N_2/N_1 = (g_2/g_1) \exp(-hc\Delta E/kT) \tag{4-22}$$

$$A_{ij} = (8\pi h v_{ij}^3/c^3)B_{ij} \tag{4-23}$$

$$B_{ij}/B_{ji} = g_j/g_i \tag{4-24}$$

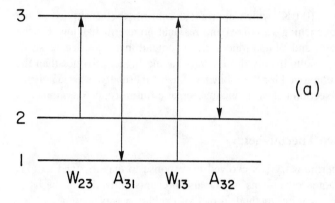

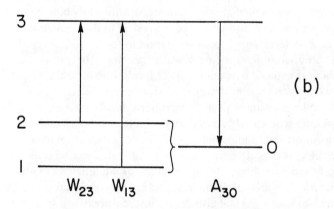

Figure 4-18 Two-line fluorescence thermometry: (*a*) atomic system, (*b*) molecular system. The zeroth level accounts for broadband emission to the ground electronic state.

Eq. (4-21) becomes

$$\frac{\Phi_{31}}{\Phi_{32}} = \zeta\left(\frac{I_{23}}{I_{13}}\right)\left(\frac{\nu_{31}}{\nu_{32}}\right)^4\left(\frac{\Delta\nu_{13}}{\Delta\nu_{23}}\right)\exp\left(-hc\Delta E/kT\right) \qquad (4\text{-}25)$$

where ΔE (cm^{-1}) is the energy difference between the two lower electronic levels. For the molecular case, a similar analysis for broadband detection gives

$$\frac{\Phi_{31}}{\Phi_{32}} = \zeta\left(\frac{I_{23}}{I_{13}}\right)\left(\frac{\Delta\nu_{13}}{\Delta\nu_{23}}\right)\left(\frac{B_{23}}{B_{13}}\right)\left(\frac{2J_2 + 1}{2J_1 + 1}\right)\exp\left(-hc\Delta F/kT\right) \qquad (4\text{-}26)$$

where ΔF is the energy difference (cm^{-1}) between the two lower rovibronic levels. As for an excitation scan, the narrow range of excitation wavelengths tends to make

I_{23}/I_{13} and $\Delta\nu_{13}/\Delta\nu_{23}$ approximately unity. Equations (4-25) and (4-26) provide direct relationships between the ratio of fluorescence signals and the desired temperature. By calibrating at a known temperature, both expressions also give a measured temperature that is independent of any systematic effects on the fluorescence signal. For atomic species, laser beam attenuation and radiative trapping can prove problematic at high atomic number densities; for molecular species, self-absorption is usually unimportant because the same fluorescence band is used for both excitation wavelengths.

The basic principles and accuracy of two-line atomic fluorescence (TLAF) were demonstrated by Omenetto et al. in 1972 (Omenetto, Benetti, and Rossi 1972; Omenetto, Browner, and Winefordner 1972). Because atomic species such as indium, gallium, and thallium must be vaporized, the technique is obviously unsuitable for measurements below approximately 500 K. Using a xenon arc lamp as the excitation source, Haraguchi et al. (1977) and Haraguchi and Winefordner (1977) demonstrated the feasibility of spatially resolved TLAF thermometry in premixed C_2H_2/air flames. These early measurements were obtained under steady laminar conditions, and thus no temporal resolution was needed to obtain a useful average temperature. The most suitable atomic additive proved to be indium because the gap between its two lower energy levels is such that high sensitivity is available for temperatures between 500 and 2500 K; moreover, its two excitation wavelengths lie in a spectral regime well suited for dye lasers (Haraguchi et al. 1977; Haraguchi and Winefordner 1977). In particular, excitation occurs at 410 and 451 nm, which corresponds to transitions from the ground $5p\ ^2P_{1/2}$ state to the excited $6s\ ^2S_{1/2}$ state, and from the metastable $5p\ ^2P_{3/2}$ state to the same excited state, respectively.

Following exploratory work by Joklik and Daily (1982), Dec and Keller (1986) subsequently demonstrated temporally resolved TLAF measurements in both laminar and turbulent flames. The accuracy of the method was substantiated by comparing the TLAF data with thermocouple measurements at 1700–2000 K. Temporal resolution was obtained by rapidly modulating between the excitation beams generated via two CW argon-ion pumped dye lasers with a mechanical chopper. The resulting fluorescence was split by a dichroic mirror into the 410- and 451-nm beams required for detection of indium. Each beam was directed to an appropriately filtered PMT, and its laser power was monitored by a pin photodiode. The fluorescence and laser signals were amplified and low-pass filtered on a shot-by-shot basis, with on-line division of each fluorescence signal by the relevant laser power signal. Indium was added to the flames by using a dilute solution of indium chloride.

Figure 4-19 shows the results for TLAF thermometry in a premixed laminar flame (Dec and Keller 1986). Calibration was effected at the center of the burner by using the thermocouple measurements. Each TLAF data point is the mean of 2000 individual temperature realizations. The average TLAF temperatures fall close to the thermocouple readings except near the burner edge, where flame flicker impedes accurate measurements. The rms temperatures indicate a precision of ± 6–9% at data rates of 1–5 kHz; the Rayleigh measurements taken for comparative purposes possess a higher uncertainty of over ± 12%. As expected, the precision of the TLAF measurements decreases at higher sampling rates, and the Rayleigh data suffer in comparison with the TLAF measurements because of the greater shot noise associated with lower photon count

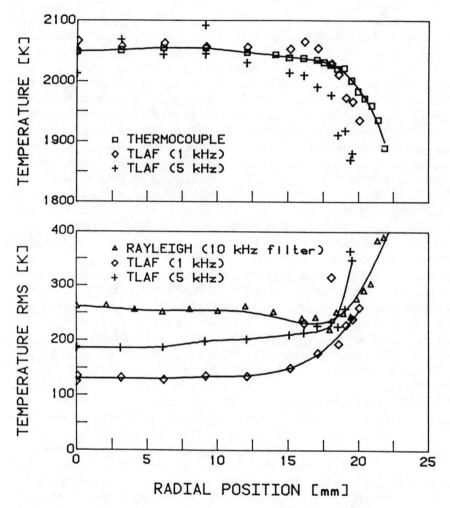

Figure 4-19 TLAF measurements in a laminar premixed flame *(Dec and Keller 1986)*: (top) radial profiles of TLAF and thermocouple temperatures, (bottom) rms deviation for TLAF and Rayleigh measurements of temperature.

rates. Under favorable conditions, Dec and Keller (1986) claim that a precision of ±3–5% can be obtained at data rates up to 10 kHz.

Two-line molecular fluorescence (TLMF) has been employed by Cattolica (1981) to monitor OH flame temperatures and utilized by Gross and McKenzie (1983, 1985) and Gross, McKenzie, and Logan (1987) to measure ambient temperatures by single- and two-photon excited fluorescence from nitric oxide. Cattolica (1981) ensured that broadband fluorescence from OH would be totally independent of quenching by pumping two rovibronic levels in the $X^2\Pi(v'' = 0)$ state to the same rovibronic level in the $A^2\Sigma^+(v' = 0)$ state. He successfully demonstrated OH thermometry near 1760 K in a premixed flame and convincingly showed that the method is insensitive to radiative trapping. Unfortunately, the sensitivity and dynamic range of the method are limited because the separation between pumped rovibronic levels in the ground electronic state is restricted by selection rules.

Utilization of stable NO has allowed TLMF thermometry at temperatures below 500 K (Gross and McKenzie 1983, 1985; Gross, McKenzie, and Logan 1987). By exciting two rotational levels in the $X^2\Pi(v'' = 0)$ state to appropriate rovibronic levels in the $A^2\Sigma^+(v' = 0)$ state, broadband fluorescence can be observed within the γ (0, 0) band (235–300 nm) with little or no interference from the excitation wavelength (225 nm for single-photon excitation and 450 nm for two-photon excitation). This method is similar to that of Cattolica (1981) except that two upper rovibronic levels are employed instead of one. These two upper levels are presumed to have similar broadband radiative lifetime and electronic quenching characteristics, an assumption that is surely more acceptable for NO than OH due to the fine rotational structure of the former (McKenzie and Gross 1981). Another advantage of using nitric oxide is the potential for accurate thermometry in low-temperature, turbulent flows owing to the high photon count rate when using broadband fluorescence. This potential, of course, deteriorates with increasing pressure because of the deleterious effects of collisional quenching.

Temporally resolved thermometry using TLMF was demonstrated by Gross and McKenzie (1983) using a cell containing 300 ppm NO in N_2 at 0.5 atm. Calibration of the method and normalization to account for pulse-to-pulse variations in laser power and bandwidth were accomplished using a reference cell containing NO at a known temperature. The experimental arrangement is shown in Fig. 4-20. Two dye lasers are simultaneously pumped at 10 pps by the third-harmonic output (355 nm) from an Nd:YAG laser. An optical delay line consisting of two plane mirrors provides a

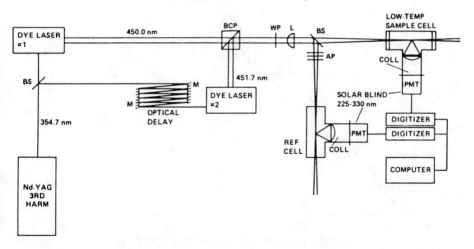

Figure 4-20 Experimental schematic for TLMF thermometry of NO *(Gross and McKenzie 1983)*. BCP, beam-combining polarizer cube; WP, half-wave plate; L, lens; BS, beam splitter; M, mirror; AP, attenuation plate; COLL, collection system; PMT, photomultiplier.

temporal resolution of 110 ns between each dye-laser pulse. In these first experiments, two-photon excitation pulses were tuned to the $J'' = 7.5$ and $J'' = 19.5$ rotational lines in the $S_{11} + R_{21}$ transitions. The two beams were then combined colinearly and partitioned into the sample and reference cells by a half-wave plate and a beam splitter. The reference cell contained 1200 ppm NO in 0.5 atm N_2 at room temperature. The broadband fluorescence was filtered and monitored by solar-blind photomultipliers. Each PMT waveform contained two pulses, from which an appropriate ratio was determined to calculate the rotational temperature. A comparison between measured rotational and thermocouple data for cell temperatures between 155 and 295 K is shown in Fig. 4-21. The average rotational temperatures were determined from 50 laser pulses; agreement with the thermocouple temperatures is within $\pm 2\%$. The precision of the single-shot measurements is approximately 3–4%, depending on the frequency and power stability of the lasers.

Gross and McKenzie (1985) and Gross, McKenzie, and Logan (1987) later applied their TLMF procedure to a two-dimensional turbulent boundary layer under supersonic

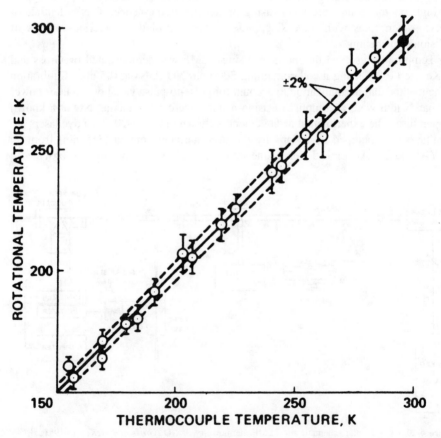

Figure 4-21 Comparison of measured TLMF and thermocouple temperatures *(Gross and McKenzie 1983)*. The cell mixture is 300 ppm NO in 0.5 atm N_2.

conditions (Ma = 2). A comparison between single- and two-photon excitation of NO indicated that the single-photon method is more accurate because line broadening caused by Stark effects limits the laser irradiance that can be utilized for two-photon excitation. For the single-photon case, the two laser pulses were tuned to the $J'' = 7.5$ rotational line in the R_{21} transition and the $J'' = 18.5$ line in the $Q_{11} + P_{21}$ transitions. The N_2 flow was doped with only 100 ppm of NO to avoid laser beam attenuation and fluorescence trapping. A band-pass filter was added to further reject scattered laser light. The high fluorescence signals for single-photon versus two-photon excitation enhanced the S/N ratio by approximately a factor of 10. The agreement with pitot tube temperatures at 150–300 K was again $\pm 2\%$; however, the single-shot precision increased to an excellent $\pm 1\%$.

Fletcher and McDaniel (1987) recently applied the TLMF method to a steady compressible flow field at 0.3–2.4 atm by employing an I_2 seed in air and a ring dye laser. Broadband fluorescence signals obtained from excitation scans of two selected transitions near 543 nm were used to determine temperatures ranging from 165 to 195 K. Corrections for collisional broadening were required to compensate for inadvertent excitation of nearby transitions because of the density of the rovibronic I_2 lines, especially at the higher pressures. Suitable corrections were effected by determining the pressure from the absolute fluorescence signal. The resulting deviation between measured and known temperatures was $\pm 2\%$.

Rea and Hanson (1988) modified a single-mode, frequency-doubled ring dye laser to obtain repetitive 300 μs spectral scans across the $R_1(7)$ and $R_2(11)$ lines in the (0, 0) band of the $A^2\Sigma^+ - X^2\Pi$ system of OH. This wavelength modulation technique was then used to demonstrate temporally resolved TLMF temperature measurements in a CH_4/air flame. The fluorescence measurements tended to be several hundred degrees greater than accompanying thermocouple measurements because of a combination of laser beam absorption, fluorescence trapping, and the influence of J-dependent quenching rates on the spectral distribution of the fluorescence signal. However, alternative trace species such as NO or I_2 could potentially alleviate many of these experimental problems, thus making wavelength modulation spectroscopy a useful tool for thermometry in turbulent reacting flows.

The dynamic range and sensitivity of two-line fluorescence methods can be enhanced without sacrificing the required independence from collisional quenching by employing two-line laser-saturated fluorescence (TLSF). This methodology has been demonstrated by Lucht, Laurendeau, and Sweeney (1982) for OH thermometry in laminar premixed flames. Two different rotational transitions were saturated in the (0, 0) band of the $A^2\Sigma^+ - X^2\Pi$ electronic system; fluorescence was monitored only from the two laser-pumped upper rovibronic levels. In this case, $W_{ji} \gg A + Q$, so that for a single transition, Eq. (4-14) becomes

$$N_j = (W_{ij}/W_{ji})N_i = (g_j/g_i)N_i \qquad (4\text{-}27)$$

Hence, under saturated conditions, the population of the upper level is related to that of the lower level through a ratio of degeneracies; the dependence on quenching is eliminated. For a pulse duration of a few nanoseconds, the total population of the two-level system formed by the laser-coupled rotational levels is constant, so that

$$N_i + N_j = N_{vJ} \qquad (4\text{-}28)$$

where N_{vJ} is the initial population of the rovibronic level in the ground electronic state.

Since quenching effects are eliminated by saturating the transition, a more flexible methodology can be developed for TLSF thermometry, as shown by the dual three-level model in Fig. 4-22. Combining Eqs. (4-2), (4-16), (4-27), and (4-28), we obtain

$$\frac{\Phi_{32}^b}{\Phi_{32}^a} = \zeta \left(\frac{\nu_{32}^b}{\nu_{32}^a} \right) \left(\frac{A_{32}^b}{A_{32}^a} \right) \left(\frac{g_{1b}g_{3b}}{g_{1a}g_{3a}} \right) \left(\frac{g_{1a} + g_{3a}}{g_{1b} + g_{3b}} \right) \exp\left(-hc\Delta F/kT \right) \qquad (4\text{-}29)$$

where the emissive power Φ has been corrected for any variation in detection efficiency, the total degeneracies of levels 1 and 3 are given by $g_J(2J + 1)$, and ΔF is the energy difference (cm^{-1}) between the two directly excited rovibronic levels in the ground electronic state. Given the measured ratio of fluorescence signals, the rotational tem-

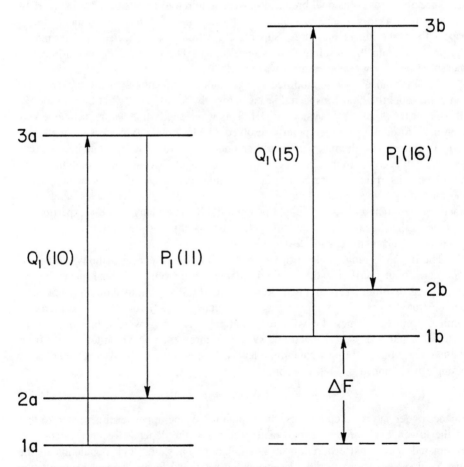

Figure 4-22 Dual three-level system for two-line saturated-fluorescence thermometry of OH.

perature is easily obtained from Eq. (4-29). The flexibility of the dual three-level scheme also allows a useful compromise between sensitivity and dynamic range. For saturated conditions, attenuation of the laser beam before it reaches the probe volume is negligible. Fluorescence trapping can be minimized by prudent selection of the observed fluorescence transitions.

TLSF thermometry has been demonstrated in premixed $H_2/O_2/Ar$ flat flames at 72 torrs (Lucht, Laurendeau, and Sweeney 1982). A Nd:YAG-pumped dye laser provided sufficient focused spectral irradiance [$\approx 5 \times 10^8$ W/(cm^2 cm^{-1})] to achieve about 95% saturation of the selected OH resonances at the peak of the laser pulse. The two fluorescence signals were monitored by placing the window of a sampling oscilloscope at the peak of both fluorescence waveforms. A representative temperature profile obtained at stoichiometric conditions is shown in Fig. 4-23. Comparisons are given with thermocouple and OH absorption measurements, and with a theoretical temperature profile obtained from a one-dimensional chemical kinetics code. The excitation and fluorescence transitions used in Fig. 4-23 are shown in Fig. 4-22. Both the precision and accuracy of the TLSF method are approximately ± 3–4%. Application of this technique to turbulent flames would require a rapid sequence of two laser pulses, as utilized by Dec and Keller (1986), Gross and McKenzie (1983, 1985), and Gross, McKenzie, and Logan (1987).

Monochromatic Fluorescence Thermometry

The utilization of two-line fluorescence thermometry for temporally resolved measurements is complicated by the need for two laser beams at different wavelengths. Considering the high differential cross section for fluorescence compared with Raman and Rayleigh scattering, fluorescence thermometry would be considerably advanced by a method that uses a single laser beam. Such monochromatic methods could also be easily extended to planar thermometry (Hanson 1986).

The most straightforward monochromatic method is thermally assisted fluorescence (THAF). In this technique the laser-induced populations of those energy levels higher than the laser-excited level are presumed to be collisionally equilibrated, so that an electronic, vibrational, or rotational temperature can be extracted from the resulting fluorescence spectrum (Zizak, Horvath, and Winefordner 1981; Zizak, Bradshaw, and Winefordner 1981; Zizak et al. 1981; Elder et al. 1984). Similar methods for measurement of the rotational and vibrational temperatures of the hydroxyl radical have been proposed by Chan and Daily (1980) and by Crosley and Smith (1980). Unfortunately, as mentioned previously, because of the large rotational constant of the OH radical, successful thermometry requires knowledge of the collisional dynamics occurring in the excited electronic state (Chan and Daily 1980; Zizak, Horvath, and Winefordner 1981; Furuya, Yamamoto, and Takubo 1985; Zizak et al. 1981). However, it must be pointed out that the THAF method has not been attempted for more favorable molecular species, such as NO and I_2.

Winefordner and coworkers (Zizak, Bradshaw, and Winefordner 1981; Zizak et al. 1981; Elder et al. 1984) have successfully demonstrated the efficacy of the THAF approach by using atomic seeds such as thallium or gallium in $C_2H_2/O_2/Ar$ flames.

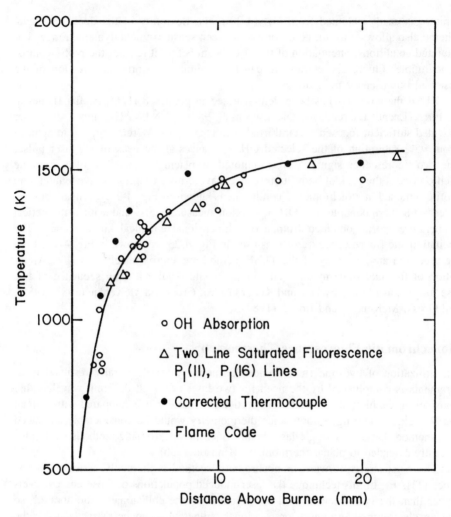

Figure 4-23 Temperature profile for TLSF thermometry of OH in a flat stoichiometric $H_2/O_2/Ar$ flame at 72 torrs with $Q_1(10, 15)$ excitation and $P_1(11, 16)$ fluorescence *(Lucht, Laurendeau, and Sweeney 1982)*. Comparisons are shown with temperatures obtained with OH absorption, a radiation-corrected thermocouple, and a one-dimensional chemical kinetics code.

For such species, a partial Boltzmann distribution can be maintained in the upper electronic states if the ratio of collisional rate coefficients for deexcitation to the directly excited and to the ground electronic states is approximately constant among the upper energy levels of interest. For the THAF experiments, a nitrogen-laser-pumped dye laser at a repetition rate of 20 pps was used to excite thallium atoms at 377.6 nm. The thallium was introduced to the flame by ultrasonic nebulization of a 3000 ppm $TlNO_3$ aqueous solution. The temperature can be determined by Eq. (4-20), except that the rotational degeneracy, $2J' + 1$, is replaced by the degeneracy of each electronic energy level. Early results gave a single-flame temperature of 2499 ± 50 K compared with 2470 ± 15 K via sodium D line reversal (Zizak et al. 1981).

Refinement of the experimental methodology eventually allowed single-shot THAF thermometry with excellent accuracy and precision (Elder et al. 1984). For such measurements, two fluorescence transitions are monitored in real time by using two monochromators, gated boxcar averagers, and a laboratory minicomputer. The signal strength is such that calibrated neutral density filters are needed to prevent saturation of the photomultiplier tubes. The temporally averaged results shown in Fig. 4-24 were obtained to verify the equilibrium assumption for THAF thermometry; note that six

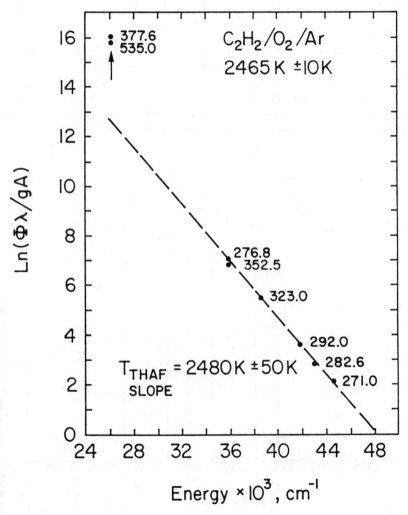

Figure 4-24 Boltsmann plot for temporally averaged THAF measurements in a thallium-doped $C_2H_2/O_2/$ Ar flame at a sodium reversal temperature of 2465 K *(Elder et al. 1984)*. The arrow indicates the laser-excited level; the wavelength (nm) for each transition is shown along the dashed line.

electronic transitions at wavelengths below 377.6 nm are equilibrated, while the directly excited level remains overpopulated. The important conclusion is that the collisional mixing rates among the thermally assisted levels are indeed much faster than the collisional rates to or from the laser-enhanced or ground electronic levels.

Subsequent single-shot measurements maximized the precision by utilizing only the two strongest fluorescence signals at 276.8 and 292.0 nm. A typical histogram of 100 consecutive single-shot temperature measurements is shown in Fig. 4-25. The uncertainty is <0.5%, and the agreement with sodium reversal measurements is excellent. For three different $C_2H_2/O_2/Ar$ flames, 100 separate single-shot measurements in each flame gave an average precision of ± 15 K and an average accuracy of ± 5 K, well within the ± 10–15 K precision of the line reversal measurements. These are outstanding results, which immediately suggest that attempts should be made to extend the THAF method to molecular species such as NO and I_2. However, even if molecular THAF thermometry is successful, the accompanying precision will suffer in comparison with atomic THAF because of the lower S/N ratio expected for molecular species.

A second monochromatic method relies on the inherent temperature dependence of the fluorescence signal. Equation (4-17) shows that the influence of temperature occurs through the population of the lower rovibronic level and the rate coefficient for collisional quenching, which is related to temperature by

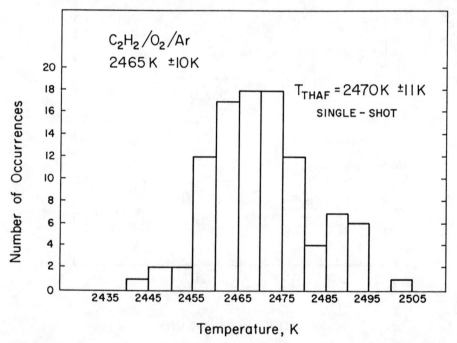

Figure 4-25 Probability histogram for 100 single-shot THAF temperature measurements in a $C_2H_2/O_2/Ar$ flame at a sodium reversal temperature of 2465 K using the 277/292 nm line pair of thallium (*Elder et al. 1984*).

$$Q = N_T \sigma_c \bar{v}(T) \qquad (4\text{-}30)$$

where N_T is the total number density of the gas mixture, σ_c is the collisional cross section, and $\bar{v}$ is the mean relative speed of the molecule. If the cross section is assumed to be independent of temperature, then combining Eqs. (4-2) and (4-30) gives

$$\Phi(T) = (x_s/T^{1/2}) \exp\left[-hcE(v, J)/kT\right] + \text{const} \qquad (4\text{-}31)$$

as $\bar{v}$ is proportional to $T^{1/2}$ from kinetic theory. This absolute fluorescence method (ABF) requires that the mole fraction x_s of the absorbing species be constant; hence best results will be obtained for nonreactive flows. Such experiments have received relatively little attention, although the technique has been used in one form or another for planar thermometry by employing either NO or O_2 as the absorbing species (Seitzman, Kychakoff, and Hanson 1985; Lee, Paul, and Hanson 1987).

An alternative approach, especially suited for nonreactive flows at lower temperatures ($T < 1000$ K), employs the so-called anomalous or dual fluorescence observed for selected polynuclear aromatic hydrocarbons. This method has been demonstrated by Peterson, Lytle, and Laurendeau (1986, 1988) for measurement of temperature in rich C_2H_4/O_2 and H_2/O_2 flames at 70–80 torrs. Because of rapid internal vibrational conversion, the fluorescence spectrum of most aromatic hydrocarbons displays a single broad feature attributable to emission from the lowest excited singlet state (S_1) to the ground electronic state (S_0). Some aromatics, such as pyrene, are characterized by a rather small energy gap between the first and second (S_2) excited singlet states. Therefore, as shown in Fig. 4-26, the broad thermal vibrational distribution in the ground state is transferred to both excited electronic states via intramolecular vibrational coupling upon laser excitation. Since the $S_2 \rightarrow S_0$ emission increases relative to the $S_1 \rightarrow S_0$ emission with increasing temperature, the ratio of the two emission bands can be directly related to the ground state vibrational temperature of the pyrene molecule.

Demonstration experiments were performed by adding trace amounts of pyrene to flat premixed flames and then exciting the pyrene with a Nd:YAG-pumped dye laser at a wavelength of 309 nm, which corresponds to excitation into both the S_1 and S_2 states (Peterson, Lytle, and Laurendeau, 1986, 1988). The fluorescence intensities were averaged over 300 pulses, and the emission spectra were recorded from 330 to 450 nm in intervals of 5 nm. Figure 4-27 shows examples of the normalized emission spectra of pyrene at three different flame temperatures. Emission from the S_1 state appears in the 380- to 410-nm region, while that from the S_2 state appears in the 340- to 370-nm region. Figure 4-28 shows the area ratio of the S_2 and S_1 emissions versus height above the burner compared with a temperature profile determined by radiation-corrected, silicon-coated Pt/Pt–10% Rh thermocouples. The two profiles are overlayed by employing a least squares procedure; the average deviation between the S_2/S_1 and temperature profiles in the postflame region is only 30 K. The agreement between these two profiles demonstrates the potential applicability of the anomalous fluorescence method. Since a ratio of fluorescence signatures is employed, the technique should be insensitive to changes in laser power and temporal pulse shape. Higher photon fluxes could be generated by employing an excimer laser with detection via

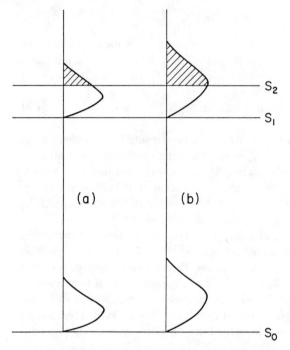

Figure 4-26 Simplified schematic displaying the influence of the ground state vibrational distribution on the S_2 character of the fluorescence spectrum for pyrene: (*a*) low temperature, (*b*) high temperature.

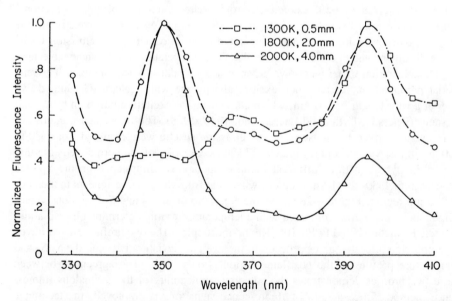

Figure 4-27 Temperature dependence of pyrene fluorescence spectra *(Peterson, Lytle, and Laurendeau 1986)*. Spectra taken at three positions in a C_2H_4/O_2 flame at an equivalence ratio of 1.7. S_1 emission at 380–410 nm and S_2 emission at 340–370 nm.

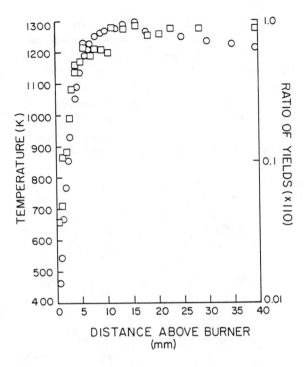

Figure 4-28 S_2/S_1 fluorescence (squares) as a function of height above the burner compared with temperatures (circles) determined by thermocouples (± 30 K) for an H_2/O_2 flame at an equivalence ratio of 1.5 *(Peterson, Lytle, and Laurendeau 1988)*.

filtered PMTs or a linear diode array. The monochromatic nature of the method also makes it suitable for temporally resolved, planar thermometry in turbulent flows.

4-6 PLANAR THERMOMETRY

In the past 5 years, thermometry by light-scattering methods has been extended to linear and planar imaging of both laminar and turbulent flow fields. Raman scattering has generally proved to be too weak for such applications, but temporally resolved measurements are possible via Rayleigh and fluorescence techniques (Hanson 1986). In general, sheet illumination is employed with right angle detection of scattered light using a two-dimensional array detector, with or without a microchannel plate image intensifier. Current detectors include photodiode arrays, vidicon cameras, and charge-coupled device (CCD) arrays, with the latter offering superior performance in terms of noise, linear dynamic range, and spatial resolution (Hanson 1986). Temporal resolution is normally set by the laser source (≈ 5 ns), while spatial resolution is determined by the pixel density of the array detector. The detector is usually coupled directly to a laboratory computer for fast storage, processing, and display of the imaged data.

Fourgette, Zurn, and Long (1986) extended Rayleigh thermometry to single-shot, planar measurements in a nonpremixed CH_4/H_2/air flame. The Rayleigh signal at each pixel was converted to temperature via Eq. (4-12). The burner consisted of a fuel nozzle surrounded by a laminar coaxial air flow; both the fuel and air were filtered before use to minimize Mie scattering. In order to maintain a constant effective cross section throughout the flow, the total Rayleigh cross section of the fuel was made equal to that of air by using a 31/69% volumetric ratio of CH_4 to H_2. To obtain quantitative measurements, the data must be corrected for both background level and spatial nonuniformities. The background level, which includes scattered light from the surroundings and electronic noise associated with the detection system, was determined by replacing both the fuel and air with helium. A room temperature calibration was effected at each pixel location; assuming steady state operation, this procedure automatically accounts for nonuniformities in both the illumination sheet and the response of the detector.

A schematic of the experimental apparatus is shown in Fig. 4-29 (Fourgette, Zurn, and Long 1986). The beam is the second harmonic (532 nm) from a Nd:YAG laser. To provide sufficient illumination, a sheet of light was formed using a multipass cell composed of two cylindrical reflectors. The Rayleigh scattered light was filtered and then imaged on a silicon-intensified target vidicon detector. The intensifier was gated to the firing of the laser to essentially eliminate contributions from flame luminosity. Each image was digitized in an 80 × 98 pixel format, yielding a resolution of 0.12 ×

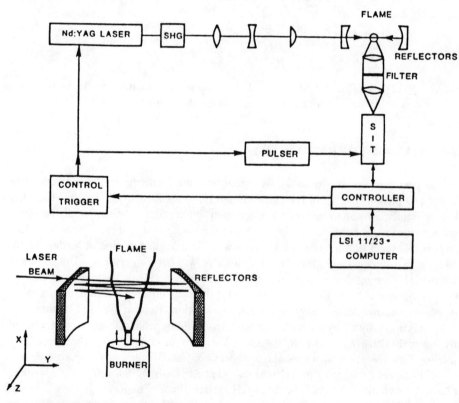

Figure 4-29 Experimental apparatus for instantaneous two-dimensional Rayleigh thermometry of a non-premixed CH_4/H_2/air flame *(Fourgette, Zurn, and Long 1986).*

0.12×0.30 mm^3 for each point. The planar measurements gave both instantaneous and average temperatures ranging from 300 to 2000 K. Regions displaying large temperature gradients and large-scale turbulent structures are easily identified, as shown in Fig. 4-30. The precision of these temperature measurements is $\pm 6\%$ from shot to shot. Average temperatures were obtained by employing 800–900 shots; in this case the maximum variation found for room temperature air was $\pm 4\%$.

Dibble, Long, and Masri (1985) combined two-dimensional Rayleigh thermometry with planar laser-induced fluorescence measurements of the C_2 radical to identify the location of the flame front in turbulent nonpremixed flames. Combining various types of planar imaging methods should eventually provide a detailed understanding of the mixing characteristics of turbulent flows. In this case the C_2 and temperature images both indicated the existence of "holes" in the flame front, thus suggesting local flamelet extinction upon turbulent mixing.

The typical experimental setup for planar laser-induced fluorescence (PLIF) is shown in Fig. 4-31 (Hanson 1986). Compared with Rayleigh thermometry, sufficient scattered signal is available to allow formation of a sheet of light by manipulation of the laser beam with cylindrical lenses rather than mirrors. The first spatially resolved temperature measurements were performed by Alden et al. (1983) in a laminar CH_4/air flame via the TLAF method. In these experiments the wavelength of a single dye laser pumped by a XeCl excimer laser was shifted between the two transitions (410 and 451 nm) of an indium dopant. The spatially resolved fluorescence signals were obtained along a line rather than in a plane, and the average temperature profile across the flame was determined by integrating the fluorescence corresponding to each transition for 100 laser shots. The precision in these early measurements was only $\pm 10\%$, but the feasibility of temporally and spatially resolved planar thermometry with TLAF was clearly demonstrated.

Planar TLMF thermometry using the hydroxyl radical has been evaluated in a premixed flat-flame burner by Cattolica and Stephenson (1983). The wavelength of a

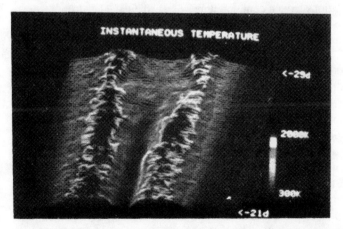

Figure 4-30 Instantaneous two-dimensional Rayleigh thermometry of a nonpremixed CH_4/H_2/air flame (Re = 6500) at 21–29 nozzle diameters downstream (*Fourgette, Zurn, and Long 1986*).

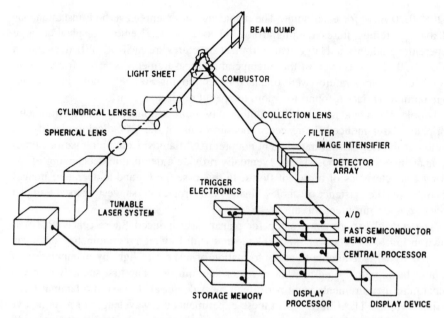

Figure 4-31 Typical experimental setup for planar laser-induced fluorescence *(Hanson 1986)*.

single Nd:YAG-pumped dye laser was again shifted between two transitions, this time corresponding to the $Q_1(5)$ line in the (1, 0) and (1, 1) vibrational bands of the OH $X^2\Pi–A^2\Sigma^+$ system. The broadband fluorescence for each transition was gated to minimize flame emission and detected with the aid of an intensified vidicon camera. After correcting for the intensity distribution of the laser sheet, the ratio of the two fluorescence signals was formed to each pixel on the 100×100 array detector. Calibration was effected by using a known temperature at the center of the flame. Since the pumped transitions have a common upper level, the calculated fluorescence ratio is independent of both quenching and radiative trapping; hence this ratio is directly related to the vibrational temperature of the hydroxyl radical (Cattolica 1981). Unfortunately, due to the weak fluorescence signal from the $v'' = 1$ level, only temperatures greater than 1800 K could be measured. Furthermore, the repeatability of the results was again limited to $\pm 10\%$. By using NO or I_2, the same methodology should provide better precision at lower temperatures.

Seitzman, Kychakoff, and Hanson (1985) obtained the first temporally resolved, planar measurements of temperature by utilizing the ABF method. In this investigation, 2000 ppm of NO was added to a rod-stabilized, fuel-lean, premixed CH_4/air flame. Based on calculated temperature sensitivities at 500–1500 K, the $Q_1(22)$ line of the NO $X^2\Pi(v = 0)–A^2\Sigma^+(v = 0)$ transition was chosen for laser excitation. The required wavelength of 225.6 nm was generated by using an Nd:YAG-pumped dye laser, and summing the doubled output of the dye laser with the residual fundamental from the Nd:YAG laser at 1.06 μm in a KDP crystal. Broadband fluorescence (235–300 nm) was detected by a time-gated (1 μs), intensified 100×100 photodiode array. A UV transmitting filter was used to reject flame emission and scattered laser light. The raw data were corrected for both nonuniformities in the laser sheet and background radiation.

Equation (4-31) shows that for a constant mole fraction of the additive, a temperature can be extracted from the absolute fluorescence signal by calibrating to a known flow field temperature. The resulting NO temperature contours are shown in Fig. 4-32 for a 9.6 × 9.6 mm region (Seitzman, Kychakoff, and Hanson 1985). The spatial and temporal resolution are 0.12 mm and 10 ns, respectively. The V shape of the rod-stabilized flame is reproduced by the temperature contours, but scattered light prevented measurements immediately downstream of the rod. The uncertainty in the temperature is approximately ± 100–200 K, arising mostly from noise in the photocathode of the image intensifier and uncertainties associated with the dependence of the collisional cross section on temperature and composition.

More recently, Lee et al. (Lee, Paul, and Hanson 1986, 1987; Lee and Hanson 1986) made planar thermometric measurements using laser-induced fluorescence from O_2. Utilization of O_2 fluorescence follows the earlier work of Massey and Lemon (1984), in which a single-shot precision of ± 1% was postulated for single-point TLMF thermometry. In the work of Lee et al. (Lee, Paul, and Hanson 1986, 1987; Lee and

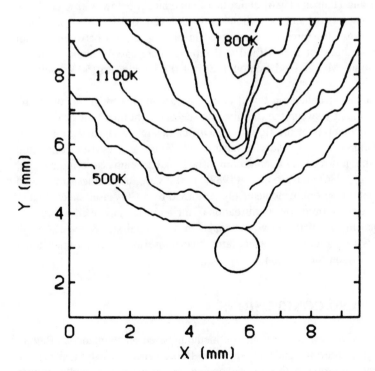

Figure 4-32 Instantaneous temperature contours using planar laser-induced fluorescence of NO *(Seitzman, Kychakoff, and Hanson 1985)*. The image represents a 9.6 × 9.6 mm section in a rod-stabilized CH_4/air flame.

Hanson 1986), rovibronic lines in the Schumann-Runge band system were excited with a broadband (0.5 nm full width at half maximum) ArF excimer laser at 193 nm. This laser produces about 100 mJ/pulse at repetition rates up to 250 pps. The resulting broadband fluorescence (200–400 nm) was first passed through a KBr filter to reject flame luminosity and scattered light, and then focused on an image-intensified 100×100 photodiode array camera. At a constant O_2 mole fraction and pressure, the fluorescence signal increases strongly with temperature for $T > 700$ K (Lee, Paul, and Hanson 1986; Lee and Hanson 1986). This dramatic increase results from excitation of the stronger $v'' = 1, 2, 3$ bands rather than the weak $v'' = 0$ band at higher temperatures. A theoretical analysis of the changes in the absorption coefficient and fluorescence signal as a function of temperature showed very good agreement with experimental calibration curves (Lee, Paul, and Hanson 1987; Lee and Hanson 1986). An important advantage of this method is that the excited B state of O_2 predissociates; thus the measured fluorescence is independent of collisional quenching at moderate pressures.

Lee, Paul, and Hanson (1987) demonstrated planar ABF thermometry with O_2 by passing a 2 cm $\times$ 0.5 mm sheet of light through a jet of heated air flowing past a 3-mm-diameter rod. Despite the strongly reduced fluorescence owing to predissociation, the S/N ratio was approximately 20. The uncertainty in the data was about ± 50 K at 500–1100 K. Lee and Hanson (1986) claim that temperatures below 300 K could be measured by using a narrowband excitation source coupled to favorable transitions. Although detailed flow structures were observed in mixing regions between the hot and cold air, the accuracy of the method will require more careful attention to atmospheric attenuation (Massey and Lemon 1984) of the laser beam (50% every 35 cm at 193 nm).

Three-dimensional visualization could clearly be accomplished in the future by rotating a mirror to rapidly sweep the illuminating sheet of light on a time scale faster than the thermal fluctuations of the flow field. Difficulties related to detection speed and data manipulation/storage would also have to be overcome. Nevertheless, initial work in this direction has recently been accomplished via Rayleigh scattering by Yip, Fourgette, and Long (1986) and Yip et al. (1987). In this case, a freon jet surrounded by a coaxial flow of air was photoacoustically perturbed by a CO_2 laser sheet. Planar images obtained at a fixed time after perturbation of the jet were convolved to produce a three-dimensional image of the flow. Although in this initial study, the Rayleigh scattered signal was used to monitor density rather than temperature, three-dimensional thermometry is clearly on the horizon.

4-7 SUMMARY AND CONCLUSIONS

In this chapter, I have reviewed thermometric methods based on spontaneous Raman scattering, Rayleigh scattering, and laser-induced fluorescence. The typical performance of these light-scattering methods based on experiments to date in steady laminar flows is summarized in Table 4-1. The differential cross sections for both vibrational and rotational Raman scattering are too low for useful applications to planar ther-

Table 4-1 Performance of light-scattering methods for measurement of temperature

Method	Accuracy/precision, %	Range, K
Vibrational Raman	2	300–2500
Rotational Raman	0.3	100–1000
Rayleigh scattering	1	300–2500
Fluorescence	1	150–2500

mometry. Even for local measurements, care must be exercised to avoid laser-induced luminosity and scattered light. Similar problems can arise when employing Rayleigh scattering, although here a larger cross section can support the extension to planar thermometry under "particle-free" conditions.

Temporally resolved planar thermometry favors the fluorescence methods because of their intrinsically high S/N ratio. Such methods are often required in turbulent flows because of the need for nonintrusive measurements with high spatial and temporal resolution. Unfortunately, the usual two-line excitation schemes are plagued by the experimental complications accompanying utilization of two laser beams. Monochromatic methods, such as the thermally assisted or absolute fluorescence techniques, show great potential in this regard, but further development is necessary. Local, thermally assisted thermometry utilizing stable molecular species should be pursued to assess the extent to which the accuracy and precision of the atomic case carry over to the molecular case.

Measurements at lower temperatures ($T < 500$ K) are not feasible using atomic seeds because of their low vapor pressure. Precision and accuracy to ± 1 K are possible near room temperature and to ± 5 K at temperatures to 1000 K using rotational Raman scattering. Similar results appear possible for fluorescence thermometry by using stable molecular species such as NO or I_2. Thus far, low-temperature fluorescence thermometry has only been attempted with the two-line excitation approach. Low-temperature planar thermometry may be possible using thermally assisted fluorescence of NO or I_2, or perhaps anomalous fluorescence from pyrene. Accuracy and precision to at least $\pm 1\%$ appear feasible for steady laminar flows.

4-8 NOMENCLATURE

A_{ij}	spontaneous emission rate coefficient (s^{-1})
B_{ij}	Einstein absorption coefficient [$cm^3/(J\ s^2)$]
c	speed of light (cm/s)
$E(v, J)$	energy (cm^{-1})
$f(J)$	rotation-vibration correction factor
$F_v(J)$	rotational energy (cm^{-1})
g	degeneracy
G	vibrational energy (cm^{-1})
h	Planck constant (J s)

I	irradiance (W/cm^2)
I_L	irradiance of laser beam (W/cm^2)
J	rotational quantum number
k	Boltzmann constant (J/K)
N_i	number density of ith energy level (cm^{-3})
N_0	Loschmidt number (cm^{-3})
N_s	number density of scattering species (cm^{-3})
N_T	total number density (cm^{-3})
P	pressure (atm)
Q	quenching rate coefficient (s^{-1})
Q_{rot}	rotational partition function
Q_{vib}	vibrational partition function
$S(J)$	line strength
t	time (s)
T	temperature (K)
v	vibrational quantum number
$\bar{v}$	mean relative speed (cm/s)
V_c	collection volume (cm^3)
W_{ij}	rate coefficient for absorption (s^{-1})
W_{ji}	rate coefficient for stimulated emission (s^{-1})
x_i	mole fraction of ith species
β	detection efficiency
γ_v	Raman correction factor
ΔE	energy difference (cm^{-1})
ΔF	rovibronic energy difference (cm^{-1})
$\Delta \upsilon$	laser spectral bandwidth (s^{-1})
ζ	correction factor for detection efficiency
η	complex refractive index
θ	scattering angle
λ	wavelength (cm)
υ	frequency (s^{-1})
σ	cross-section (cm^2)
σ_c	collisional cross section (cm^2)
σ_{eff}	effective Rayleigh cross section (cm^2)
ϕ	polarization angle
Φ	signal power (W)
Ω_c	solid angle of collection optics (sr)

Subscripts

AS	anti-Stokes
F	fluorescence
i	level/species
j	level index
J	rotational quantum number
k	level index

L   laser
R   Rayleigh
S   Stokes
v vibrational quantum number

REFERENCES

Aeschliman, D. P., Cummings, J. C., and Hill, R. A. 1979. *J. Quant. Spectrosc. Radiat. Transfer* 21:293.

Alden, M., Grafstrom, P., Lundberg, H., and Svanberg, S. 1983. *Opt. Lett.* 8:241.

Anderson, W. R., Decker, L. J., and Kotlar, A. J. 1982. *Combust. Flame* 48:179.

Arshinov, Yu. F., Brobrovnikov, S. M., Zuev, V. E., and Mitev, V. M. 1983. *Appl. Opt.* 22:2984.

Bechtel, J. H. 1979. *Appl. Opt.* 18:2100.

Breiland, W. G., and Ho, P. 1984. *Proceedings of the ninth international conference on chemical vapor deposition*, eds. McD. Robinson, C. H. J. van den Brekel, G. W. Cullen, J. M. Blocher, and P. Rai-Choudhury, p. 44. Pennington, N.J.: The Electrochemical Society.

Breiland, W. G., Coltrin, M. E., and Ho, P. 1986. *J. Appl. Phys.* 59:3267.

Cattolica, R. J. 1981. *Appl. Opt.* 20:1156.

Cattolica, R. J., and Schefer, R. J. 1982. *Nineteenth symposium (international) on combustion*, p. 311. Pittsburgh, Pa.: The Combustion Institute.

Cattolica, R. J., and Stephenson, D. A. 1983. *Dynamics of flames and reactive systems*, eds. J. R. Bowen, N. Manson, A. K. Oppenheim, and R. I. Soloukhin, p. 714. Progress in Astronautics and Aeronautics, vol. 95. Washington, D.C.: American Institute of Aeronautics and Astronautics.

Chan, C., and Daily, J. W. 1980. *Appl. Opt.* 19:1963.

Chandran, S. B. S., Komerath, N. M., Grissom, W. M., Jagoda, J. I., and Strahle, W. C. 1985. *Combust. Sci. Technol.* 44:47.

Cooney, J. A. 1983. *Opt. Eng.* 22:292.

Crosley, D. R., and Smith, G. P. 1980. *Appl. Opt.* 19:517.

Crosley, D. R., and Smith, G. P. 1982. *Combust. Flame* 44:27.

Dec, J. E., and Keller, J. O. 1986. *Twenty-first symposium (international) on combustion*, p. 1737. Pittsburgh, Pa.: The Combustion Institute.

Dibble, R. W., and Hollenbach, R. E. 1981. *Eighteenth symposium (international) on combustion*, p. 1489. Pittsburgh, Pa.: The Combustion Institute.

Dibble, R. W., Long, M. B., and Masri, A. 1985. *Dynamics of reactive systems*, vol. II, eds. J. R. Bowen, J.-C. Leyer, and R. I. Soloukhin, p. 99. Progress in Astronautics and Aeronautics, vol. 105. Washington, D.C.: American Institute of Aeronautics and Astronautics.

Drake, M. C., and Hastie, J. W. 1981. *Combust. Flame* 40:201.

Drake, M. C., and Rosenblatt, G. M. 1978. *Combust. Flame* 33:179.

Drake, M. C., Bilger, R. W., and Starner, S. H. 1983. *Nineteenth symposium (international) on combustion*, p. 459. Pittsburgh, Pa.: The Combustion Institute.

Drake, M. C., Lapp, M., and Penney, C. M. 1982. *Temperature: its measurement and control in science and industry*, vol. 5, ed. J. F. Schooley, p. 631. New York: American Institute of Physics.

Drake, M. C., Rosasco, G. J., Schneggenburger, R., and Nolen, R. L., Jr. 1979. *J. Appl. Phys.* 50:7894.

Drake, M. C., Lapp, M., Penney, C. M., Warshaw, S., and Gerhold, B. W. 1981. *Eighteenth symposium (international) on combustion*, p. 1521. Pittsburgh, Pa.: The Combustion Institute.

Drake, M. C., Asawaroengchai, C., Drapcho, D. L., Viers, K. D., and Rosenblatt, G. M. 1982. *Temperature: its measurement and control in science and industry*, vol. 5, ed. J. F. Schooley, p. 621. New York: American Institute of Physics.

Eckbreth, A. C. 1981. *Eighteenth symposium (international) on combustion*, p. 1471. Pittsburgh, Pa.: The Combustion Institute.

Eckbreth, A. C. 1987. *J. Propulsion Power* 3:210.

Eckbreth, A. C., Bonczyk, P. A., and Verdieck, J. F. 1979. *Prog. Energy Combust. Sci.* 5:253.

Elder, M. L., Zizak, G., Bolton, D., Horvath, J. J., and Winefordner, J. D. 1984. *Appl. Spectrosc.* 38:113.

Fletcher, D. G., and McDaniel, J. C. 1987. *Opt. Lett.* 12:16.

Fourgette, D. C., Zurn, R. M., and Long, M. B. 1986. *Combust. Sci. Technol.* 44:307.

Furuya, Y., Yamamoto, M., and Takubo, Y. 1985. *Jpn. J. Appl. Phys.* 24:455.

Gross, K. P., and McKenzie, R. L. 1983. *Opt. Lett.* 8:368.

Gross, K. P., and McKenzie, R. L. 1985. *AIAA J.* 23:1932.

Gross, K. P., McKenzie, R. L., and Logan, P. 1987. *Exper. Fluids* 5:372.

Hanson, R. K. 1986. *Twenty-first symposium (international) on combustion*, p. 1677. Pittsburgh, Pa.: The Combustion Institute.

Haraguchi, H., and Winefordner, J. D. 1977. *Appl. Spectrosc.* 31:195.

Haraguchi, H., Smith, B., Weeks, S., Johnson, D. J., and Winefordner, J. D. 1977. *Appl. Spectrosc.* 31:156.

Hill, R. A., Peterson, C. W., Mulac, A. J., and Smith, D. R. 1976. *J. Quant. Spectrosc. Radiat. Transfer* 16:953.

Joklik, R. G., and Daily, J. W. 1982. *Appl. Opt.* 21:4158.

Kohse-Höinghaus, K., Perc, W., and Just, Th. 1983. *Ber. Bunsenges. Phys. Chem.* 87:1052.

Kreutner, W., Stricker, W., and Just, Th. 1987. *Appl. Spectrosc.* 41:98.

Lapp, M. 1974. *Laser Raman gas diagnostics*, eds. M. Lapp and C. M. Penney, p. 107. New York: Plenum.

Lapp, M., and Penney, C. M. 1977. *Advances in infrared and Raman spectroscopy*, chap. 6, vol. 3, ed. R. J. H. Clark and R. E. Hester. London: Heyden and Sons.

Lapp, M., Penney, C. M., and Goldman, L. M. 1972. *Science* 175:1112.

Lee, M. P., and Hanson, R. K. 1986. *J. Quant. Spectrosc. Radiat. Transfer* 36:425.

Lee, M. P., Paul, P. H., and Hanson, R. K. 1986. *Opt. Lett.* 11:7.

Lee, M. P., Paul, P. H., and Hanson, R. K. 1987. *Opt. Lett.* 12:75.

Leyendecker, G., Doppelbauer, J., Baucile, D., Geittner, P., and Lydtin, H. 1983. *Appl. Phys.* A30:237.

Lucht, R. P., Laurendeau, N. M., and Sweeney, D. W. 1982. *Appl. Opt.* 21:3729.

Massey, G. A., and Lemon, C. J. 1984. *IEEE J. Quant. Electr.* QE-20:454.

McKenzie, R. L., and Gross, K. P. 1981. *Appl. Opt.* 20:2153.

Measures, R. M. 1984. *Laser remote sensing*. New York: John Wiley.

Michael-Saade, R., Sawerysyn, J. P., Sochet, L.-R., Buntinx, G., Crunell-Cras, M., Grase, F., and Bridoux, M. 1983. *Dynamics of flames and reactive systems*, eds. J. R. Bowen, N. Manson, A. K. Oppenheim, and R. I. Saloukhin, p. 658. Progress in Astronautics and Aeronautics, vol. 95. Washington, D.C.: American Institute of Aeronautics and Astronautics.

Namer, I., and Schefer, R. W. 1985. *Exper. Fluids* 3:1.

Omenetto, N., Benetti, P., and Rossi, G. 1972. *Spectrosc. Acta* 27B:453.

Omenetto, N., Browner, R., and Winefordner, J. 1972. *Anal. Chem.* 44:1683.

Penney, C. M. 1969. *J. Opt. Soc. Am.* 59:34.

Peterson, D. L., Lytle, F. E., and Laurendeau, N. M. 1986. *Opt. Lett.* 11:345.

Peterson, D. L., Lytle, F. E., and Laurendeau, N. M. 1988. Flame temperature measurements using the anomalous fluorescence of pyrene. *Appl. Opt.* 27:2768.

Rea, E. C., Jr., and Hanson, R. K. 1988. Rapid laser wavelength modulation spectroscopy applied as a fast temperature measurement technique in hydrocarbon combustion, *Appl. Opt.* 27:4454.

Schefer, R. W., Robben, F., and Cheng, R. K. 1980. *Combust. Flame* 38:51.

Seitzman, J., Kychakoff, G., and Hanson, R. K. 1985. *Opt. Lett.* 10:439.

Smith, G. P., and Crosley, D. R. 1981. *Eighteenth symposium (international) on combustion*, p. 1511. Pittsburgh, Pa.: The Combustion Institute.

Smith, J. R., and Giedt, W. H. 1977. *Int. J. Heat Mass Transfer* 20:899.

Sochet, L.-R., Lucquin, M., Bridoux, M., Crunelle-Cras, M., Grase, F., and Delhaye, M. 1979. *Combust. Flame* 36:109.

Stephenson, D. A. 1979. *Seventeenth symposium (international) on combustion*, p. 993. Pittsburgh, Pa.: The Combustion Institute.

Yip, B., Fourgette, D. C., and Long, M. B. 1986. *Appl. Opt.* 25:3919.

Yip, B., Lam, J. K., Winter, M., and Long, M. B. 1987. *Science* 235:1209.

Zizak, G., Bradshaw, J. D., and Winefordner, J. D. 1981. *Appl. Spectrosc.* 35:59.

Zizak, G., Horvath, J. J., and Winefordner, J. D. 1981. *Appl. Spectrosc.* 35:488.

Zizak, G., Horvath, J. J., Van Dijk, C. A., and Winefordner, J. D. 1981. *J. Quant. Spectrosc. Radiat. Transfer* 25:525.

FIVE

JOINT LASER RAMAN SPECTROSCOPY AND ANEMOMETRY

D. R. Ballal

5-1 INTRODUCTION

At present, there is considerable interest in applying joint laser Raman spectroscopy and anemometry for combustion measurements. This interest stems from the growing need to (1) use computer models for predicting combustor performance, (2) develop combustion theory for explaining experimental findings, and (3) test simplified models of complex combustor hardware in a laboratory.

Computer modeling of turbulent combustion processes is in an active state of development. Some of these models are used in industry in a qualitative or a semi-quantitative way for predicting flow fields, heat release rates, and pollutant formation in combustors. These models provide finite difference solutions of nonlinear, time-dependent conservation equations of mass, momentum, species, and energy. For solving these and other relevant equations, closure assumptions are normally employed in the description of turbulent chemistry, turbulent transport, and turbulent dissipation effects. In the absence of combustion measurements, modelers are often forced to make empirical, unjustifiable, or even erroneous assumptions. For example, (1) turbulent chemistry closure requires complete information on the joint velocity-scalar probability

This work was supported by the U.S. Air Force Wright Research and Development Center, Aero Propulsion and Power Laboratories under contracts F33615-82-C-2255 and F33615-87-C-2767. The author wishes to thank W. M. Roquemore, the Air Force technical monitor, for providing the photograph in Fig. 5-8. Both R. W. Dibble of Sandia National Laboratories, Livermore, California, and J. F. Driscoll of University of Michigan, Ann Arbor, Michigan, provided many comments that led to an improved manuscript. I am most grateful to these researchers and to my coworkers A. J. Lightman and P. P. Yaney for help, advice, and many fruitful discussions.

density function (pdf), $P(u, \phi)$, but in many models this is chosen somewhat arbitrarily, (2) the turbulent transport process is often cast in the form of turbulent eddy diffusivity, a concept increasingly under attack, and finally, (3) modeling of the turbulent dissipation term is empirical at best. Clearly, the utility, applicability, and potential design capabilities of the models can be fully realized only if combustion measurements are available to assist closure formulations.

Although the conservation equations that describe turbulent reacting flows are well known, discrepancies between combustion theory and experimental results arise for three main reasons. First, the simplifying closure assumptions used in the development of the theory compromise the accuracy of final predictions. Second, measurements of velocity-scalar correlation and joint pdfs are very difficult to perform in a combusting flow. As a consequence, submodels for such correlations have been neither fully evaluated nor refined by direct comparison with their measurements. Third, the solutions of these equations produce predictions of a variety of fluid and thermodynamic quantities, only a few of which can be measured even with the most sophisticated diagnostics available to date. Indeed, some statistical terms (e.g., entropy) are introduced to facilitate analysis but are not directly observable, or have very intricate physical interpretation.

Finally, combustion measurements using laser or mechanical probes are extremely difficult to perform in large-scale practical combustors. Several factors, such as large density gradients, heat transfer, flame luminosity, soot particulates, flame quenching, and two-phase quasi-steady combustion, complicate measurements in practical devices. Also, the inlet flow boundary conditions are ill defined. Therefore an alternate approach is to gain sound fundamental knowledge from small-scale laboratory experiments and then extend it to evaluate and refine computer models of practical combustor hardware. For this reason, "benchmark quality" data from well-conceived laboratory experiments that simulate the combustion process in practical combustors are required.

This chapter begins by outlining the requirements of laser diagnostic systems. Next, we describe the integration of laser Raman spectroscopy with the laser Doppler anemometer (LDA), giving details of individual instruments, their optical and electronic integration, sources of error, and data reproducibility. Then, combustion measurements in a variety of premixed and diffusion flames are presented. These data are interpreted in light of their contribution to our knowledge of combustion mechanisms and model solutions. We conclude with some thoughts on future developments in advanced laser diagnostics for combustion measurements.

5-2 REQUIREMENTS OF COMBUSTION MEASUREMENTS

To define the specifications of the laser diagnostic system, we must first review the range of combustion and fluid flow parameters over which laboratory and practical combustion systems operate. Goulard, Mellor, and Bilger (1976) categorized many turbulent combustion processes in terms of their characteristic mean velocity U, integral length scale L, and Kolmogoroff scale η. Also, the characteristic time τ_u for the Kolmogoroff scale to be convected past a point in the flow is $\tau_u = \eta/U$, and the dissipation rate of concentration fluctuation is $\tau_c = \lambda^2/6\nu_0$, where λ is the Taylor microscale, and ν_0 is the characteristic kinematic viscosity. These authors have identified regimes corresponding to the operation of gas turbine combustors, afterburners, small-scale fires, and furnace combustion in terms of these parameters.

Referring to Goulard, Mellor, and Bilger (1976), at one extreme are high-speed, highly turbulent combustion processes in the aviation gas turbine combustor and afterburner, for which (under atmospheric pressure and room temperature inlet conditions) typical values of the characteristic parameters are $U = 100$ m/s, $L = 3$–5 cm, $\eta = 100$ μm, $\tau_u = 1$ μs, and $\tau_c = 50$ ms. At the other end are the buoyancy-dominated fires and furnace combustion in which $U = 2$–5 m/s, $L = 2$ m, $\eta = 1$ mm, $\tau_u = 200$–500 μs, and $\tau_c = 3$ s. Intermediate between these two extremes are regions associated with turbulent combustion in laboratory and small-scale model combustors.

Thus, for measurements in high-speed reactive flows, a laser diagnostic system should have a spatial resolution at least equal to the smallest turbulence scale (i.e., Kolmogoroff eddy size) $\eta = 100$ μm and a temporal resolution equal to the characteristic time $\tau_u = 1$ μs. If high inlet pressure and temperature characteristic of practical combustors are taken into account, then the resolution requirement of a laser diagnostic system increases dramatically, over and above these already difficult-to-attain numbers. Now, for many laboratory flames and model combustors, typical values of the characteristic parameters are $U \leq 25$ m/s, $L = 2$ cm, $\eta = 250$ μm, and $\tau_u = 10$ μs. As described in the next section, some modern laser diagnostic systems in use today can meet these requirements of spatial and temporal resolutions in laboratory experiments. However, significant improvements in high-power pulsed lasers and data processing electronics are necessary before the laser diagnostic systems can fully meet or exceed the resolution requirements of the high-speed reactive flows.

For gaining even a rudimentary understanding of turbulent combustion, time-resolved measurements of velocity, temperature, and concentration are necessary. These measurements can be performed by operating LDA and Raman spectroscopy systems independently of each other. However, as discussed in detail in Chapter 1 and also pointed out by Libby and Williams (1980), and Chigier (1981), the development of closure formulation and sophisticated combustion theory requires the simultaneous measurements of velocity-scalar correlations $\overline{uc}$, $\overline{vc}$, $\overline{uT}$, $\overline{vT}$, and joint pdfs $P(u, \phi, x)$ in turbulent flames. Such data facilitate the following.

1. Calculation of mean reaction rate (required in the turbulent chemistry closure) in a premixed flame from the measured joint pdfs.
2. Evaluation of conserved scalar approach for turbulent diffusion flame calculations.
3. Computation of turbulent Schmidt/Prandtl number, which defines a ratio between the propagation of shear stress and scalar flux.
4. Verification of the well-known gradient transport assumption, or the turbulent eddy viscosity concept.
5. Computation of cross-dissipation terms in the modeling equations.

For successfully performing such combustion measurements, both Raman spectroscopy and LDA systems have to be spatially and temporally integrated. These individual diagnostic systems, their calibration, and their optical and electronic integration are described below.

5-3 RAMAN SPECTROSCOPY–ANEMOMETRY INTEGRATION

Advances in lasers and electrooptics in the last two decades have made possible the development and application of several advanced laser diagnostic techniques such as LDA, spontaneous Raman spectroscopy, coherent anti-Stokes Raman spectroscopy, and Rayleigh scattering to combustion measurement. To date, these instruments have yielded independent measurements of velocity, temperature, concentration, and density fluctuations in both laboratory and practical combustors. As the next logical step in widening the application of these techniques to combustion measurements, LDA has been integrated with a variety of spectroscopic techniques since the early 1980s. This section describes the individual laser Raman spectroscopy and LDA system hardware and then discusses their integration. The fundamentals of laser anemometry and Raman scattering are covered in Chapters 2 and 4, respectively.

The pioneering investigations of Lapp and Penney (1974), Lederman (1977), and Warshaw et al. (1980) established spontaneous Raman scattering for the time-resolved measurements of temperature and species concentration in flames. Recently, Eckbreth (1988) made a valuable addition to this literature. Both rotational-vibrational (Q branch) and pure rotational Raman scattering can be used for combustion measurements. The Q branch spectra are generally preferred for flame application because of the relative freedom from spectral overlap of different molecular species. In contrast, a pure rotational Raman system has the main advantage that the Stokes rotational Raman cross section for a single transition is generally an order of magnitude stronger than that for an entire vibrational branch. This is a great asset in combustion measurements, where the Raman signal is usually extremely weak and yet high sampling rates are desired. Thus pure rotational Raman scattering can be effectively employed for simple mixtures or where one constituent, e.g., molecular nitrogen, is dominant. These two main types of spontaneous Raman spectroscopy systems are now described.

Vibrational Raman Spectroscopy (VRS)

A single-pulse VRS system has been extensively employed for combustion measurements by Drake et al. (1980, 1983; Drake, Pitz, and Lapp 1986), Dibble, Kollmann, and Schefer (1984), Johnston, Dibble, and Schefer (1986), and Magre and Dibble (1988). This system is ideally suited for the time-resolved measurements of scalar pdfs in combustion measurements.

Figure 5-1 illustrates the general arrangement of the principal optical components of the combined VRS-LDA system as employed by Dibble, Kollmann, and Schefer (1984). The brief description of the VRS system given below follows from the papers of Dibble, Kollmann, and Schefer (1984) and Magre and Dibble (1988). A dye laser pumped by a flash lamp provides pulses (1 J/pulse, 3 pulses/s, 2 μs pulse width, $\lambda = 532.1$ nm, $\Delta\lambda = 0.3$ nm) to excite the Raman scattering. The scattered signal is

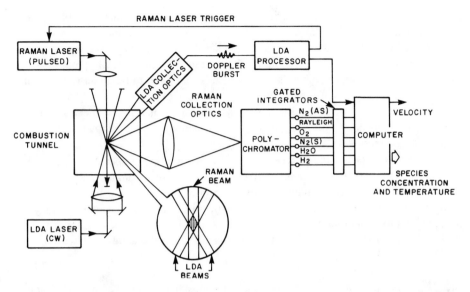

Figure 5-1 Arrangement of the principal optical components of the joint VRS-LDA system *(Dibble, Kollmann, and Schefer 1984)*.

collected at right angles by a six-element, 30-cm focal length, $f/2$ collection lens and is relayed at $\times 3$ magnification to the entrance slit of a 3/4-m spectrometer. The 3-mm width of the entrance slit determines the length of the Raman probe volume, while the height of the probe volume is decided by the laser beam diameter (<1 mm). With this slit, the measured Raman signal is insensitive to laser beam steering in the turbulent flame. At the exit plane of the spectrometer, five photomultiplier tubes are positioned to receive Raman-scattered light from N_2, O_2, H_2, H_2O, and the anti-Stokes vibrational Raman scattering from N_2. In addition, the Rayleigh and Mie scattering are also measured. The electrical outputs of the phototubes are connected to a 12-channel charge integrator (CAMAC), which is gated for the duration of the laser pulse. As an indication of the overall efficiency of the Raman collection system, Magre and Dibble (1988) state that 6000 photoelectrons per joule of laser light are collected from vibrational Raman scattering of N_2 in room air.

In calibrating the VRS system, it should be recognized that the integrated charge Q_i from a photomultiplier tube is linearly related to laser energy Q_1 and species concentration $[N_i]$ as follows:

$$Q_i = k_i Q_1 [N_i] f_i(T) \tag{5-1}$$

In Eq. (5-1) the proportionality constant k_i is dependent on the vibrational Raman cross section, wavelength, geometry, and optical collection efficiency. This constant is ultimately determined by calibration at room temperature and pressure. The bandwidth factor $f_i(T)$ accounts for the temperature-dependent distribution of molecules in their allowed quantum state. Also, the spectral location, shape, and bandwidth of the spectrometer and the bandwidth of the laser are included in this factor. Therefore the of the bandwidth factor are experimentally determined in laminar, adiabatic, and premixed hydrogen-air flat flames. Next, the fuel flow rate determines the equivalence ratio from which equilibrium values of the species concentration are calculated. These concentrations are used for calibration of the Raman signal. Finally, background fluorescence in the Stokes Raman signal, which is measured to be about 0.5% of the Raman signal from gases at room temperature and pressure, is subtracted from all the Raman data. Following these calibrations, measurement of a given species concentration from each laser pulse requires that laser pulse energy, Raman signal, and temperature be recorded.

For each laser pulse, the concentrations of individual major species (H_2, N_2, O_2, H_2O) are determined from their respective Stokes vibrational Raman intensities. The temperature is determined from the ratio of anti-Stokes-to-Stokes N_2 vibrational Raman scattering intensities. Due to the weakness of the anti-Stokes signal, the accuracy of this approach to temperature measurement is low. For better accuracy the temperature can be calculated from the total number density N_{total} and the ideal gas law:

$$T = T_0(N_0/N_{total}) = T_0N_0/([H_2O] + [H_2] + [O_2] + [N_2]) \qquad (5\text{-}2)$$

where N_0 is the total number density of air at atmospheric pressure and temperature T_0 ($N_0 = 2.687 \times 10^{19}$ molecules/cm^3 when $T_0 = 273$ K). Extensive calibrations of this Raman system in laminar, premixed flat flames indicate the measurement accuracy to within 100 K for temperature and ± 1 mole % for species mole fraction. The laser pulse width limits the temporal resolution of the system to 2 μs, and the spectrometer entrance slit and collection optics magnification together provide a spatial resolution of 300 μm $\times$ 300 μm $\times$ 700 μm.

Rotational Raman Spectroscopy (RRS)

The unavailability of a powerful and rapidly pulsed laser source for the VRS system does not permit measurement of the power spectral density function or autocorrelation function in combusting flows. Such data are required for calculating temperature or concentration scales in turbulent reactive flows. These functions can be obtained from a real-time record of fluctuations if the sampling rate is sufficiently high. In the RRS arrangement, a high-power continuous wave (CW) laser and a multipass cell optics configuration are used to detect pure rotational Raman transitions. This combination provides signals 1000 or more times stronger than those typically observed from the N_2 Q branch using a 1-W CW laser.

In recent years, Yaney et al. (1982, 1985, 1987) developed the RRS system for combustion measurements. Figure 5-2 shows the layout of the principal optical components of the joint RRS-LDA system. The left side optics table shown in the figure carries a three-channel, photon-counting RRS system capable of simultaneously measuring the Rayleigh intensity, and either the concentration of two gases in cold flow, or the temperature and concentration fluctuations of a single gas species, such as N_2 in a flame. This system uses the 9-W, 488-nm line of a CW argon-ion laser. The choice of laser wavelength, 488 nm, is appropriate since, if the 514.5-nm line is used, a C_2 Swan band overlaps the Stokes rotational Raman spectrum. A retroreflecting

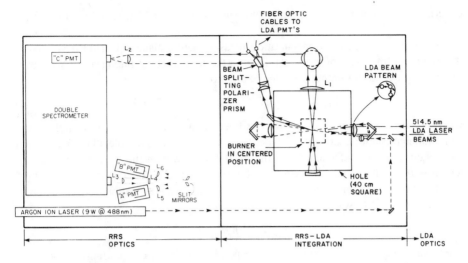

Figure 5-2 Layout of the principal optical components of the joint RRS-LDA system.

multipass cell designed after Hill, Mulac, and Hackett (1977) multiplies the laser power incident upon the observed volume by a factor of 13 over a single pass. The bowtie-shaped measurement volume produced by the multipass cell optics has the dimensions of 50 μm × 300 μm × 1000 μm. A double Spex spectrometer is modified to provide two exit slits, one for each photon-counting channel. As shown in Fig. 5-2, the third, Rayleigh channel ("C" photomultiplier tube), is located in the first half of the double spectrometer. The wavelength spacing of the two Raman channels has a range of 0.76 nm, designed for measurements on the pure rotational Raman lines. The RRS system performs combustion measurements using a 200- to 250-μs window and a signal-sampling rate up to 5 kHz in cold flows, and over 100 Hz in flames.

In the RRS system the signal detection is accomplished by two cooled RCA C31034A photomultiplier tubes. The 3-ns output pulses of these tubes are amplified and standardized to 20-ns fast nuclear instrument module (NIM) pulses using LeCroy model 821 quad discriminators. These signals are recorded by a MODCOMP computer using a Raman interface unit (RIU).

Yaney et al. (1985) described in detail the computation of scalar quantities from the RRS signals. The temperature T of a diatomic gas is related to the ratio R of the intensities of two pure rotational Raman lines specified by quantum numbers J_A and J_B by the relation

$$T = K_1/\ln (K_2 R) \tag{5-3}$$

where $K_1 = -171.75$, $K_2 = 1.5182$, $J_A = 11$, and $J_B = 8$ for the N_2 molecule. Theoretical temperature calibration curves, resulting from the J ratio of 11/8, and three other line pairs, 7/4, 9/6, and 14/10, are given by Yaney et al. (1985). To apply Eq. (5-3) successfully in practice, the background signals under the two lines must be measured. If the background spectrum is reasonably steady in magnitude and shape, then its contribution to the two Raman channels A and B can be measured in a separate experiment. Its contribution can then be subtracted to yield true temperature measurement.

For constant-pressure concentration measurements, Yaney et al. (1985) derived that the mole fraction c of a gas is related to the Raman count rate $\dot{n}_j$ of the rotational line having quantum number J by the expression

$$c = n_j\{\lambda_j^3\lambda_L T^2 \exp\ [1.4388B_0 J(J\ +\ 1)/T]/(\epsilon P_e \ell p)\} \tag{5-4}$$

where λ_j is the wavelength of the Raman line, λ_L the wavelength of laser light, ϵ the calibration constant dependent on the collection geometry, the molecular species, and the system detection efficiency, P_e the excitation power in the observed volume of length ℓ, p the atmospheric pressure, and the rotational Raman constant $B_0 = 1.9895$ cm^{-1} for N_2. For concentration measurements, an absolute calibration is required. However, in cold flows, Eq. (5-4) simplifies to

$$c = \dot{n}_j/\dot{n}_{jr} \tag{5-5}$$

where $\dot{n}_{jr}$ is the reference count rate corresponding to 100% concentration ($c = 1$) of the particular gas species in the observed volume. Thus in cold flows the use of two parallel Raman channels allows simultaneous measurements of concentration of two gases.

Laser Doppler Anemometry System

LDA is a well-known technique for measuring velocity fluctuations in turbulent flows. In Chapter 2 the principles, applications, and optical layouts of a variety of LDA systems are fully described. Two different types of LDA systems were used in the integration with VRS and RRS systems. These LDA systems are briefly described below.

The LDA used by Dibble, Kollmann, and Schefer (1984) and Johnston, Dibble, and Schefer (1986) is a commercially available two-color, real-fringe, forward scattering system. Its general setup is shown in Fig. 5-1. The 488-nm, 1-W laser beams are focused to produce an optical volume of diameter 250 μm × 300 μm in length. A 5-MHz Bragg-cell shift is used to eliminate directional ambiguity in the radial velocity component. Coincidence of radial and axial velocity measurements is verified using a multichannel interface, and a variable time window set at 10 μs ensures that the velocity measurement in each direction is from the same seed particle. Both the fuel and air flows are seeded with 0.3-μm diameter alumina particles using commercial seeders.

The LDA system used for integration with the RRS system is fully described by Ballal et al. (1987; Ballal, Chen, and Schmoll 1988). Briefly, this is a two-component, real-fringe system assembled on the right side of the table, which carries optical components of the RRS-LDA integrated system as shown in Fig. 5-2. It uses the 2-W, 514.5-nm green line of a CW argon-ion laser as a light source. The two measurement channels are separated by polarization because this LDA cannot use the 488-nm laser beam reserved for the RRS system. The optical train provides three parallel beams with the orientation and polarization as shown in Fig. 5-2 (inset). This LDA system has three distinguishing features. First, it incorporates Bragg-cell frequency shifting for unambiguous measurements in recirculatory flows; second, for detecting very low velocities (~ 1 cm/s), a phase-locked mixer is designed to electronically downshift the signal frequency from 40 MHz to 5 or 10 MHz; and finally, a channel-coincidence circuit is incorporated for verifying beam crossing in combusting flows. The LDA-scattered signal is collected in a forward direction $10°$ off-axis. This scheme maximizes signal strength, eliminates any potential crosstalk between the two channels, and preserves the polarization in the scattered radiation. The LDA measurement volume dimensions are 50 μm $\times$ 200 μm $\times$ 650 μm. Typical sampling rates for the LDA measurements are 3 kHz in cold flows and 1 kHz in combusting flows.

The integration of VRS and RRS systems with their respective LDA systems has to be carried out at two levels: optical integration and electronics/software integration. In the optical integration, spatial coincidence of the two systems is achieved, while in the electronic integration their temporal coincidence is obtained. Descriptions of VRS-LDA and RRS-LDA systems follow.

Integrated VRS-LDA System

Figure 5-1, reproduced from Dibble, Kollmann, and Schefer (1984), shows the schematic of the VRS-LDA optical integration. As illustrated in this figure (inset), Raman and LDA beams are brought to focus in the same spatial location having a waist diameter of 0.5 mm and a length of 2 mm. A major problem created by the spatial coincidence of Raman and LDA systems is that the presence of an LDA seed particle in the measurement volume produces a stray Mie-scattered signal, which is very strong in comparison with the Raman-scattered signal. To reject this stray Mie scattering and thereby improve the signal-to-noise ratio, Dibble et al. (1984) mount colored glass and interference filters in front of their Raman photomultiplier tubes, which detect the Stokes-shifted scattering. Further, to increase the dynamic range of the VRS-LDA system, some of the phototube outputs are connected to two channels, one of them with an attenuation factor of 10.

In the electronic integration of the VRS-LDA systems, the dye laser of the VRS system is delay triggered by an LDA event. Dibble, Kollmann, and Schefer (1984) found that if the delay is too long, the velocity and scalar measurements become uncorrelated, whereas if it is too short, the laser pulse collides with the seed particle present in the measurement volume and produces unusually large Rayleigh signals. From various trial-and-error experiments, an optimum value of time delay was found to be 40 μs. For this optimum value, only 1% of the Raman data is rejected as unsuitable for scalar calculations.

In this VRS-LDA system the LDA realizes 15 events/min, and the interval between velocity and scalar measurement is 40 μs. Thus frequency components up to 12 kHz can contribute to the velocity-scalar correlation. Although high event and frequency rates are not necessary for measuring correlations and joint pdfs accurately, it is interesting to know that this joint system provides adequate time resolution for combustion measurements in high-speed flows.

Integrated RRS-LDA System

Figures 5-2 and 5-3 show the optical beam paths of the 488-nm blue RRS laser and the 514.5-nm green LDA laser. The blue RRS beams forming the multipass cell are confined to a horizontal plane parallel to the optical table top, as shown clearly in Fig. 5-3. The green LDA beams are brought around the outer periphery of the multipass cell, above and below the RRS folding mirrors and through a common lens, so as to focus them to exactly the same location in a flame. As shown in Fig. 5-3, during combustion measurements, the LDA signal is collected 10° off-axis in a forward scattering mode, and the RRS signal is collected at 90° to the input axis to minimize the interrogation volume.

The electronic-hardware integration scheme of the RRS-LDA systems is illustrated in Fig. 5-4. The LDA phototube signals are processed by a TSI, Inc., counter processor (CP) and sent to the LDA interface (one for each channel), which also records the time of arrival (32-bit clock, 100-ns resolution) and the number of cycles in the LDA burst. The CP data and the clock data are then sent to a MODCOMP computer through a direct memory access channel. The Raman signals are detected by two RCA C31034A photomultiplier tubes. After their amplification and discrimination, these signals are

Figure 5-3 Photograph illustrating the optical integration of RRS-LDA systems for performing measurements in a turbulent jet diffusion flame. Numerous passes of the blue (488 nm) Raman beam form a multipass cell. The outer green (514.5 nm) LDA beams are focused to produce spatial coincidence within the flame.

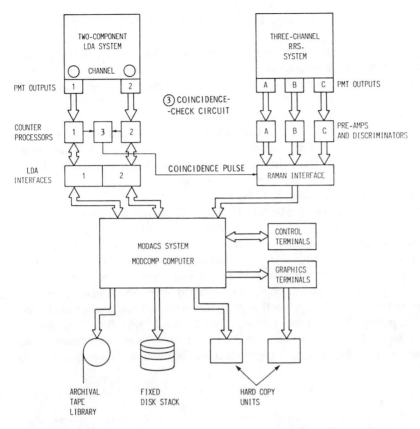

Figure 5-4 Diagram illustrating the electronic-hardware integration of the RRS-LDA systems.

sent to a Raman interface unit (RIU). The RIU provides a choice of three simultaneous data acquisition modes (discussed below), a choice of two- or three-channel operation, and control and monitoring of the spectrometer wavelength setting, Raman window setting, and any digital signal on the data bus. Following the detection of a valid LDA event by the coincidence circuit, the Raman pulses are counted for 200–250 µs (user selectable). Then, all the data are routed to the 16-bit-wide data bus via a 4096-word, first-in, first-out (FIFO) buffer memory. Finally, to calculate the scalar quantities, these data are transferred to a computer through a direct memory access (DMA) channel.

The RIU has wider applications besides providing a variety of joint measurement schemes for the RRS-LDA system. For example, the Rayleigh photomultiplier tube can be reconfigured so that it can detect CH-stretch Raman vibrational bands of the fuel species or an oxygen Raman rotational line. Alternately, the three RRS channels can be rearranged to make laser-induced fluorescence measurements, or one channel can be used for background signal measurement. Finally, RIU also provides a trigger that can initiate integration between physical probes, such as thermocouples and pressure transducers, with optical probes, such as the RRS system. Such a capability not only permits a wide variety of simultaneous measurements in an experimental run, but also provides an excellent means for checkout and calibration of the optical probes against mechanical ones.

Figure 5-5 shows one independent and two basic modes of the software-based integration scheme for simultaneous velocity-temperature data acquisition. First, in the independent mode, both LDA and RRS signal acquisitions are controlled by a 32-bit reference clock within the RIU, but the start and continuation of data acquisition proceed independently. Second, in the clock mode the signal acquisition clock of the Raman interface is locked onto the LDA clock. Both LDA and Raman take data independently, and the software searches for near-simultaneous events. Third, in the sync-to-LDA mode, RRS takes data within a preset window (usually 200–250 μs) after the LDA coincident data-ready pulse is received.

The integration modes described above provide the choice of a software-based approach or a hardware-based approach. Each has advantages and disadvantages. In the software-based approach, a large quantity of data must be recorded by the RRS-LDA system to obtain an adequate number of valid, simultaneous velocity-scalar measurements. However, these data are available for unlimited trials of a variety of criteria or algorithms for defining simultaneous measurements. In the hardware-based approach, the only data recorded are those determined by the fixed delay and window width set on the RIU. This approach is preferred, provided that it has already been determined, perhaps by using the software-based approach, what delays and window widths are acceptable. Also, the hardware-based approach saves on computer storage and is most accurate but requires a higher data acquisition time. Depending upon the

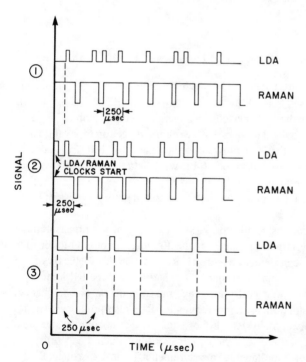

Figure 5-5 Diagram illustrating the RRS-LDA software integration schemes: 1, independent mode; 2, LDA clock mode; and 3, sync-to-LDA mode.

objectives of the testing, accuracy desired, time available, and computer resource capabilities, these different data acquisition choices can be used.

5-4 SOURCES OF ERROR

In designing individual VRS, RRS, and LDA systems, several sources of error arise. These errors affect the operation of the combined VRS-LDA and RRS-LDA systems in performing accurate joint measurements of correlations and pdfs. Thus an error analysis should be performed, errors eliminated, and correction schemes developed. Since the Raman-scattered signal is very weak, it is susceptible to interferences from flame luminosity, LDA seed, and soot particulate scattering. Also, the statistics of photon-counting data acquisition produce uncertainties. Likewise, the LDA instrument suffers from errors principally due to nonuniform seeding and particle biasing. These errors are now discussed.

Background Luminosity

Both VRS and RRS systems are strongly susceptible to errors produced by background luminosity from soot, high-power laser light, and fluorescence.

Since the scattered Raman signals are situated in the visible or near-ultraviolet portion of the spectrum, flame luminosity due to chemiluminescent emission and soot smears the signals. Various combinations of interference and polarizing filters may be used to reduce background flame luminescence. In premixed near-stoichiometric flames, chemiluminescent emissions are the dominant source of luminosity, whereas in diffusion flames, soot emissions are predominant. Eckbreth (1988) demonstrated the severity of this problem by calculating the radiation energy density emitted by 40-nm-diameter spherical soot particles at a temperature of 2000 K. He found that this luminous background will easily swamp the Raman-scattered signals from a 10-W CW argon-ion laser. In a simulated combustor environment, Eckbreth (1977) noted that background luminosity is suppressed to some extent if it is slowly modulated by combustion instability, aerodynamic effects, swirl, or turbulent fluctuations. In such cases, electronic filtering can be successfully used to prevent the luminosity component from entering the signal-processing equipment.

In their studies of hydrogen-air diffusion flame using the VRS-LDA system, Dibble, Kollmann, and Schefer (1984) measured, by scanning the spectrometer away from the Raman line, the contribution of background fluorescence to the Raman signal to be about 0.5% of the Raman signal from N_2 at NTP. This constant background is subtracted from all the Raman data. Recently, Masri, Bilger, and Dibble (1987) reported a detailed investigation of the fluorescence interference with Raman measurements in diffusion flames of methane.

In their application of the RRS-LDA system to premixed methane-air flames, Yaney et al. (1987) noted that errors due to background luminosity may be alleviated by choosing a narrow bandwidth, viz., 0.023–0.07 nm, and by recording a large number of samples, e.g., >4096. Another solution is in measuring background levels by shifting the spectrometer down by 0.075 nm just before each test run and subtracting it from the data. However, this corrective scheme assumes that the background levels are flat and statistically quiet as compared with Raman signal fluctuations. This would generally be the case, at least in steady laboratory hydrocarbon-air flames. However, in practical combustors, background level may fluctuate widely from sample to sample. In such cases, the scalar measurements can be smeared unpredictably.

Particulate Scattering

Scattering from LDA seed particles and soot can easily surpass the strength of the molecular Raman-scattered signal. This is such a severe problem for both VRS-LDA and RRS-LDA systems, that exact simultaneity of velocity and temperature measurements is not possible, and appreciable time delay is necessary between the detection of LDA and subsequent collection of Raman signals.

LDA seed particles produce Mie scattering. For example, the Mie-scattering cross section of a 0.1-μm-diameter particle is 20 orders of magnitude greater than the N_2 Raman cross section. This scattering can be suppressed by employing various spectrometer slit-filter combinations or by a specially designed transient-overload circuit, which desensitizes the photomultiplier tube just before the seed signal overloads it, and quickly returns the tube to full operational status afterward. In spite of these precautions, Mie scattering is still large enough to compete with the Raman signal. Therefore, and for the flow velocities of their experiment, Dibble, Kollmann, and Schefer (1984) introduced an optimum time delay of 40 μs between velocity and scalar measurements to clear the measurement volume of the seed particles. Further, 1% of the Raman data was rejected presumably because of the random excursion of some seed particles through the Raman volume.

The RRS-LDA system has been found remarkably tolerant to the presence of LDA seed particles in the measurement volume. Ballal, Lightman, and Yaney (1987) report that even for a heavily seeded flow, the LDA seed particles contribute only up to 3% of the total Raman signal. One reason for this low signal loss is undoubtedly the use of a double Spex monochromator with a very high rejection ratio (10^{12}) and a narrow slit opening. The seed particles can also increase the opacity to the laser light and displace the gas volume at the measurement location. Again, in actual tests, little deterioration in Raman measurements was observed due to this effect.

Soot particulates produce background flame luminosity, and in interacting with the Raman laser source, they can fluoresce, Raman scatter, and incandesce. Leonard (1974) observed strong hydrocarbon fluorescence that prevented his Raman system from detecting NO from aircraft engine emissions. Alden et al. (1982) encountered fluorescence interferences in an atmospheric flame. However, these interferences were successfully subtracted out by sampling them with the incident laser tuned off the OH species being probed. Similarly, particulate Raman scattering is spectrally specific and generally does not pose a problem. Wright and Krishnan (1973) report that carbon has broad Raman bands at 1360 and 1580 cm^{-1} and these merge. Therefore errors due to particulate Raman scattering are negligible. Finally, when soot particles absorb incident radiation from a very powerful laser source, they can heat up to temperatures well above that of the flame and emit blackbody-like radiation. This is a broadband interference and cannot be avoided easily. Thus significant errors can arise when using a high-power laser Raman system to perform combustion measurements in highly sooty flames.

Statistical Uncertainty

The RRS is essentially a photon-counting system. This type of measuring process introduces intrinsic random fluctuations characterized by Poisson statistics superimposed on the intensity fluctuations of the measured Raman signal. Therefore this Poisson uncertainty has to be subtracted from the measured signal to obtain correct values for temperature and concentration fluctuations. Assuming negligible background signal levels, it is possible to predict the Poisson uncertainty in the temperature and concentration data as measured by the RRS system. Yaney et al. (1985) have done these calculations and show that the Poisson standard deviation in temperature is

$$S_T \approx \overline{T^2} \sqrt{f_r[(1/\bar{n}_A) + (1/\bar{n}_B)]}/|K_1| \tag{5-6}$$

and in concentration it is

$$S_c \approx \bar{c} \sqrt{(f_r/\bar{n}_B) + [(K_B/\overline{T}) - 2]^2 (S_T/\overline{T})^2} \tag{5-7}$$

In Eq. (5-6), f_r is the sampling rate. This equation clearly shows the desirability of having high Raman count rates ($\dot{n}_A$ and $\dot{n}_B$), particularly when a high sampling rate is used. In Eq. (5-7), $\bar{c}$ is the mean mole fraction, and $K_B = 1.4388 B_0 J_B (J_B + 1)$.

Figure 5-6 shows the Poisson standard deviation in both temperature and concentration measurements performed in a methane-air jet diffusion flame by Yaney et al. (1985). In this study the desired accuracy of each sample, as set by the Poisson uncertainty, limited the maximum sampling rate to about 100 Hz in flames. However, newer statistical data analysis techniques, as proposed by Yaney et al. (1987), promise to deliver unambiguous measurements of temperature and concentration at much higher sampling rates.

Optical Alignment

The optical integration of VRS-LDA and RRS-LDA systems poses several alignment difficulties. For example, finer alignment of the pulse laser beams is extremely difficult to make, and beam steering through thick glass windows and unmixedness of the flame environment pose two major hurdles to aligning the multipass cell. Misalignment can lower the Raman signal gain and produce spurious signals. A degree of correction is possible, for example, by detecting the changes in the position of the laser beam after it exits the multipass cell and using this information to correct for the loss or gain. Absorption of flame radiation by the optical mounts also causes frequent alignment drifts. The orientation of the multipass cell must be chosen to optimize the light collected by the spectrometer. The aspect ratio of the light source should also match the aspect ratio of the spectrometer entrance slit. However, this requirement may conflict with the need to have a wider slit opening for the Raman signal gain or a

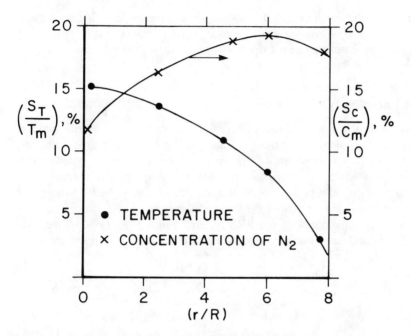

Figure 5-6 Poisson uncertainty in the measurement of temperature and concentration fluctuations in a methane-air diffusion flame using the RRS-LDA system.

finer slit opening to reduce background luminosity due to laser light scattered inside the spectrometer. All these factors restrict choices and produce optical alignment difficulties resulting in loss in Raman signal.

Turbulence, flame curvature, and thermal gradient effects lead to defocusing and steering of the laser beams due to refractive index gradient effects. This poses serious problems, particularly at elevated pressure. For example, Setchell (1978) noted that in a reciprocating engine combustion, there are crank angle regions, associated with the combustion wave passage, where the signal is lost. By arranging for the laser beams to travel parallel to the gradient of density fields, errors due to beam steering and defocusing can be reduced, if not eliminated altogether.

The next two sections examine the errors present in LDA measurements. Reference should also be made to Chapter 2 for a detailed discussion of these and other errors associated with using an LDA.

LDA Seed Rate

A uniform and controlled seeding of combusting flow is a difficult and frustrating task. If the seed rate is nonuniform or uneven for the fuel and airstreams, LDA bias errors increase rapidly. Therefore, to seed multiple-fluid streams with different densities, Ballal, Lightman, and Yaney (1987) employed separate seeders for each stream.

The relative seeding level of the two seeders is adjusted such that the visible laser-scattered intensity remains uniform across the entire plane of measurement. This procedure ensures minimum particle bias errors in the LDA data. A different procedure is adopted by Dibble, Kollmann, and Schefer (1984), who use alternate seeding of fuel jet and coflowing air in their experiments. By conditionally sampling the data, velocity statistics for fuel and airstreams are obtained. Such a sampling process can yield upper and lower bounds to the seeding bias; the actual bias is somewhere inter-mediate between the two bounds.

LDA Particle Bias

When particles are fed and distributed uniformly in the flow field, the measured velocity statistics will be biased because of the dependence of LDA sampling rate on velocity. Since the combustion process produces large density and velocity gradients, the final measured velocity is statistically biased toward higher mass flux (velocity × density) when number weighted averages are used to calculate stationary statistics. Recently, Edwards (1987) reviewed the various particle bias corrections. One correction rec-ommended is due to Chen and Lightman (1985) as follows:

$$P_u = P_b r/r(U) \tag{5-8}$$

where the subscript u refers to an unbiased pdf and b to a biased velocity pdf, r is a norm of $r(U)$, the conditioned measurement rate of velocity, i.e., the average mea-surement rate for a certain bin velocity. Figure 5-7 shows the measured and corrected (by using Eq. (5-8)) velocity pdf data in the turbulent mixing layer of a bluffbody. The biasing of the measured data toward higher velocity values is clearly evident from this figure.

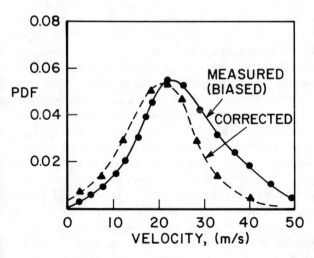

Figure 5-7 Effects of particle ar-rival statistics on the LDA mea-surement of velocity pdf in a bluffbody combustor.

5-5 ERROR ESTIMATES AND DATA REPRODUCIBILITY

Several checks on the data must be performed for assessing the accuracy and reproducibility of the combustion measurements. To verify LDA measurement accuracy,
mass conservation balance should be carried out by integrating the velocity and mass
fraction profiles across the flow field. For the VRS-LDA and RRS-LDA systems
described earlier, a survey of the published literature (e.g., Ballal and Chen 1987;
Dibble, Kollmann, and Schefer 1984; Yaney et al. 1987) indicates that the mass balance
agrees to within 5%. In addition, conservation of momentum is verified to within 4–
6%. For rms velocity fluctuations below 15% of the mean velocity, LDA bias errors
have been found to be about 5%. Dibble, Kollmann, and Schefer (1984) estimated
the errors due to velocity gradient broadening to be about 0.3%, and the seed particle
velocity lag was negligible for flow frequency up to 8 kHz.

The main errors in the VRS and RRS systems arise during calibration of the
Raman signal detection, shot noise, background luminosity, and Poisson uncertainty.
With due care and attention to these factors, Raman measurements have been found
accurate to 3% at 300 K, 7% at 2000 K, and ± 1 mole % for species mole fraction.

Finally, in the published literature (e.g., Ballal and Chen 1987; Dibble, Kollmann,
and Schefer 1984; Yaney et al. 1985) the long-term repeatability of individual LDA,
VRS, and RRS instruments is quoted to be within 5%, but this value deteriorates to
around 7% for the joint VRS-LDA and RRS-LDA systems.

5-6 COMBUSTION MEASUREMENTS USING JOINT RAMAN-LDA SYSTEMS

This section describes measurements in one nonreactive and several reactive flows
using the VRS-LDA and RRS-LDA systems. As a first step toward understanding a
jet diffusion flame, we discuss the mixing of round jets of CO_2 and nonreacting propane
in airstreams with and without free stream turbulence. Next, we present measurements
in turbulent premixed flames, followed by simultaneous velocity-scalar measurements
in turbulent jet diffusion flames. Finally, for each case, we interpret these measurements
and the extent to which they have contributed to our knowledge of combustion mechanisms.

Before proceeding with quantitative point measurements using the sophisticated
systems described here, it is highly desirable to perform flow visualization to examine
the details of the flow field. Once the qualitative and global features of the flow field
are understood, quantitative measurements can be made at select locations to reveal
the most important details of the flow. Flow visualization also helps in the interpretation
of quantitative data. Recently, Roquemore et al. (1987) developed a reactive Mie-
scattering technique in conjunction with a laser sheet lighting arrangement to photograph a variety of nonreactive and reactive flows. Although the description of the
technique is beyond the scope of this chapter, Fig. 5-8 shows a typical photograph of
the turbulence structure in the axial plane of a turbulent jet diffusion flame of methane
and air. In this figure, details of the random turbulent motion and eddy structures in

Figure 5-8 Flow visualization photograph illustrating the structure of turbulence in a methane–air jet diffusion flame, stabilized on a 1-cm-diameter nozzle *(Roquemore et al. 1987)*.

the flame are revealed with impressive clarity. Such photographs add greatly to our understanding of the flow physics and the interpretation of quantitative combustion measurements.

Nonreactive Round Jet

Schefer et al. (1986) examined the mixing of a propane jet $D = 0.5$ cm, $U_j = 53$ m/s, $Re_D = 68,200$ with an airstream of velocity, $U_a = 9.5$ m/s. These measurements were made in the Sandia Turbulent Combustion Test Facility, which has a test section 30-cm^2 cross section and 200-cm long. This facility is a forced draft, vertical wind tunnel with an axisymmetric fuel jet located at the upstream end of a test section. The test section empties into an exhaust hood. It can also be traversed in x, y, z directions.

These authors report measurements of velocity-mixture fraction joint pdfs and correlations using the VRS-LDA system. Figure 5-9(a) shows pdf correlation plots of axial velocity-concentration product in the far-field region ($x/D = 30$). Only the coflowing airstream was seeded for these tests. This result shows that a positive

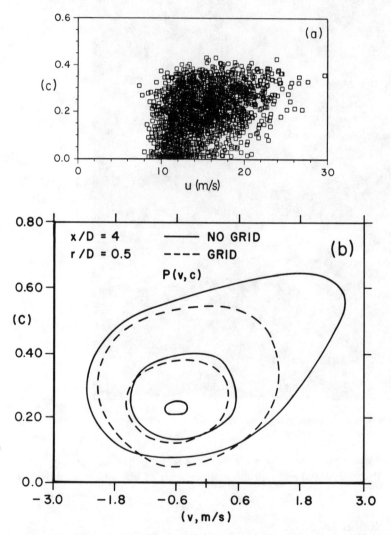

Figure 5-9 Correlation plots of joint pdfs of velocity and concentration. (a) Nonreacting propane jet spreading into seeded coflowing air, $x/D = 30$, $y/D = 2.3$ *(Schefer et al. 1986)*. (b) CO_2 jet spreading into coflowing air without and with free stream turbulence, $x/D = 4$, $r/D = 0.5$.

correlation exists between axial velocity and concentration. These workers also observed a negative correlation between radial velocity and concentration. Similar measurements on the centerline showed that the mixture fraction was uncorrelated with either velocity component.

These measurements suggest that the jet fluid is being transported to the outer airstream by the axial velocity, whereas air is being entrained radially inward into the jet. At the centerline, fluid originating from both streams is well mixed, and hence no correlation is observed between velocity and concentration.

It is interesting to compare the above observations in the far-field region of the jet ($x/D = 30$) with the results of Ballal and Chen (1987) in the near-field region ($x/D = 4$) obtained using the RRS-LDA system. The latter results are shown in Fig. 5-9(b). These joint pdfs were computed from 4096 pairs of two components of velocity data, and N_2 and CO_2 concentrations, respectively. It is observed that in the near-field region, large-scale eddies with outward positive radial velocity are mainly responsible for transporting the jet fluid CO_2 to the outer airstream. Further, in a later paper, Ballal and Chen (1988) showed that in the presence of free stream air turbulence, a weak correlation is observed between radial velocity and concentration. This weak correlation is shown as near-circular (Gaussian) correlation contours of joint pdf in Fig. 5-9(b). Clearly, this result suggests that the outward transport of jet fluid is no longer controlled solely by the large-scale eddies. Rather, both increased jet entrainment due to random, small-scale free stream turbulence, and outward motion of the large-scale eddies contribute to the spreading of jet in the airstream. Thus the effect of free stream turbulence is to disrupt the radial outward transport of concentration solely by large-scale eddies, and also to increase jet entrainment. In this manner, such joint measurements have helped to reveal these most interesting changes in the mixing patterns of a jet at different locations, and in the presence of appreciable free stream turbulence.

Turbulent Premixed Flames

The development of computer models of turbulent chemistry, scalar transport, and scalar dissipation in a turbulent premixed flame requires the measurements of intensity, length scales, and dissipation rate of scalar fluctuations. Accordingly, Ballal, Chen, and Yaney (1986) and Yaney et al. (1987) employed the RRS-LDA system and performed independent and joint measurements of velocity and temperature in turbulent premixed methane-air flames. In the experiments of Ballal, Chen, and Yaney (1986), a 1-cm-diameter, and in the measurements of Yaney et al. (1987), a 4-cm-diameter, Bunsen burner were employed. The results of their investigations are presented in Figs. 5-10 to 5-12 and are discussed below.

Figure 5-10 illustrates typical autocorrelation and power spectral density functions (psdf) of temperature fluctuations in a turbulent premixed flame. These tests and photographic evidence (see Ballal, Chen, and Yaney 1986) showed that a wrinkled, wavy flame front repeatedly sweeps the RRS measurement volume. This produces, in Fig. 5-10, periodic oscillations in autocorrelation function corresponding to sweep amplitude, and a discrete peak at 200 Hz in the psdf curve corresponding to sweep

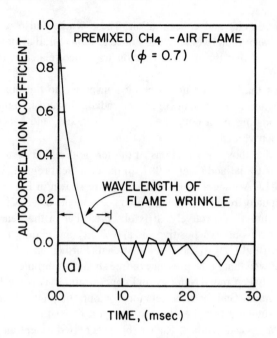

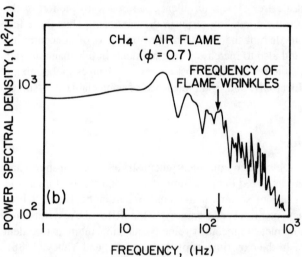

Figure 5-10 Typical autocorrelation and power spectrum of temperature fluctuations on the centerline of turbulent methane-air premixed flame measured at $x/D = 1.4$, using the RRS-LDA system.

frequency. Now, if a purely bimodal temperature distribution is assumed, then the wavelength of this flame wrinkle will be exactly twice the integral length scale of temperature fluctuations defined by

$$L_T = U\tau_T = U \int_0^\infty R_T(t) \, dt \tag{5-9}$$

From a physical viewpoint, an integral temperature scale represents the mean diameter of a pocket of gas of uniform temperature. For the premixed flame of Ballal, Chen, and Yaney (1986), measurements revealed that $L_T = 6$ mm and the wavelength of flame wrinkle, $L_f = 2L_T = 12$ mm. These values are very close to those measured in similar Bunsen burner turbulent premixed flames by Yoshida and Tsuji (1982) and Yanagi and Mimura (1981), who used fine-wire compensated thermocouples for their temperature fluctuation measurements. This agreement illustrates the ability of the RRS system to successfully perform measurements of scalar-length scales in turbulent combusting flows.

In its simplest form, a laboratory premixed flame represents a thin and sharp interface between the reactants and the products. In contrast, flames in practical combustors are highly turbulent and, as described by Ballal and Lefebvre (1975), combustion is sustained solely by the reactions taking place at the interfaces formed between the combustion products and small entrained eddies of fresh mixture. Further, these entrained eddies are completely consumed as they progress downstream through a thick "combustion in-depth" region. These differences between thin laboratory and thick practical flame fronts are presented in Fig. 5-11, in the form of their respective dissipation rate ratios as computed by Ballal, Chen, and Yaney (1986) from conditioned measurements. Here, $\bar{\chi}$ represents the dissipation rate and the subscripts u, uc, and T refer to axial velocity, velocity-progress variable (mixture fraction), and temperature, respectively. These data show the following.

1. The variation of dissipation rate ratios $(\bar{\chi}_u/\bar{\chi}_{uc})$ and $(\bar{\chi}_u/\bar{\chi}_T)$ with the progress variable $\bar{c}$, (zero for unburned, 1 for burned gases) is qualitatively similar for thin and thick flames.
2. In the flame zone the magnitudes of the dissipation rates rank as $\bar{\chi}_T > \bar{\chi}_{uc} > \bar{\chi}_u$.
3. Both the ratios $(\bar{\chi}_u/\bar{\chi}_{uc})$ and $(\bar{\chi}_u/\bar{\chi}_T)$ exhibit maximum values in the flame zone, approximately corresponding to the location of the flame tip.
4. Scalar dissipation rates are comparatively higher in a thick flame than in a thin flame. As described above, a thick flame comprises small packets of fresh mixture eddies, entrained and burning inside the flame front. These packets represent regions of very high temperature gradients and hence high scalar dissipation rates. In contrast, a thin flame is associated with large-scale wrinkling effects, i.e., large gas packets in which the dissipation rates are comparatively lower. These physical differences in the structure of the two types of turbulent flames translate into important differences in their relative dissipation rates. Clearly, such differences call for drastic changes in the modeling strategies for thin and thick turbulent premixed flame fronts.

Figure 5-12 shows measurements of normalized turbulent fluxes $\overline{uT}$, $\overline{vT}$ along the axial direction in a turbulent premixed flame. It is observed that the correlation $\overline{vT}$ decreases monotonically with downstream distance. This can be understood by recognizing that as the flame spreads downstream from the nozzle exit, it accelerates the gas motion in the axial direction and its radial momentum and heat are rapidly dissipated by entrainment of the cold surrounding air. Thus the radial velocity-temperature cor-

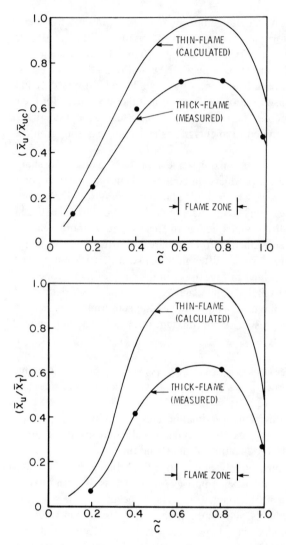

Figure 5-11 Variation of dissipation rate ratios for a thin and a thick turbulent methane-air premixed flame measured at $x/D = 1.4$.

relation weakens with downstream distance. As for the axial velocity-scalar correlation $\overline{uT}$, it first falls rapidly with downstream distance in the unburned mixture region, then rises and reaches a peak in the flame zone, and finally decreases monotonically with downstream distance in the hot gas region. Clearly, in the flame zone the flame-generated acceleration of burning mixture is mainly responsible for arresting the monotonic decrease in this correlation with downstream distance. Otherwise, the entrainment of cold air, radial spreading of the flame, and dissipation of axial momentum all would contribute to a weakening of this correlation with downstream distance.

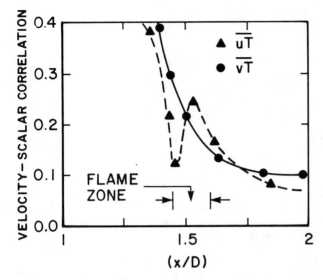

Figure 5-12 Centerline variation of normalized axial and radial turbulent scalar fluxes in a premixed methane-air flame measured using the RRS-LDA system.

Turbulent Diffusion Flames

An essential feature of many practical combustors is that the rate of chemical reactions is limited by the rate of turbulent mixing of the reactants. In most devices, reactants enter the combustion zone of interest in two separate streams. For such a situation, Bilger (1980) showed that an assumption of equal diffusivity for all species leads to the state of mixing being uniquely determined by one conserved scalar variable ξ. The conserved scalar variable is an algebraic combination of reactive scalars, e.g., fuel and product atomic concentrations, such that ξ is independent of the progress of chemical reaction but changes due only to turbulent and molecular transport. Linking the mean and higher moments of species and temperature to the pdf of the conserved scalar, ξ becomes the central problem in the theory whose solution is sought in the analysis of turbulent diffusion flames.

In turbulent nonpremixed flames, the fuel-air mixing rates are several orders of magnitude higher than in laminar flames. These higher rates are a consequence of rapid transport induced by turbulent fluxes or correlations between velocity and temperature, density, or species concentration. At present, computer models use untested and unrefined assumptions to describe these correlations. Clearly, what is needed are direct and unambiguous measurements of these correlations in turbulent diffusion flames.

Figure 5-13 shows joint pdf plots of velocity-atomic mixture fraction and velocity-density correlations as measured by Dibble, Kollmann, and Schefer (1984), using the VRS-LDA system. These measurements were performed in a horizontal turbulent jet diffusion flame of hydrogen (containing 0.22 mole fraction of argon) issuing out of a

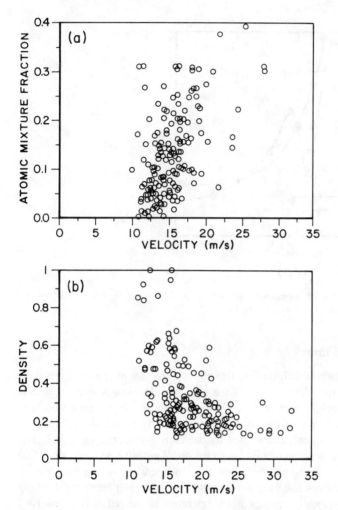

Figure 5-13 Joint pdf plots constructed from 160 simultaneous VRS-LDA events collected at $x/D = 50$, $y/D = 4$ in a hydrogen-air jet diffusion flame. (*a*) Velocity-conserved scalar correlation and (*b*) Velocity-density correlation (*Dibble, Kollmann, and Schefer 1984*).

fuel nozzle 0.53 cm in diameter, at 154 m/s ($Re_D = 24,000$), into a coflowing airstream of velocity $U_a = 8.5$ m/s. This diffusion flame was enclosed in a 30-cm^2 cross section and 200-cm-long working section provided with glass windows.

Components of a typical $u - \xi$ matrix are plotted in Fig. 5-13(*a*), where each data point represents a single VRS-LDA event. The scatter shown here is due to random fluctuations and not to experimental errors. A typical experimental run produced a data matrix of 160×2 velocity and conserved scalar pairs by using a data reduction method described by Dibble, Kollmann, and Schefer (1984). This $u - \xi$ matrix has been transformed into a $u - \rho$ matrix plotted in Fig. 5-13(*b*).

The correlations shown in Fig. 5-13 are in part a result of mean gradients of the scalar interacting with the fluctuating velocity field. A better insight into the physics of the flow is obtained when the correlations are converted to radial plots of $\overline{\rho u}$, $\overline{u\xi}$, and $\overline{uT}$, shown in Figs. 5-14(a) and 5-14(b). The negative values of density-axial velocity correlations observed in Fig. 5-14(a) suggest that, on average, the lighter burned fluid is moving faster than the denser unburned fluid, i.e., the low-density fluid has higher velocity fluctuations, and vice versa. In Fig. 5-14(b), a close agreement between $\overline{u\xi}$ and $\overline{uT}$ is evident. This is a consequence of the axial measurement station $x/D = 50$ being just beyond the flame tip. At this location the flow is characterized primarily by the mixing of hot combustion products with the surrounding air.

While the $\overline{u\xi}$ correlation appears reasonably symmetric, the $\overline{\rho u}$ profile is asymmetric. Dibble, Kollmann, and Schefer (1984) attribute this asymmetry to the effect

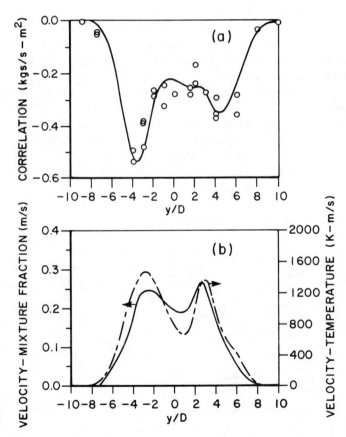

Figure 5-14 Radial profiles of (a) velocity-density correlation and (b) velocity-temperature correlation $\overline{uT}$ (dashed line) and velocity-mixture fraction correlation $\overline{u\xi}$ (solid line) at $x/D = 50$ (*Dibble, Kollmann, and Schefer 1984*).

of buoyancy in the horizontal flame. This asymmetry is not predicted by the common gradient diffusion model for the correlations

$$\overline{\rho u} = -D_T \, \partial\overline{\rho}/\partial x \qquad \overline{u\xi} = -D_T \, \partial\overline{\xi}/\partial x \qquad (5\text{-}10)$$

since the gradients themselves are symmetric. Further investigations by Dibble, Kollmann, and Schefer (1984) reveal that the correlation $\overline{\rho u}$ is not determined by the axial variation of mean density as in Eq. (5-10), but is generated by the radial derivatives of density and velocity with the destructive mechanism provided by the pressure correlation (see Bilger 1980). Since the buoyancy forces will result in an asymmetric pressure distribution, a similar asymmetry of pressure correlation can be expected, thus leading to the observed asymmetry of the $\overline{\rho u}$ profile.

Figure 5-15 illustrates the temporal velocity-scalar correlations for density, conserved scalar, and temperature as measured by Dibble, Kollmann, and Schefer (1984) in a hydrogen-air diffusion flame using the VRS-LDA system. From these data, Dibble, Kollmann, and Schefer (1984) computed that the correlation half-life is ~1.5 ms, and the measured mean velocity at this location is 15 m/s. The product of these quantities yields an integral scale ~2 cm. This value of the length scale is consistent with the often-used order-of-magnitude assumption (see Goulard, Mellor, and Bilger 1976) that it is of the order of the turbulent jet diameter.

Gradient diffusion relations such as Eq. (5-10) have been used in many first-order closure models of turbulent combustion to describe turbulent eddy diffusivity. This model assumption is important and should be tested by comparing it directly with the correlation measurements of $\overline{\rho u}$ and $\overline{\rho v}$ in a turbulent diffusion flame. Driscoll, Schefer, and Dibble (1982) employed a combined Rayleigh-LDA rather than a VRS-LDA system to perform such measurements. Strictly, the Rayleigh signal is invalid when an LDA seed particle is present in the measurement volume. This poses special dif-

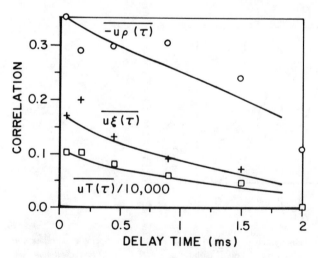

Figure 5-15 Measured temporal velocity-scalar correlation for density $\overline{u\rho}(\tau)$, atomic mixture fraction $\overline{u\xi}(\tau)$, and temperature $\overline{uT}(\tau)$ plotted as a function of delay time at $x/D = 50$, $y/D = 4$. These measurements used the VRS-LDA system *(Dibble, Kollmann, and Schefer 1984)*.

ficulties in measuring velocity and density simultaneously using a Rayleigh-LDA technique. Driscoll, Schefer, and Dibble (1982) resolved many of these difficulties and employed a time delay of 40 μs between velocity and density measurements, which was much shorter in comparison with the characteristic time of turbulent combusting flow in their experiments. Therefore these measurements can be considered essentially simultaneous.

Figure 5-16 presents axial velocity-density correlation and axial-density gradient, whereas Fig. 5-17 shows radial velocity-density correlation and radial density gradient. The data of Fig. 5-16 clearly show that the correlations $\overline{\rho u}$ are negative where the measured $\partial\bar{\rho}/\partial x$ also are negative. To explain this result physically, it should be recognized that at any given position in a jet diffusion flame, the existing fluid is convected from upstream locations. The fluid convected from larger radii will, on average, have large density, low axial velocity, and negative radial velocity, i.e., $\rho >$

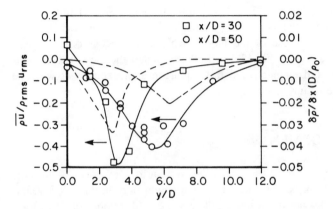

Figure 5-16 Radial profiles of axial velocity-density correlations $\overline{\rho u}/(\rho_{rms}u_{rms})$ and normalized axial density gradient $\partial\bar{\rho}/\partial x(D/\rho_0)$. Dashed line, $\partial\bar{\rho}/\partial x(D/\rho_0)$ at $x/D = 30$; semi-solid line, at $x/D = 50$ *(Driscoll, Schefer, and Dibble 1982)*.

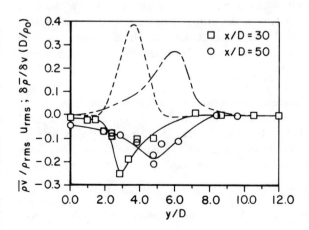

Figure 5-17 Radial profiles of radial velocity-density correlations and normalized radial density gradient. Dashed line, $\partial\bar{\rho}/\partial y(D/\rho_0)$ at $x/D = 30$; semi-solid line, at $x/D = 50$ *(Driscoll, Schefer, and Dibble 1982)*.

0, $u < 0$, and $v < 0$. Also, fluid convected from smaller radii will, on average, have low density, high axial velocity, and positive radial velocity, i.e., $\rho < 0$, $u > 0$, and $v > 0$. Therefore the corresponding product of the above terms will yield the observed negative values of the density-velocity correlations.

The result of Fig. 5-16 contradicts the relationship expressed in Eq. (5-10) by suggesting a negative value for the turbulent eddy diffusivity. Thus we have here direct experimental evidence of countergradient diffusion in this nonpremixed flame. However, Driscoll, Schefer, and Dibble (1982) pointed out that measurements along the centerline in the same flame show that there are regions where the axial density gradient and the correlation $\overline{\rho u}$ have opposite signs. Hence gradient diffusion is not always an unreasonable assumption. Furthermore, and as shown in Fig. 5-17, the measured radial mass flux $\overline{\rho v}$ appears to follow a gradient diffusion relation throughout the flow field.

Further analysis of the data presented in Figs. 5-16 and 5-17 shows that both $\overline{\rho u}$ and $\overline{\rho v}$ correlations are dominated by the radial gradients, which are an order of magnitude larger than axial gradients. Thus closure models for these correlations should almost always include both radial and axial gradients. Finally, Driscoll, Schefer, and Dibble (1982) recommend the use of the more sophisticated second-order closure models to replace relations such as Eq. (5-10). In these models the local flux depends upon the history of the fluid element, and this feature is more consistent with the actual "nonlocal" nature of turbulence.

5-7 FUTURE DEVELOPMENTS

Serious applications of the combined Raman-LDA system to correlations, joint pdf, and spectral measurements in combusting flow have begun only in the last 5 years. At present, these techniques are in an intermediate stage of development. Yet, their spatial and temporal resolutions are approaching the requirements, their measurement accuracy appears acceptable, and the prospects for their future development appear good, at least in laboratory situations. From an experimental viewpoint, both Rayleigh and Raman techniques are well understood and simpler to use in relation to the more sophisticated coherent anti-Stokes Raman scattering (CARS), inverse Raman scattering, or planar imaging techniques. In the future, Raman techniques could be used for combustion measurements in practical combustors. However, this type of application poses several challenges, as follows.

1. Since flames in practical combustors are confined, the Raman signal may be contaminated by the fluorescence originating from the impurities in the window glass. Dibble, Kollmann, and Schefer (1984) report that in their experiments, the fluorescent background was significantly reduced when the test section windows were removed. Eckbreth (1988) suggests the use of compound windows with opaque baffling. However, the use of such a glass window has yet to be demonstrated on high-temperature combustors.
2. Flames in practical combustors produce soot that attenuates laser beams by scat-

tering and absorption. In some devices, e.g., fuel sprays and coal-laden streams, extinction is likely to limit detectivity and potentially preclude combustion measurement. Soot particles also absorb incident laser radiation. This heat may be transferred back to the gas and in this way lead to errors in the temperature measurement.

3. In practical combustors, a Raman signal well above the level of background flame luminosity can be obtained by using a powerful laser source. However, such high-intensity laser radiation may break down the medium under examination, particularly in the presence of large soot particles. Also, a window coated with soot is susceptible to optical damage from powerful laser beams. These difficulties limit usable laser intensities, a serious drawback in practice.

Recently, other techniques for making simultaneous velocity-scalar measurements are also being developed. For example, Goss, Trump, and Roquemore (1988) integrated the CARS and LDA techniques and performed combustion measurements in a diffusion flame, and Long, Levin, and Fourguette (1985) obtained simultaneous two-dimensional concentration-temperature measurements in turbulent flames. Now, the planar-imaging techniques (see Hanson 1986; Eckbreth 1988) for the scalar are being successfully integrated with quantitative velocity measurement systems. Soon, experimentalists will have a laser diagnostic scheme that provides the breadth, depth, and volume of detail now available only through computer simulation. The challenge for the future lies in successfully integrating experiments with models and thereby obtaining even better understanding of the turbulent combustion process.

We have discussed ways in which some of the shortcomings in our current knowledge of turbulent combustion can be overcome with the applications of joint Raman-LDA technique for performing combustion measurements. For example, by making available benchmark quality data sets on joint pdfs and correlations, the need to guess pdfs is redundant. Rather, exact representation of certain processes, such as finite-rate chemistry and a more rigorous treatment of turbulent diffusion and chemical reactions, i.e., pdf modeling, can become possible (see Pope 1985). In contrast, the moment approach to modeling, which has many engineering applications, requires closure assumptions of varying levels of arbitrariness. These closure formulations can be evaluated and refined by comparing them with direct and unambiguous measurements of various correlation terms in reactive flows. This opens up new possibilities such as sophisticated or general closure schemes for a variety of flows, better accuracy of prediction, and finally, the availability of modeling codes that can be used with a sufficient degree of realism, confidence, and certainty to design practical combustion systems of the future.

5-8 SUMMARY

In this chapter, integrated Raman-LDA systems capable of performing simultaneous measurements of velocity, species concentration, and temperature are described. Both vibrational Raman and rotational Raman spectroscopic systems were combined optically and electronically with an LDA system. Details of each individual instrument, various schemes for integrating them, their calibration, data processing, operational capabilities, and limitations are all discussed.

In the design and successful operation of VRS-LDA and RRS-LDA systems, several sources of error arise, such as Raman signal contamination due to background flame luminosity, interference from particulate Mie scattering (both LDA seed and soot particles), uncertainties due to Poisson statistics, beam steering, beam defocusing, and LDA particle bias errors. All these errors are carefully analyzed, and correction schemes are developed for them.

The VRS-LDA and RRS-LDA systems are used to measure joint pdfs, velocity-scalar correlations, scalar power spectra, and scalar length scales in variable-density nonreactive round jets, turbulent premixed flames, and turbulent diffusion flames. The round jet measurements reveal the mixing process between jet fluid and outer airstream. In the near-field region, large-scale eddies with positive radial velocity spread the jet outward, whereas in the far-field region, it is the axial velocity that is responsible for the transport of jet fluid. Combustion measurements in turbulent premixed flames demonstrate the wrinkled, wavy nature of the flame front, and the relatively high scalar dissipation rates associated with practical thick flames in comparison with laboratory thin flames. Finally, in turbulent diffusion flames, measurements of conserved scalar fluxes and mass fluxes are presented. These data demonstrate the interaction between mean gradients of the scalar and fluctuating velocity field, effects of buoyancy, countergradient diffusion at selected locations in a flame, and gradient transport at other locations.

Finally, future developments in laser diagnostics and their application in combustion modeling are discussed.

5-9 NOMENCLATURE

B_0	rotational Raman constant
c	concentration, progress variable
D	jet exit diameter, turbulent diffusivity
f_i	bandwidth factor
f_r	data sampling rate
J	quantum number
k_i	proportionality constant, Eq. 5-1
L	integral length scale
n_j	Raman count rate of rotational line
N	number density
p	pressure
P	probability density function
P_e	excitation power
Q_1	laser energy
Q_i	photomultiplier tube integrated charge
R	correlation coefficient

Re	Reynolds number
S	Poisson standard deviation
t	time
T	temperature
u, v	axial and radial fluctuating velocities
U, V	axial and radial mean velocities
$u\phi, v\phi$	velocity-scalar correlation ($\phi = c$ or T)
x, r	axial and radial directions
ϵ	Raman calibration constant, Eq. 5-4
η	Kolmogoroff scale
λ	Taylor microscale, wavelength
ν	kinematic viscosity
ξ	conserved scalar
ρ	density
τ	characteristic time, delay time
ϕ	equivalence ratio, mixture fraction
χ	turbulent dissipation rate

Subscripts

a	coflowing airstream
A, B	channel numbers
c	concentration
j	jet
r	radial direction
rms	root mean square value
T	temperature
u	velocity

Superscripts

~	Favre-averaged value
—	Reynolds-averaged value

REFERENCES

Alden, M., Edner, H., Holmstedt, G., Svanberg, S., and Hogberg, T. 1982. Single-pulse laser-induced OH fluorescence in an atmospheric flame, spatially resolved with a diode array detector. *Appl. Opt.* 21:1236–40.

Ballal, D. R., and Chen, T. H. 1987. Studies of CO_2 round jet using an integrated Raman-LDA system. American Institute of Aeronautics and Astronautics paper AIAA-87-0377.

Ballal, D. R., and Chen, T. H. 1988. Effects of freestream turbulence on the development of a CO_2 round jet. American Institute of Aeronautics and Astronautics paper AIAA-88-0536.

Ballal, D. R., and Lefebvre, A. H. 1975. The structure and propagation of turbulent flames. *Proc. R. Soc. London, Sev. A*, 334:217–34.

Ballal, D. R., Chen, T. H., and Yaney, P. P. 1986. Scalar fluctuations in turbulent combustion—An experimental study. American Institute of Aeronautics and Astronautics paper AIAA-86-0367.

Ballal, D. R., Chen, T. H., and Schmoll, W. J. 1988. Laser diagnostics for gas turbine combustion research. American Society of Mechanical Engineers paper 88-GT-21.

Ballal, D. R., Lightman, A. J., and Yaney, P. P. 1987. Development of test facility and optical instrumentation for turbulent combustion research. *AIAA J. Propul. Power* 3:97–104.

Bilger, R. W. 1980. Turbulent flows with nonpremixed reactants. In *Turbulent reacting flows*, eds. P. A. Libby and F. A. Williams, pp. 65–113. New York: Springer-Verlag.

Chen, T. H., and Lightman, A. J. 1985. Effects of particle arrival statistics on laser anemometer measurements. American Society of Mechanical Engineers paper ASME-FED-5.

Chigier, N. A. 1981. *Energy, combustion and environment*. New York: McGraw-Hill.

Dibble, R. W., Kollmann, W., and Schefer, R. W. 1984. Conserved scalar fluxes measured in a turbulent nonpremixed flame by combined laser Doppler velocimetry and laser Raman scattering. *Combustion Flame* 55:307–21.

Drake, M. C., Pitz, R. W., and Lapp, M. 1986. Laser measurements on nonpremixed H_2—air flames for assessment of turbulent combustion models. *AIAA J.* 24:905–17.

Drake, M. C., Lapp, M., Penney, C. M., Warshaw, S., and Gerhold, B. W. 1980. Measurements of temperature and concentration fluctuations in turbulent diffusion flames using pulsed Raman spectroscopy. In *Eighteenth symposium (international) on combustion*, pp. 1521–30. Pittsburgh, Pa.: The Combustion Institute.

Drake, M. C., Lapp, M., Pitz, R. W., and Penney, C. M. 1983. Dynamic measurements of gas properties by vibrational Raman scattering: application to flames. In *Time resolved vibrational spectroscopy*, ed. G. H. Atkinson, pp. 83–95. New York: Academic.

Driscoll, J. F., Schefer, R. W., and Dibble, R. W. 1982. Mass fluxes $\overline{\rho u}$ and $\overline{\rho v}$ measured in a turbulent nonpremixed flame. In *Nineteenth symposium (international) on combustion*, pp. 477–85. Pittsburgh, Pa.: The Combustion Institute.

Eckbreth, A. C. 1977. Laser Raman thermometry experiments in simulated combustor environments. *Progr. Aeronaut. Astronaut.* 53:517–47.

Eckbreth, A. C. 1988. *Laser diagnostics for combustion temperature and species*. Cambridge, Mass.: Abacus Press.

Edwards, R. V. 1987. Report of the special panel on statistical particle-bias problems in laser anemometry. *ASME J. Fluids Eng.* 109:89–93.

Goss, L. P., Trump, D. D., and Roquemore, W. M. 1988. A combined CARS-LDA instrument for simultaneous temperature and velocity measurement. *Exp. Fluids* 6:189–98.

Goulard, R., Mellor, A. M., and Bilger, R. W. 1976. Combustion measurements in air breathing propulsion engines—survey and research needs. *Combust. Sci. Technol.* 14:195–219.

Hanson, R. K. 1986. Combustion diagnostics: planar imaging techniques. In *Twenty-first symposium (international) on combustion*, pp. 1677–91. Pittsburgh, Pa.: The Combustion Institute.

Hill, R. A., Mulac, A. J., and Hackett, C. E. 1977. Retroreflecting multipass cell for Raman scattering. *Appl. Opt.* 16:2004–6.

Johnston, S. C., Dibble, R. W., and Schefer, R. W. 1986. Laser measurements and stochastic simulations of turbulent reacting flows. *AIAA J.* 24:918–37.

Lapp, M., and Penney, C. M., eds. 1974. *Laser Raman gas diagnostics*. New York: Plenum.

Lederman, S. 1977. Use of laser Raman diagnostics in flow fields and combustion. *Progr. Energy Combust. Sci.* 3:1–34.

Leonard, D. A. 1974. Field test of a laser Raman measurement system for aircraft engine exhaust emissions. Air Force Aero Propulsion Laboratories technical report AFAPL-TR-74-100, Wright-Patterson Air Force Base, Ohio.

Libby, P. A., and Williams, F. A., eds. 1980. *Turbulent reacting flows*. New York: Springer-Verlag.

Long, M. B., Levin, P. S., and Fourguette, D. C. 1985. Simultaneous two-dimensional mapping of species concentration and temperature in turbulent flames. *Opt. Lett.* 10:267–69.

Magre, P., and Dibble, R. W. 1988. Finite chemical kinetic effects in a subsonic turbulent hydrogen flame. *Combust. Flame* 73:195–206.

Masri, A., Bilger, R. W., and Dibble, R. W. 1987. Fluorescence interference with Raman measurements in nonpremixed flames of methane. *Combust. Flame* 68:109–19.

Pope, S. B. 1985. Pdf methods for turbulent reactive flows. *Progr. Energy Combust. Sci.* 11:119–50.

Roquemore, W. M., Chen, L.-D., Goss, L. P., and Lynn, W. F. 1987. Structure of jet diffusion flames. Paper read at the United States–France Joint Workshop on Turbulent Reactive Flows, 6–10 July 1987, Rouen, France.

Schefer, R. W., Johnston, S. C., Dibble, R. W., Gouldin, F. C., and Kollmann, W. 1986. Nonreacting turbulent mixing flows: a literature survey and data base. Sandia National Laboratories technical report SAND-86-8217, Livermore, Calif.

Setchell, R. E. 1978. Initial measurements within an internal combustion engine using Raman spectroscopy. Sandia National Laboratories technical report SAND-78-1220, Livermore, Calif.

Warshaw, S., Lapp, M., Penny, C. M., and Drake, M. C. 1980. Temperature-velocity correlation measurements for turbulent diffusion flames from vibrational Raman-scattering data. In *Laser probes for combustion chemistry.* ACS Symposium Series 134, ed. D. R. Crosley, pp. 239–46. Washington, D.C.: American Chemical Society.

Wright, M. L., and Krishnan, K. S. 1973. Feasibility study of in situ source monitoring of particulate composition by Raman or fluorescence scatter. Environmental Protection Agency technical report EPA-R2-73-219, Washington, D.C.

Yanagi, T., and Mimura, Y. 1981. Velocity-temperature correlations in premixed flames. In *Eighteenth symposium (international) on combustion*, pp. 1031–39. Pittsburgh, Pa.: The Combustion Institute.

Yaney, P. P., Becker, R. J., Magill, P. D., and Danset, P. 1982. Dynamic temperature measurements of flames using spontaneous Raman scattering. In *Temperature, its measurement and control in science and industry*, ed. J. F. Schooley, vol. 5, pp. 639–48. New York: American Institute of Physics.

Yaney, P. P., Becker, R. J., Danset, P. T., Gallis, M. R., and Perez, J. I. 1985. The application of rotational Raman spectroscopy to dynamic measurements in gas flowfields. *Progr. Aeronaut. Astronaut.* 95:672–99.

Yaney, P. P., Chen, T. H., Spicer, M. R., and Schmoll, W. J. 1987. Kilohertz measurements of temperature in a premixed methane-air flame using two-channel rotational Raman spectroscopy. American Institute of Aeronautics and Astronautics paper AIAA-87-0301.

Yoshida, A., and Tsuji, H. 1982. Characteristic scales of wrinkle in turbulent premixed flames. In *Nineteenth symposium (international) on combustion*, pp. 403–11. Pittsburgh, Pa.: The Combustion Institute.

SIX

SPRAY DIAGNOSTICS BY LASER DIFFRACTION

J. Swithenbank, J. Cao, and A. A. Hamidi

6-1 INTRODUCTION

The spray combustion process depends on various physical phenomena such as fluid dynamics, chemical kinetics, and particle dynamics. Since the governing equations of these processes include the particle diameter as a parameter, it is clear that any fundamentally based industrial design or operation procedure must have this basic information. Indeed, the specification of many industrial products includes the particle size distribution as an important parameter.

6-2 PARTICLE SIZING METHODS

A large number of drop-sizing techniques have been developed since the significance of drop size distribution and its control was realized. The ideal diagnostic technique should provide a complete size distribution. It should also cover a wide range with a good spatial resolution and require a minimum of additional information about the properties of the particles or the dispersion medium. The need for a wide size range is particularly important in sprays, since the drop sizes containing most of the mass of the fluid usually extend over about 2 orders of magnitude (i.e., a range of 1–100 from the smallest to the largest drops). A further, practical characteristic of the instrument is the requirement of minimum time for data generation and analysis.

Support for some parts of this work by the MoD, USAF, and British Council is gratefully acknowledged.

In general, drop and particle sizing instruments can be divided into two main categories, intrusive and nonintrusive techniques. These categories are indicated in Figs. 6-1 and 6-2.

Intrusive Techniques

The intrusive diagnostic methods, sometimes referred to as mechanical techniques, were the first to be developed. These techniques require the insertion of a solid probe into the flow field, and they can thus interfere with the particle field before it is measured. This interference effect should be quantified and minimized. In the case of liquid sprays, the removal of a sample from the flow is useful for patternation measurements. This can be achieved by isokinetic sampling in which the velocities of the drops and the fluid are closely matched. It should be mentioned that when true isokinetic sampling cannot be achieved, then the sampling errors are smaller if one samples above the isokinetic velocity rather than below. Most intrusive techniques are sampling techniques, i.e., a sample of particles is collected, preserved, and taken away to be measured, usually by microscopic analysis. The sampling techniques differ according to the method of collection of the sample, the method of preservation of the sample, and the method of analysis of the sample.

One way in which the liquid sampling problem has been solved in sprays is to solidify the drops. Two methods have been used: in the first, a hot liquid wax that solidifies upon cooling in the atmosphere was used; in the second, the liquid spray was collected and solidified in liquid nitrogen. The solidified particles were then sized using a sieve system.

The electrical properties of the collected particles can be utilized to measure their size. These techniques use the Coulter effect in which, if a nonconducting particle suspended in a conducting medium is placed within a small aperture, an increase in the resistance across the orifice is produced, relative to that of the medium alone. The magnitude of this increase in the resistance can be related to the size of the particle.

Other properties of the dispersed phase such as inertial separation and the impact of a spray on a coated glass slide have also been employed to measure drop and particle size distribution.

Nonintrusive Techniques

Nonintrusive techniques require no solid objects to be inserted into the particle field, so that the errors resulting from interference with the particles before or after sampling no longer occur. However, other errors are present that must be quantified. All nonintrusive techniques in common use the properties of light beams traversing the particle field. As seen from Fig. 6-1, optical techniques can be clearly divided into imaging techniques and light-scattering techniques. We shall outline some of the more common of these techniques here.

However, before describing these techniques, it is important to make a distinction between two types of size distributions measurable, depending on the technique employed (Tishkoff 1984). The distinction is between temporal and spatial size distribution.

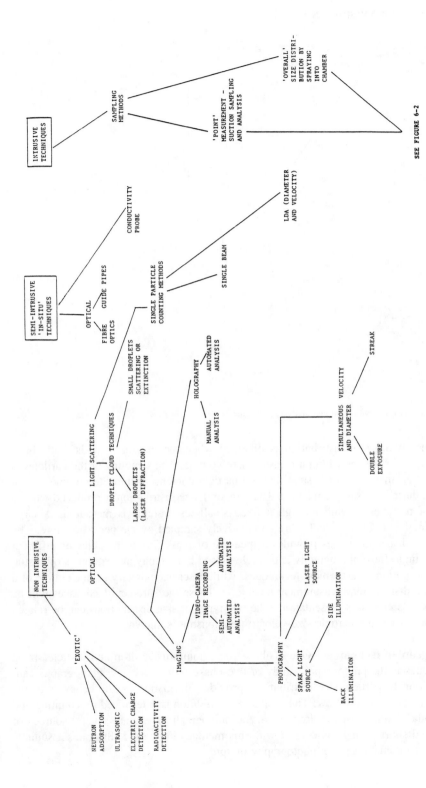

Figure 6-1 Droplet-sizing techniques.

181

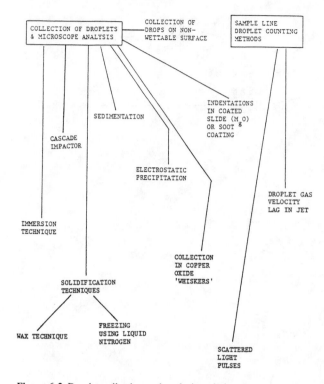

Figure 6-2 Droplet collection and analysis techniques.

The spatial size distribution is obtained when the measurement is made on a collection of particles within a measurement volume. The sampling of the particles is made at an instant of time such that the particle contents within the volume do not change during the sample interval. The temporal size distribution is produced when a flow of particles passing through a fixed small test volume is measured. A large number of individual particles are cumulatively sampled over a period of time. The temporal distribution thus provides a measure of mass flux of particles of different sizes. Single-flash photography and single-pulsed holography are examples of spatial size distribution measurement techniques, while laser anemometry is an example of a temporal distribution measurement technique. These measurement techniques will be discussed later in this presentation. The two types of size distributions can be related to each other if the particle size-velocity relationship is known.

Photographic techniques. One of the earliest nonintrusive drop and particle sizing techniques is the photographic study of two-phase flow processes. Photographs can, in addition to particle size distribution, provide information on the behavior of the dispersed phase in the flow. The drop size distribution is determined by counting and sizing the diameter of the drops from the photograph. There are several sources of error and shortcomings associated with this method. Figure 6-3 illustrates a suitable optical geometry for spark photography of sprays.

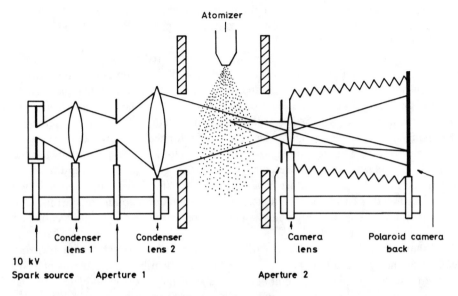

Figure 6-3 Optical setup for photography of sprays.

The technique can only be applied to relatively dilute regions of the particle field. Also, as only one plane of the field at a time can be photographed, care must be taken not to include the image of the drops out of focus into the size analysis. The consequence of this ambiguity tends to be a biasing toward the larger sizes, as larger particles are in focus over a greater depth of field.

A further disadvantage of this method is the large error inherent in counting; for example, counting errors in the range -17% to $+13\%$ have been reported in the measurement of drop size distribution in sprays (Watson and Mulford 1954).

To reduce the counting error, it is required that large numbers of particles be examined. Since this is time consuming, automatic (Morgan and Meyer 1959) and semiautomatic (McCreath, Roett, and Chigier 1974) systems have been developed to carry out this task. Some of these units can be interfaced with minicomputers, which means that the data reduction task is much easier. Some of these units are even capable of discriminating between an image of a particle in focus and one that is not.

Photographic negatives have usually been used to process the data (Simmons 1977). A strobe light has also been used to store an image on a vidicon tube. The image is scanned to obtain particle size data, and the cycle is repeated 10 times per second. The smallest size measurable by these photographic methods is approximately 5 μm.

Holographic techniques. Photographs are a two-dimensional representation of a three-dimensional space. However, holograms can be used to obtain three-dimensional pictures of an object. This three-dimensional image can, after recording, be examined in much the same way as a photograph; however, the viewing plane can be traversed in depth. Alternatively, a laser diffraction particle sizing technique can be employed to analyze the hologram by utilizing its diffraction pattern (Ewan, Houdard, and Bastenhof 1983).

A hologram is the result of interference of the scattered and unscattered coherent monochromatic laser light from a particle field. The illumination time of the particle field depends on the particle velocity and diameter and must be such that the particle is not displaced by more than a tenth of its diameter during the exposure. Therefore the employment of a Q-switched pulse laser is essential in the holographic measurement of moving sprays, since the light pulse can then be limited to a few nanoseconds.

In general, there are two possible configurations for a holographic setup. For the in-line holographic technique the reference beam and the beam producing the diffraction pattern are the same. On the other hand, the off-axis holographic setup has these two beams separated. The in-line technique is the more practical configuration for many spray measurement studies. Figure 6-4 shows the optical setup for the production and reconstruction of a hologram.

Thompson and Dunn (1980) discussed the merits of holographic particle-sizing techniques and applications to particle size and velocity distribution measurement using double-pulse holography. The minimum distance of the holographic plate from the spray is an important factor and depends on the smallest particle diameter in the particle field and the wavelength of the illuminating light.

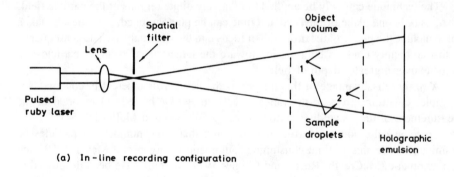

(a) In-line recording configuration

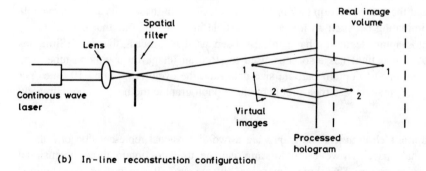

(b) In-line reconstruction configuration

Figure 6-4 Optical setup for holographic technique.

While attempts have been made to measure submicron particles using holograms, it has been shown that the minimum size that can be measured by this technique is 2 μm. The upper limit is in the region of 1 mm.

Particle-sizing interferometry. Particle-sizing interferometry is an extension of laser Doppler anemometry (LDA) in which the size and the velocity of the drops can be recorded. Two nonparallel laser beams originating from the same source are focused at the measurement volume in the spray or particle field, as shown in Fig. 6-5. At the intersection of the two beams an interference fringe pattern is set up. A particle traveling through this fringe will scatter light, and using the visibility of the signal, defined as

$$V = \frac{I_{max} - I_{min}}{I_{max} + I_{min}} \tag{6-1}$$

the particle size can be determined over a limited range. The frequency of the Doppler signal is dependent on fringe spacing and the velocity of the particle, and hence the velocity of the particle crossing the measuring volume can also be determined (Farmer 1972; Hong and Jones 1976). This visibility technique needs careful choice of collection angle and aperture. A diagrammatic representation of the Doppler signal for a particle traversing the measurement volume is shown in Fig. 6-6.

The visibility V as a function of particle diameter d and fringe spacing δ has been shown to be (Robinson and Chu 1975)

$$V = \frac{2J_1(d/\lambda)}{(\pi d/\delta)} \tag{6-2}$$

where J_1 is the first-order spherical Bessel function and λ is the light wavelength. However, this LDA system has several drawbacks that make its use difficult.

It is essential that at any one time only one particle be present in the measuring volume, otherwise the data acquired cannot be interpreted. The measurement volume is usually 1 mm³, and most sprays and particle fields have concentrations that exceed this restriction. For small particles the visibility function is resolution limited, and for

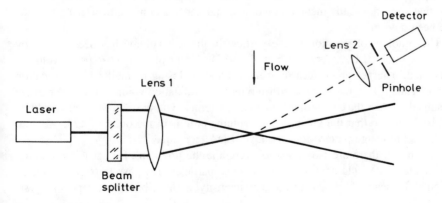

Figure 6-5 Optical setup for laser Doppler anemometry.

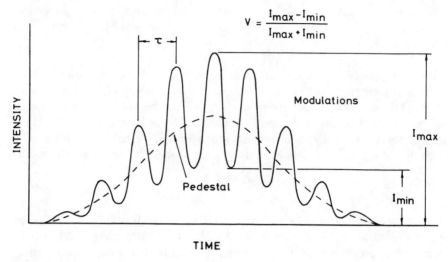

$$V = \frac{I_{max} - I_{min}}{I_{max} + I_{min}}$$

Figure 6-6 Typical Doppler burst.

large particles the limitation is due to the ambiguities that arise when the particle diameter exceeds the fringe spacing. A lower limit of 300 nm on particle diameter has been reported (Jones 1977). The upper limit can be removed as long as it is possible to increase the test volume.

It must be noted that due to the Gaussian distribution of light intensity of a laser beam (Fig. 6-7), it is critical that the particle pass through the center of the test volume. A large particle passing through the edge of the fringe pattern can have the same Doppler signal as a smaller particle passing through the fringe pattern center. This problem exists irrespective of the particle field concentration. A great deal of effort has been put into overcoming this problem. Thin wires have been used for calibration purposes, since the method works for cylindrical as well as spherical objects.

Hong and Jones (1976) used an oscilloscope to observe each Doppler signal individually and hence discriminated manually between signals that were right and those that were not. This method is tedious, and hence automatic measuring systems have been devised.

The problem of position of the particle in the test volume has been solved by using a two-color LDA (Azzopardi et al. 1982). A pair of concentric probe volumes are created, a larger green measurement volume enclosing a smaller blue validation volume. Thus if a signal is received from the blue validation volume, then the particle is assumed to cross the center of the green measurement volume. The validation volume must be small, so that the green-to-blue volume ratio is of the order of 125:1. Figure 6-8 shows the fringe pattern for this novel LDA technique.

In general, there are two ways to overcome the problem of a particle not going through the center: (1) analysis of only those particles that pass through a selected portion of the beam of known and constant intensity, as described above, or (2) analysis

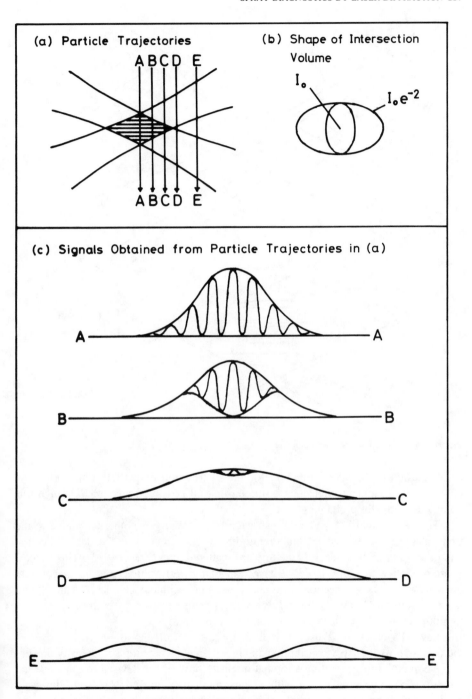

Figure 6-7 Properties of the Gaussian LDA intersection volume and signals.

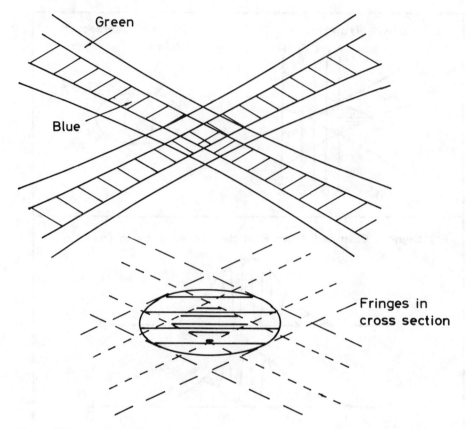

Figure 6-8 Probe volume for two-color LDA.

of all particles and later correction for the distribution of particle trajectories and corresponding incident intensities assuming that all possible trajectories through this volume are equally probable.

The two-color LDA technique is of the first category. Another technique in the first category uses a second photomultiplier at right angles to the data collection photomultiplier, which receives the forward scattered light (Ereaut 1983). This "gate" photomultiplier can only "see" the section of the test volume that has a constant intensity in the axial direction. In order to use this technique, the laser beam should have a "top hat" intensity profile. Only the signals received by both the gate photomultiplier and the collection photomultiplier are analyzed.

Several workers have used a small aperture to cut off the low-intensity edges of the Gaussian light energy distribution and hence create a near top hat intensity profile. A novel method of generating a top hat intensity profile was devised by Belvaux and Verdi (1975), who used a filter with a Gaussian density profile. The use of this filter produced uniform illumination. The size range of the particles that can be measured by this technique is 3 μm to 5 mm, depending on the fringe spacing. Particle velocities up to approximately 1000 m/s can be measured using LDA.

Another important and interesting technique based on LDA for drop size and velocity measurement has been developed (Bachalo 1980, Bachalo and Houser 1984) and produced as the phase Doppler particle analyzer by Aerometrics. The technique has an optical configuration similar to that of a conventional LDA system except that three detectors are used to receive the scattered light signals from three different angles. The drop velocity is determined from the Doppler burst as in an LDA system. The phase shift between the Doppler signals received from the three detectors is used to determine the drop size. With the off-axis light-scattering detection, the measurement results are unaffected by laser beam attenuation; the technique can therefore be used for fairly high-density spray measurement. Particle sizes over a very large size range (3 μm to 5 mm) can be measured by the technique.

Scattering methods. Scattering techniques use the diffraction pattern generated as a result of the interaction of monochromatic light with the particle field, to deduce the size distribution of the particles. As with most particle-sizing methods, the technique normally requires that the particles be spherical. A brief account of diffraction theory and its application in particle sizing are presented here.

6-3 DIFFRACTION THEORY AND PARTICLE SIZING

When a parallel beam of light falls on a circular (or noncircular) disc or aperture, hereafter referred to as a particle or drop, it can be absorbed, scattered, or transmitted, depending on the physical dimensions and properties of the particle. Application of the conservation of energy necessitates that the sum of the three effects be equal to the energy of the incident light in a reference area. C_{ext}, called the extinction cross section, is defined as the difference between the incident and transmitted light energy. It is therefore evident that

$$C_{ext} = C_{sca} + C_{abs} \qquad (6\text{-}3)$$

Cross-section efficiencies for scattering, absorption, and extinction can be defined as

$$Q = \frac{C}{\pi a^2} \qquad (6\text{-}4)$$

where a is the radius of the spherical particle. The significance and application of the extinction cross-section efficiency Q_{ext} will be noted below when the Fraunhofer diffraction theory is outlined.

The general theory for the evaluation of the extinction cross section is the Lorenz-Mie theory, which is fully discussed in many standard texts on optics (van de Hulst 1981; Kelvin 1970). The theory starts from Maxwell's electromagnetic equations and proceeds to exact answers for the cross-section efficiencies. It shows that the cross-section efficiencies depend on the particle diameter expressed as a size parameter X, the ratio of the refractive indices of the dispersed and continuous media m, and the angle of observation. The refractive index term can be a complex number, of which the real part describes the refraction, and the imaginary part the absorption.

The solution of Lorenz-Mie theory produces an exact answer, but in general, the solution is not a simple relation between the particle size and the optical measurements. Therefore well-established limiting cases for ''small'' and ''large'' particles are used wherever applicable. There are several criteria as to the applicability of these limiting cases. The simplest one relates the particle diameter d to the wavelength of the light λ such that

$$d \ll \lambda, \quad d \simeq \lambda, \quad d \gg \lambda \tag{6-5}$$

for Rayleigh scattering, Lorenz-Mie theory, and diffraction theory, respectively.

Fraunhofer Diffraction Theory

When a large opaque particle $(d \gg \lambda)$ is illuminated by a beam of parallel mono-chromatic light, the Fraunhofer diffraction pattern can be observed at infinity, super-imposed on the geometrical image of the particle. This pattern is very large in comparison with the particle image.

From the evaluation of the Fraunhofer diffraction integral, the diffracted light intensity distribution for a particle of radius a can be shown to be

$$I = I(0)\left(\frac{2J_1(X)}{X}\right)^2 \tag{6-6}$$

where $I(0)$ is the intensity at the center of the diffraction pattern, J_1 is the first-order spherical Bessel function, and X is a dimensionless size parameter given by

$$X = \frac{2\pi aS}{\lambda f} \tag{6-7}$$

where f is the focal length of the collection lens and S is the radial distance in the detection plane (focal plane) as measured from the optical axis.

Equation (6-6), plotted as $I/I(0)$ versus X in Fig. 6-9, is known as the Airy function. The pattern consists of a series of maxima and minima that relate to the location of peak and zero intensities, physically appearing as a series of bright and dark concentric rings.

As the diffraction pattern is unique for a particle or a collection of particles of different sizes, it is therefore possible to deduce the size distribution of a collection of particles from their diffraction pattern signature. It was shown that accurate mea-surement of light intensity is difficult for the purpose of particle sizing. Therefore measurement and analysis of the light energy distribution over a finite area of the detector plane are employed in a commercial instrument such as the Malvern particle sizing instrument. Integrating Eq. (6-6), the fraction of light energy scattered by a particle or a monosize particle field contained within a circle of radius S on the detector is given by

$$L = 1 - J_0^2(X) - J_1^2(X) \tag{6-8}$$

where J_0 is the zero-order spherical Bessel function. Thus the light energy within any ring of radii S_1 and S_2 is derived as

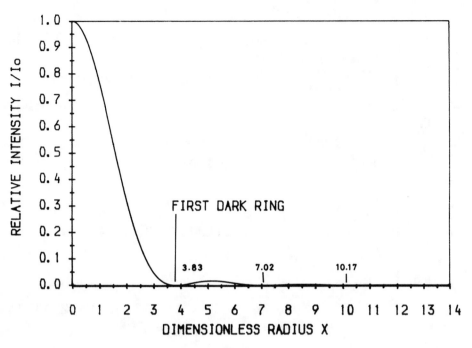

Figure 6-9 Intensity distribution of Fraunhofer diffraction.

$$E_{S1,S2} = E[J_0^2(X_1) + J_1^2(X_1) - J_0^2(X_2) - J_1^2(X_2)] \tag{6-9}$$

where X_1 and X_2 are equal to $2\pi a S_1/\lambda f$ and $2\pi a S_2/\lambda f$, respectively.

E is the energy falling on the particle, which is proportional to the cross-section area of the particle of radius a and the intensity of the incident beam I_0. For N particles of the same size, the total energy falling on the particles can be written as

$$E = NI_0\pi a^2 \tag{6-10}$$

The number of particles N is related to the total particle weight W as

$$N = \frac{3W}{4\pi a^3 \rho} \tag{6-11}$$

where ρ is the particle density. Figure 6-10 shows the plots of Eqs. (6-8) and (6-9). Points of inflection on the total energy curve correspond to the maxima and minima on the energy distribution curve, representing the bright and dark bands. If, instead of a single particle, a collection of particles of different sizes is considered, then the light energy falling on any ring in the focal plane is the sum of the contributions from individual particles:

$$E_{S1,S2} = C \sum_{j=1}^{M} \frac{W_j}{a_j} [J_0^2(X_{j,1}) + J_1^2(X_{j,1}) - J_0^2(X_{j,2}) - J_1^2(X_{j,2})] \tag{6-12}$$

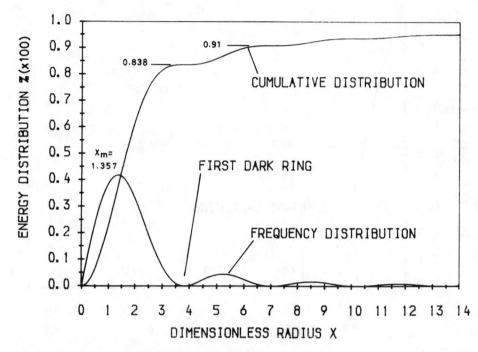

Figure 6-10 Energy distribution of Fraunhofer diffraction.

where $X_j = 2\pi a_j S/\lambda f$, C is a constant, W_j is the weight fraction of the particles of radius a_j, and M is the total number of particle size ranges.

If the particle size distribution is classified into a number of size ranges, then it is preferable to make measurements at radii where the first maximum of the diffracted energy distribution occurs. The location of the first maximum of the light energy distribution for each size can be calculated by differentiating Eq. (6-8). This location is given by

$$2\pi a S/\lambda f = 1.357 \qquad (6\text{-}13)$$

If a detector is used that is divided into a set of circular rings, then each of these rings will define a characteristic particle size range. The total light energy distribution is the sum of the product of the energy distribution for each size range and weight fraction in that range. This can be expressed as

$$E(I) = W(J)T(I, J) \qquad (6\text{-}14)$$

where $E(I)$ and $W(J)$ are the light energy and weight distributions, respectively, and $T(I, J)$, known as the T matrix, contains the coefficients that define the light energy distribution for each particle size range. The weight distribution is then deduced in such a way that

$$\sum_{I=1}^{N} [E_m(I) - W(J)T(I, J)]^2 = \min \qquad (6\text{-}15)$$

where $E_m(I)$ is the measured light energy distribution and N is the total number of detector rings.

The ratio of light intensity measured at the center diode, before and after the particle field is introduced in the beam, gives the fraction of the light transmitted by the particles. The transmission is related to the total projected area of the particles by the Beer-Lambert law:

$$\ln (I/I_0) = -\tau l \tag{6-16}$$

where I_0 and I are the light intensities before and after the introduction of the sample, l is the optical path length, and τ is given by

$$\tau = Q_{\text{ext}} \sum_{j=1}^{M} N_j \pi a_j^2 \tag{6-17}$$

For Fraunhofer diffraction the extinction cross-section efficiency Q_{ext} takes the value of 2. The diffraction theory outlined above has been used by many researchers to develop instruments for the measurement of particle size distribution in a variety of applications.

Cornillault (1972) developed a technique for the measurement of size distribution of solid particles suspended in some suitable liquid. The measurement range was from 2 to 100 μm. The diffracted light distribution as a function of radial distance was measured using a photodetector and a rotating screen with windows placed radially on the screen. Prior to Cornillault's technique, Dobbins, Crocce, and Glassman (1963) had developed a similar technique to determine the Sauter mean diameter in sprays. A normal white light was used as the source, and a narrow-band filter was used to obtain monochromatic light. The scattered light was collected using a photomultiplier.

An apparatus developed by Swithenbank et al. (1977) has been used extensively in two-phase flow applications. The technique collects the light diffracted by a particle field using a set of concentric photodetector rings. Each ring corresponds to a size band, depending on the focal length of the collection lens employed. Using different focal lengths for the receiving lens, drops or particles in the size range 0.2–2000 μm can be measured. The optical components for this apparatus are schematically shown in Fig. 6-11.

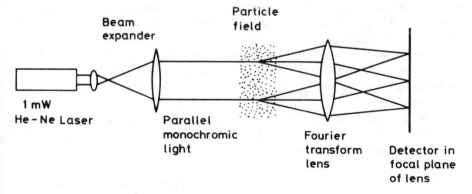

Figure 6-11 Optical setup for laser diffraction technique.

Drop Sizing in Low-Density Sprays

The measurement of drop/particle size distribution and concentration in low drop density systems such as wind-dispersed aerosols poses particular problems. At Sheffield University (in collaboration with Malvern Instruments) a portable instrument has been developed for aerosol detection "in the field." The aerosols to be measured are in the range 1–30 μm at concentrations down to about 20 mg/m³.

The developed instrument is based on a standard Malvern particle sizer. In order to detect the scattered light signal from aerosols, which is usually weak due to low particle concentration, a 20-mW solid state laser is used in place of the conventional low-power HeNe laser, and the detector/amplifier sensitivity is considerably increased. The portable battery-powered system includes a detector control, data acquisition and storage computer with machine code software, and a remote control unit. The data stored during a series of measurements are later processed on a standard PC to give the particle size distribution and concentration.

Problems arising during the application of the instrument include the effect of daylight noise that can be present when the instrument is used in the field. A light baffle has been designed with the FLUENT flow simulation system, which physically blocks the daylight from coming into the measurement volume but does not obstruct the passage of particles. Two-phase flow through the light baffle is illustrated in Fig. 6-12.

The problems of the anomalous diffraction effect raised by small and transparent particles, the lens vignetting effect raised when the particles may not all be close to the receiving lens, and the measurement of particle concentration from the scattered light energy have been solved and are discussed below.

Anomalous Diffraction Effect

When applying Fraunhofer diffraction theory to particle size distribution measurement, one of the assumptions made is that the particles are opaque, or the ratio of the refractive index of the particles to that of the surrounding medium (RI ratio) is much larger than unity. The scattered light pattern is predominantly contributed by the diffracted light described by Fraunhofer diffraction theory.

However, in many applications, such as aerosol detection in the field, the particles may be small and optically transparent or the RI ratio of the particles to the medium may be close to unity. In these cases the "diffraction" pattern is a result of interference between the diffracted and the transmitted light from the particles. This effect of interference between the diffracted and the transmitted light is described by van de Hulst (1981) as the anomalous diffraction effect. Obviously, the relation between the diffraction pattern and the particle sizes is different from that derived from Fraunhofer diffraction theory.

In order to use the Malvern particle sizer to measure the size distribution of small and transparent particles, a new algorithm must be employed that involves generating the matrix and particle size bands covered by each detector ring for size distribution calculation.

The effect of anomalous diffraction on the light energy pattern has been experi-

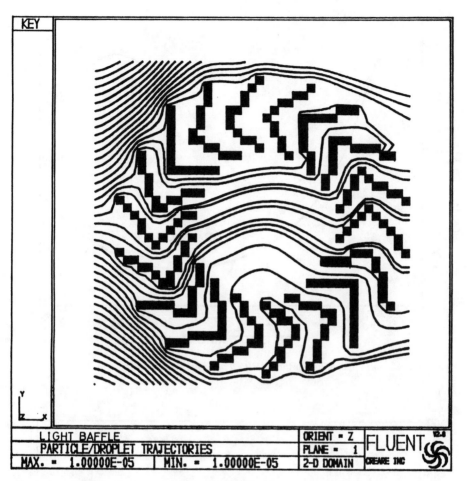

Figure 6-12 Modeling of particle trajectories through light baffle.

mentally investigated by Brown, Alexander, and Cao (1990). The study was carried out by using a standard Malvern particle sizer to measure the diffracted light energy distribution produced by standard glass sphere samples suspended in liquid. The liquid was prepared by mixing various portions of two organic liquids of different refractive indices, so that variation of the liquid refractive index could be obtained. Thus the RI ratio of the particles to the liquid could be changed.

It was shown that as the RI ratio changed to close to unity from both directions, there was an increase in the light energy arriving at the outer rings of the detector. This increased energy could give a fallacious indication of small particle size if the analysis procedure based on Fraunhofer diffraction theory was used. Figure 6-13 shows a typical comparison of the measured light energy distributions for both Fraunhofer and the anomalous diffraction. Figure 6-14 shows the derived size distributions for both cases by using the algorithm based on Fraunhofer theory.

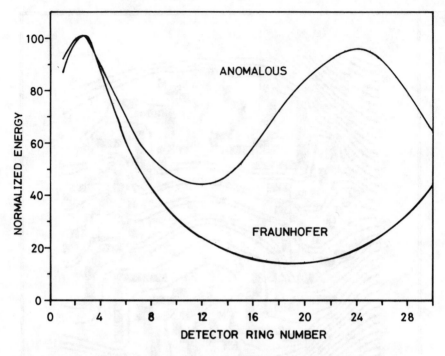

Figure 6-13 Measured light energy distributions for Fraunhofer and anomalous diffraction.

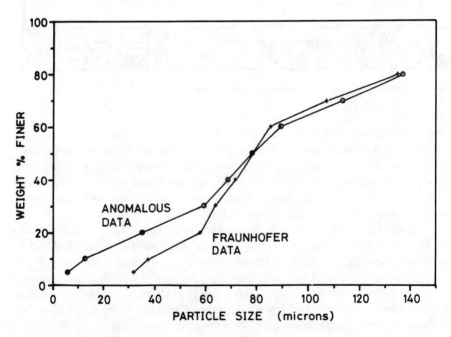

Figure 6-14 Size distribution determined by Fraunhofer theory.

To derive the expression for anomalous diffraction, a parameter ρ is introduced, which has a physical meaning of phase lag suffered by the light wave passing through the particle along the full diameter, and can be written as

$$\rho = \frac{4\pi a}{\lambda}|m - 1| \tag{6-18}$$

where a is the radius of the particle, m is the RI ratio of the particle to the medium, and λ is the light wavelength in the medium.

Consider a spherical nonabsorbing particle of radius a: the amplitude of the scattered light at any point P on the diffraction pattern produced by the particle can be written as

$$A(P) = C'\pi a^2 \int_0^{\pi/2} [1 - \exp(-i\rho \sin \alpha)]J_0(X \cos \alpha) \sin \alpha \cos \alpha \, d\alpha \tag{6-19}$$

which is a complex function, where X is the size parameter defined by Eq. (6-7) and C' is a constant. The expansion of the integral shows that as ρ approaches infinity, the imaginary part of the function vanishes, and the result for Fraunhofer diffraction is obtained. The scattered light intensity and energy distribution can then be derived in a manner similar to that used for Fraunhofer diffraction.

Unfortunately, an analytic expression for the intensity and the energy distribution cannot be given due to the difficulties in the expansion of the integrals involved. Instead, numerical methods have been used to calculate the intensity and the energy distributions for different values of ρ and m. Figure 6-15 shows a calculation result of the intensity distribution, the fraction of energy within a circle of radius X, and the energy distribution as a function of X for the case $\rho = 3.0$. The differences in these patterns from those of Fraunhofer diffraction, reflected by the changes in the zero and peak positions, can easily be seen.

The first peak position X_m in the energy distribution is of more interest, since it is used in the determination of the particle size bands covered by each detector ring and the mean size in each size band. The values of X_m as a function of ρ are calculated and plotted in Fig. 6-16. It can be seen that the value of X_m oscillates around and converges to the value of 1.357 (which is the result for Fraunhofer diffraction) as the value of ρ increases. It is also found that for the cases where the value of ρ is larger than 30, the Fraunhofer diffraction approximation can be applied. For $\rho \ll 1$, Rayleigh scattering theory must be used.

The new T matrix and the particle size bands used for the size distribution calculation are generated and interfaced with the standard Malvern particle sizer, so that the correct size distribution can be obtained. Figure 6-17 shows a comparison of the derived size distributions from the same light energy distribution by both Fraunhofer and anomalous diffraction theories.

Particle Field Concentration

The conventional Malvern particle sizer can be used for the measurement of concentration as well as particle size distribution.

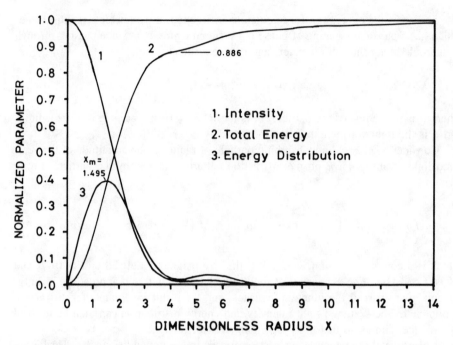

Figure 6-15 Intensity and energy distributions of anomalous diffraction for the case of $\rho = 3$.

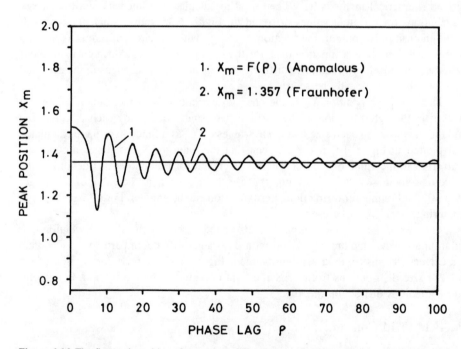

Figure 6-16 The first peak position of anomalous diffraction as a function of ρ.

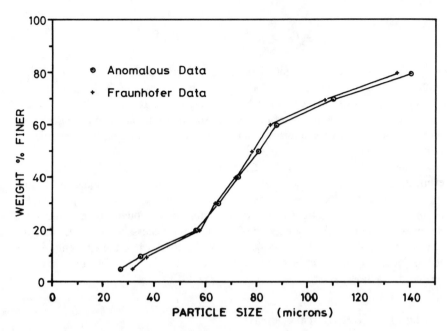

Figure 6-17 Size distribution determined by anomalous theory.

For a particle field with an obscuration between 5% and 50%, the concentration can be calculated from the measured obscuration and size distribution. The equation used in the calculation can be derived from Beer-Lambert law (Eq. (6-18)) as

$$C_v = -2 \ln (I/I_0) \left[3l \sum_{j=1}^{M} (-V_j Q_{ext,j}/d_j) \right]^{-1} \tag{6-20}$$

where d_j is the mean drop diameter in the size range j, $Q_{ext,j}$ is the extinction cross-section efficiency for the drops in size range j, and V_j is the drop volume distribution that results from the size distribution measurement. The summation is carried out over all size ranges.

For Fraunhofer diffraction, a value of $Q_{ext,j} = Q_{ext} = 2$ applies. For anomalous diffraction, $Q_{ext,j}$ can be calculated by (van de Hulst 1981)

$$Q_{ext,j} = 2 - \frac{4 \sin \rho}{\rho} + \frac{4(1 - \cos \rho)}{\rho^2} \tag{6-21}$$

where ρ is defined by Eq. (6-18).

For a low-density particle field, Eq. (6-20) cannot be used, since the obscuration is far too low to be detected by the central diode due to the limited number of bits available in the signal analog-to-digital (A/D) conversion. A new equation has been derived that allows the concentration to be calculated from the total scattered light energy measured by the detector. The equation is given by

$$C_v = C'' \sum_{i=1}^{N} E_i \left(l \sum_{i=1}^{N} \sum_{j=1}^{M} V_j T_{i,j} \right)^{-1} \tag{6-22}$$

where ΣE_i is the total scattered light energy collected from all detector rings (excluding the central diode), C'' is the constant concerned with the power of the laser transmitter and the efficiency of the optics, which can be determined by calibration, and $T_{i,j}$ is the matrix used in the calculation of the particle size distribution and can be generated by using Fraunhofer or anomalous diffraction theory, depending on the properties of the particles concerned.

It is noted from Eq. (6-22) that C_v is a linear function of ΣE_i. Results from experiments confirm that at low density the concentration is proportional to the sum of the scattered energy collected by the detector, but this does not apply for higher concentrations due to the effect of multiple scattering. Although proper corrections can be made for the multiple-scattering effect, it is not necessary to do so, since the method based on Beer-Lambert law can always be used when the concentration is high, provided that the effect of multiple-light scattering is not sufficient to severely distort the diffraction pattern.

Drop Sizing in High-Density Sprays

Since the original development of this instrument, several of the initial limitations of the technique have been studied extensively by many researchers. As a result, it is now possible to employ the instrument in a wide variety of applications.

A major improvement to the instrument has resulted from the study of the effect of multiple scattering (Cao and Brown 1990; Gomi 1986; Hirleman 1988; Felton, Hamidi, and Aigal 1985; Hamidi and Swithenbank 1986). The study of scattering of light by particle fields and sprays occurring both naturally and artificially had previously been restricted to the study of single scattering. A mathematical model has therefore been constructed that simulates the effect of multiple scattering of light by dense particle fields (Felton, Hamidi, and Aigal 1985). These studies thus extend the application of particle-sizing techniques based on Fraunhofer diffraction, to dense particle fields.

The effect of multiple scattering on the generated light energy distribution for a monosize, $d = 100$ μm, particle field and a particle field that can be described by the Rosin-Rammler distribution, $\overline{X} = 26$ and $N = 2.9$, is shown in Figs. 6-18 and 6-19, respectively. The choice of the Rosin-Rammler parameter used was due to the availability of National Bureau of Standards (NBS) glass spheres that could be characterized by these Rosin-Rammler distribution parameters. Hence comparison between experimental and theoretical data could be made. Each diagram shows the light energy distribution produced by a given particle field at several different concentrations, this being expressed as the percentage obscuration. The effects of multiple scattering are clearly shown. First, the peak of the light energy distribution shifts toward the outer detector ring, hence indicating a smaller apparent size. This is due to the fact that when the incident light is multiply scattered, the overall scattering angle increases, and hence the resultant energy distribution appears to originate from particles smaller than those actually present in the particle field.

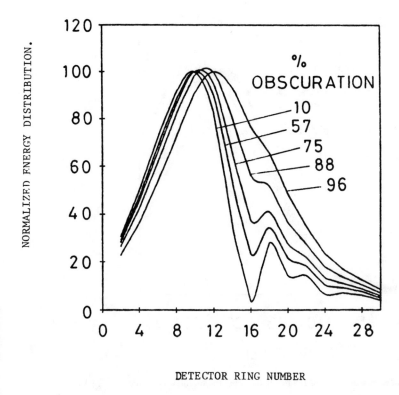

Figure 6-18 Light energy distribution as a function of obscuration for monosize (d = 100 μm) particle field.

The second effect is a general broadening of the light energy distribution. Again this is easily explained by the fact that the light is scattered at larger and larger angles. This broadening of the energy distribution results in an apparent broadening of the size distribution. The secondary maxima and minima, characteristic of the light energy distribution for monosize particles, disappear as the obscuration increases, hence highlighting the broadening effect due to multiple scattering.

Figures 6-20 and 6-21 show the effect of multiple scattering on the measured Rosin-Rammler size distribution parameters. The experimental work was carried out using a dense suspension of NBS standard glass spheres in a magnetically stirred cell. The effect of multiple scattering is clearly illustrated in these diagrams. The measured size is reduced, and the distribution is broadened as the obscuration of the light increases.

Parametric studies showed that the multiple-scattering correction factor is both a function of the sample obscuration and the actual particle dispersion. As a result of these studies, correction equations have been derived for two parameter size distribution models. One such set of equations applicable to the Rosin-Rammler distribution model is given by (Hamidi and Swithenbank 1986)

$$\overline{X}/\overline{X}_{app} = 1.0 + [0.036 + 0.49(OB)^{9.0}]N_{app}^{[1.9 - 3.33(OB)]} \tag{6-23}$$

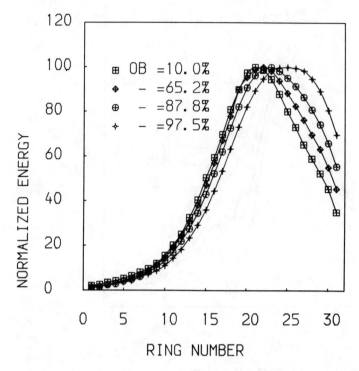

Figure 6-19 Light energy distribution as a function of obscuration for $\overline{X} = 26$ and $N = 2.9$ Rosin-Rammler distribution.

$$N/N_{app} = 1.0 + [0.035 + 0.11(OB)^{8.65}]N_{app}^{[0.35 + 1.45(OB)]} \qquad (6\text{-}24)$$

where OB is light obscuration and $\overline{X}_{app}$ and N_{app} are the apparent Rosin-Rammler parameters.

Although convenient, two-parameter size distribution models cannot adequately characterize all particle size distributions. Therefore the so-called model independent option is employed in such cases, where the size distribution is described by 15 independent size bands. The multiple-scattering study has therefore been extended to be employed with the model independent option size distribution (Cao and Brown 1990; Hamidi and Swithenbank 1986).

Near-Monodisperse Drop-Sizing Theory

When using the laser diffraction method to determine drop size distribution, various algorithms are available that characterize the size distribution, that is, by Rosin-Rammler parameters, lognormal parameters, or model independent parameters. However, for certain special applications the size distribution is near monodisperse, in which case the method described below gives an accurate assessment of the mean size and size distribution for spherical particles.

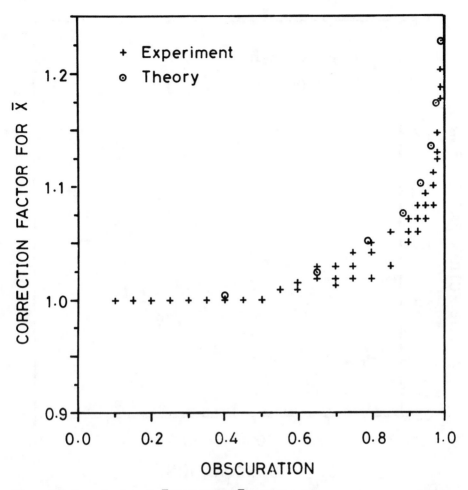

Figure 6-20 Correction factor for $\overline{X}$ for distribution $\overline{X} = 21$, $N = 2.9$.

As shown previously, the light energy distribution from exactly monodisperse spherical particles varies as

$$2J_1^2(X)/X \tag{6-25}$$

where X is the dimensionless size parameter defined in Eq. (6-7).

It follows from the Bessel function that this light energy pattern has an infinite number of maxima with zeros between.

If a number of particle sizes are present, the peaks and zeros due to each size are displaced in radius, and the combination of the light energies results in the minima increasing from zero until, with a wide spread of sizes, the intervening minima vanish, as shown in Fig. 6-22. The ratio of the height of the second peak above the first minimum to the height of the first peak above this minimum is therefore a unique

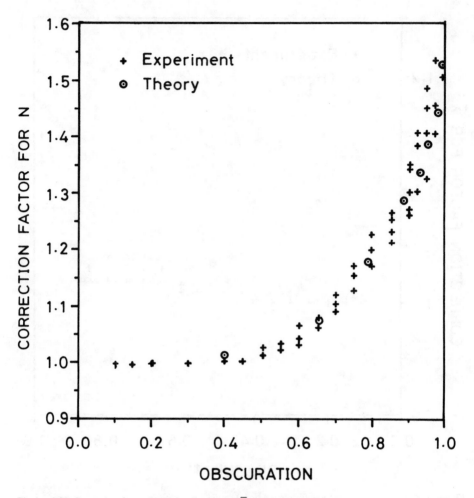

Figure 6-21 Correction factor for N for distribution $\overline{X} = 21$, $N = 2.9$.

function of the spread of the particle size for any given distribution function. Similarly, the mean size can be determined very accurately from the location of the first maximum.

Considering the first estimate of the mean size, as shown by Swithenbank et al. (1977), this is given by $X_{max} = 1.357$; hence the mean diameter of the particle is

$$d_1 = 1.357\lambda f/\pi S_{max} \qquad (6\text{-}26)$$

e.g., for $\lambda = 0.6328$ μm and $f = 300$ mm,

$$d_1 = 82.00/S_{max} \quad \mu m$$

where S_{max} is the radius of the first maximum in millimeters. Similarly, a second estimate of the mean particle diameter may be made from the first minimum:

$$d_2 = 3.84\lambda f/\pi S_{min} \qquad (6\text{-}27)$$

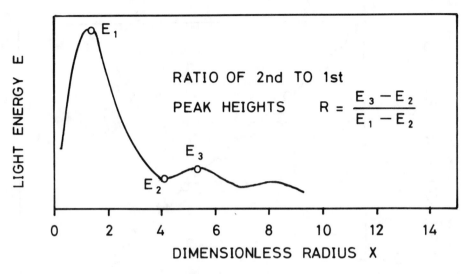

Figure 6-22 Dimensionless diffracted light energy distribution.

In order to obtain the radius of the first maximum and minimum it is important that these are not taken directly from the readout of the photodetector, since the detector rings are not of constant width, and therefore the actual readout is a distorted version of the true light energy distribution. (This arrangement is used to reduce the dynamic range of the signal to the A/D converter, thus improving the instrument precision.) The actual readings from the rings must therefore be divided by their respective widths to yield the true light energy distribution. For example, the true widths and mean radii of the rings in the Malvern instrument could be used.

To obtain an accurate value of the radius at the first maximum and first minimum of the true light energy distribution, a parabolic curve can be fitted to the adjacent group of points using a standard least squares error curve-fitting procedure. The resulting equations are then differentiated to give the required radii and the corresponding values of the light energy at the first two maxima and intervening minimum. An accurate mean value for the diameter of the particles can be obtained by substituting these radii into the three respective equations.

Considering next the determination of the size distribution about the mean, the convolution of the light energy distribution equation with the rectangular and Gaussian distribution functions has been integrated numerically to obtain the corresponding light energy distributions for a range of deviation parameters. The results for the rectangular distribution are shown in Fig. 6-23, expressed as the ratio of the second peak to the first peak, where the peaks are measured with respect to the intervening trough. The rectangular distribution is characterized by the dimensionless ratio of its half width to the mean. In this way, the curve is applicable for any mean size of particles. In order to use this curve in an automated analysis of size distribution it is convenient to fit it with a polynomial.

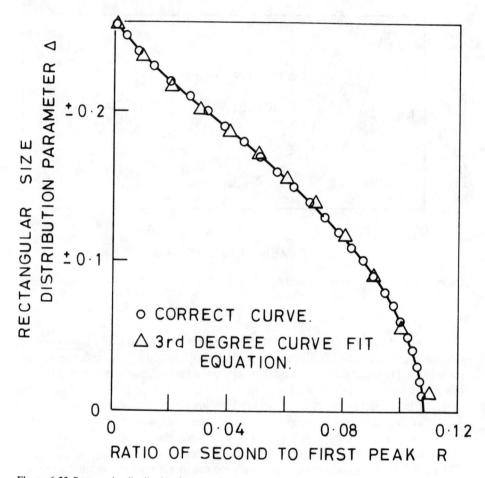

Figure 6-23 Rectangular distribution Δ versus R.

The following equation can be used:

$$\Delta = 0.262 - 2.782R + 32.707R^2 - 255.646R^3 \qquad (6\text{-}28)$$

where Δ is the spread of particle size ($\bar{d} \pm \Delta$) and R is the ratio of second peak to first peak. It should be noted from Fig. 6-23 that Δ is insensitive to R in the range $0 < \Delta < 0.04$, and it is recommended that experimental results be interpreted with care in this region. It will also be noted from Fig. 6-23 that the method is applicable up to the relatively wide range of $\pm 26\%$.

The Gaussian distribution is more generally encountered, and the standard deviation σ is used to characterize the width of this distribution curve. The relation between the standard deviation and the ratio of the second to first peaks in the diffracted light energy distribution to the standard deviation is shown in Fig. 6-24. It can be seen that the standard deviation is a sensitive unique function of the peak height ratio throughout

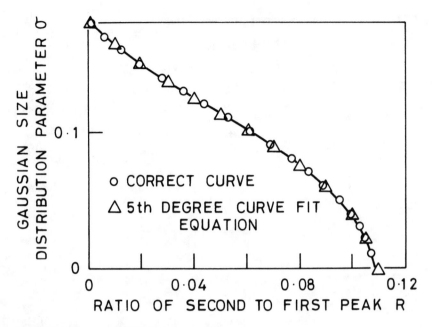

Figure 6-24 Gaussian distribution σ versus R.

the range of σ from 0 to 18.3%. Again, the curve was calculated by numerical integration of the Bessel functions weighted by the distribution function. For convenience of automated size distribution analysis, the curve has been fitted by a polynomial, and the use of the fifth degree is recommended. The relation is given by

$$\sigma = 0.1856 - 3.0951R + 105.102R^2 - 2499.2R^3$$
$$+ 26,407.5R^4 - 102,956R^5 \qquad (6\text{-}29)$$

The third-degree equation may be used, but it is slightly less accurate at large values of R.

$$\sigma = 0.1837 - 2.127R + 23.1R^2 - 166R^3 \qquad (6\text{-}30)$$

Summarizing, the procedure to evaluate the mean particle size and narrow size distribution using the Malvern or a similar instrument is as follows.

1. Measure the background light distribution on the photodetector array.
2. Measure the light distribution diffracted by the particles and subtract the corresponding background readings.
3. Divide each reading by the corresponding ring width and represent the resulting true light energy distribution as a function of the mean ring radius also given in Table 6-1. (Change the transform lens if the peaks are near the ends of the range.)
4. Locate the first maximum and first and subsequent minima and, by means of a quadratic curve fit, use the radii at which these occur to calculate the mean size using Eqs. (6-26) and (6-27).

Table 6-1 Detector ring dimensions of Malvern particle sizer

| | Radius, mm | | | Width, |
Ring no.	Inner	Outer	Mean	mm
1	0.149	0.218	0.184	0.067
2	0.254	0.318	0.286	0.064
3	0.353	0.417	0.385	0.064
4	0.452	0.518	0.485	0.066
5	0.554	0.625	0.589	0.071
6	0.660	0.737	0.699	0.076
7	0.772	0.856	0.814	0.084
8	0.892	0.986	0.939	0.094
9	1.021	1.128	1.074	0.107
10	1.163	1.285	1.224	0.122
11	1.321	1.461	1.391	0.140
12	1.496	1.656	1.576	0.160
13	1.692	1.880	1.786	0.188
14	1.915	2.131	2.023	0.216
15	2.167	2.416	2.291	0.249
16	2.451	2.738	2.595	0.287
17	2.774	3.101	2.938	0.328
18	3.137	3.513	3.325	0.376
19	3.549	3.978	3.763	0.429
20	4.013	4.501	4.257	0.488
21	4.536	5.085	4.811	0.549
22	5.121	5.738	5.429	0.617
23	5.773	6.469	6.121	0.696
24	6.505	7.282	6.894	0.777
25	7.318	8.184	7.751	0.866
26	8.219	9.185	8.702	0.965
27	9.220	10.287	9.754	1.067
28	10.323	11.501	10.912	1.179
29	11.537	12.837	12.187	1.300
30	12.873	14.300	13.586	1.427
31	14.336	15.900	15.188	1.565

5. Calculate the ratio of the second to first peak, and using Eq. (6-28) or (6-29), calculate the distribution half width or standard deviation as appropriate.

To illustrate the application of this method, experimental data for the light diffracted by near mono-sized particles is plotted in Fig. 6-25. The data are first corrected to produce a true light energy distribution. To determine the mean particle size, the values of S/f for the first peak and first three minima are read. These are shown in Table 6-2.

Clearly, because the diffracted light energy contains a series of peaks and minima, the particles have a narrow size distribution. Assuming the distribution is Gaussian, the standard deviation σ may be evaluated from the ratio R as follows.

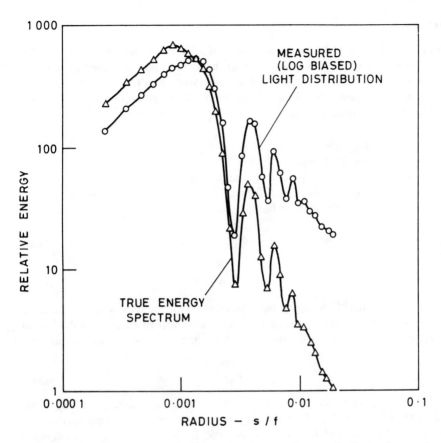

Figure 6-25 Comparison between the measured and corrected energy distribution for near-monosized particles.

First peak relative energy value 700
First minimum relative energy value 8
Second peak relative energy value 50

Hence the ratio is 0.06, and the standard deviation (σ) for the spread of size is 0.1, or 26.9 µm.

Table 6-2 Values used to determine mean particle size

	S/f	X_m	a, µm
First maximum	0.001	1.357	138
First minimum	0.0029	3.84	133
Second minimum	0.0059	7.02	136
Third minimum	0.0078	10.17	131

Here, the average particle radius is 134.5. Hence mean diameter is 269 µm.

Vignetting

When the particle field to be measured is located too far from the collection lens, the light scattered at large angles is lost due to the limited lens aperture, causing the measured size distribution to be biased toward large particles. This is called the vignetting effect. Figure 6-26 (Wild and Swithenbank 1986) shows an optical system where vignetting is taking place, with

R_L = radius of the collection lens.
R_b = radius of the laser beam.
R_d = the radial position of point P on the detector.
Z_m = the maximum recommended distance of the particle field from the lens.
Z_p = the position of the particle field.

The onset of vignetting is given by $Z_p > Z_m$, where Z_m is defined as (Dodge 1984)

$$Z_m = f(R_1 - R_b)/R_{dm} \qquad (6\text{-}31)$$

R_{dm} is the outer radius of the largest detector ring and f is the focal length of the lens.

An efficiency, η_R, is introduced to correct the measured energy distribution for the effect of vignetting; η_R is defined for a detector ring as (Wild and Swithenbank 1986)

$$\eta_R = E_R/E'_R \qquad (6\text{-}32)$$

where E_R is the energy collected on a ring using limited lens aperture and E'_R is the energy that could be collected on the ring when using an infinite lens aperture. In evaluation of the scattered light energy received by a detector ring, the Gaussian intensity profile of the laser beam should be considered, since the beam dimension plays an important part in introducing the efffect of vignetting.

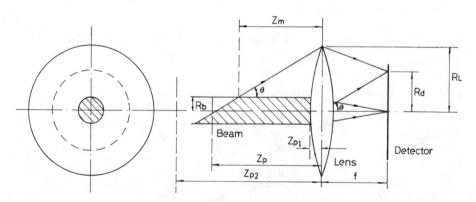

Figure 6-26 Optical configuration with vignetting.

The evaluation of the efficiency can be carried out in different regions of the measurement volume. A few special cases can be noted.

1. If the measurement volume V is outside the maximum recommended measurement volume V_m, i.e.,

$$Z_{p2} > Z_{p1} \geq f(R_1 + R_b)/R_d$$

then $\eta_R = 0$, i.e., no correction can be made.
2. If the apparent measurement volume V' is inside the volume V_m, i.e.,

$$Z_{p1} < Z_{p2} \leq f(R_1 - R_b)/R_d$$

then $\eta_R = 1$, i.e., no correction is needed.
3. When the apparent measurement volume V' is overlapped with the volume V_m, then $0 < \eta_R < 1$.

Once the correction efficiency is obtained for each ring, the measured energy distribution can be corrected by dividing the energy on each ring by its associated efficiency.

A result of using the correction efficiency to correct the vignetting effect for rings 29 and 30 is given in Fig. 6-27 (Wild and Swithenbank 1986). Corrections would not normally be required below these ring numbers.

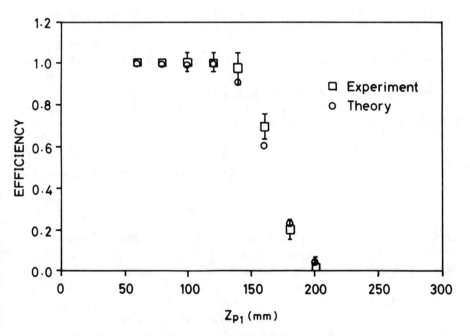

Figure 6-27 Vignetting correction efficiency for rings 29 and 30.

Drop Size and Velocity Measurement

In spray characterization, knowledge of droplet velocity as well as size distribution is often required. Several techniques have been developed for drop size and velocity measurement. Among them are photography, holography, and LDA and its variations. Each has its restrictions as well as advantages for different applications.

The drop size distribution measured by the Malvern particle sizing instrument is spatially averaged; the results therefore provide information on the number density of drops in the spray at an instant of time. In spray characterization, as the drops traverse at different velocities, the information on mass flux of drops of different sizes is often of interest. This is given by the temporal size distribution. The measured spatial distribution can be converted to temporal distribution if the drop velocity distribution can be determined. A technique developed at Sheffield University can provide an on-line measurement of both the drop size and velocity distributions, thus permitting both spatial and temporal size distributions to be obtained.

In the technique the scattered light energy is measured as time series signals from each of the 31 detector rings of the Malvern particle sizer. Figure 6-28 shows the measured scattered light signals from several detector rings. According to diffraction theory, the radial position of the peak of the scattered light energy distribution on the detector plane is inversely proportional to the drop diameter. The light energy received from each detector ring is therefore mainly related to the drops in a certain size range. Information related to the droplet formation process and velocity can be obtained by analyzing the measured time series signals using autocorrelation and cross-correlation techniques.

The success of applying correlation techniques in spray signal analysis depends on the nonuniformity of the drop concentration, which is essential for providing fluctuating signals for the correlation. In practice, drops are formed by the instability of thin liquid sheets, which necessarily results in the required fluctuating signals as shown in Fig. 6-28.

Assuming that the drops have a Gaussian velocity profile and the intensity of the laser beam is also of Gaussian distribution, the autocorrelation function of the scattered light energy signal can then be expressed as

$$R(\tau) = \int\limits_{-\infty}^{+\infty}\!\!\int c(t) \exp\left[\frac{-2v^2(t + \tau/2)^2}{r_0^2}\right] \exp\left[\frac{(v - V_0)^2}{-2\sigma_v^2}\right] \Bigg/ (\sqrt{2\pi}\ \sigma_v)\ dv\ dt$$

$$(6\text{-}33)$$

where r_0 is the $1/e^2$ beam radius, V_0 is the drop mean velocity, σ_v is the velocity deviation, and $c(t)$ is the drop concentration within the measurement volume. If the drops have a constant concentration, the autocorrelation function of the signal will have a Gaussian shape. If the drop concentration varies periodically, the resulting correlation function will show a certain degree of oscillation, and the energy spectrum of the signal, which is the Fourier transform of the autocorrelation function, will show a peak at a certain frequency. Figures 6-29 and 6-30 show the measured autocorrelation

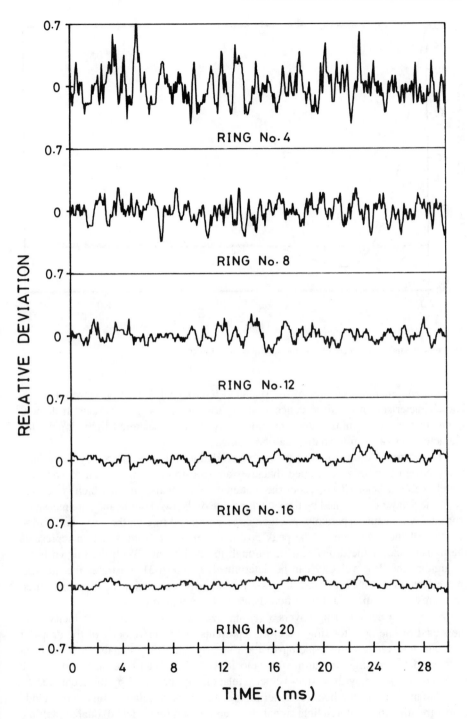

Figure 6-28 Measured scattered light energy signals from several detector rings.

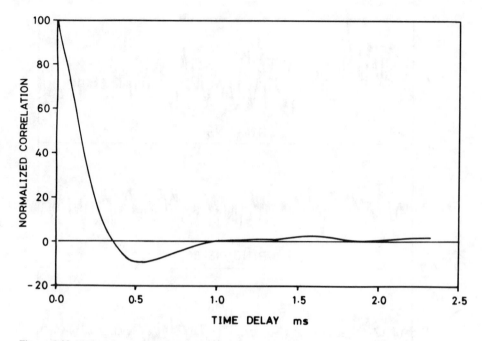

Figure 6-29 Measured autocorrelation function of scattered light signal.

function and energy spectrum of the scattered light signal from a pressure jet spray. The characteristic of the drop concentration variation can be clearly seen. It demonstrates that by performing autocorrelation analysis of the scattered light signals, the characteristics of drop formation can be studied.

To determine drop velocity, the cross-correlation technique is employed. A dual-beam system has been developed that uses an acousto-optic Bragg cell to split and switch the laser beam. Two sets of the scattered light signals, one of which is delayed by a time delay τ determined by the switching rate of the two laser beams, are measured from the two beams separately and then cross correlated to give the cross-correlation function of the light signal. The peak position of the correlation function represents the transit time of the drops passing through the two beams. With the known beam separation, the drop velocity can be determined. Figure 6-31 illustrates the diagram of the technique, and Fig. 6-32 shows a typical cross-correlation function measured for a pressure jet spray and the fitted theoretical correlation curve.

Since each detector ring covers a certain size range, the measured velocity from the signal of one detector ring therefore represents the mean velocity of the drops of a certain size range corresponding to the ring. By analyzing signals received from different detector rings, the drop size–velocity correlation can be established. Figure 6-33 shows a jet spray drop size–velocity relationship measured by the technique.

Variants of the method for determining the cross-correlation function include autocorrelating the scattered light signals from continuous (unswitched) dual or multiple laser beams.

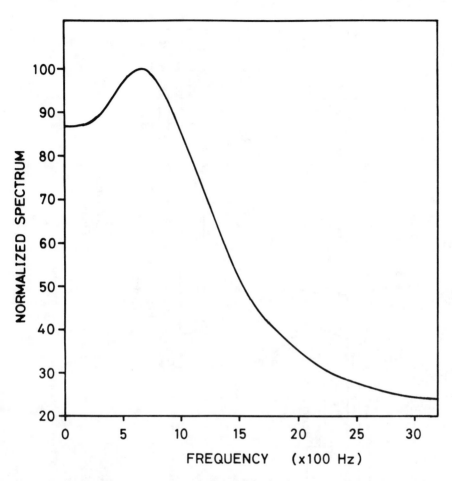

Figure 6-30 Measured energy spectrum of scattered light signal.

Detailed results on this modulated beam correlation method for the measurement of both drop size and velocity distributions are reported by Cao and Swithenbank (1990).

Drop Position Determination

The investigation, prediction, and modeling of the vaporization and combustion of fuel sprays may require detailed information on the local spray structure, as specified by spatial distributions of volume concentrations, drop size distributions, and volume fluxes. All of this detailed information is concerned with the positions of the drops that are of interest. The standard Malvern particle sizer cannot provide this information directly, since the information received by the detector is line integrated along the laser beam traversing the spray.

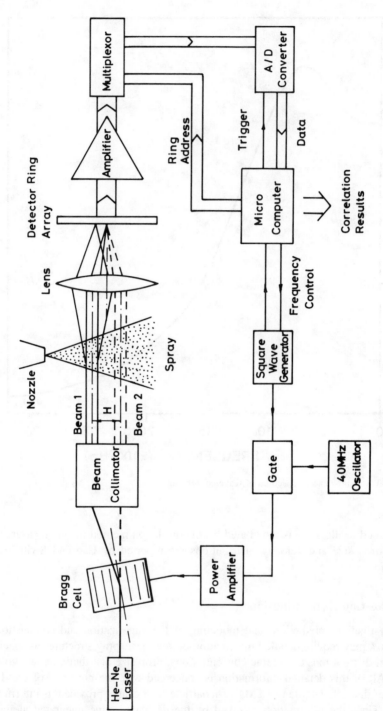

Figure 6-31 Diagram of spray drop and velocity distribution measurement system.

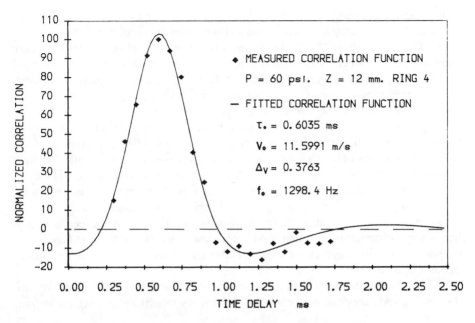

Figure 6-32 Comparison between the measured and the fitted correlation function.

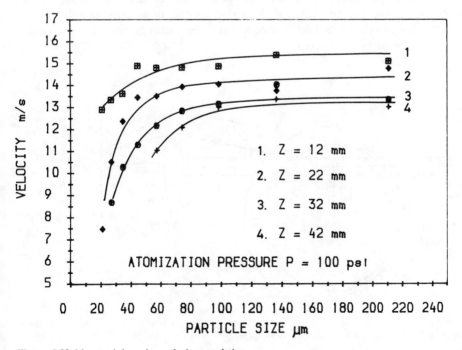

Figure 6-33 Measured drop size–velocity correlation.

Tomography. A tomography technique has been developed by Yule et al. (1979) for the measurement of drop size distribution and concentration in volume elements within an axisymmetrical spray. The technique uses the same concept as the medical X-ray brain and body scanner. A standard Malvern particle sizer is used to "scan" a cross section of a spray. The measured scattered light data from different parts of the spray are Abel transformed to give the two-dimensional distribution of particle sizes and concentrations in the plane across the spray.

Figure 6-34 (Yule et al. 1979) shows a diagram of light scattering while passing through a spray. The beam position is defined by (p, ψ) referred to the axis of the spray. The position of a volume element in the spray is defined by (r, θ). The light extinction coefficient and the scattered light energy distribution on the focal plane can be measured and expressed as a function of beam position when the measurements are taken at different parts of the spray. A Fourier transformation is carried out to transform the data obtained in the (p, ψ) domain into distributions of local scattered light energy distribution and extinction coefficient in the (r, θ) domain. From these distributions the local particle volume size distribution and the local volume concentration can be obtained.

The technique has been used successfully for the measurement of local drop size distribution and vaporization rate in a kerosene spray vaporizing in a hot gas stream.

Optical techniques. Apart from the technique mentioned above, there are other optical techniques that can be used in conjunction with laser diffraction instruments to determine the drop positions in the spray. For example, the diffracted light from particles of the same size may be "filtered" by a ring aperture or an imaging optic, so that only the light from the particles at a certain position within the measurement volume can be collected by the detector.

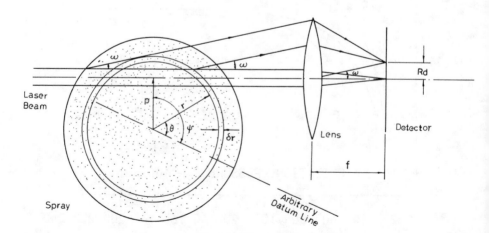

Figure 6-34 Drop position determination by tomography.

"Drop" Shape Effects

Although fuel sprays normally contain spherical drops, there are a number of important cases where nonspherical shapes are present. This occurs, for example, in the region near the fuel nozzle, where ligaments of liquid are formed from a thin fuel sheet, which then break up into the spray. In the case of coal water slurries and other slurry fuels, the ligaments tend to be long lived. Once a liquid fuel pyrolyzes to form a char, the resulting particle is often nonspherical. Elliptically shaped particles have also been observed when drops are spinning rapidly. It is therefore relevant to assess the effect of nonsphericity on the interpretation of the output of instruments based on laser diffraction. For simplicity, diffraction by cylindrical or rectangular shapes will be discussed.

When a parallel beam of monochromatic light falls on a cylindrical particle, the diffraction pattern is formed, superimposed on the geometrical image, similar to that described above for a spherical particle.

From the evaluation of the Fraunhofer diffraction integral, the diffracted light intensity distribution in the focal plane of the Fourier transform lens for a cylindrical particle of width a and length l can be shown to be

$$I = C_0 \left(\frac{\sin \beta}{\beta}\right)^2 \left(\frac{\sin \gamma}{\gamma}\right)^2 \tag{6-34}$$

where C_0 depends on the intensity of the incident beam and the particle area, $\beta = 2\pi a S'/\lambda f$, $\gamma = 2\pi l S''/\lambda f$, f is the focal length of the collection lens, and S', S'' are the orthogonal radial distances in the detection plane as measured from the optical axis.

A long cylinder representation of Eq. (6-34) is plotted as I versus X in Fig. 6-35. This long cylinder (or slot) diffraction pattern is equivalent to the Airy function. The pattern consists of an equidistant series of maxima and minima that relate to the location of peak intensity and zero intensity, physically appearing as a series of bright and dark parallel lines of diminishing intensity.

If such a group of "long fibers" were to be randomly aligned, the diffraction light energy would vary with radial distance from the axis, at the focal plane of the lens, proportional to

$$l^2 \frac{\sin^2 (\beta)}{\beta} \tag{6-35}$$

This function is also plotted in Fig. 6-35, and it can be seen that the energy reaches a maximum at approximately $X = \beta = 1.2$. This value compares with a value of 1.357 for the first maximum for circular objects.

Much of the scattered light energy is contained before the first minimum located at $X = \beta = \pi = 3.14$, whereas Fig. 6-10 shows that the first minimum for circular objects lies at $X = 3.84$. However, the light energy does not fall off as rapidly as in the case with a circular object or sphere. Thus, as expected, there is a close correspondence between the size indications for circular and long noncircular diffracting

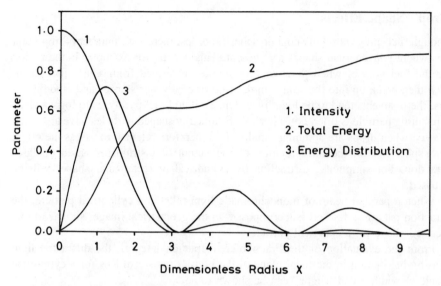

Figure 6-35 Diffraction by cylindrical particles.

objects. If the output of the Malvern machine were to be interpreted to deduce the width of the fibers, or long cylinders, the size given by the instrument would have to be reduced by about 12% to yield the true diameter of the cylinder.

If the cylinders have a length l equal to 3 times the width a, then the diffraction pattern for a single cylinder (or rectangular opening) appears as shown in Fig. 6-36. To obtain the light energy as a function of radius, we must integrate the light intensity illustrated in Fig. 6-36 for all angles. This intensity distribution is illustrated in Figs. 6-37(a) and 6-37(b) as a function of the angle for a few ranges of radial distances X. If we again take account of the random orientation of many such particles, the diffraction information may be represented as shown in Fig. 6-38. Here it can be seen that representing the diffraction pattern by the energy distribution effectively introduces a radial weighting factor proportional to the radius in the focal plane of the lens. The effect of this is to attenuate the signal at smaller radii, corresponding to the diffraction information from the longer dimension of the particles relative to that from the smaller dimension. An instrument such as the Malvern machine therefore tends to indicate the smaller dimension. If we take the center of gravity of the light energy distribution to the first minimum for the cylindrical (rectangular) particles, this is located at $X = 1.2$. Again this would imply that we must reduce the size readings given by the machine by about 12% to yield the true diameter of the cylinders.

Bearing in mind that cylindrical particles will usually be arranged in a random orientation with respect to the axial direction, the projected length will on average be reduced from the total length by a factor of

$$2l/\pi \int_{0}^{\pi/2} \cos \theta \, d\theta = 2l/\pi = 0.636l \qquad (6-36)$$

Figure 6-36 Diffraction pattern from a rectangular opening (after A. Kohler).

As the diffraction pattern is unique for a cylinder or a collection of cylinders of different sizes, it is therefore possible, in principle, to deduce the size distribution of a collection of particles from their diffraction pattern provided that the length/diameter ratio is known. In general, particles have a range of length/diameter ratio, and their shape cannot be characterized accurately. The most useful conclusion that can be drawn is that the Malvern (or similar) particle sizing instrument will tend to overestimate the width by about 12% and take relatively little account of the length when used with an algorithm that assumes the particles are spherical.

The ratio of light intensity measured at the center diode, before and after the particle field is introduced in the beam, gives the fraction of the light transmitted by the particles. As above, the transmission is related to the total projected area of the particles by the Beer-Lambert law, and using $Q_{ext} = 2$, the concentration can be deduced.

6-4 TWO-PHASE REACTING FLOW CALCULATIONS

The purpose of the spray size and velocity measurement is for process equipment design or design validation. Modeling/calculation procedures therefore form an integral part of spray technology. Combustion systems represent typical applications of this complementary aspect of technology, and it may be noted that the important spray data required are the initial drop size and velocity distributions. In the case of a pressure

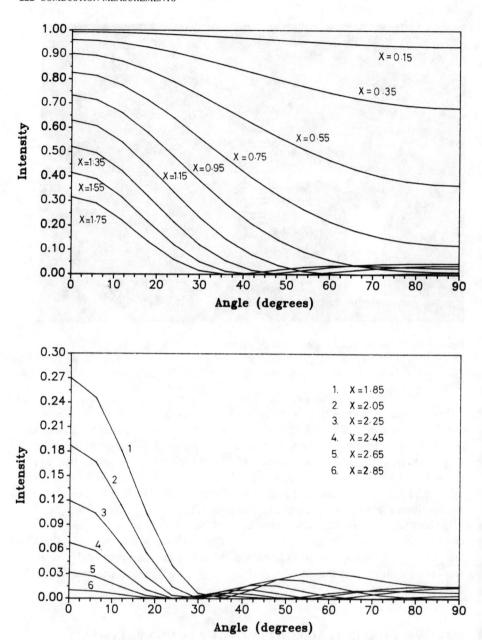

Figure 6-37 Diffraction by rectangular objects ($L/D = 3$).

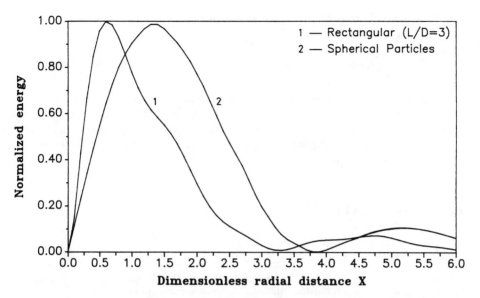

Figure 6-38 Radial diffraction energy for rectangular shapes ($L/D = 3$) and spherical particles.

jet atomizer, the initial velocities are almost equal for all drop sizes. However, for an air blast atomizer, the initial velocities may vary with drop size.

As a result of modern spray diagnostics and computational fluid dynamic techniques, the rational design and operation of spray systems is advancing rapidly.

6-5 CONCLUSIONS

Modern spray combustion system design methods depend on accurate information on the spray characteristics, and especially on the drop size distribution. Spray diagnostics therefore play an essential role in combustion system design technology. Similarly, knowledge about spray characteristics is of little use unless there is an integration of this information with combustion system designs.

In this study a range of drop-sizing instruments has been outlined, and the advantages of nonintrusive instruments highlighted. The features of diagnostic instruments based on Fraunhofer diffraction are particularly emphasized. It has been shown how this type of instrument can be used to obtain information on the following.

Drop size distribution.
Low-density aerosols.
Very small drops (anomalous diffraction regime).
High-density sprays.
Near-monosize sprays.

Sprays relatively distant from the lens (vignetting).
Drop velocity distribution.
Drop position.
"Drop" shape effects.

   Mathematical modeling, which uses finite difference techniques to evaluate the governing differential equations relevant to liquid atomization and spray systems, has now evolved to the point where, in conjunction with laser diffraction spray diagnostics, the level of prediction of flow fields, drop trajectories, and volatilization rates is adequate for many engineering purposes.

6-6 NOMENCLATURE

a	drop or particle radius
a	width of cylindrical particle, Eq. (6-34)
a_j	drop or particle radius in size band j
$A(P)$	amplitude function of the scattered light
$c(t)$	drop concentration within the measurement volume
C	cross-section area, Eqs. (6-3) and (6-4)
C, C', C'', C_0	constant, Eqs. (6-12), (6-19), (6-22), and (6-34), respectively
C_v	particle volume concentration
d	drop or particle diameter
d_1, d_2	mean drop diameters, Eqs. (6-26) and (6-27)
E	total light energy falling on the particles
$E(I)$	scattered light energy distribution
$E_m(I)$	measured light energy distribution
$E_{s1,s2}$	scattered light energy within a ring of radii S_1 and S_2
E_R	light energy received on a ring using limited lens aperture
$E_{R'}$	light energy received on a ring using infinite lens aperture
f	focal length
I	scattered light intensity from an interference fringe pattern, Eq. (6-1)
I	scattered light intensity, Eq. (6-6)
$I(0)$	light intensity at the center of the diffraction pattern
I_0	light intensity of the incident beam
J_0	zeroth-order spherical Bessel function
J_1	first-order spherical Bessell function
l	optical path length, Eq. (6-16)
l	length of cylindrical particle, Eq. (6-34)
L	fraction of the scattered light energy within a circle
m	particle relative refractive index
M	number of particle size bands
N	number of particles, Eqs. (6-10) and (6-11)

N	number of detector rings, Eq. (6-15)
N	Rosin-Rammler parameter
P	atomization pressure, Figs. 6-32 and 6-33
Q	cross-section efficiency
r_0	$1/e^2$ beam radius
R	ratio of second peak to first peak, Eqs. (6-28), (6-29), and (6-30)
R_L	radius of lens
R_b	radius of laser beam
R_d	radial location on the detector plane
$R(\tau)$	correlation function of the scattered light signal
S	radial location on the detector plane
S_1, S_2	detector ring dimensions
t	time
$T(I, J)$	elements of diffraction matrix, Eqs. (6-14) and (6-15)
$T_{i,j}$	elements of diffraction matrix, Eq. (6-22)
v	drop velocity, Eq. (6-33)
V	visibility, Eq. (6-1)
V	measurement volume
V'	apparent measurement volume
V_j	particle volume distribution, Eqs. (6-20) and (6-22)
V_m	maximum measurement volume
V_0	mean drop velocity, Eq. (6-33)
W	total particle weight
W_j	weight fraction of particles of radius a_j
$W(J)$	particle weight distribution, Eqs. (6-14) and (6-15)
X	dimensionless particle size parameter
$\overline{X}$	Rosin-Rammler parameter
X_m	first peak position in scattered light energy distribution
Z	distance from the nozzle, Figs. 6-32 and 6-33
Z_m	maximum recommended distance of the particle field from the lens
Z_p	distance of the particle field from the lens
δ	interference fringe spacing
Δ	spread of particle size, Eq. (6-28)
Δ_v	σ_v/V_0
η_R	correction efficiency for vignetting
λ	light wavelength
ρ	phase lag, Eq. (6-18)
ρ	particle density, Eq. (6-11)
σ	standard deviation, Eqs. (6-29) and (6-30)
σ_v	drop velocity deviation, Eq. (6-33)
τ	total extinction cross section, Eq. (6-17)
τ	time delay, Eq. (6-33)

Subscripts

abs	absorption
app	apparent
ext	extinction
max	maximum
min	minimum
sca	scattering

REFERENCES

Azzopardi, B. J., Yeoman, M. L., White, H. J., Bates, C. J., and Roberts, P. J. 1982. *Optical development and application of a two-colour LDA system for the simultaneous measurement of particle size and particle velocity*. Atomic Energy Research Establishment report AERE RL0468. Harwell, England.

Bachalo, W. D. 1980. Method for measuring the size and velocity of spheres by dual-beam light scattering interferometry. *Appl. Opt.* 19(3):363–70.

Bachalo, W. D., and Houser, M. J. 1984. Phase/Doppler spray analyzer for simultaneous measurements of drop size and velocity distributions. *Opt. Eng.* 23(5):583–93.

Belaux, Y., and Verdi, S. P. S. 1975. A method for obtaining a uniform non-Gaussian laser illumination. *Opt. Commun.* 15:193.

Brown, D. J., Alexander, K., and Cao, J. 1990. Anomalous diffraction effects in the sizing of solid particles in liquids. *Particle and Particle System Characterization* (in press).

Cao, J., and Brown, D. J. 1990. Accurate laser diffraction particle size measurement in dense suspensions and dense sprays by correction for multiple scattering. Report No. HIC 493, Dept. of Mechanical and Process Engineering, University of Sheffield, England.

Cao, J., and Swithenbank, J. 1990. Drop size and velocity measurement using laser diffraction. Report No. HIC 494, Dept. of Mechanical and Process Engineering, University of Sheffield, England.

Cornillault, J. 1972. Particle size analyser. *Appl. Opt.* 11(2):265–68.

Dobbins, R. A., Crocce, L., and Glassman, I. 1963. Measurement of mean particle size of sprays from diffractively scattered light. *AIAA J.* 1(8).

Dodge, L. G. 1984. Calibration of the Malvern particle sizer. *Appl. Opt.* 23(14):2415–19.

Ereaut, P. R. 1983. Simultaneous size and velocity measurement of droplets in fuel sprays using laser Doppler anemometry. Ph.D. thesis, Sheffield University, Sheffield, England.

Ewan, B. C. R., Houdard, T., and Bastenhof, D. 1983. The holographic measurement of diesel sprays. Paper read at Conference on Instrumentation in Aerospace Simulation Facilities, September 1983, at Institute Saint Louis, France.

Farmer, W. M. 1972. Measurement of particle size, number density and velocity using a laser interferometer. *Appl. Opt.* 11:2603.

Felton, P. G., Hamidi, A. A., and Aigal, A. K. 1985. Measurement of drop size distribution in dense sprays by laser diffraction. Paper read at International Conference on Liquid Atomization and Spray Systems, at the Institute of Energy, London.

Gomi, H. 1986. Multiple scattering correction in the measurement of particle size and number density by the diffraction method. *Appl. Opt.* 25(19):3552–58.

Hamidi, A. A., and Swithenbank, J. 1986. Treatment of multiple scattering of light in laser diffraction measurement technique in dense sprays and particle fields. *J. Inst. Energy* (June 1986):101–5.

Hirleman, E. D. 1988. Modeling of multiple scattering effects in Fraunhofer diffraction particle size analysis. *Particle Particle Syst. Characterization* 5:57–65.

Hong, N. S., and Jones, A. R. 1976. A light scattering technique for particle sizing based on laser fringe anemometry. *J. Phys. D: Appl. Phys.* 9:1839.

Jones, A. R. 1977. A review of drop size measurement: the application of techniques to dense fuel sprays. *Prog. Energy Combust. Sci.* 3:225–34.

Kelvin, M. V. 1970. *Optics*. New York: John Wiley.

McCreath, C. G., Roett, M. F., and Chigier, N. 1974. A technique for measurement of velocities and sizes of particles in flames. *J. Phys. E: Sci. Instrum.* 5:601–4.

Morgan, B. B., and Meyer, E. W. 1959. Multichannel photographic scanning instrument for sizing microscopic particles. *J. Sci. Instrum.* 1(36):492.

Robinson, D. M., and Chu, W. P. 1975. Diffraction analysis of Doppler signal characteristic for a cross beam laser Doppler anemometer. *Appl. Opt.* 14:2177.

Simmons, H. C. 1977. The correlation of drop size distribution in fuel nozzle sprays. *J. Eng. Power* 99:309–14.

Swithenbank, J., Beer, J. M., Taylor, D. S., Abbott, D., and McCreath, G. C. 1977. A laser diagnostic technique for the measurement of droplet and particle size distribution. *Prog. Astronaut. Aeronaut.* 53:421–47.

Thompson, B. J., and Dunn, P. 1980. Advances in far-field holography: theory and application. *Recent Advances in Holography* 215:102–11.

Tishkoff, J. M. 1984. Spray characterization: practices and requirements. *Opt. Eng.* 23(5):557–60.

van de Hulst, H. C. 1981. *Light scattering by small particles*. New York: Dover.

Watson, H. H., and Mulford, D. F. 1954. A particle profile test strip for assessing the accuracy of sizing in regularly shaped particles with a microscope. *Brit. J. Appl. Phys. Suppl.* 3:5105.

Wild, P. N., and Swithenbank, J. 1986. Beam stop and vignetting effects in particle size measurement by laser diffraction. *Appl. Opt.* 25:3520.

Yule, A. J., Seng, C. A., Felton, P., Ungut, A., and Chigier, N. A. 1979. A laser tomographic investigation of liquid fuel sprays. In *Eighteenth symposium (international) on combustion*. Pittsburgh, Pa.: The Combustion Institute.

DIAGNOSTICS FOR FUEL SPRAY
CHARACTERIZATION

W. D. Bachalo, A. Breña de la Rosa, and S. V. Sankar

7-1 INTRODUCTION

In situ measurements of the fuel spray size distribution and the drop dynamics are of importance in enhancing our understanding of spray combustion phenomena. The quality of atomization and subsequent development of the size and velocity distributions within the combustion chamber have a significant effect on the performance of the combustion system, which is generally evaluated in terms of ignition, flame stability, combustion efficiency, combustor durability, and pollutant emissions. These factors can be related directly to fuel spray characteristics and air-fuel mixing. For example, drop spacing, spatial oxygen distribution, ambient temperature, the relative velocity between drops and air, drop size, and drop interactions have been found to influence the production of pollutants. Furthermore, inlet air swirl and air preheat have a direct effect on energy release and pollutant formation in turbulent flames. Both drop number density and mass flux also contribute to the processes leading to soot formation. These latter quantities are especially difficult to characterize due to the complexities of the flow field, which may include high turbulence levels, recirculation, and swirl.

Spray combustion environments, typically characterized by turbulent flows, pose a number of measurement difficulties to the experimentalist. Laser based optical diagnostics offer the greatest potential for obtaining fluid flow data in such complex environments. In spite of the advantages, laser-based diagnostics have their drawbacks: Optical access to the flowfield of interest can be limited and the windows of the vessel housing the flow under investigation become contaminated frequently. In sprays, drop number densities are usually high and coincident particle occurences in the measurement volume pose a detection problem for single particle counter instruments. Beam extinction will limit the accuracy for both line of sight and point measurement particle sizing techniques. Under burning conditions, the reacting flow will cause laser beam steering and spreading due to the steep density gradients, which in turn will induce gradients in the index of refraction. These phenomena preclude the use of certain optical methods in reacting flows.

Because of the limitations of the available instrumentation and the difficulties associated with obtaining in situ measurements, spray characterization has normally been carried out in quiescent laboratory environments. However, recent results have shown that even for simple pressure atomizers, spray formation and the development of the size distribution downstream are complicated by entrainment and interaction with the induced airflow, relaxation of the drop velocity of the various size classes, drop-drop interactions, and coalescence. When the spray is being injected into turbulent flows with swirl and recirculation, the local drop size distributions are significantly influenced by the interaction of the spray with the gas phase flow. Thus conclusions about the influence of local drop size distributions on combustion performance may not be complete without considering the effects of gas dynamics on the spray.

Over the past decade, a great deal of development has occurred in the optical diagnostics for spray field characterization. The line-of-sight measurements of particle size using small-angle forward light scatter detection methods (e.g., Malvern Instruments) have gained acceptability as a means for spray characterization. With the development of calibration techniques, the method has achieved a reasonable level of confidence in terms of measurement accuracy. Unfortunately, the method furnishes limited spatial resolution, although an Abel inversion scheme can be used with the assumption of an axisymmetric spray to procure spatially resolved measurements. The method fails in flow fields with minor fluctuations in the refractive index, which can produce significant beam deflections. Furthermore, the method cannot be combined with other methods such as laser Doppler velocimetry to produce simultaneous size and velocity measurements of the drops.

A number of other techniques have been proposed in an effort to fulfill the need to acquire simultaneous drop size and velocity measurements in practical environments. For example, hybrid methods combining light-scattering cross-section methods with the laser Doppler velocimeter have been used with success limited to relatively dilute sprays. The signal visibility method was also advanced but was limited by both a small dynamic range and a sensitivity to optic distortions introduced by the measurement environment. However, theoretical analysis of the visibility led Bachalo and Houser (1984a) to the derivation of the phase Doppler concept.

The phase Doppler method uses light-scattering interferometry to acquire simultaneously the size and velocity of spherical particles. As such, the measurements are dependent upon the wavelength of the scattered light as the measurement scale, which is not easily affected by environmental conditions. That is, the beam intensity may be attenuated by spray or intervening optics, whereas the wavelength is unaffected. After a number of years of development and refinement, the method has received wide acceptance as a reliable means for completely characterizing sprays in practical environments, including spray combustion. It has the advantage of being robust and is also relatively easy to use. The phase Doppler method can provide size and velocity data, number density, volume flux, drop angle of trajectory, and time-resolved information.

In the following sections the basic light-scattering theory involved in the phase Doppler methods will be presented and subsequently described in sufficient detail to provide an appreciation for the technique and its capabilities and limitations. A series of measurement examples will then be given to demonstrate the measurement accuracy and capabilities of the method. The results, which were obtained from measurements in complex flow field environments, such as turbulent recirculating flows, spray combustion, and diesel sprays, include drop size–velocity correlations, particle/gas phase turbulent interaction, and particle number densities.

7-2 BRIEF REVIEW OF LIGHT-SCATTERING INTERFEROMETRY

Light-Scattering Theory

The scattering of light by homogeneous dielectric spheres of arbitrary size is described by the well-known Lorenz-Mie theory. However, even with large computers, the time required to calculate the scattering coefficients for a range of drop sizes can be prohibitive. The number of terms needed to be computed in the series solution is proportional to the size parameter $\alpha = \pi d/\lambda$. Fortunately, for drop sizes greater than the light wavelength λ, the simpler theory of geometrical optics produces accurate results.

Van de Hulst (1957) showed that for spheres much larger than the light wavelength and with refractive index sufficiently different from the surroundings, the amplitude functions derived from the geometrical optics approach were, in the asymptotic limit, equal to the Mie amplitude functions. Comparisons have been made to the exact Mie theory by a number of researchers (Born and Wolf 1975; Gouesbet, Mahen, and Grehan 1988; Hodkinson and Greenleaves 1963) to demonstrate the accuracy of the geometrical optics method. Very good agreement was demonstrated by Glantschnig and Chen (1981) for drops as small as 5 μm. Van de Hulst demonstrated that for $\alpha > 10$, the scattering of light can be separated into the simplified theories of diffraction, refraction, and reflection.

Light scattered by diffraction, which is described by the following expression,

$$S_d(\alpha, \theta) = \frac{\alpha^2}{4\pi} \lambda \left[\frac{J_1(\alpha \sin \theta)}{\alpha \sin \theta} \right] \tag{7-1}$$

where J_1 is the Bessel function of the first kind, order 1 and θ is the scattering angle, is concentrated in a lobe centered about the transmitted beam. This forward scattered light becomes more intense in proportion to the diameter squared of the particle, and smaller in angular distribution with increasing particle diameter. Hodkinson and Greenleaves (1963) showed that for $\alpha > 15$, diffraction becomes insignificant at scattering angles greater than 10°. For spheres of refractive index $m = 1.5$ and diameters as small as 2 μm, diffractive scatter is less than 10% of the total light scattered at 45°. Thus the light scattered by diffraction can be avoided, when desired, by proper placement of the receiver optics.

Light scattered by reflection and refraction is best described in terms of rays and the use of the simple laws of reflection and refraction. When a ray impinges on the surface of a transparent sphere, it produces a reflected and refracted ray (Fig. 7-1). The direction of the refracted ray follows from Snell's law:

$$\cos \tau = m \cos \tau' \tag{7-2}$$

where τ and τ' are the angles between the surface tangent and the incident and refracted rays, respectively. The partitioning of the energy into two rays follows from the Fresnel coefficients.

The emerging rays are characterized by two parameters: the angle τ of the incident ray and the integer p of the interface from which it emerges. That is, $p = 0$ for the first surface reflection, $p = 1$ for the transmitted ray, and $p = 2$ for the ray emerging after one internal reflection. The energy in the remaining reflections is insignificant. The angle between the incident ray and the pth emergent ray is given by

$$\theta = 2(p\tau' - \tau) \tag{7-3}$$

The fraction of the incident intensity contained in the emergent rays can be obtained from the Fresnel coefficients. The details of derivation of the closed form θ-dependent intensity functions are given by Glantschnig and Chen, and only the results will be quoted here.

In the notation of van de Hulst, the scattered light is described in terms of two amplitude functions, $S_1(\alpha, m, \theta)$ and $S_2(\alpha, m, \theta)$ for the perpendicular (to the scattering

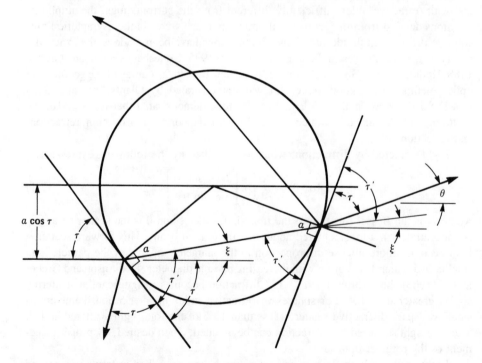

Figure 7-1 Geometrical ray trace of an incident light ray through a sphere.

plane) and parallel incident polarizations, respectively. Van de Hulst demonstrated that a large sphere scatters over 90% of the incident light in the forward direction with 99.5% of the forward scattered light emerging from the first two interfaces ($p = 0$) and ($p = 1$). Thus, only the first two terms of the series describing the rays emerging from the various interfaces need be considered for most practical applications.

Glantschnig and Chen obtained expressions for the amplitude function describing the rays reflected from the first surface of the sphere, $S_j^1(\alpha, m, \theta)$, where superscript 1 implies the $p = 0$ reflection and subscripts 1 and 2 represent the perpendicular and parallel polarizations, respectively. He also obtained expressions for the amplitude functions describing the rays emerging from the second surface after refraction by the sphere, $S_j^2(\alpha, m, \theta)$. Summing the amplitude functions, the total forward light scattering is given as

$$S_j(\alpha, m, \theta) = S_d(\alpha, \theta) + S_j^1(\alpha, m, \theta) + S_j^2(\alpha, m, \theta) \qquad j = 1, 2 \quad (7\text{-}4)$$

The dimensionless intensities i_j are simply

$$i_j(\alpha, m, \theta) = |S_j(\alpha, m, \theta)|^2 \qquad (7\text{-}5)$$

As indicated by van de Hulst, these expressions produce accurate results with the exception of scattering at grazing or nearly grazing incidence and for scattering into the rainbow angles. The geometrical optics results can be used for values of $\alpha > 15$, depending upon the information (scattering angles) and accuracy required. Comparisons of a sample calculation between geometrical optics and Mie theory (from Glantschnig and Chen 1981) are shown in Fig. 7-2. The agreement for the 5-μm-diameter droplets is very good, even though this is near the minimum size wherein the geometrical optics approximation applies.

For spheres large enough ($d > 3$ μm) to satisfy the above criteria, the scattered light intensity appears as an obvious means for size analysis. The square of the diameter of the spherical scatterer can be assumed to be proportional to the scattered light intensity,

$$I_{sc}(d, m, \theta) = C(m, \theta)d^2 \qquad (7\text{-}6)$$

This relationship is correct, provided that the receiver $f/$(number) is small enough to form an average over the angular fluctuations resulting from the interference between the reflected ($p = 0$) and refracted ($p = 1$) light. The measured light intensity is also dependent upon the incident intensity on the particles I_0 which is multiplied by the scattering coefficient, Q_{sc}. In practice, the optical collection efficiency and losses in the system are determined by calibrations using spherical particles or drops of known size. Unfortunately, in realistic environments the drops will attenuate the transmitted beams and the scattered light by an indeterminate amount.

Interferometry

Phase shift of scattered light. The light rays emerging from the spherical scatterer will have different path lengths, depending upon both the angle of scatter and the path lengths through the sphere. For this reason, the intensities of the outgoing beams cannot be added directly. The beams that all originated from the same coherent incident

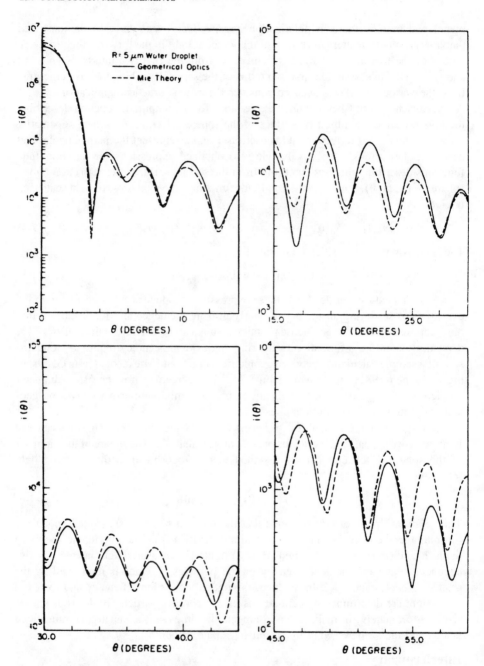

Figure 7-2 Comparisons of angular scattering diagrams for 5-μm drops computed from geometrical optics and Mie approaches (*Glantschnig and Chen 1981*).

wave must have their complex amplitudes added, and the squared modulus then becomes the correct intensity.

In order to compute the phases from ray optics, it is practical to reference the actual ray to a hypothetical ray scattered without a phase lag at the center of the sphere. Neglecting phase shifts of π at reflection and phase shifts of $\pi/2$ at focal lines (all of which cancel from the subsequent analysis), a simple geometric analysis results in the expression

$$\eta = 2\alpha(\sin \tau - pm \sin \tau') \tag{7-7}$$

for the phase shift due to the length of the optical path. It is of interest to note that the phase is directly proportional to $\alpha = \pi d/\lambda$, which implies that the number of extrema in the scattering pattern is also proportional to the dimensionless size, α. The change in phase, which is independent of the incident intensity or scattering amplitudes but is directly proportional to the particle diameter, is a more practical means for obtaining size information.

Dual-beam light scattering. One approach for extracting size information from the phase shift is to utilize the dual-beam scattering arrangement of the familiar laser Doppler velocimeter (LDV), Fig. 7-3. Assuming linearly polarized light, the amplitude functions associated with scattering from beams 1 and 2 are

$$S_{11}(m, \theta, d) = \sqrt{i_1} \exp (j\sigma_1) \tag{7-8}$$

$$S_{12}(m, \theta, d) = \sqrt{i_2} \exp (j\sigma_2) \tag{7-9}$$

where the double subscript indicates only polarization 1 is considered, j is the imaginary value, $j^2 = -1$, and $\sigma = \eta$ if phase shifts of π due to reflection, or of $\pi/2$ due to focal lines and the Fresnel coefficients are neglected. In Eqs. (7-8) and (7-9), σ_1 and

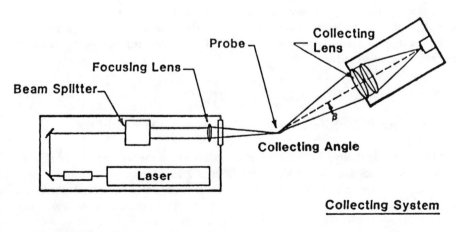

Figure 7-3 Schematic of the optical system for an LDV and particle-sizing system (top view).

σ_2 are $\pi/2$ plus the advance in phase of the actual ray with respect to a hypothetical ray scattered without phase lag at the center of the sphere.

When a spherical particle passes through the intersection of the two beams, it will scatter light from each beam as if the other beam was not there. Assuming linearly polarized light, i.e., $n = 1$, and droping this subscript, the scattered light waves may be described as

$$E_1(m, \theta, d) = S_{11}(m, \theta, d) \frac{\exp(-jkr + j\omega_1 t)}{jkr} \qquad (7\text{-}10)$$

$$E_2(m, \theta, d) = S_{12}(m, \theta, d) \frac{\exp(-jkr + j\omega_2 t)}{jkr} \qquad (7\text{-}11)$$

where $k = 2\pi/\lambda$ is the wave number, r is the polar coordinate, and ω_1, ω_2 are the light frequencies of beams 1 and 2, respectively.

The total energy scattered is obtained by summing the complex amplitudes from each beam. The intensity can then be determined by

$$I(m, \theta, d) = (|E_1|^2 + |E_2|^2 + 2|E_1||E_2| \cos \sigma) \qquad (7\text{-}12)$$

where σ is the phase difference between the scattered fields. In this expression the cross-product term $2|E_1||E_2| \cos \sigma$ corresponds to the sinusoidal intensity variation of the fringe pattern, while the $|E_1|^2 + |E_2|^2$ terms are the dc or pedestal components. The visibility of the scattered fringe pattern is simply equal to the ratio of these two terms:

$$V = \frac{2|E_1||E_2| \cos \sigma}{|E_1|^2 + |E_2|^2} \qquad (7\text{-}13)$$

Confusion can occur here by the use of the terms visibility and signal visibility. In the former case, Michelson's definition is implied, which is a measure of the distinctness of the fringes formed in the space surrounding the drop (Fig. 7-4). If a point detector was moved normal to the resultant fringe pattern, a signal proportional to the local light intensity would be produced, and the visibility of the resultant sinusoidal signal would be

$$V = \frac{V_{max} - V_{min}}{V_{max} + V_{min}} \qquad (7\text{-}14)$$

where V_{max} and V_{min} are the maximum and minimum voltages of the Doppler signal, respectively.

The signal visibility, $\mathcal{V}$, is the relative modulation of the signal received through the detection optics and the photodetector. This is usually produced by integration of the scattered fringe pattern over the area of the receiver lens. In general, the signal visibility depends on the receiver aperture, drop size, beam intersection angle, and light wavelength as well as other parameters and will always be less than or equal to the scattered interference fringe visibility.

Signal visibility has been used by Farmer (1972) as a means for determining the drop size. When the drop moves through the intersecting laser beams, the fringe pattern

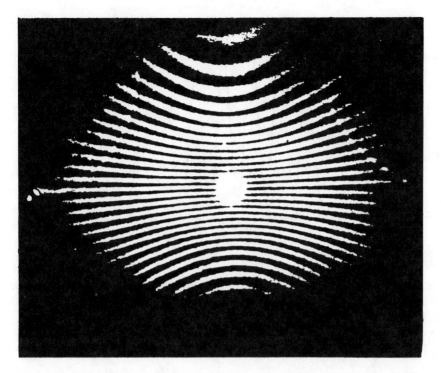

Figure 7-4 Far-field scattered light interference fringe pattern produced by a spherical particle.

that is produced appears to move at the Doppler difference frequency. The Doppler difference frequency is a function of the beam intersection angle, light wavelength, and velocity of the drop. The spatial frequency of the fringe pattern is dependent upon the angle of observation, drop index of refraction, beam intersection angle, laser wavelength, and drop diameter. Placement of a receiver lens to collect the scattered light will produce a Doppler burst signal, as shown in Fig. 7-5. Bachalo (1980a) showed that for off-axis light scatter detection, the lens acts as a length scale to measure the spatial frequency of the scattered fringe pattern. Integration of the fringe pattern over the receiver aperture produces a Doppler burst signal with a signal visibility that may be related to drop size. This relationship is shown in Fig. 7-6.

Although the method based on measuring the signal visibility partially eliminated the problems associated with intensity measurements and provided a means for the simultaneous measurement of the drop size and velocity with high spatial resolution, it has several shortcomings. As previously mentioned, the visibility of the scattered fringe pattern will be equal to unity if the scattered intensities from each beam are equal. Since the measured signal visibility depends on this being true, the incident laser beams must be of equal intensity in addition to having the same linear polarization direction and to being coherent. These requirements may be frustrated by optical imperfection and alternate attenuations of the beams by large drops. Because of the Bessel function relationship between the measured signal visibility and the dimensionless drop size, the error produced can be relatively large at the small size end of

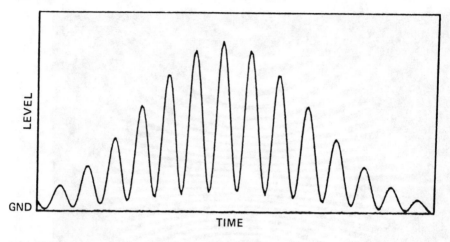

Figure 7-5 Laser Doppler burst signal.

the measurement range. Another difficulty associated with the instrument response function is the limited size range of a factor of 10 or less. The size range of typical particle size distributions extends over a factor greater than 10. Because it is required to have equal intensities from each beam incident upon the drop, the alignment of the system is also critical.

7-3 THE PHASE DOPPLER METHOD

Description of the Concept

The need to measure particles to sizes larger than approximately 50 μm in diameter, with good resolution, suggested the analysis of light scattered by reflection or refraction rather than by diffraction. The angular distribution of the light scattered by reflection and refraction is independent of the drop size except for the higher frequency resonant lobes produced by interference between the scattering components. The measurement of the energy scattered at angles away from the forward direction (greater than 20°) can be used to avoid the light scattered by diffraction. A detailed treatise of the scattering phenomena is given by van de Hulst (1957).

Bachalo (1980a) derived a theory for drop sizing utilizing the phase shift of the light transmitted through or reflected from spherical particles. Based on the analysis given by van de Hulst, the optical path length of a light ray passing through a sphere relative to a reference ray deflected at the center of the sphere was previously given as

$$\eta = 2\alpha(\sin \tau - pm \sin \tau') \tag{7-15}$$

where p is a parameter that characterizes the emerging rays and relates to the interface from which it emerges. For example, $p = 0$ for the first surface reflection, $p = 1$ for the transmitted ray, $p = 2$ for the ray emerging after one internal reflection, m is

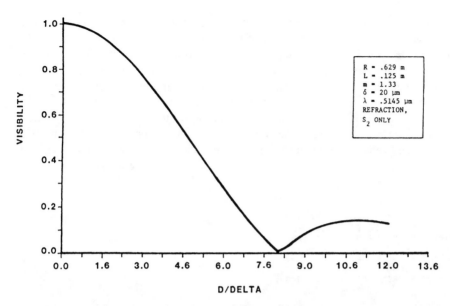

Figure 7-6 Signal visibility versus dimensionless particle size; $f/5$ and collection angle $30°$.

the relative index of refraction of the particle to the medium, and τ and τ' are the angles between the surface tangent and the incident and refracted rays, respectively.

The phase shift can be obtained using light-scattering interferometry produced with a standard dual-beam LDV. The rays from each beam are incident upon the particle at different optical angles and therefore reach the common points on the receiver by different optical paths. Neglecting the phase shifts at reflection and focal lines, the relative phase shift due to the differing optical paths is described as

$$\phi = \frac{2\pi d}{\lambda} [(\sin \tau_1 - \sin \tau_2) - pm(\sin \tau_1' - \sin \tau_2')] \qquad (7\text{-}16)$$

where the subscripts represent beams 1 and 2. Since the angles τ are fixed by the receiver geometry, the phase difference only changes as a result of the particle diameter d. These phase differences produce an interference fringe pattern that can be analyzed to obtain the size and velocity of the spherical particle. The temporal frequency of the fringe pattern is the Doppler difference frequency, which is a function of the beam intersection angle, laser wavelength, and velocity of the particle. The spatial frequency of the fringe pattern, which is linearly related to drop size, is dependent on the beam intersection angle, laser wavelength, drop diameter, angle of observation, and drop index of refraction. Therefore the mathematical description of the interference fringe pattern, which includes the effects of all the optical parameters, is required to infer particle size. With the complete theoretical description of the scattered fringe patterns for the appropriate parameters, there is no longer any need to calibrate the system for each measurement task. The drop size measurement can subsequently be obtained from the accurate measurement of the spatial frequency of the interference fringe pattern.

Three scattering regions of practical interest were considered to compute the fringe patterns for any selected set of optical parameters: forward scatter $10° \leq \beta \leq 50°$, backscatter $130° \leq \beta \leq 170°$, and $\beta = 90°$, where β is measured with respect to the transmitted beam direction. Light scattering by a combination of refraction and reflection at similar intensity will occur at some angles and under certain parametric conditions. Where this occurs, the spatial fringe pattern is no longer a pure sinusoidal intensity variation because of the multicomponent scattering interference. That is, additional interference between the refracted and reflected rays can occur and will produce significant errors in the measurements. Such errors can be minimized or eliminated with the proper selection of detection and processing methods.

```
DROPLET DIAMETER=          160.0000
RECEIVER FOCAL LENGTH= 636.0000
DIAMETER RECEIVER =     125.0000
THETA        =             0.0000
FRINGE SPACING    =        50.0000
LASER WAVELENGTH  =          .6328
INDEX OF REFRACTION =      1.3300
```

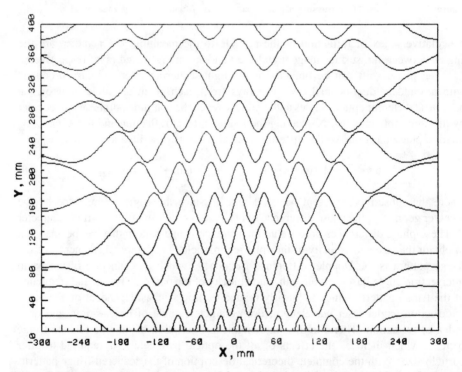

Figure 7-7 Interference fringe pattern computed for a spherical particle using geometrical optics theory. Droplet diameter, fringe spacing, and laser wavelength expressed in μm; receiver focal length and diameter in mm.

An example of the computed interference fringe pattern is given in Fig. 7-7. The sinusoidal fringe pattern was computed for a plane normal to the beam directions. Only the upper half of the symmetric fringe pattern was reproduced. As is observed in the figure, the fringes are hyperbolic curves showing a decreasing spatial frequency with distance from the beam axis at $x = 0$, $y = 0$. The pattern is also symmetric about $x = 0$. Thus the spatial frequency of the fringe pattern is dependent on where the measurement is made.

Measurement of the fringe pattern would be straightforward if the fringes were of relatively high intensity and of low temporal frequency. Unfortunately, the temporal frequency of the fringe pattern is essentially the Doppler difference frequency that will vary according to the speed of the drop. This frequency may be as high as 20 MHz in spray environments. The scattered light intensities will be low and will vary over several orders of magnitude. A receiver lens and either photomultiplier tubes or solid state detectors are required to provide the necessary sensitivity.

Measurement of the Interference Pattern

A direct means of measuring the interference pattern was developed by Bachalo and Houser (1984*a*). The scheme uses pairs of detectors located at known angles to the laser beam and separated by fixed spacings (Fig. 7-8). As the particle passes through the beam intersection region, it produces a scattered interference fringe pattern that appears to move past the receiver. Doppler burst signals are produced by each detector with a phase shift between them, as illustrated in Fig. 7-9. The signals in this figure have been high-pass filtered to remove the pedestal component. The phase difference between the detectors, ϕ_{12}, is then determined by measuring the time, τ_{12}, between the zero crossings of the signals from detectors 1 and 2, and dividing by the measured Doppler period. That is,

$$\phi_{12} = \frac{\tau_{12}}{\tau_D} 360° \qquad (7\text{-}17)$$

where the measurements are averaged over all the cycles in the Doppler burst signal.

Measurements of the phase shift are then related to particle size using the linear relationships derived from the calculations and shown in Fig. 7-10. In this figure the effect of changing the optical parameters, which include the laser beam intersection angle, collection angle, drop index of refraction, laser wavelength, and scattering component detected, is to simply change the slope of the linear response curves. That is, only the size scale is changed for the same range of phase angles, since all curves must pass through the origin.

It is possible to know if a spherical scatterer is a particle or a bubble from the movement of the interference fringes. Theoretical analysis of light scattering interferometry has revealed that the movement of the spatial interference pattern, as observed by the receiver lens, is in opposite direction for a liquid particle and a bubble (Bachalo and Sankar 1988; Breña de la Rosa et al. 1989).

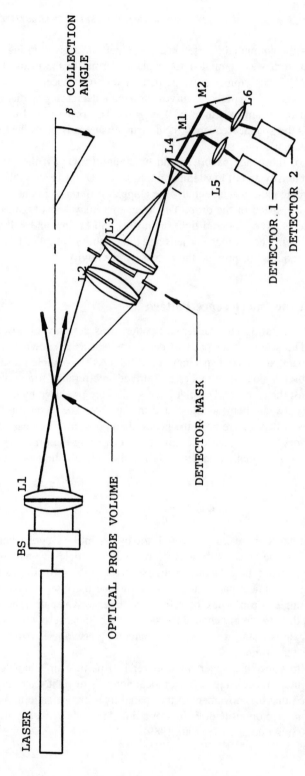

Figure 7-8 Optical component layout of the phase Doppler particle analyzer (top view). The direction of a monosize stream of drops is perpendicular to the plane of the figure.

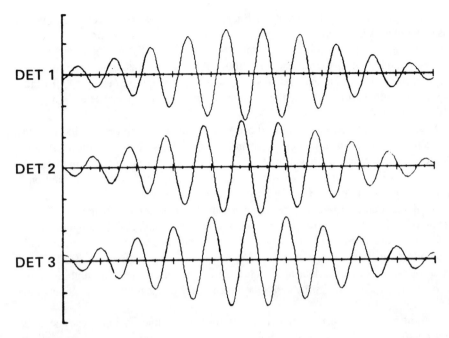

Figure 7-9 High-pass filtered signals.

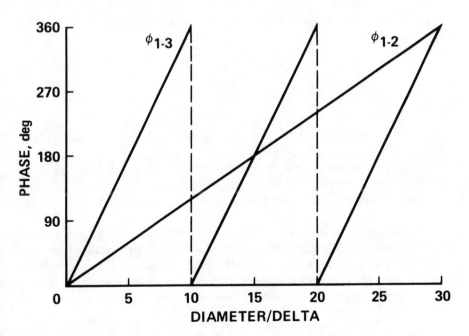

Figure 7-10 Theoretical prediction showing the phase variation with the dimensionless particle size.

The original ideas and concepts of Bachalo and Houser (1984a) evolved into the phase Doppler particle analyzer. This instrument uses three detectors to extend the measurement size range while maintaining good sensitivity. The two phase angles also serve as a redundant measurement for additional testing of the signals and to extend the size range sensitivity at one optical setting to a factor of approximately 100. Because drops scatter light in proportion to their diameter squared, the required detector response is 10^4 for a size range of 35.

Practical liquid sprays, such as those present in rocket or gas turbine engines, require for their study information regarding the spatial distribution of drop size, drop velocity, and interparticle arrival time. These spray systems exhibit very high particle number densities, which require processing, transferring, and storing information in computer memory at rates of the order of 100,000 samples/s.

Such high data rates are required to ensure that the drop arrivals, which are Poisson distributed, are not missed when they pass the sample volume at close intervals. For mass flow measurements it is important that the processing rate be sufficient to handle the smallest interparticle arrival times. Average data rates as high as 5000–15,000 particles/s have been experienced in actual spray measurements. The highest arrival rate or interparticle arrival time can vary by an order of magnitude from the average value.

Assuming that the airflow and spray are thermodynamically in a steady state (or using conditional sampling in unsteady flows), the time averages are used to obtain the mean and rms fluctuating velocities for each size class. That is, all of the measurements of particle size, velocity, and time of arrival are stored in memory. The data can be processed to form velocity probability density functions (pdfs) for each size class. A mean and rms velocity can then be obtained for each particle size class i using

$$u(d_i) = \sum_{j=1}^{N(d_i)} \frac{U_i(d_i, t)}{N(d_i)} \tag{7-18}$$

and

$$\langle U'(d_i) \rangle = \left(\frac{1}{N(d_i)} \sum_{i=1}^{N(d_i)} [U_i(d_i, t) - u(d_i)]^2 \right)^{1/2} \tag{7-19}$$

where the summation is taken over all the particles in size class i, $N(d_i)$ is the number of particles sampled in size class i, and $U_i(d_i, t)$ are the velocities of individual drops of size class i. Typically, 10,000 or more particle measurements should be acquired at each point in the flow field.

The measurement of the transverse velocity components and in regions with recirculation usually requires frequency shifting. Frequency shifting causes the interference fringe patterns to appear to move at the shift frequency. This frequency offset allows the measurement of the small transverse velocity component and the resolution of the directional ambiguity in recirculating flows. With frequency shifting, the velocity is

$$U_i(d_i, t) = \lambda \frac{f_D(d_i, t) - f_s}{2 \sin \gamma/2} \tag{7-20}$$

where $f_D(d_i, t)$ is the Doppler difference frequency, f_s the shift frequency, and γ the beam intersection angle. Frequency shifting also serves to compress the Doppler frequency bandwidth, which allows the processing of an increased turbulence frequency bandwidth.

Subsequently, the individual particle size and velocity measurements taken by the instrument are accumulated in histogram form to obtain the overall size and velocity distributions and the various mean diameters.

Modeling of the Light-Scattering Phenomena

It is important to recognize that, unlike the LDV, the phase Doppler technique requires calibration curves to convert the phase differences between the outputs of the detectors to particle diameters. Furthermore, if the phase Doppler technique is to function properly, these calibration curves should be linear over the entire range of particle diameters of interest to the investigator. Since the scattered interference fringe pattern is dependent on the transmitting and receiving optical configuration of the phase Doppler technique as well as on the refractive index of the scatterer, it may not be possible to obtain linear calibration curves for any arbitrary setup. Ideally, if we can identify a scattering angle for which only one of the scattering components (for example, reflection or refraction) is present, then we are assured of linear calibration curves. In practice, this is seldom possible, for different scattering components are invariably simultaneously present and the best that one can hope for is that one of the scattering components is dominant. In order to identify the optical configurations for which the phase Doppler technique will operate optimally and also to generate calibration curves at such configurations, theoretical models have been developed to simulate the various features of the phase Doppler technique as closely as possible.

Mie theory versus geometrical optics theory. Two independent light-scattering models of the phase Doppler technique were developed using both Lorenz-Mie theory and geometrical optics theory, each having their individual advantages and disadvantages. Mie theory is exact and completely describes the scattering of plane electromagnetic waves by spherical particles. It is based on the solution of complete electromagnetic wave equations along with appropriate boundary conditions (van de Hulst 1957; Born and Wolf 1975). On the other hand, geometrical optics (or ray optics) is approximate, and its solutions asymptotically approach those of the Mie solutions for particles very much larger than the incident light wavelength. Van de Hulst (1957) showed that for $\pi d/\lambda \gg 1$ it is possible to approximate Mie scattering by the interference of diffracted, refracted, and reflected rays, which is the basis for the geometrical optics approach.

The geometrical optics approach, though approximate, provides one with a much greater physical insight into the complex problem of light scattering than does the Mie theory, where one tends to get lost within the mathematics. Furthermore, the computational efficiency of the geometrical optics method is far superior to that of the Mie theory. Mie theory is based on series solutions where the number of terms required for convergence is directly proportional to the size of the particle. This implies that the computation time increases with particle size. On the other hand, the computation time of the geometrical optics theory is independent of particle size. The higher computation speed of the geometrical optics theory therefore makes it economically viable to compute the amplitude and phase of the scattered light over a fine grid on the receiver aperture, especially for large sized particles.

In several practical applications of the phase Doppler technique, the laser beam diameters at the probe volume are required to be of the order of the particle diameter itself. In such situations, the nonuniform (Gaussian) illumination of the laser beam cannot be ignored. The geometrical optics approach possesses a significant advantage over Mie theory in handling such cases. Nonuniform illumination, on the other hand, violates a fundamental assumption of Mie theory, and therefore the theory needs modification. Recently, such modifications to Mie theory have been made by some researchers (Glantschnig and Chen 1981).

The geometrical optics theory, in spite of all its advantages over Mie theory, does have certain disadvantages. For instance, it is not valid for the analysis of light scattering by small particles, and the predictions of the theory when used for such purposes is open to question. Furthermore, the geometrical optics theory solutions fail near the regions of rainbow and glory and therefore are generally not very suitable for backscatter angles. After careful consideration of the advantages and disadvantages of the two approaches, theoretical models of the phase Doppler technique were developed based on both approaches. Of these two, the Mie theory based model was generally used for small sized particles (<20 μm) and the one based on the geometrical optics approach was used for larger particles. The models basically differ only in the way the scattering amplitude functions $S_1(m, \theta, d)$ and $S_2(m, \theta, d)$ are computed for each of the incident beams.

In order to demonstrate, theoretically, the creation of spatial fringe patterns by the interference of scattered light, the developed models were used to map out the light intensity distribution as seen by the receiver lens (the interested reader is referred to Bachalo and Sankar (1988) for a detailed analysis and discussion on this topic).

Figure 7-11 shows a computer-generated spatial fringe pattern that is formed on the receiving lens due to the scattering of light by a 40-μm water droplet. The assumption of pure refraction ($p = 1$ only) has been made in the calculations. The circle in Fig. 7-11 delineates the receiving lens. This lens collects the scattered light, and a system of optics images the collected light from three different areas of the lens on to three separate photodetectors. The parallel vertical lines in Fig. 7-11 show the three areas of the receiving lens. Apart from the presence of distinct fringes in the x' direction (see Fig. 7-12), a gradual intensity variation in the z' direction can also be seen in this figure. These secondary fringes are responsible for causing oscillations in the calibration curves, and their presence can be physically understood as arising due to the interference of one scattering order with another, for example, refraction with reflection. Also observable in the figures is a certain randomness in the intensity distributions, which also contributes to oscillations. These undesirable contributions can be treated as "noise," and the receiving lens performs a reasonable job of integrating them away and yielding linear calibration curves for certain optical configu-

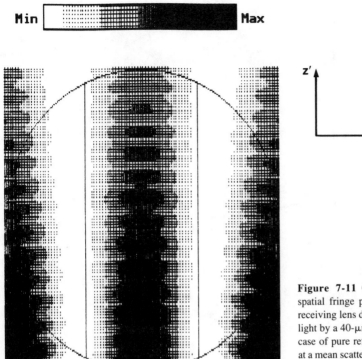

Min [ ] Max

z'

x'

Figure 7-11 Computer-generated spatial fringe pattern formed on the receiving lens due to the scattering of light by a 40-μm water droplet. Ideal case of pure refraction ($p = 1$ only) at a mean scattering angle of 30°. The circle delineates the receiving lens.

rations. The wavelength of the spatial intensity variation in the x' direction is directly and linearly related to the size of the particle. Furthermore, it is this wavelength that we are attempting to measure indirectly by determining the phase differences between the outputs of the different detectors.

The simultaneous presence of external reflection, refraction, and second internal reflection at 30° results in a degradation of the fringe pattern. This effect can be seen in Fig. 7-13, which is a computed fringe pattern assuming interference by reflection, refraction, and second internal reflection. It is clear from this figure that the presence of the additional scattering components, namely, reflection and second internal reflection, has led to modulation of the intensity in the z' direction. Note, however, that the intensity variation in the x' direction remains distinct and therefore still carries particle size information that can be easily retrieved.

Results similar to those discussed above were obtained for various scattering angles between 20° and 80°. For all angles that were investigated in this forward scatter region, it was possible to obtain useful linear calibration curves by integrating the fringe pattern over the lens surface.

Backscatter angles ($\approx 150°$) provide another region of significant practical applications. For water droplets, however, this region corresponds to the main rainbow

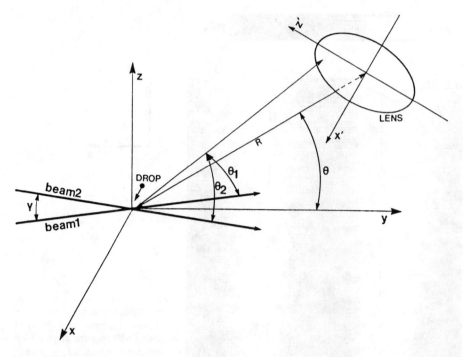

Figure 7-12 Coordinate system used in the model.

region, where external reflection and two components of first internal reflection are present. Our experimental investigations at the backscatter (150°) angles have resulted in good agreement with the forward scatter (30°) data (Fig. 7-14).

Calibration curves were initially generated using just three points on the receiver lens for two different cases, namely, first internal reflection ($p = 2$) only and combined external and internal reflections ($p = 0$ and $p = 2$). The interference of external reflection with the first internal reflection gives rise to high-frequency oscillations with phase varying from 0° to 360°.

The effect of performing spatial and temporal integrations is shown in Fig. 7-15. An integration mesh of size 100×20 was used for these spatial integrations, and five time steps were used for performing the time averaging. It is interesting to note that the process of integration has indeed been able to damp out high-frequency oscillations, especially for the larger particles, leaving behind only small oscillations in the region of small particles (<10 μm in diameter). Comparing the integrated calibration curves with the linear regression fits presented in Fig. 7-15, it is clear that light-scattering collection at 150° is certainly adequate for sizing water droplets using the phase Doppler method.

The small oscillations seen in the calibration curves for the smaller sized particles could be due to the fact that the mesh size used for performing the spatial integrations

Min ▭ **Max**

Figure 7-13 Computer-generated spatial fringe pattern for a 40-μm water droplet. Simultaneous presence of external reflection, refraction, and second internal reflection at a mean scattering angle of 30°.

was not sufficiently fine. A more probable cause is that the geometrical optics is not a good approximation of Mie theory for very small particles, especially near the rainbow region. It remains to be seen whether integration of the Mie-scattering solutions over the receiver surface can remove these oscillations.

Basic Evaluation of the Method

The particles necessary to evaluate the phase Doppler method and to verify the theoretical predictions were generated from three different devices: the monodisperse drop generator, a spray nozzle, and a spinning disc atomizer.

The range of drop sizes obtainable with the monodisperse drop generator was roughly 90–160 μm. Not only could this device produce a known, variable monodisperse drop field, but the streamwise nature of the drops also allowed them to be selectively directed through specific parts of the probe volume.

The highly controllable nature of the monodisperse drop generator was necessary for the verification of the theory but did not represent the true environment of most drop fields. Polydispersions of drop sizes and a certain randomness of drop trajectory, velocity, and arrival at the probe volume are more characteristic of a spray environment.

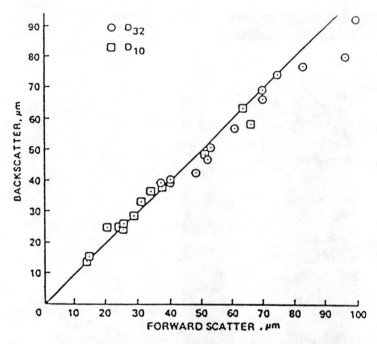

Figure 7-14 Experimental comparison of backscatter and forward scatter results.

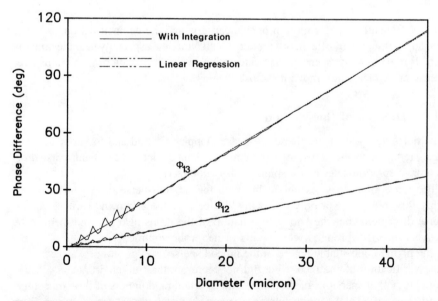

Figure 7-15 Computed calibration curves for 150° mean scattering angle showing the effect of performing spatial and temporal integration of the spatial intensity pattern.

Unfortunately, there are no "calibrated" drop sources of the type described, due in part to the lack of instrumentation capable of performing such calibrations.

To better simulate the actual spray environment, several spray nozzles were used. The water supply used in the laboratory had a maximum line pressure of 551 kPa. The pressure was regulated to produce varying spray characteristics. The nozzles used were low to moderate flow rate (1.9–190 L/h), solid cone type. High flow rate sprinkler nozzles were sometimes used to produce larger drops in a hollow-cone spray pattern.

A spinning disc atomizer was utilized for one set of tests. This device is used for room humidification and is known to generate very small drops (approximately 4–30 μm in diameter) in a very narrow distribution.

7-4 PRACTICAL APPLICATIONS OF THE PHASE DOPPLER METHOD

Sample Volume Characterization

The accurate determination of the sampling cross section of a single particle-counter instrument using laser beams with Gaussian beam intensity distributions is difficult. By now, it is well known that the sample cross section changes with particle size, measurement conditions, and instrument setup parameters (Fig. 7-16).

This effect on the sampling statistics was recognized a number of years ago, and corrections to the particle size distribution were made based upon a theoretical description of the light intensity distribution in the sampling cross section. That is, the beam intensity distribution orthogonal to the direction of propagation z is given by

$$I_s = I_0 \exp \left(-2 \frac{a^2}{b_0^2} \right) \tag{7-21}$$

where a is the diameter variable and b_0 is the diameter at which the intensity falls to $1/e^2$ of the peak intensity, assuming an ideal Gaussian beam. Recognizing that the particles scatter light in proportion to their diameter squared, and that there is a minimum signal amplitude that can be detected, has been used with the above expression to theoretically define the sample cross-section diameter $y(d_i)$ as a function of particle diameter. The resulting equation is

$$y(d_i) = \left\{ y_{min}(d_{min})^2 + \frac{b_0^2}{2} \ln \left[\left(\frac{d_i}{d_{min}} \right)^2 \left(\frac{\mathcal{V} + 1}{\mathcal{V}_{min}} \right) \right] \right\}^{1/2} \tag{7-22}$$

where the subscript min implies the values at the minimum detectable limit and $\mathcal{V}_{min}$ is the signal visibility corresponding to d_{min}. This result is essentially the theoretical curve shown in Fig. 7-16. As previously noted, this approach is influenced by the measurement conditions.

Early in the development of the phase Doppler method, Bachalo and Houser (1984a, b) recognized that the light-scattering interference fringes formed an intrinsic scale with which the diameter of the sampling cross section could be measured as a

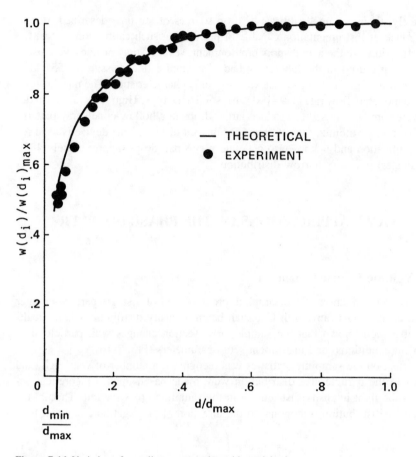

Figure 7-16 Variation of sampling cross section with particle size.

function of the particle diameter (Fig. 7-17). By counting the number of cycles in the Doppler burst signal, the desired information can be obtained. Although the particles will pass the beam on random trajectories, the greatest number of them will pass through the trajectory with the maximum number of fringe crossings $N_{max}(d_i)$, followed by a lower likelihood for those particles crossing $[N_{max}(d_i) - 1]$ fringes, and so on.

The maximum number of fringe crossings $N_{max}(d_i)$ for particles of diameter d provides the desired beam diameter; that is,

$$D(d_i) = N_{max}(d_i) \, \delta \qquad (7\text{-}23)$$

where the fringe spacing $\delta = \lambda/2 \sin(\gamma/2)$ and γ is the beam intersection angle. Since there is a minimum number of fringe crossings required for signal processing, the actual width $w(d_i)$ of the measurement cross section that forms the sampling area is

$$w(d_i) = \{y(d_i)^2 - [N_{min}(d_i) \, \delta]^2\}^{1/2} \qquad (7\text{-}24)$$

where $N_{min}(d_i)$ is the minimum number of cycles required for processing the signals.

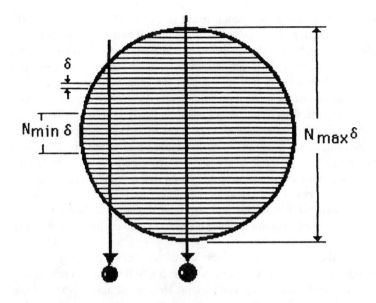

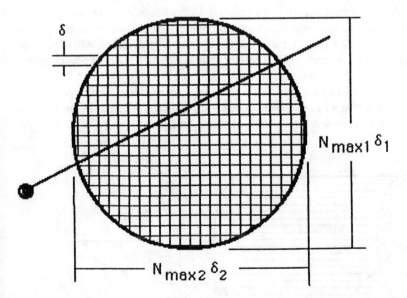

Figure 7-17 Schematic of the sample volume showing the implicit fringe pattern. (Top) Orthogonal particle trajectories. (Bottom) Fringe pattern for a two-component system for oblique trajectories.

This method, shown as the experimental result in Fig. 7-16, works very well, since it does not require a priori the knowledge of the optical parameters determining the beam waist diameter except for the intersection angle and laser wavelength, and accounts for all the variances in the beam due to the measuring environment. The main disadvantage of the method is that it is dependent upon the flow angle being orthogonal to the fringe pattern. In most applications, this does not represent a serious problem. In cases where the flow angle is unknown and may be quite different for each particle size class, the flow direction must be measured. With a two-component phase Doppler instrument, this is done automatically. Figure 7-17 shows the orthogonal fringe patterns that are used to generate the resultant particle path lengths through the beam. The measured sampling cross section is then used in the calculations of number density and volume flux. For a one-component system, each of the two velocity components can be measured separately, and the flow direction deduced for each particle size class. The sample volume is then corrected for the effects of the flow angle.

A slit aperture in the receiver unit, which serves to control the length of the sampling cross section (Fig. 7-18), is imaged across the focused beams. In this figure, β is the detection angle, and s the slit aperture width. How accurately this dimension can be defined depends largely upon the resolution of the receiver lens system. The $f/5$ receiver lens used in the phase Doppler instrument is a custom-made, air-spaced triplet with a nominal resolution of 15 μm. Hence the blur spot or point spread of this lens will produce an uncertainty of approximately 15% when using a 100-μm aperture. However, the effective spot size will vary with particle size, since the energy distribution in the blur circle is nearly Gaussian and the radius of detection will depend on the amplitude of the scattered light. Therefore smaller particles will produce smaller uncertainty in the length of the sampling cross section.

The effect of the lens blur spot on the sampling cross section can be seen in Fig. 7-19 (Payne, Crowe, and Plank 1986). In this study a monodispersed stream of drops of known size was traversed through the sampling cross section using a micrometric

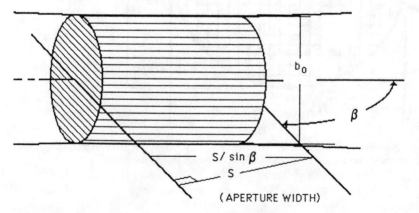

Figure 7-18 Schematic showing the definition of the sample volume length.

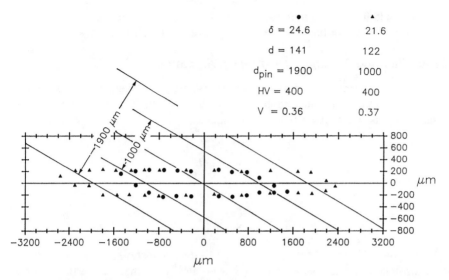

Figure 7-19 Direct measurement of the sampling cross section showing the effect of the receiver blur spot. The x and y coordinates represent the dimensions along and across the sampling cross section, d is the diameter of the particle in μm, d_{pin} the diameter of the pinhole in μm, HV the photomultiplier high voltage, and $\mathcal{V}$ the signal visibility.

traversing system. Theoretical dimensions of the sampling cross section are delineated on the data plot and given by the x and y coordinates, which represent the dimension along and across the sampling cross section, respectively. The diameters of the circular pinholes used for the experiments were d_{pin} = 1900 and 1000 μm, for the two cases studied here. The photomultiplier high voltage was set at HV = 400 V, and the signal visibility yielded $\mathcal{V}$ = 0.36 and 0.37. The edge of the sampling cross section was defined as the point where the data rate dropped to approximately 10% of the maximum. Detection of particles passing on trajectories beyond the limits set by the slit aperture is due to the blur circle. These data represent the worst case, since the drops were at the end of the selected size range. Since it is the large particles that produce the greatest uncertainty, the mass flux, which is heavily weighted toward the large drops, will be most seriously affected.

In actual sprays the particles cover a wide range of sizes; therefore the effective size of the probe volume cross section is determined for each size class, and the individual bin count is normalized to the largest probe volume cross section found. This results in a bin count increase for some size classes. The basis for this normalization lies in the fact that larger particles can be detected over a larger spatial area more easily by the signal processor because they scatter much more light than do smaller particles. As a result, larger particles that transit the probe near the lower intensity edges of the Gaussian intensity profile may produce adequate signal levels, whereas smaller particles must transit nearer the high-intensity center of the probe to generate adequate signals. This disparity in effective probe volume cross section is removed by the normalization criteria. The raw (uncompensated) bin counts are shown as solid bars within the histogram bars, whereas the corrected bin counts for each size class are given by the ordinate in the size histogram of Fig. 7-20.

Volume Flux and Number Density Determination

Volume flux is determined from the measured volume mean diameter D_{30}, sampling cross sectional area A, and the number of particles counted N, in the sampling time interval t. That is, the volume flux is calculated as

$$F = \frac{\pi}{6} D_{30}^3 \frac{N}{At}$$
(7-25)

where

$$N = \sum_{i=1}^{n} N(d_i)$$
(7-26)

$N(d_i)$ is the number of particles in size class i and the summation is taken over all the particle size classes. The mass flux can be determined from the volume flux and the density of the particles. The quantity D_{30} is influenced most by the largest particles in the distribution. Because size distributions are highly skewed with a relatively small population in the larger size classes, it is important that a large number of particles be measured to obtain good statistical representation. Furthermore, D_{30} is cubed, so the accurate determination of this quantity is most important in realizing good flux measurements.

Local particle number density is also determined with the instrument. However, uncertainties in the number density are not as dependent on the lens blur spot size, since the small particles that constitute the greatest population have the smallest blur circle. The number density is calculated from the measured number of particles N passing the sampling cross section. If the particles can be assumed to be traveling in the same direction (no reversed flow), the swept volume for particle size class i, $V(d_i)$ may be determined from

$$V(d_i) = |u(d_i)| \, tA$$
(7-27)

where $u(d_i)$ is the velocity of the particles in size class i, t is the total sample time, and A is the normalized sampling area.

The particle number density is then obtained from

$$N_d = \frac{1}{A} \sum_{i=1}^{n} \frac{N(d_i)}{|u(d_i)|} \left(\sum_{j=1}^{N(d_i)} \Delta t_{ij} \right)^{-1}$$
(7-28)

where the summation is taken over all particle size classes n and Δt_{ij} is the interarrival time between particles of size j within the ith velocity bin.

One of the major advantages of using optical probes to measure mass flux is that the data can be obtained with good spatial and temporal resolution in flows with high swirl, recirculation, and pulsations, and in reversing flows. Unlike continuous phase flows, two-phase flows can have particles of different sizes moving in different directions. With frequency shifting, the velocity of the particles can be measured in a

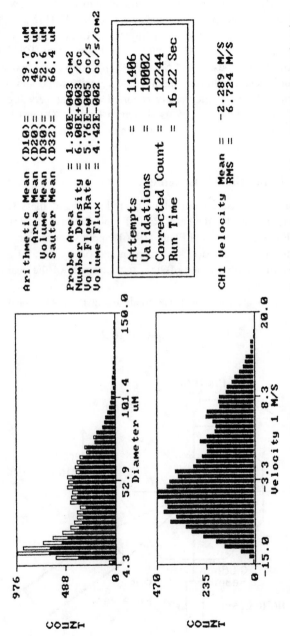

Figure 7-20 Particle size distributions showing the corrected and uncorrected particle counts to account for variations in the sample volume size.

plane orthogonal to the beams (a three-component system is needed in three-dimensional flows). The flux for each particle size class or for all size classes can be determined in the direction of interest. For the number density, it is the average spacing between particles that is needed to establish the swept volume V. However, the particles can be moving in a wide range of angles of trajectory. In fact, particles in the same size class have been found (McDonell, Cameron, and Samuelson 1987; Jackson and Samuelson 1987; Bachalo, Houser, and Smith 1987) to move in opposite directions in some cases. Under these conditions, the average velocity could approach zero and cause the number density to be incorrectly reported as being very high or even to diverge. Therefore the absolute velocity as shown in Eq. (7-28) must be used in the calculations. Once again, motion in the third dimension can produce measurement error when only two components of velocity are measured.

Methods for Verifying the Measurements

To provide comparative measurements of number density and mass flux, a simple light extinction system, used by Bachalo, Rudoff, and Breña de la Rosa (1988) and fully described by Payne, Crowe, and Plank (1986), and a sampling probe were used. The schematic of the light extinction system mounted on the spray rig is shown in Fig. 7-21. This system acquired a line-of-sight measurement of the beam attenuation on the optical path.

Determination of the transmittance was made using Beer's law, namely,

$$T = \frac{I}{I_0} = \exp\left(-N_d \sigma_e L\right) \tag{7-29}$$

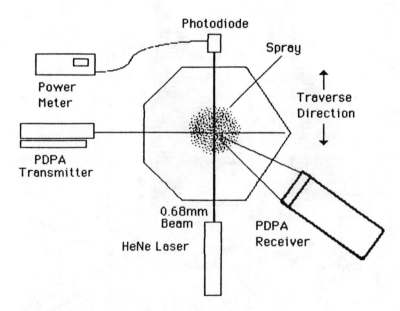

Figure 7-21 Schematic of the laser light beam extinction system.

where N_d is the particle number density, σ_e the extinction cross section, and L the optical path length. The extinction cross section may be related to the measured drop area mean diameter D_{20}, which is the diameter of a drop that would have the same area as the entire spray, via the following relation:

$$\sigma_e = \frac{\pi}{4} \overline{Q}_e D_{20}^2 \tag{7-30}$$

where $\overline{Q}_e$ is the mean extinction efficiency. For large spheres, $\overline{Q}_e = 2$, which was the case considered here, since the minimum drop size in the distribution was 5 μm.

The phase Doppler instrument provided D_{20} at points along the optical path that could be integrated to estimate the transmittance. This value was then compared with the measured transmittance value of the extinction system as a consistency check on the number density determined by the phase Doppler instrument.

Mass flux comparisons were made using a simple sampling probe having a sharp-edged orifice area of 0.24 cm². The samples were collected immediately below the optical sample volume. Direct measurements of the particle size and velocity of the drops near the mouth of the probe showed only minor deflection of relatively few small drops due to nonisokinetic sampling. Thus probe perturbation was insignificant to the flux measurement.

Measurements of particle number density and mass flux. Figure 7-22 shows the radial distributions of particle number density and volume flux obtained at 10 cm below the exit plane of a Hago 3-gph, 45° solid-cone pressure atomizer, spraying water and operating at 415 kPa. Also shown in Fig. 7-22 (bottom) is the volume flux obtained with a sample probe having a sharp-edged orifice area of 0.245 cm² and located immediately below the optical sample volume. The results showed an agreement between the volume flux measurements within 10%.

A high number density water spray was produced by an array of four identical nozzles, Hago 3-gph, 45° solid cone also operated at 415 kPa. Each nozzle was directed at an inward angle to produce a uniform impinging spray with a peak number density of approximately 7500 particles/cm³. Figure 7-23 shows the radial distributions of particle number density, D_{30}, and volume flux at 10 cm from the nozzle exit plane. It should be noted that the size of the largest drops being measured in the distribution dictates the minimum allowable size of the focused beam diameter. Care should be exercised not to choose a beam diameter that gives large sample cross sections, since coincident particle occurrences can take place in the sample volume, especially in high-density sprays like the one depicted here.

Liquid nitrogen (LN_2) sprays were investigated as a convenient alternative for the study of evaporating sprays in nonreactive conditions. The tests used a pressure atomizer Hago 10-gph, 45° solid cone, operated at 690 kPa. Comparative measurements of number density were provided by a light extinction system described by Bachalo, Rudoff, and Breña de la Rosa (1988). This system acquired a line-of-sight measurement of the beam attenuation on the optical path. The transmittance was determined from Beer's law of light extinction, the drop area mean diameter D_{20}, the particle number

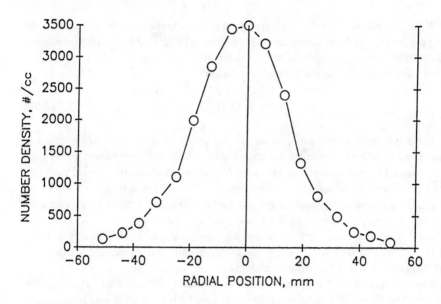

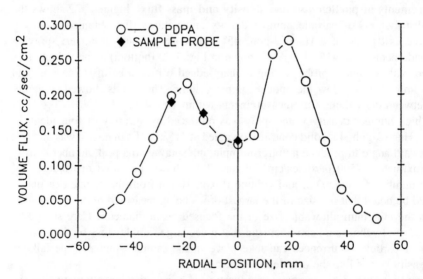

Figure 7-22 Measurements for a simple pressure atomizer operated at 415 kPa. (Top) Radial distribution of number density. (Bottom) Radial distribution of volume flux.

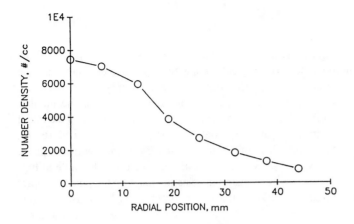

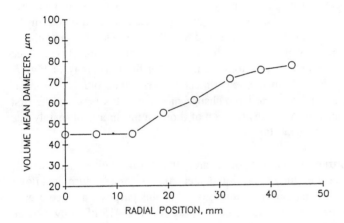

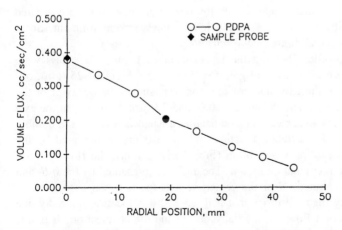

Figure 7-23 Results from the four-nozzle array of pressure atomizers operated at 415 kPa and measured at $X = 10$ cm below face of atomizer. (Top) Radial distribution of particle number density. (Center) Radial distribution of D_{30}. (Bottom) Radial distribution of volume flux.

density, and the extinction cross section. The phase Doppler instrument provided the D_{20} and the particle number density at points along the optical path that could be integrated to estimate the transmittance. This value was then compared with the measured transmittance of the extinction system as a consistency check on the number density determined by the phase Doppler instrument. Figure 7-24 (top) shows the radial distribution of number density taken at three axial locations, $X = 5$, 10, and 16 cm, from the nozzle and a table with values of the transmittance obtained with the phase Doppler instrument and the extinction system. It should be noted that the table in Fig. 7-24 (top) shows that at $X = 16$ cm the transmittance estimated with the phase Doppler technique was 0.95 compared with 0.76 of the extinction system. The reason for this lower value was that the cold LN_2 had chilled the air and condensed the water vapor in the air surrounding the nitrogen spray, causing further extinction and hence a lower transmittance at this location (the extinction system produces a line-of-sight measurement). In this experiment the measurement of the condensed water droplets in the surrounding air of the LN_2 spray was not performed with the phase Doppler instrument.

Figure 7-24 (bottom) is a plot of the radial distribution of volume flux at three axial locations $X = 5$, 10, and 15 cm from the nozzle for the LN_2 spray of Fig. 7-24 (top). No attempt was made in this experiment to verify the volume flux measurements with the sample probe. The large change in volume flux between the first two axial stations was caused by a redistribution of drops in the spray but mainly due to the high evaporation rate of the liquid.

Turbulent air-liquid interactions. One of the areas that has been given considerable attention at Aerometrics is the study of air-liquid interaction in complex turbulent flows typical of liquid rocket and gas turbine cumbustors. For this purpose, a Hago 3-gph, 45°, hollow-cone pressure atomizer was used to spray water at 415 kPa downstream of a bluff body. The bluff body consisted of a circular disk 76 mm in diameter mounted on the nozzle supply tube. This assembly was placed in a wind tunnel with an air velocity of 21 m/s. The results obtained with the nozzle-bluff body were compared with those obtained with the nozzle spraying into a quiescent environment. Some important points of this research are now discussed.

Detailed velocity profiles showing the response of each drop size class were measured for the nozzle-bluff body configuration (Fig. 7-25). Figure 7-25 (top), obtained at $X = 5$ cm from the exit plane of the atomizer, shows the behavior of the spray (9.0- to 25.5-μm drops) and the airflow (0.6- and 2.0-μm drops). Near the axis the average drop velocities were near zero, indicating a balance between the number of drops injected in the downstream direction and those being recirculated by the airflow. Further downstream at $X = 10$ cm (Fig. 7-25, bottom), the flow was approaching the end of the recirculation region. The airflow represented by the 0.6- and 2-μm particles was decelerated. Near the centerline axis the reversed airflow had a greater negative velocity than the drops that had reached this location, partially due to the external recirculation flow and partially due to the direct injection. It is this difference between the airflow velocity and the drop velocity that provides the drag force that returned the drops toward the atomizer.

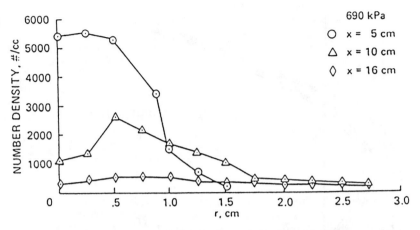

(cm)	Transmittance (%)		
Axial	Extinction		Percent
Position	System	PDPA	Difference
5	0.82	0.83	-1.2
10	0.90	0.90	0.0
16	0.76	0.95	-20.0

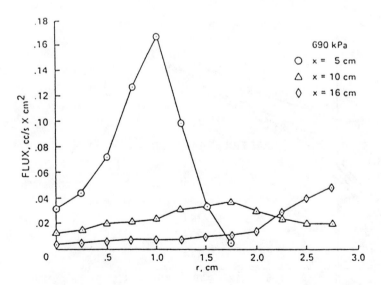

Figure 7-24 Results of liquid nitrogen spray measurements. (Top) Radial distribution of number density. (Bottom) Radial distribution of volume flux.

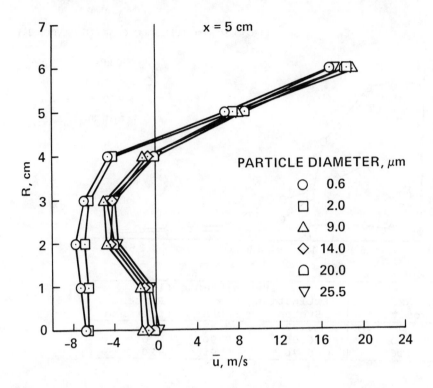

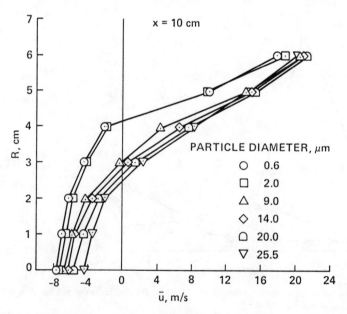

Figure 7-25 Mean axial velocity for discrete size classes in the wake of a disk-shaped bluff body. (Top) $X = 5$ cm. (Bottom) $X = 10$ cm.

264

Figures 7-26 and 7-27 show the drop size and drop velocity distributions and the size-velocity correlations for a single nozzle spraying into quiescent environment and the nozzle-bluff body configuration operating at the same previous conditions. The results shown correspond to $X = 5$ and 10 cm from the exit plane of the nozzle at the centerline. The effect of the recirculation caused by the bluff body is appreciable, with the drop sizes showing a bimodal distribution due to the combination of injection and recirculation. Further downstream at $X = 10$ cm (Fig. 7-27) the nozzle-bluff body configuration yielded a narrow distribution of velocities with practically constant average velocity of $u = -5$ m/s, while the single nozzle showed, as expected, a wider velocity distribution with the larger drops having the higher velocities.

Figure 7-28 is a plot similar to the previous two figures, except that the measurements were taken at $X = 5$ cm, $R = 1.5$ cm for the nozzle spraying into a quiescent environment, and $X = 5$ cm, $R = 2$ cm for the nozzle-bluff body configuration. Both of these conditions represent locations close to the periphery of the spray sheath. It is observed that the effect of the recirculation induced by the bluff body is not as strong as it appears at the centerline. Further downstream, at $X = 10$, $R = 4$ cm (not shown) the average velocity yielded, practically, a constant value for the nozzle-bluff body configuration.

Figures 7-26 through 7-28 clearly show the drastic differences in the drop size/drop velocity distributions and in the size-velocity correlations in a practical spray environment, and the errors that could be incurred in predicting the aerodynamics of actual flows by analyzing the atomization characteristics of simplistic single-nozzle configurations operating in a quiescent environment.

Particle Drag

The drag coefficient can be determined by performing a balance between the pressure drag forces and the inertial forces acting upon a decelerating drop. In terms of C_D, the well-known equation representing this balance is

$$C_D = \frac{4}{3}\left(\frac{\rho_d}{\rho_g}\right)\left(\frac{d}{U_r^2}\right)\left(\frac{dU}{dt}\right) \tag{7-31}$$

where ρ_d and ρ_g are the drop and gas densities, respectively, d is the drop diameter, dU/dt is the drop acceleration, and U_r is the relative velocity between the drop and the surrounding gas. The assumptions that must be made to obtain this equation are as follows.

1. No change in the state of the gas due to drop injection.
2. The drops remain spherical during their lifetime.
3. Gravitational effects are negligible.
4. Momentum flux due to evaporation is negligible.

With respect to the experiment presented here, these assumptions were believed to be reasonable due to the relatively low accelerations, velocities, and volatilities of the test fluids.

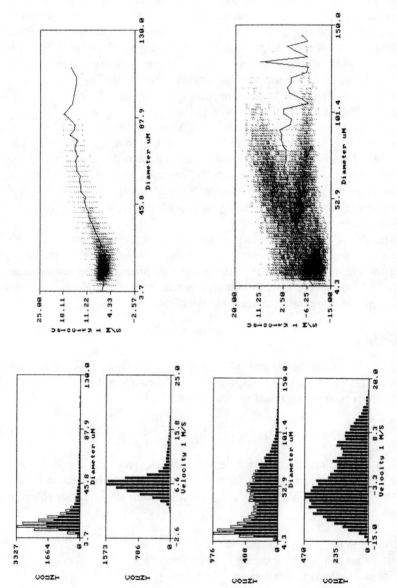

Figure 7-26 Drop size and drop velocity distributions and size-velocity correlations at $X = 5$ cm, $R = 0$. (Top): Nozzle spraying into quiescent air. (Bottom) Nozzle-bluff body configuration.

266

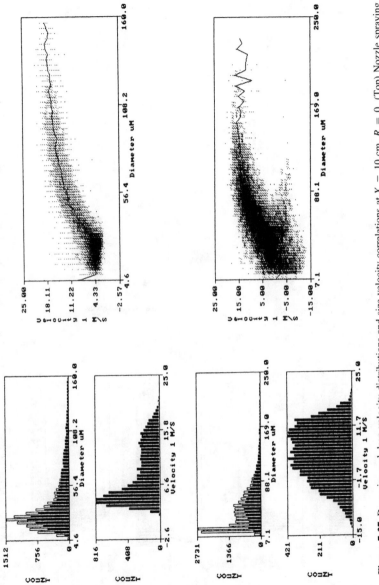

Figure 7-27 Drop size and drop velocity distributions and size-velocity correlations at $X = 10$ cm, $R = 0$. (Top) Nozzle spraying into quiescent air. (Bottom) Nozzle-bluff body configuration.

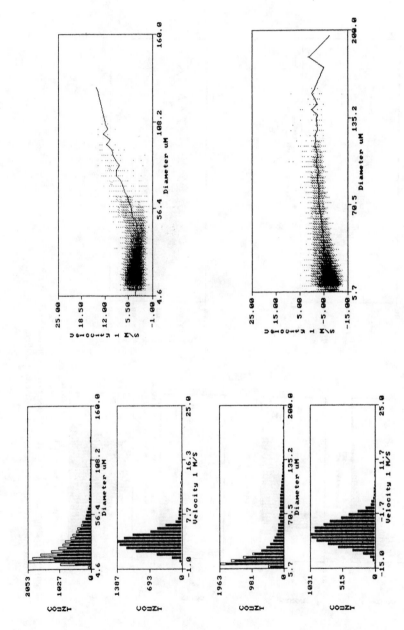

Figure 7-28 Drop size and drop velocity distributions and size-velocity correlations. (Top) Nozzle spraying into quiescent air at $X =$ 5 cm, $R = 1.5$ cm. (Bottom) Nozzle-bluff body configuration at $X = 5$ cm, $R = 2$ cm.

In this study the polydispersed spray from nozzles located 1 m upstream of a polished cylinder 63.5 mm in diameter with an afterbody was utilized to make the drag measurements (Fig. 7-29). The cylinder was anodized black to minimize the reflections of the laser beams. The afterbody was fitted to reduce the oscillations of the stagnation streamline.

The cylinder was located at the aft end of the two-phase flow wind tunnel test section, which allowed a uniformly turbulent coflow to interact with the spray (Fig. 7-30). The wind tunnel has a test section 2 m long with clear acrylic walls for optical access. Axial position was determined with a dial gage accurate to 0.025 mm. All measurements were performed at midspan of the cylinder.

Different screens to produce various degrees of turbulence can be inserted at the upstream end of the test section to generate an isotropic turbulent flow. The grid chosen for this work was rectangular, with a 25-mm-square cross section, spaced 50 mm apart from edge to edge. This gave an open area of 56% and turbulence intensity levels of the order of 8–12% based upon the free stream velocity. The results of this turbulent flow were compared with the results of the unmodified low turbulence flow with a turbulence intensity of 3%.

The size and velocity of the drops were determined with the phase Doppler instrument as they traveled along the cylinder stagnation streamline. This well-defined uniform deceleration was imperative to provide the lowest possible scatter in the measurements. For each size class the velocity was determined at a series of axial points, typically 10 mm apart, approaching the cylinder. The deceleration of each size class was obtained from the velocity measurements, the distance traveled, and the average velocity between the two measurement stations. The deceleration between each measurement point was used along with the measured drop and air velocities to determine C_D versus Reynolds number, Re, for each size class. Relative velocity was easily determined from the difference between the air and drop velocities. Seeding with water drops of the order of a few microns in diameter, which the phase Doppler

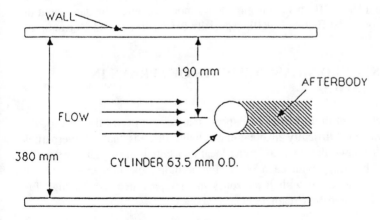

Figure 7-29 Cylinder with afterbody.

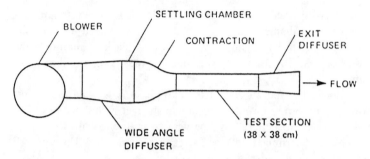

Figure 7-30 Two-phase flow wind tunnel used in the experiments.

instrument was able to discriminate from the larger drops, was utilized to determine the local air velocity (Bachalo, Rudoff, and Houser 1987). Radial velocity was also measured to ensure that the stagnation streamline was followed and hence that the drops had only a single velocity component.

Results of the drop drag measurements. The results of C_D depicted in Figures 7-31 and 7-32 were obtained following the procedures described by Kamemoto (1989). Figure 7-31 presents C_D versus Re for a single nozzle spraying in a low turbulence environment, with a free stream velocity of $U_{fs} = 19$ m/s. The values of Re were varied from approximately 1 to 100 for the flow conditions of this test. The data show that the drops tend to be grouped by size class but they all follow closely the correlation given by Torobin and Gauvin (1960), which is shown as the straight line on Figure 7-31. The correlation is

$$C_D = \frac{24}{Re} (1 + 0.15\ Re^{0.687})$$

(7-32)

Figure 7-32 is a plot of C_D versus Re for a single nozzle atomizing in a low turbulence environment with a $U_{fs} = 10$ m/s. This plot shows more scatter of the data about the correlation of Torobin and Gauvin (1960), especially for the 10 μm and 20 μm drops.

7-5 TIME-RESOLVED MEASUREMENTS OF SPRAYS IN TURBULENT FLOWS

Another important aspect in the analysis of sprays that has not been fully investigated because of the inherent difficulty in obtaining reliable data, is that of interparticle spacing. It is well known that in a spray the local concentration of fuel is related to the proximity of the drops with each other. Furthermore, the propensity for soot formation has been identified with local zones in the spray that exhibit high fuel concentration (Chigier and McCreath 1974).

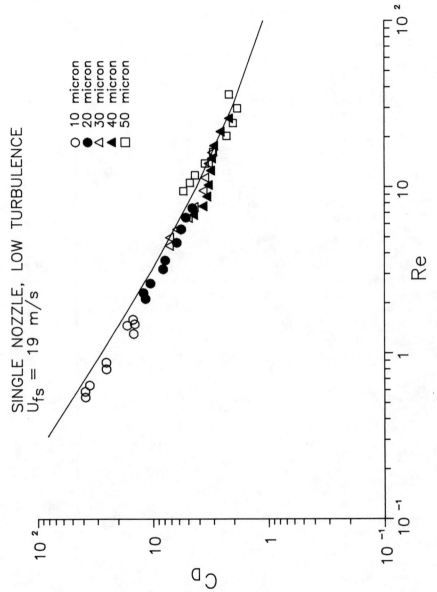

Figure 7-31 Drop drag coefficient C_D versus Reynolds number for various drop sizes.

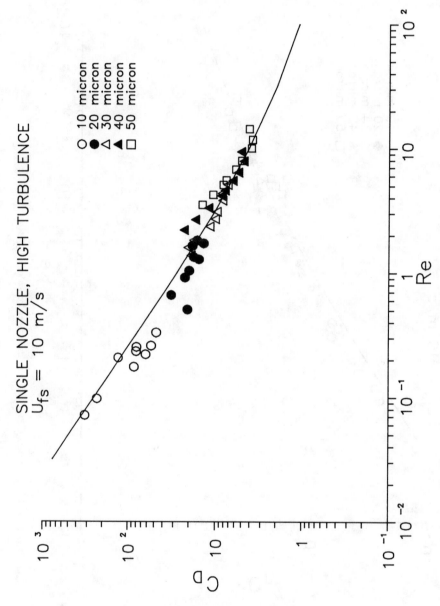

Figure 7-32 Drop drag coefficient C_D versus Reynolds number for various drop sizes.

For a particle-counter instrument, this information can be obtained by keeping track of the particle arrival times as they sweep through the probe volume at a particular location. Raw data taken at a local region in the spray can be processed to yield the time intervals of particle arrivals for specified drop size and drop velocity ranges. The distance between two particles can be readily obtained from the time elapsed between particle arrival times and the instantaneous velocity of the particles, assuming a constant velocity during this short interval. It is to be noted that very fast data acquisition rates are necessary for a time analysis of sprays, since valuable information could be omitted in a few milliseconds, as will be seen in a later section.

Current hardware and software capabilities allow timing the events to an accuracy of 1 μs. It is also possible to obtain mean data acquisition rates of 100,000 particles/s and a processing time of only 3 μs per particle, allowing an equivalent rate of 300,000 particles/s in short-duration bursts.

For these experiments a Hago 3-gph, 45° solid-cone pressure atomizer was used in both quiescent flow and in the wind tunnel (U_{fs} = 21 m/s) with a bluff body. Figure 7-33 is a plot of velocity versus a total run time of 0.01 s and includes all particle sizes. This is a very interesting figure, since it is readily seen that drops tend to form clusters that vary widely in number and concentration. The local number density of these clusters can be calculated from these plots (by expanding the time window to get a better resolution), since spacing between the drops is obtained from the velocity and the time interval. The measurement depicted in Fig. 7-33 was taken at X = 5 cm on-axis from the atomizer in quiescent flow.

The data of Fig. 7-34 were taken in quiescent conditions at X = 5 cm near the periphery of the spray in a more diluted region of the spray. It suggests also that some drops travel in clusters; however, the average spacing among the drops appears more uniform than that shown in the previous figure. The total run time represented in Fig. 7-34 is 0.19 s, and it is seen that the drops appear to be collected in six clusters, which contain 7, 7, 5, 8, 5, and 11 particles. Local number densities were calculated for the second and fifth clusters, giving N_d = 1664 particles/cm^3 and N_d = 2835 particles/cm^3, respectively. The time-averaged particle number density measured for a collected sample of 10,000 particles at this particular location was N_d = 397 particles/cm^3, which is significantly less than the previous "instantaneous" N_d values.

The data obtained with the nozzle-bluff body configuration show some differences when compared with the quiescent case. Figure 7-35 represents data taken at X = 2 cm from the atomizer at the centerline and plotted for a Δt = 0.0036 s. Drops seemed to be grouped in larger clusters with a wide range of velocities due to the effect of the recirculation induced by the bluff body.

Figure 7-36 depicts the time analysis for the nozzle-bluff body configuration over a run time of 0.04 s at X = 2 cm, R = 1.25 cm, i.e., near the periphery of the spray cone. It shows mainly three clusters of closely packed drops that exhibit a wider range of velocities due to the strong recirculation zone close to the edge of the spray.

Current research with the time analysis capability is focused on the mapping of the particle arrivals at different regions of the spray to study local particle concentrations and interparticle spacing that can be related to a given nozzle-bluff body and nozzle-swirler configurations (Breña de la Rosa, Wang, and Bachalo 1990).

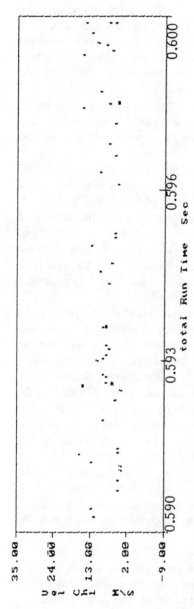

Figure 7-33 Time analysis of a spray in quiescent environment over a run time of 0.01 s showing drops grouped in clusters.

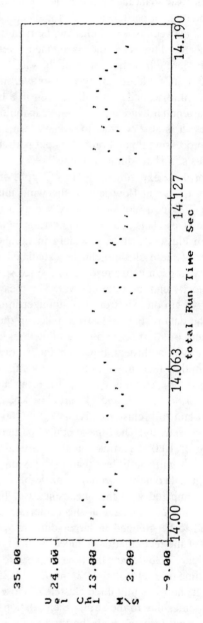

Figure 7-34 Time analysis of spray in quiescent environment over a run time of 0.19 s close to the periphery of the spray cone, $X = 5$ cm, $R = 1.75$ cm.

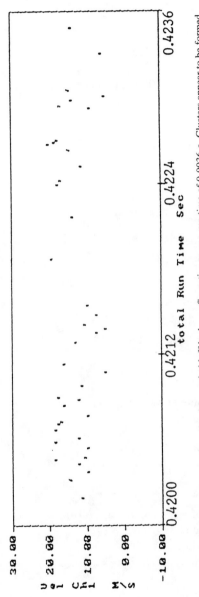

Figure 7-35 Time analysis of spray for the nozzle-bluff body configuration over a run time of 0.0036 s. Clusters appear to be formed of larger number of particles than in the quiescent case, $X = 2$ cm, $R = 0$.

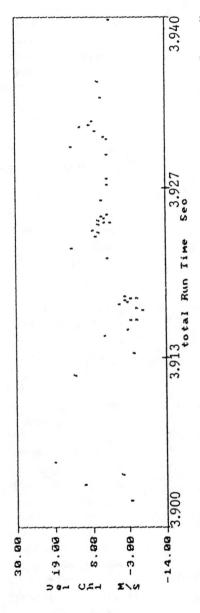

Figure 7-36 Time analysis of spray for the nozzle-bluff body configuration over a run time of 0.04 s, near periphery of spray, $X = 2$ cm, $R = 1.25$ cm.

7-6 NOMENCLATURE

A	sampling cross-sectional area
b_0	beam radius
C_D	drag coefficient of drop
d	diameter of particle
$D_b(d_i)$	cross-section diameter of drop size of class i
D_{30}	volume mean diameter
$\|E_{jk}(m, \theta, d)\|$	amplitude of scattered light, Eqs. (7-10) and (7-11)
f_s	shift frequency, Eq. (7-20)
$i_j(\alpha, m, \theta)$	dimensionless intensities, Eq. (7-5)
I	light intensity
I_{sc}	intensity of scattered light
k	wave number ($= 2\pi/\lambda$)
L	optical path length
m	relative index of refraction of particle to medium
N	number of particles passing sample cross section
N_d	particle number density
$N(d_i)$	number of particles in size class i
p	parameter, Eq. (7-3)
Q_e	mean extinction efficiency
Q_{sc}	scattering coefficient
r	polar coordinate, Eqs. (7-10) and (7-11)
R	radial position from centerline of nozzle
R	distance from receiver to probe volume, Fig. 7-12
Re	Reynolds number ($= \rho d U/\mu$)
s	slit aperture width, Fig. 7-18
$\|S\|$	amplitude of scattered light
$S_d(\alpha, \theta)$	amplitude function of light scattered by diffraction, Eq. (7-1)
$S_j(\alpha, m, \theta)$	total amplitude function, Eq. (7-4)
t	time
u	mean particle velocity
$u(d_i)$	mean velocity of drops of class i
U	instantaneous velocity
U_{fs}	free stream velocity
$U_i(d_i, t)$	velocities of individual drops of size class i, Eq. (7-18)
$\langle U'(d_i) \rangle$	rms velocity of particles of size class i, Eq. (7-19)
V	visibility of the fringe pattern, Eq. (7-13)
$V(d_i)$	volume swept by particles of size class i, Eq. (7-27)
$\mathcal{V}$	signal visibility, Eq. (7-22)
$w(d_i)$	actual width of sample cross section
x, x'	spatial coordinates, Fig. 7-12
X	axial distance from nozzle
$y(d_i)$	sample cross-section diameter, Eq. (7-22)
z, z'	spatial coordinates, Fig. 7-12

α	particle size parameter $(= \pi d/\lambda)$
β	collection angle
γ	beam intersection angle
δ	fringe spacing in probe volume, Eq. (7-23)
η	phase shift due to optical path length, Eq. (7-7)
θ	scattering angle
λ	wavelength of light
μ	dynamic viscosity
ρ	density
σ	phase shift of scattered light, Eqs. (7-8) and (7-9)
τ, τ'	angles between surface tangent and the incident and refracted rays
τ_e	extinction cross section, Eq. (7-29)
ϕ	phase difference between detectors, Eq. (7-17)
ω	frequency of light

Subscripts

d	drop
D	Doppler difference
g	gas
i	drop size class, incident beam
j	direction of polarization, Eq. (7-4)
k	referring to either beam 1 or 2, Eqs. (7-10) and (7-11)
p	particle
r	relative value of parameter

REFERENCES

Bachalo, W. D. 1980a. Method for measuring the size and velocity of spheres by dual-beam light-scatter interferometry. *Appl. Opt.* 19(3):363.

Bachalo, W. D., and Houser, M. J. 1984a. Phase/Doppler spray analyzer for simultaneous measurements of drop size and velocity distributions. *Opt. Eng.* 23(5):583.

Bachalo, W. D., and Houser, M. J. 1984b. NASA Lewis report 174636, NASA Lewis Research Center.

Bachalo, W. D., and Sankar, S. V. 1988. Analysis of the light scattering interferometry for spheres larger than the light wavelength. Paper read at the Fourth International Symposium on Applications of Laser Anemometry to Fluid Mechanics, 11–14 July 1988, Lisbon, Portugal.

Bachalo, W. D., Houser, M. J., and Smith, J. N. 1987. Behavior of sprays produced by pressure atomizers as measured using a phase Doppler instrument. *Atomization Spray Technol.* 23:53–72.

Bachalo, W. D., Rudoff, R. C., and Houser, M. J. 1987. Particle response in turbulent two-phase flows. Paper read at ASME Winter Annual Meeting Symposium on LDA, November 1987, Boston.

Bachalo, W. D., Rudoff, R. C., and Breña de la Rosa, A. 1988. Mass flux measurements of a high number density spray system using the phase Doppler particle analyzer. Paper AIAA-88-0236 read at the AIAA 26th Aerospace Sciences Meeting, 11–14 January 1988, Reno, Nevada.

Born, M., and Wolf, E. 1975. *Principles of optics*, 5th ed. New York: Pergamon.

Breña de la Rosa, A., Sankar, S. V., Weber, B., Wang, G., and Bachalo, W. D. 1989. A theoretical and experimental study of the characterization of bubbles using light scattering interferometry. Paper presented at the Third International Symposium on Cavitation Inception, 10–15 December 1989, San Francisco, California.

Breña de la Rosa, A., Wang, G., and Bachalo, W. D. 1990. The effect of swirl on the velocity and turbulence fields of a liquid spray. Paper presented at the 35th ASME Gas Turbine and Aeroengine Congress and Exposition, 11–14 June 1990, Brussels, Belgium.

Chigier, N. A., and McCreath, C. G. 1974. Combustion of droplets in sprays. *Acta Astronaut.* 1:687–710.

Farmer, W. M. 1972. Measurement of particle size, number density, and velocity using a laser interferometer. *Appl. Opt.* 11:2603.

Glantschnig, W. J., and Chen, S. 1981. Light scattering from water droplets in the geometrical optics approximation. *Appl. Opt.* 20(14):2499.

Gouesbet, G., Maheu, B., and Grehan, G. 1988. Light scattering from a sphere arbitrarily located in a Gaussian beam, using a Bromwich formulation. *J. Opt. Soc. Am.* 5:1427.

Hodkinson, J. R., and Greenleaves, I. 1963. Computations of light-scattering extinction by spheres according to diffraction and geometrical optics, and some comparisons with the Mie theory. *J. Opt. Soc. Am.* 53(5):577.

Jackson, T. A., and Samuelsen, G. S. 1987. Droplet sizing interferometry: a comparison of the visibility and phase Doppler techniques. *Appl. Opt.* 26(11):2137.

Kamemoto, D. Y. 1989. A study of seeding particulates for laser velocimetry applications. Masters thesis, Mechanical Engineering Department, George Washington University.

McDonell, V. G., Cameron, C. D., and Samuelsen, G. S. 1987. Symmetry assessment of a gas turbine air-blast atomizer. Paper read at the AIAA/SAE/ASME/ASEE 23rd Joint Propulsion Conference, 29 June–2 July 1987, San Diego, California.

Payne, A. L., Crowe, C. T., and Plank, D. R. 1986. *Laser light attenuation for determining loading ratio in a gas-solid flow.* ISA paper 86-0138.

Torobin, L. B., and Gauvin, W. H. 1959. Fundamental aspects of solids–gas flow. *Can. J. Chem. Eng.* 37(4):129–141.

van de Hulst, H. C. 1957. *Light scattering by small particles.* New York: John Wiley.

IN SITU PARTICLE MEASUREMENTS IN COMBUSTION ENVIRONMENTS

Donald J. Holve and Patricia L. Meyer

8-1 INTRODUCTION

Research on new fossil energy processes and research directed at improving existing power plant operations require instrumentation that accurately characterizes operational conditions. The availability of reliable data is crucial to basic research, design, and process control. In both conventional and advanced systems the problems of slagging, fouling, and erosion are primary limitations on the efficient use of coal. The ability to measure particle size, concentration, and velocity distributions is a key step in controlling these adverse effects.

There is a variety of commercially available particle measurement techniques. This range of methods is indicative of the range of environments and conditions where particle measurements are required. Figure 8-1 shows the basic relationship between traditional "off-line" methods and the newer "in-line" techniques. Off-line techniques require the collection of a representative sample for analysis. In-line techniques have the advantage of making measurements under the actual system conditions. No collection or handling of the sample is required. For both in-line and off-line techniques the analysis may be essentially real-time, or more typically, the analysis may require minutes or even hours to complete.

Optical techniques, based on light scattering, have the capability to provide in-line measurements of particle-laden flows. Instruments incorporating these techniques can provide real-time measurements of the process stream. Since this measurement technique is based on the analysis of light scattered by airborne particulate, it avoids

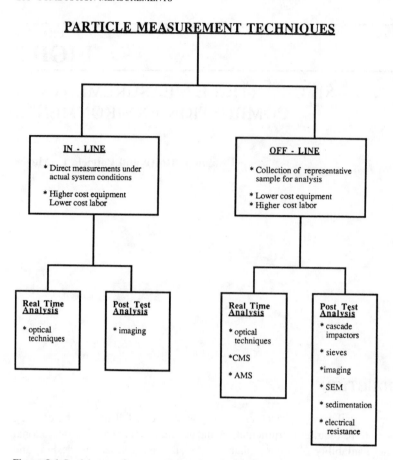

Figure 8-1 Particle measurement techniques can be divided into in-line and off-line methods.

the sample-handling problems encountered by off-line techniques that must collect particle samples for analysis.

The class of light-scattering techniques can be divided into ensemble particle-counting and single-particle-counting techniques. Ensemble particle counters, such as transmissometers, opacity meters, and ensemble diffraction techniques, measure the light scattered from a group, or ensemble, of particles. Single-particle counters, such as interferometric, transit timing, and absolute intensity techniques, measure light scattered from individual particles as they pass, one by one, through a focused laser beam. Although particles are individually measured, a completed single-particle counter measurement will be based on information from thousands to millions of particles.

While light-scattering techniques can provide the required in-line, real-time particle measurements, there are problems applying such techniques to large-scale combustion applications. The key problems are due to beam steering and aerosol opacity effects. Thermal gradients in the process stream cause changes in the refractive index that deflect (or steer) the laser beam. Maintaining the alignment of an optical system becomes difficult when there are large refractive index variations over the laser beam path length. Aerosol opacity refers to the problem of transmitting the light signals through high particle concentrations over the long distances in industrial systems. It is difficult to maintain clean windows for optical access in particle-laden process streams. Also, it is difficult to avoid degradation of the light signals through secondary scattering by other particles in the flow. A third concern is that the measurement technique itself must minimize interference with the process flow. A device that causes a major disturbance in the flow will not accurately measure or represent the flow.

In many combustion applications the particles' material and shape properties can also vary widely. The particles to be measured may include liquids, solids, or slurries. These particles may also be regular or irregular in shape. The particles' material, phase, and shape all affect the optical properties (refractive index and surface reflectivity) on which optical techniques are based.

This chapter presents Insitec's solution to these design and application challenges. Insitec's particle counting, sizing, velocity (PCSV) instruments are single-particle counters designed for applications in laboratory and industrial combustion environments. This chapter describes the instrument configuration and specifications, principle of operation, design criteria, and applications.

8-2 INSTRUMENT CONFIGURATION AND SPECIFICATIONS

PCSV instrument specifications are shown in Table 8-1. The PCSV instruments are unique in incorporating all of the following capabilities.

In-line measurements: Direct measurements of the particle-laden flow stream. No sample collection or sample conditioning is required.
Multiple measurements: Measurements of particle size, concentration, and speed.

Table 8-1 PCSV system specifications

	Specification
Size range	General capability from 0.20 to 200 μm; a typical specific instrument configuration is 0.30–100 μm
Concentration	Absolute particle concentrations up to $10^7/cm^3$ for submicron range or up to 10 g/m^3 for supermicron range (assuming a particle material density of 1 g/cm^3)
Speed	0.1–400 m/s
Particle type	Solid, liquid, composite, volatile or nonvolatile
Particle environment	No sample conditioning required; capability to make in-line measurements of particles in gas or liquid carriers over a wide range of pressures and temperatures
Accuracy	Typically $\pm 10\%$ of indicated size
Calibration	Factory calibrated with monodispersed polystyrene latex spheres, standard polydispersed aerosols, and Insitec Reference Reticle; no further calibration required
Particle pulse rate	Up to 500 kHz

Point measurements: Spatially resolved measurements can be obtained, since the PCSV makes measurements at a point in space.

Real-time measurements: Particle measurements are made at count rates up to 500,000 particles/s. Data analysis requires less than 1 s.

Measurements of different types of particles: Particles can be regular or irregular in shape. The PCSV is not limited to measurements of spherical particles. Particles can have different refractive indices. Absorbing particles such as coal or fly ash can be measured as well as nonabsorbing particles such as water drops or latex spheres.

Measurements in high-temperature, high-pressure environments: Since the PCSV-E hardware is external to the particle flow field, direct measurements of particles under high pressures and temperatures can be made. The PCSV-P is water cooled for operation at temperatures up to 1400°C.

Bench-top to large-scale systems with remote operation: The PCSV-E is designed for laboratory-scale to industrial systems, where the particle-laden flow stream measures up to a meter in diameter. The PCSV-P is designed for large-scale systems and offers the advantage of built-in, gas-purged windows. Fiber optic signal transmission and a computer-driven motor system allow both instruments to be remotely operated.

Figure 8-2 shows the main components of a PCSV instrument and an overview of the signal acquisition and processing system. The three main components of the PCSV are optical head, signal processor, and computer and software. The optical head can be in the form of a completely nonintrusive external system (PCSV-E). The PCSV-E can span ducts or flow streams up to about 1 m in diameter. This upper limit is imposed by beam stability requirements, beam attenuation in high-density flow streams, and instrument alignment requirements.

For large-scale combustion and industrial systems it is necessary to place the optical system in the end of a cooled probe (Fig. 8-3). This configuration, known as the PCSV-P, is designed to meet the significant space and durability constraints on the optical system hardware. The PCSV-P optical head must be compact and rugged to withstand high temperatures and the mechanical stresses due to thermal expansion.

The PCSV-P uses a window purge system to keep the optical head free from deposits. The windows are recessed to minimize the chance for window obscuration and to reduce the aerodynamic disturbance of the purge air on the primary flow through the flow access region (Fig. 8-3). The exposed windows are made of fused silica for reliability and stable optical quality at high temperatures. Purge air flow rates can be adjusted for the various measurement conditions to avoid perturbing the sample volume region.

The optical heads of both the PCSV-E and PCSV-P require continuous in situ alignment in order to maintain alignment under changing thermal conditions. This design constraint is solved by using a large-diameter, multimode fiber optic cable to transfer the scattered light signals from the optical head to the signal processor for detection and analysis. Computer keyboard controls allow the fiber optic system to be

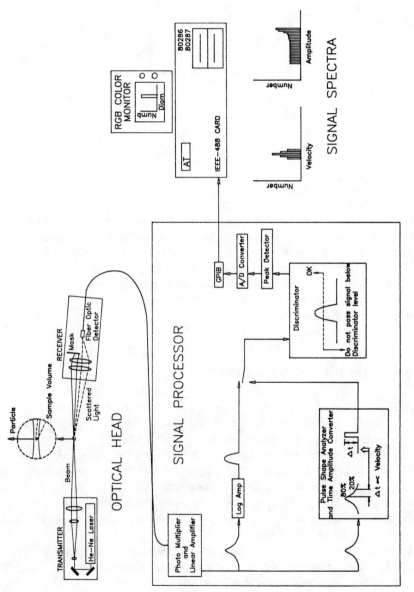

Figure 8-2 Schematic of PCSV system showing signal acquisition and processing.

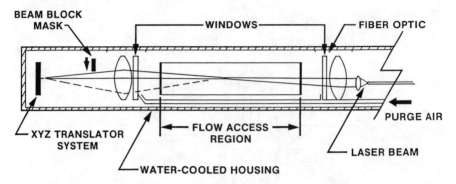

Figure 8-3 Schematic of the PCSV-P optical head. The probe measures 9-cm OD with a 10 × 4 cm flow access region.

translated within the optical head to ensure proper alignment of the optical system during measurements in hostile environments.

In order to make direct measurements in combustion applications, it is necessary to separate the laser light signals from the background radiation in the system. PCSV instruments use a narrow band-pass interference filter centered at 632.8 nm to screen out light at wavelengths other than the illuminating HeNe laser light.

The signal processor, computer, and software can be used with either or both of the optical heads. Both the PCSV-E and PCSV-P systems can be remotely operated at distances up to 400 feet (122 m).

8-3 PRINCIPLE OF OPERATION

The PCSV instruments are part of a group of light-scattering instruments whose principle of operation is based on single-particle counting. More specifically, these instruments measure the peak intensity and the width of scattered light signals produced by single particles moving through the sample volume of a single, focused laser beam.

The sample volume is defined by the dimensions of the laser beam at the focus and by the receiver aperture. Figure 8-4 shows a schematic (not to scale) of the sample volume region. Particles may pass anywhere along the length of the laser beam. However, only those particles passing through the sample volume will scatter light, which is collected at the receiver.

As particles pass through the sample volume, light scattered in the near-forward direction is collected by the receiver lens and focused onto a fiber optic cable. The fiber optic cable conducts the scattered light pulses from the optics to the detector in the signal processor, where they are analyzed.

For each scattered light pulse the signal processor measures the peak signal intensity and the signal width. Knowing the laser beam focus diameter allows particle velocities to be derived from the pulse width of each scattered light signal (Holve 1982). Particle size information is derived through measurement of the peak intensity

TRANSMITTER   **RECEIVER**

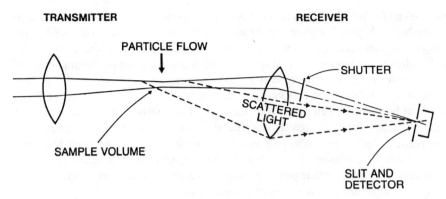

Figure 8-4 Schematic of PCSV sample volume. Focused laser beam and receiver optics define the sample volume.

of each scattered light signal. However, the amplitude of the scattered light signal depends not only on the particle size, but on its trajectory, as shown in Fig. 8-5. While off-line instruments based on single-particle counting can direct the particle trajectory through a portion of the sample volume having a known intensity, this type of sample manipulation is not possible in an in-line particle counter such as the PCSV.

Because particle trajectory is not controlled in the PCSV, the particles can pass anywhere through the sample volume. Since the laser light intensity varies across the measurement volume, a particle trajectory through the center of the measurement volume results in a much higher signal intensity than does a particle trajectory near the boundary (Fig. 8-5). A graph of scattered light pulses from many particles shows

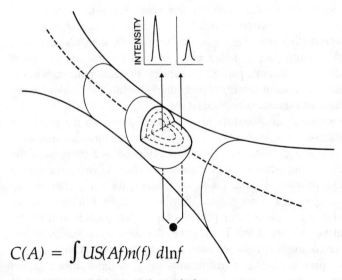

$$C(A) = \int US(Af)n(f)\, d\ln f$$

Figure 8-5 Intensity deconvolution method.

a wide range of pulse amplitudes, even if there is only one particle size passing through the laser beam. Figure 8-2 shows an example of the intensity spectrum obtained for monodisperse particles. Since the particle trajectory can occur anywhere within the sample volume, the light intensity distribution throughout the sample volume must be known in order to solve for particle size and concentration.

The sample volume size also varies with particle size. Larger particles experience larger sample volumes than do smaller particles. This means that the probability of counting a large particle is greater than that of counting a small particle. Depending on the particle size range, the sample volume cross-sectional area can vary by 1 or 2 orders of magnitude between the smallest and the largest particles measured. In order to accurately determine particle concentrations, it is essential to account for the variation in sample volume size as a function of particle size.

PCSV instruments resolve both of these issues (the unknown particle trajectory and the variation in sample volume size) through use of an intensity deconvolution algorithm (Holve and Self 1979, Part I; Holve and Davis 1985):

$$C(A) = \int US(A\mathbf{f})n(\mathbf{f})\,\mathrm{d}\ln\mathbf{f}$$

The intensity deconvolution algorithm is based on the statistical analysis of a large number of individual events, the scattered light signals from single particles passing through the measurement volume. The mathematics of the intensity deconvolution algorithm are analogous to the mathematics used to obtain CAT-scan images in the medical field. The analysis has the advantage that no assumptions are required about the shape of the size distribution. Use of the intensity deconvolution algorithm allows the absolute particle concentration and particle size distribution to be obtained directly from the experimental data.

Figure 8-6 shows an example of the results format. The graphs display the absolute number and mass concentration as a function of particle size for a pulverized coal sample. The histogram shows the frequency distribution. The integral of the frequency distribution is the cumulative distribution, which is represented by the curve.

The advantage of presenting information in the form of absolute concentration measurements is that this information is independent of both the instrument's measurement technique and the instrument's particle size range. Presenting the data in the form of absolute concentration measurements is particularly useful when making comparisons between different measurement methods.

In addition to the absolute concentration measurements shown in the graphs in Fig. 8-6, the tabulated data on the right also express the results in the form of normalized distributions (i.e., the percentage of the particle number of mass as a function of the instrument's size range). An important point is that the normalized distributions are a function of that particular instrument's size range. This is true for any particle-sizing instrument. Therefore comparisons of normalized distributions from different types of instruments will be misleading if a significant portion of the actual particle distribution lies outside the instrument's range. If this is the case, the normalized values should only be used for relative comparison between runs made by the same instrument.

To summarize, comparisons between instruments using different measurement techniques must be done on the same basis. Absolute mass or number concentration

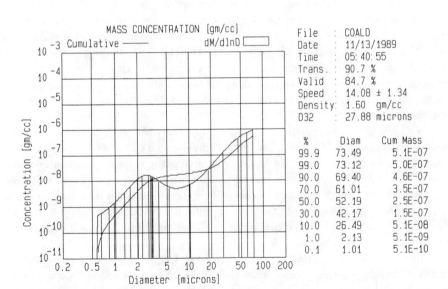

Figure 8-6 Example of PCSV data presentation.

measurements allow comparisons to be made independently of the instrument's measurement method or measurement range. On the other hand, normalized distributions are a function of the particular measurement technique and its size range. Caution is advised when making comparisons using normalized distributions, particularly when the particle size range of the material to be measured exceeds the size range capability of the measurement method.

8-4 DESIGN CRITERIA

The specific choice of PCSV optical components depends on the application requirements. The primary variables in any application are the working space and access available to the instrument and the particle size range, concentration, refractive index, and shape. The following discussion outlines the basic instrument parameters considered in optimizing the PCSV optical system.

Optical Configuration and Access

Off-axis configurations, where the angle between the incident light and the scattered light collection optics is of the order of 90°, have smaller sample volumes than near-forward configurations and therefore have the ability to make measurements at higher number densities. The trade-off is that off-axis configurations have more sensitivity to particle shape and refractive index than do near-forward configurations. Off-axis configurations are suitable for measurements of spherical particles.

For irregularly shaped particles, either transparent or opaque, it is essential to perform measurements at near-forward angles, just a few degrees off-axis from the direction of the incident light. The advantage of analyzing near-forward scattered light is that its main component is diffracted light. Therefore the intensity of the scattered light signal is mainly dependent on the particle's cross-sectional area; it is nearly independent of the particle's surface characteristics or its shape. This optical arrangement provides the flexibility to measure spherical drops as well as irregular particles, such as coal or fly ash.

Backscattering configurations, where the angle between the incident and scattered light approaches 180°, provide a convenient single-ended geometry for easy access in some applications. However, backscattering configurations have two main disadvantages. The first is that the signal level is 10–100 times lower than in forward scatter geometries. The second limitation of backscattering configurations is the nonmonotonic scattering response function. Both of these restrictions reduce the sizing sensitivity. However, for some applications the convenience and capabilities of single-ended access may override the disadvantages.

Working space is defined as the exposed beam length between the transmitter and receiver. Larger working spaces require larger lens apertures and are more sensitive to beam-pointing errors, vibration, and turbulent beam steering in high-temperature operations. The PCSV-E has an upper working space of approximately 1 m.

When system diameters exceed 1 m, the PCSV-P can be used more effectively. The PCSV-P has a fixed working space or exposed beam length of 10 cm, but the PCSV-P optical head itself may be extended to make measurements several meters into the process stream.

Particle-Scattering Response Function

Typically, PCSV instruments use a near-forward light-scattering configuration that has been shown to be minimally sensitive to particle shape and refractive index.

The typical signal response as a function of particle size is shown in Fig. 8-7 for a PCSV instrument with near-forward scattering geometry. This graph includes the response function for both transparent and opaque particles. The power dependence information in Fig. 8-7 is summarized in Table 8-2.

The size ranges in Table 8-2 are approximate and may vary by a factor of 2, depending on the optical scattering geometry chosen.

In addition, Fig. 8-7 shows that there are resonances in the response function (primarily for transparent spherical particles) that reduce the size resolution locally. Nevertheless, these tabulated power dependencies give one a general understanding of the size sensitivity in different ranges.

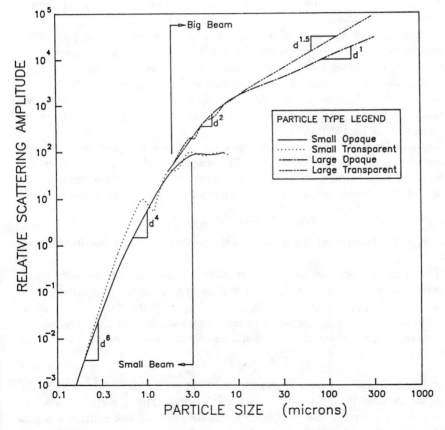

Figure 8-7 PCSV near-forward scattering response for opaque and transparent particles. Slope triangles show approximate power law dependence as a function of particle size.

Table 2 Particle size and signal power dependence for near-forward scattering geometry

Particle size range, μm	Signal Dependence	
	Opaque particles	Transparent spherical particles
<0.2	d^6	d^6
0.2–1.0	d^4	d^4
1.0–10.0	d^{2-3}	d^{2-3}
>10.0	d^1	$d^{1.5}$

From Bohren and Huffman (1983).

Sample Volume Size and Number Concentration Considerations

PCSV instruments are single-particle counters. The instrument's sample volume size must be small enough that only one particle is in the sample volume at a time. In order to obtain accurate particle size and concentration information, the instrument must measure light scattered from individual, separate particles.

In order to cover the wide dynamic range in concentration and size, the PCSV instruments typically use two beams with different beam diameters and different sample volume sizes. One beam measures the smaller particle sizes, from about 0.3 to 2–3 μm. A second beam measures the larger particle sizes, from about 3 to 100–200 μm. The total particle distribution reported results from the combination of these two independent measurements. The congruence of these two measurements is a significant consistency check for the PCSV.

The exact particle size ranges measured by each beam depend on the sample volume size and the particle concentration. The sample volume of each beam is defined by the intersection of the laser beam waist (focus) and the detector optics as shown in Fig. 8-4. The sample volume size, V_s, is given by the following equation:

$$V_s = (\text{const}) W_0^2 W_s / \sin \theta$$

where W_0 is the beam waist diameter, W_s is the detector slit width, and θ is the view angle.

The sample volume size required for a particular application is inversely related to the particle number concentration. The absolute particle number concentration (or number density) of particles in size class j, (n_j) is related to the number of counts per second in this size class (c_j) measured by the instrument, the velocity of the particles (u_j), and the sample volume cross-sectional area for this size class (s_j):

$$n_j = c_j / (u_j s_j)$$

For applications with low particle concentrations, larger sample volumes can be tolerated, and therefore the measurable size range is increased. Higher particulate loadings require smaller sample volumes in order to ensure that only one particle at a time passes through the sample volume.

Figure 8-8 shows the relation between the measurable size range and particle concentration. This figure shows that for each beam the upper bound of the measurable

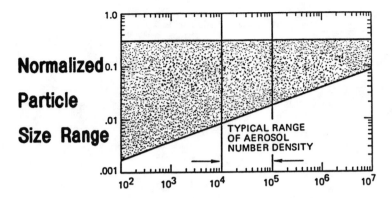

Cumulative Number Density

Figure 8-8 PCSV measurable size range as a function of particle concentration. PCSV size range decreases as particle number density increases.

size range is fixed. The upper bound for particle size measurements is determined by the laser beam waist diameter at the sample volume (the beam diameter waist must be at least twice as large as the largest particle to be measured). The upper bound does not vary as a function of particle concentration.

However, the lower bound of the measurable size range varies with particle number concentration. In order to meet the requirement that only a single particle be present in the sample volume at any one time, the lower bound moves to larger particle sizes as the number density increases. Conversely, as the particle number density decreases, the measurable size range increases. However, the measurable size range cannot expand indefinitely. For very low number density flows, the increase in the measurable size range is eventually limited by signal-to-noise considerations. Optimal selection of beam diameters at the sample volume can be made by considering the anticipated particle concentration and size distribution. Figures 8-9 and 8-10 show dashed lines that represent typical particle distribution shapes at different concentrations. In these figures the particle concentration decreases inversely with the third power of the particle diameter. Although actual particle distributions may vary greatly from these idealized ones, our experience shows that these idealized distributions generally represent worst case scenarios for estimating particle concentration as a function of particle size.

Superimposed on the graphs of the idealized particle distributions are rectangular boxes for each of the two indicated beam diameters. Each box represents the operational range for the selected beam diameter. Each operational range is constrained by the size range and concentration. For example, at high particle concentrations, smaller operational size ranges are available to the PCSV. Nevertheless, even at high concentrations, distribution information can be obtained by interpolating between the two

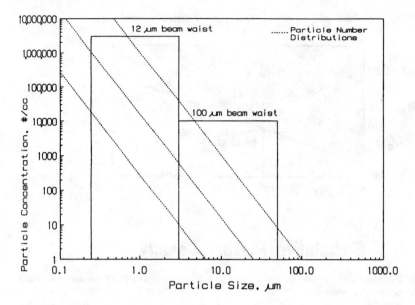

Figure 8-9 Optical geometries and beam diameters (12 and 100 μm) are chosen to measure different particle size and concentration ranges.

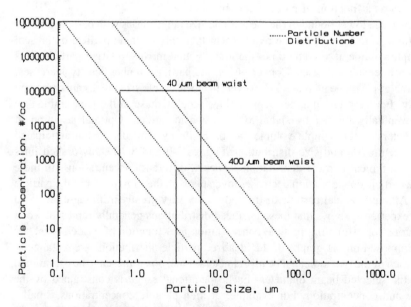

Figure 8-10 Optical geometries and beam diameters (40 and 400 μm) are chosen to measure different particle size and concentration ranges.

size ranges. The assumption here is that there are no rapid variations in the slope of the particle size distribution.

8-5 APPLICATIONS

The PCSV-E has been used to make in-line measurements of pulverized coal, coal-water slurries, liquid fuels, limestone powders, and fly ash in laboratory and industrial environments. Measurements have been made in ambient and high-temperature particle-laden flow streams (Holve 1986; Dunn-Rankin et al. 1987; Bonin and Quieroz 1989; Mescher et al. 1989). The measurements have been validated by mass balances and comparisons with other instruments (Holve and Self 1979; Holve 1986; Mescher et al. 1989).

The PCSV-P has been developed more recently to provide in-line particle measurements in large-scale combustion facilities (Holve and Milanovich 1986). Experimental results are presented here that describe in-line particle measurements in combustion applications that are not amenable to measurement by conventional instrumentation. These are unique measurements in that they provide the first experimental data describing the in-line particle size, concentration, and speed in these difficult environments.

Southern Company Services hosted a demonstration of the PCSV-P at Mississippi Power's Plant Watson unit 4 (Holve et al. 1988). Boiler 4 is an opposed wall, pulverized coal-fired unit. It is rated at 250 MW and fires a high-sulfur eastern bituminous coal with an ash content of 8–9%. A Western Precipitator, rated at 154 specific collection area (SCA), is used for particulate collection.

The PCSV-P was used to obtain particle measurements at three locations: furnace (1330°C), inlet duct to the electrostatic precipitator (125°C), and exit duct from the electrostatic precipitator (121°C). Southern Company Services also funded Southern Research Institute to make cascade impactor measurements at the electrostatic precipitator (ESP) inlet and exit ducts in parallel with the PCSV-P measurements.

Particle size calibration checks were performed prior to and following the Plant Watson tests. The primary standards used for calibrating the PCSV-P are a rotating reticle (for particles larger than 6 μm) and monodisperse latex spheres (for submicron particles). These sizing standards are used to check the instrument response at several points on the calibration curve (Fig. 8-7 shows examples of generalized calibration curves). Table 8-3 compares the known and measured values of particle size before and after the tests at Plant Watson. These results show that the instrument calibration was maintained throughout the cross-country shipment and the tests at the plant. The consistency of the calibration checks confirms the precision of measurements.

The following sections discuss some of the main features of the results.

Particle Size and Velocity Distributions

Spatial variations. Since the PCSV-P measures only those particles passing through the sample volume of the instrument, it can be used to make spatially resolved size

Table 8-3 PCSV-P particle size calibration results before and after Watson tests

	Measured diameter, μm	
Reticle diameter, μm	Before	After
6.0	4.7	6.1
9.0	8.5	—
24.0	21.0	24.6

Uncertainties in the reticle diameters are ± 1.0 μm.

and velocity measurements. The PCSV-P obtained particle velocity measurements in the furnace and at the ESP inlet and outlet. Southern Research Institute also obtained flow velocities at the inlet and outlet of the electrostatic precipitator using a type S pitot tube. The two techniques show the same trends as a function of position in the duct, and both instruments report a wide range in velocity values at the ESP exit. Table 8-4 summarizes the results from each location.

The significant feature of the furnace measurements is that the PCSV-P velocity is about twice the volumetric average velocity of the furnace (based on total gas flow and cross-sectional area at the measurement plane). Plant operators have observed increased deposition on the high-temperature superheater just downstream of the probe measurement port. This finding is consistent with the higher fly ash velocities measured by the optical probe. Measurement of actual velocity distributions can provide useful information for plant performance and future design.

PCSV-P measurements in the furnace and at the ESP inlet and outlet showed that there is little spatial variation in the shape of the particle size distribution at each location. In fact, particle size distributions at all three measurement locations have the same characteristic shape. This finding is consistent with previous PCSV measurements of pulverized coal in small-scale combustors (Holve 1986). Both the PCSV-P and the cascade impactor reported a particle concentration peak at about 10–20 μm. This peak is preserved throughout the process flow stream from the furnace to the precipitator exit.

Temporal variations. A unique feature of particle-counting techniques such as the PCSV-P is the capability to show temporal variations in particle concentration. Variations occurring over a several-minute period could be detected, since the PCSV-P required just 2 min for each measurement. These variations cannot be measured by

Table 8-4 Velocity ranges (m/s) at each measurement location

	PCSV-P	Pitot tube
Furnace	27–33 (m/s)	
Precipitator inlet	5.7–7.8	7.3–10
Precipitator exit	16–25	12–20

techniques such as the cascade impactor, which makes measurements over much longer time intervals, e.g., 1–2 hours at the ESP exit.

The PCSV-P showed some variation in the number density as a function of time at all three measurement locations. However, the location exhibiting the largest variation in particle concentration was the outlet of the ESP.

Figure 8-11 compares number density variations at the inlet and outlet of the ESP. These curves show the cumulative particle number concentration per cubic centimeter of the local gas density. The cumulative number density curve represents the total concentration for all particles larger than the indicated diameter. The upper set of curves in Fig. 8-11 represents seven individual PCSV-P measurements made over a 2-hour period at the ESP inlet. The eight ESP outlet measurements were all taken over a period of 30 min.

The spread in the number concentration curves represents variations in ESP inlet and outlet operating conditions over time. The outlet measurements show a wider variation in number density, especially for the larger particles. Figure 8-12 shows another way to represent the data at the ESP outlet. Particle count rate is plotted as a function of time for small (0.3–2.1 μm) and large (2.3–40 μm) particles. The axis on the right also shows the average particle velocity during the 30-min test period. While the small-particle count rate fluctuates over a range of 11.5–15.5 kHz, much larger, cyclic variations are observed in the large particle count rate, from 2.9 to 19.2

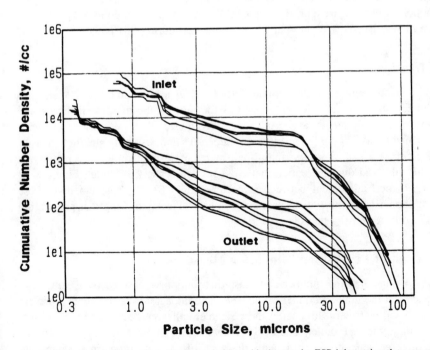

Figure 8-11 Cumulative number density variations with time at the ESP inlet and outlet as measured by the PCSV-P. Inlet measurements were made at a fixed location over a 2-hour interval. Outlet measurements were made at a fixed location over a 30-min interval.

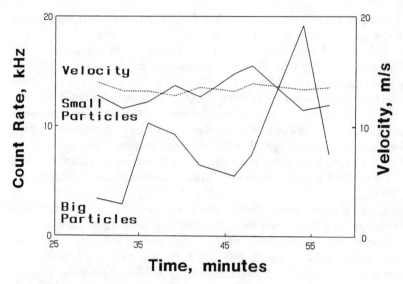

Figure 8-12 Temporal variation of the large- and small-particle count rates and the average particle velocity at the ESP outlet.

kHz. The wide spread in the number distribution curves and the fluctuations in the large particle count rate are related to rapping, i.e., the removal of collected particulate from the plates. Bursts of larger particles are released during this process.

ESP Efficiency

PCSV-P measurements of the precipitator efficiency as a function of particle size are shown in Fig. 8-13. Penetration is defined as the ratio of the mass frequency at the outlet to the mass frequency at the inlet for a given size. PCSV-P mass frequencies were obtained by averaging the results from three runs and assuming a particle density of 2.48 g/cm^3. Cascade impactor mass frequencies represent averages over traverses at the ESP inlet and over the entire duct cross section at the ESP outlet. Figure 8-13 shows very good agreement between impactor and PCSV-P measurements. These results are consistent with typical precipitator performance, showing that penetration increases with decreasing particle size.

Comparison of PCSV-P and Impactor Measurements

Figure 8-14 compares the number and mass distributions measured by the PCSV-P and impactor at the inlet and outlet of the ESP. The impactor measurements represent averages over traverses of the duct at the ESP inlet and averages over the entire duct cross section at the ESP outlet. The PCSV-P data shown in these figures represent the average of three separate measurements made at different positions in the inlet and outlet ducts. The impactor and PCSV-P measurements were also made at different times during the tests at Plant Watson.

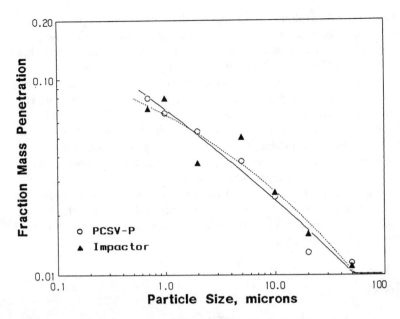

Figure 8-13 ESP efficiency as a function of particle size.

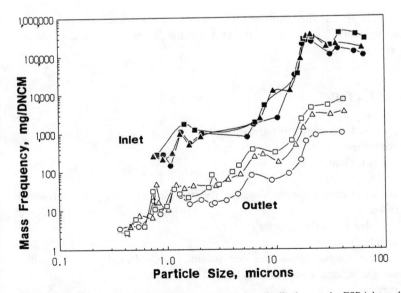

Figure 8-14 Comparing PCSV-P and impactor number distributions at the ESP inlet and outlet. PCSV-P data represent single-point measurements. Impactor data represent averages across the duct.

The number density data (plotted as number per dry normal cubic centimeter) and the mass density data (plotted as milligrams per dry normal cubic meter) show good agreement between the two techniques for particles up to 10 μm in diameter. For larger particle sizes the PCSV-P always measured higher particle concentrations than did the cascade impactor. While the measurements were made at different times, differences in the measurement techniques contribute to this discrepancy as well. The PCSV-P technique is based on the direct measure of particle counts for particles up to 100 μm in size, while the cascade impactor is based on the direct measure of particle mass for particles up to 7 μm in size. One factor affecting the results may be the curve-fitting program used by the cascade impactor to extend the measurements to particle sizes larger than the 7-μm upper limit of the impactor stages. A second factor affecting the results is that the PCSV-P used the same particle density value for the entire size range. While this value describes the dense, spherical small particles formed from molten ash, it is not appropriate to describe large particles such as cenospheres or agglomerates rapped off the collectors.

8-6 SUMMARY

This chapter has provided an introduction to PCSV measurement techniques, examples of design choices that must be made when applying the instrument in different environments, and examples of instrument applications. Unlike off-line sampling instruments, PCSV instruments have the capability to provide direct, real-time particle measurements in combustion environments. This is a significant advance, in that it allows particulate levels to be immediately and directly linked to changes in operating conditions.

In addition, sample-handling problems such as condensation, sticking, and particle agglomeration are avoided, since measurements are made directly in the process stream under actual system conditions.

8-7 NOMENCLATURE

A	amplitude of scattered light signal
C	counts/time
d	particle diameter
F	beam f number
$F(d_j, \tilde{m})$	scattered light response function for particle size class j and refractive index $\tilde{m}$
$\tilde{m}$	complex refractive index $(= n_1 + \mathbf{i}n_2)$
n	channel number
n_j	particle number density per unit volume $[= C_j/(U_j S_j)]$
S	sample volume cross-sectional area
U	particle speed
V_s	sample volume

W_0 beam waist
W_s detector slit width
x normalized particle size ($= \pi d/\mathbf{l}$)
θ light collection angle or view angle

REFERENCES

Bohren, C. F., and Huffman, D. R. 1983. *Absorption and scattering of light by small particles*. New York: John Wiley.

Bonin, M. P., and Queiroz, M. 1989. An analysis of single stream droplet combustion through size and velocity determinations using a laser-based, in situ measurement technique. Paper read at the Combustion Institute Western States Meeting, Spring 1989, Pullman, Washington.

Dunn-Rankin, D., Hoornstra, J., Gruelich, F. A., and Holve, D. J. 1987. Coal rank influences on agglomeration, swelling, and fragmentation of coal/water slurries during combustion. *Fuel* 56:1139–46.

Holve, D. J. 1982. Transit timing velocimetry (TTV) for two phase reacting flows. *Combust. Flame* 48:105–8.

Holve, D. J. 1986. In situ measurements of flyash formation from pulverized coal. *Combust. Sci. Technol.* 44:269–88.

Holve, D. J., and Davis, G. W. 1985. Sample volume and alignment analysis for an optical particle counter-sizer and other applications. *Appl. Opt.* 24(7):998–1005.

Holve, D. J., and Milanovich, F. P. 1986. A hybrid optical probe for in situ particle measurements in large scale combustion systems. Paper WSS-86-14 read at the Combustion Institute Western States Meeting, Fall 1986, Tucson, Arizona.

Holve, D. J., and Self, S. A. 1979. An optical particle-sizing counter for in situ measurements, Parts I and II. *J. Appl. Opt.* 18(10):1632–52.

Holve, D. J., Meyer, P. L., Muzio, L. J., and Shiomoto, G. H. 1988. On-line, in situ particle measurements in large scale combustion systems. Paper read at EPA/EPRI 7th Symposium on the Transfer and Utilization of Particulate Control Technology, 22–25 March 1988, Nashville, Tennessee.

Mescher, A., Essenhigh, R. H., and DeFazio, M. 1989. Comparison of batch and continuous flow particle sizing of pulverized coal. Paper read at the Combustion Institute Central States Meeting, Spring 1989, Dearborn, Michigan.

FOR FURTHER READING

Dunn-Rankin, D., Hoornstra, J., and Holve, D. J. 1986. In situ non-Doppler laser anemometer particle sizer applied in heterogeneous combustion. Paper read at the Third International Symposium on Applications of Laser Anemometry to Fluid Mechanics, 7–9 July 1986, Lisbon, Portugal.

Holve, D. J. 1980. In situ optical particle sizing technique. *J. Energy* 4(4):176–83.

Holve, D. J., and Annen, K. 1984. Optical particle counting and sizing using intensity deconvolution. *Opt. Eng.* 23(5):591–603.

Holve, D. J., and Meyer, P. L. 1987. Coal-water slurries: fuel preparation effects on atomization and combustion. *Combust. Sci. Technol.* 52:243–68.

Holve, D. J., Fletcher, T. H., and Gomi, K. 1987. Comparative combustion studies of ultrafine coal/water slurries and pulverized coal. *Combust. Sci. Technol.* 52:269–91.

Holve, D. J., Tichenor, D. A., Wang, J. C. F., and Hardesty, D. R. 1981. Design criteria and recent developments of optical single particle counters for fossil fuel systems. *Opt. Eng.* 20(4):529–39.

OPTICAL CHARACTERIZATION OF CONDENSED PHASES IN COMBUSTION SYSTEMS

A. D'Alessio, F. Beretta, and A. Cavaliere

9-1 INTRODUCTION

Combustion processes transform complex organic molecules, via fossil fuels, into CO_2 and H_2O through a sequence of gas phase reactions. Fossil fuels are generally complex mixtures of liquid hydrocarbons or solid carbonaceous material. In order to increase the specific surface area for better heat and mass transfer, they are atomized into sprays of micronic droplets or pulverized into an ensemble of micronic particles, respectively.

Furthermore, the mixing of gaseous or multiphase fuels with the combustion air produces a spatial and temporal distribution of the fuel concentration in the gas phase. In regions with high C/O atomic ratio, additional condensed phases are produced, e.g., structures with a large number of carbon atoms such as polycyclic aromatic hydrocarbons (PAH), carbon clusters, and solid submicronic soot particles. For these reasons the presence of condensed (liquid and solid) phases plays an important role in combustion systems.

Before being transformed into CO_2 and H_2O, the original condensed phases and the flame-generated particles may be involved in heterogeneous chemical reactions. For slow mass and heat transfer from the condensed phases, pyrolytic or oxidative reactions take place at the droplet surface or even inside the droplets. Under these circumstances, porous carbonaceous particles of micronic size (cenosphere) may be produced. Coal particles are transformed into char by devolatilization. In any case, the final oxidation of micronic and submicronic carbonaceous particles is mainly a heterogeneous process.

The physical and chemical properties of the micronic particles change in very short intervals of space and time. The distribution of the submicronic particles is determined by the structure of the turbulent reacting flow field. Therefore optical studies of condensed phases in combustion processes are concerned about particles widely different in size, shape, chemical nature, and concentration.

This chapter will not review the total field of optical diagnostics of condensed phases in combustion systems. It summarizes and discusses the methods developed by the authors and their colleagues in the last few years in Naples for the study of stationary gaseous flames, spray flames, and nonstationary diesel sprays. It updates previous monographs of the authors on light scattering and optical measurements in flames and sprays (D'Alessio 1981; D'Alessio, Cavaliere, and Menna 1983; D'Alessio et al. 1983).

The first section summarizes the scattering theory for particles much smaller than the wavelength of the incident radiation. Submicronic particles irradiated with visible or UV light lie within this regime of scattering. Light scattering for particles much larger than the wavelength, e.g., particles of micronic sizes such as liquid droplets, is discussed briefly.

In the following section the experimental characterization of the optical properties of submicronic soot particles is presented. Subsequently, the characterization of monodisperse droplets and of differently atomized sprays by light scattering is discussed.

The roles of the different fuel properties, the atomization procedures, and the spray-air interaction are illustrated through selected examples with emphasis on diagnostics. The light-scattering study of sprays under burning conditions points out the importance of soot formation and fuel pyrolysis. The final section reports measurements employing two-dimensional time-resolved light scattering to analyze unsteady dense diesel sprays. The relevance of this diagnostic technique for understanding atomization regimes is also briefly discussed.

9-2 THEORETICAL BACKGROUND

A physical quantum-mechanical description of the interaction between radiation and a condensed medium allows a unified treatment of absorption, emission, and different scattering processes like Rayleigh, Raman, Thomson, and Compton scattering (Chen 1981). In this chapter, only elastic effects are considered, where the interaction of radiation with the scatterer produces no frequency shift.

The interaction of particles with radiation is described in terms of the optical properties of the particles, namely, their complex refractive index $m = n - ik$, or in an equivalent way, their complex dielectric function $\epsilon = \epsilon' + i\epsilon''$. The scattering and absorption properties are classically considered as an electromagnetic problem. Quantum mechanic concepts are necessary only for interpreting the frequency dependence of the optical properties.

Within this concept the interaction of an electromagnetic field with a particle depends primarily on the ratio D/λ of the typical dimension of the particle and the wavelength of the radiation, and on the optical properties. For $D/\lambda << 1$ the elementary electrical dipoles induced by the incident electromagnetic wave produce wavelets that are almost in phase. Their combined effects sum up to give an overall scattering pattern with an angular distribution and polarization typical for a radiating dipole. When D/λ increases, the scattered wavelets show a phase difference due to scattering from different parts of the particles, and a complex scattering pattern is generated. In the limit $D/\lambda \gg 1$ the angular scattering pattern and its polarization are described in terms of geometrical optics of refracted and reflected rays, e.g., through the Fresnel and Snell equations.

Since ultraviolet and visible radiation, with wavelength ranging between 0.2 and 0.6 μm, is considered in the present study, combustion-generated submicron particles lie in the regime $D/\lambda < 1$, and micronic particles and droplets have a D/λ ratio larger than 1.

The amplitude and phase of the light scattered from a single dipole are determined by the optical properties. Their variation with wavelength is generally modeled as a series of damped Lorenz oscillators (Born and Wolf 1970). In a quantum mechanical interpretation the resonance frequency of these oscillators is proportional to the energy difference between the initial and final quantum states of the transition. The strength of the oscillator is proportional to the probability of the transition, and the damping factor is related to the probabilities of transitions to all other quantum states. Hence the imaginary part of the refractive index k becomes significant only in a frequency range where allowed transitions between the quantum states of the material in question occur. Depending on the level of the quantum states of the carbon atoms in the carbonaceous material, the resonance frequency of the damped Lorenz oscillator lies in the visible or near UV.

In this regard, structures formed by carbon atoms may take up basically three configurations (Birks 1970). Tetragonal bondings, or sp^3 hybridization, originate in saturated molecules such as methane or can be a diamond-like structure in the case of solid material. Trigonal, or sp^2 hybridization, provides the hexagonal ring structure of benzene and of the PAHs or can be a graphitic-like structure in the case of solid material, while diagonal or sp hybridization produces a configuration that occurs in the acetylenic molecules and, perhaps, in polymeric solid materials. The sp^2 configuration consists of hybrid orbitals symmetrical about their bonding axes known as σ electrons and an antisymmetrical orbit known as the π electron. Whereas σ electrons are localized, π electrons are delocalized, and their excited states are in a lower energy range with respect to σ electrons, so that resonances with radiation in the visible and near UV can take place. For aromatic molecules present in liquid phase or for graphitic structures of the solid carbonaceous phase, the imaginary part of the refractive index of carbon compounds is significantly different from zero in the visible and UV.

These qualitative considerations may be expressed quantitatively for scattering by a spherical particle. For this case, Lorenz (Logan 1965) and Mie (van de Hulst 1957; Kerker 1969) independently found the exact solutions of Maxwell's equations for diameter and complex refractive index, which appear as one main variable. The mathematical basis of the theory is given in Chapter 10, and only some results will be quoted here.

The scattering cross section of a particle, defined as the ratio between the energy flux of the light scattered from the sphere (W_s) into a unit solid angle and the incident energy flux (I_i), is given by the expression

$$C_{\text{scatt}} = \frac{W_s}{I_i} = \frac{\lambda^2}{2\pi} \sum_{n=1}^{\infty} (2n + 1)(|a_n|^2 + |b_n|^2) \tag{9-1}$$

The extinction cross section is equal to

$$C_{\text{ext}} = \frac{\lambda^2}{2\pi} \sum_{n=1}^{\infty} (2n + 1)[\text{Re}(a_n + b_n)] \tag{9-2}$$

where a_n and b_n are functions of the size parameter $\alpha = \pi D/\lambda$ and of the complex refractive index $m = n - ik$. For real m, $C_{\text{scatt}} = C_{\text{ext}}$, since no absorption occurs inside the medium. For complex m there is absorption, and the absorption cross section is given by $C_{\text{abs}} = C_{\text{ext}} - C_{\text{scatt}}$.

The dependence of C_{ext} upon the size of the scatterer at $\lambda = 500$ nm is illustrated in Fig. 9-1 for refractive indices with the real part $n = 1.5$ and the imaginary part k ranging from zero to 10^{-2}. For the transparent sphere ($k = 0$), two asymptotic regimes are present: for small particles, C_{ext} is proportional to D^6, and for very large particles, C_{ext} is proportional to D^2. For absorbing small spheres ($k \neq 0$), C_{ext} is much larger than in the transparent case and is proportional to D^3. For very large absorbing particles, C_{ext} is again proportional to D^2. Simple analytical expressions are obtained for C_{ext} and C_{scatt} in the case of small particles ($\alpha \ll 1$). Expanding the various functions in the scattering coefficients a_n and b_n in power series and retaining only the first few terms (van de Hulst 1957) yields

$$C_{\text{ext}} = \frac{\pi D^2}{4} \left\{ 4\alpha \ \text{Im} \left(\frac{m^2 - 1}{m^2 + 2} \right) + \frac{8}{3} \alpha^4 \left[\text{Re} \left(\frac{m^2 - 1}{m^2 + 2} \right) \right]^2 \right\} \tag{9-3}$$

and

$$C_{\text{scatt}} = \frac{\pi D^2}{4} \frac{8}{3} \alpha^4 \left[\text{Re} \left(\frac{m^2 - 1}{m^2 + 2} \right) \right]^2 \tag{9-4}$$

Equations (9-3) and (9-4) explain the observed dependences of C_{ext} on particle size given in Fig. 9-1. Furthermore, it can be shown that they are equivalent to those of an ideal radiating dipole with a moment $\mathbf{P}$ equal to

$$\mathbf{P} = \left[\frac{\pi D^3 (m^2 - 1)}{2(m^2 + 2)} \right] \mathbf{E}_0 \tag{9-5}$$

where $\mathbf{E}_0$ is the strength of the incident field and the term in brackets is the particle polarizability. For large particles ($\alpha \to \infty$) the incident wave can be subdivided into a number of rays, each of which corresponds to a value of n in the Lorenz-Mie series (van de Hulst 1957; Nussenzweig 1979). The rays interact with the surfaces according to geometrical optics laws given by Fresnel and Snell's equations and are attenuated inside the medium according to the Lambert-Beer law. The extinction cross section is twice the geometrical one, $C_{\text{ext}} = 2\pi D^2/4$, and the scattering cross section can be written as

$$C_{\text{scatt}} = C^{\text{diff}} + C^{\text{refl}} + C^{\text{trans}} \tag{9-6}$$

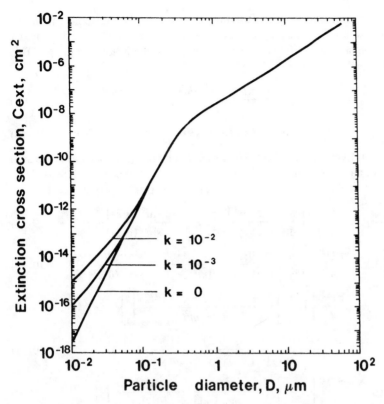

Figure 9-1 Extinction cross section versus diameter for different values of the imaginary part k of the refractive index.

where $C^{\text{diff}} = \pi D^2/4$ and takes into account diffraction, C^{refl} is the cross section for external reflection, and C^{trans} is the cross section for radiation, which is transmitted during or after internal reflections. This last term is zero for absorbing spheres ($k \neq 0$).

The Lorenz-Mie theory also provides angular distribution and polarization of scattered radiation in terms of the angular cross sections $C_{\text{VV}}(\theta)$ and $C_{\text{HH}}(\theta)$. They are the energy fluxes of light scattered into a unit solid angle for a unit incident energy flux of light polarized perpendicular or parallel to the scattering plane. They are given by the following expressions in terms of amplitude function S_1 and S_2:

$$C_{\text{VV}}(\theta) = \frac{\lambda^2}{4\pi} \sum_{n=1}^{\infty} \frac{2n+1}{n(n+1)} (a_n \pi_n + b_n \tau_n)^2 = \frac{\lambda^2}{4\pi} |S_1|^2 \qquad (9\text{-}7)$$

and

$$C_{\text{HH}}(\theta) = \frac{\lambda^2}{4\pi} \sum_{n=1}^{\infty} \frac{2n+1}{n(n+1)} (a_n \tau_n + b_n \pi_n)^2 = \frac{\lambda^2}{4\pi} |S_2|^2 \qquad (9\text{-}8)$$

where π_n and τ_n are functions of the scattering angle θ. It is also convenient to define the polarization ratio $\gamma(\theta) = C_{HH}/C_{VV}$. For small particles ($\alpha \ll 1$) the angular pattern of the angular cross sections is identical to that of the irradiating dipole. This is known as Rayleigh scattering.

$$C_{VV}(\theta) = \frac{\pi D^2}{4} \alpha^4 \left| \frac{m^2 - 1}{m^2 + 2} \right|^2 \tag{9-9}$$

$$C_{HH}(\theta) = C_{VV}(\theta) \cos^2 \theta \tag{9-10}$$

In the limit of a very large sphere the scattering angular cross section is the sum of external reflection cross sections $C^{refl}(\theta)$ and refraction cross sections $C^{refr}(\theta)$. They are given by the following expressions, which are valid for transparent spheres:

$$C_{VV}^{refl}(\theta) = \frac{\pi D^2}{4} \frac{1}{4} \left(\frac{\sin(\theta/2) - [m^2 - 1 + \sin^2(\theta/2)]^{1/2}}{\sin(\theta/2) + [m^2 - 1 + \sin^2(\theta/2)]^{1/2}} \right)^2 \tag{9-11}$$

$$C_{HH}^{refl}(\theta) = \frac{\pi D^2}{4} \frac{1}{4} \left(\frac{m^2 \sin(\theta/2) - [m^2 - 1 + \sin^2(\theta/2)]^{1/2}}{m^2 \sin(\theta/2) + [m^2 - 1 + \sin^2(\theta/2)]^{1/2}} \right)^2 \tag{9-12}$$

$$C_{VV}^{refr}(\theta) = \frac{\pi D^2}{4} \left\{ 1 - \left[\frac{1 + m^2 - 2m \cos(\theta/2)}{1 - m^2} \right]^2 \right\}$$

$$\cdot \frac{m^2 \sin(\theta/2)[m \cos(\theta/2) - 1][m - \cos(\theta/2)]}{2 \sin(\theta)[m^2 + 1 - 2m \cos(\theta/2)]} \tag{9-13}$$

$$C_{HH}^{refr}(\theta) = \frac{\pi D^2}{4} \left\{ 1 - \left[\frac{(1 + m^2) - \cos(\theta/2) - 2m}{(m^2 - 1) \cos(\theta/2)} \right]^2 \right\}$$

$$\cdot \frac{m^2 \sin(\theta/2)[m \cos \theta/2) - 1][m - \cos(\theta/2)]}{2 \sin(\theta)[m^2 + 1 - 2m \cos(\theta/2)]} \tag{9-14}$$

Note that the scattering cross section at very low scattering angles is dominated by diffraction effects. For small scattering angles the cross section of diffraction is some order of magnitude higher than the cross sections of reflection and refraction. The small scattering angle limit is given by the Fraunhofer law

$$C^{diff}(\theta) = \frac{\pi D^2}{4} \left[\frac{J_1(\alpha \sin \theta)}{\alpha \sin(\theta)} \right]^2 = \frac{\lambda^2}{4\pi} \alpha^2 \left[\frac{J_1(\alpha \sin \theta)}{\alpha \sin(\theta)} \right]^2 \tag{9-15}$$

where J_1 is the Bessel function of first order. Outside the diffraction lobe this geometrical optics approximation predicts the angular distribution and polarization properties of the scattered light only as a first approximation for infinitely large spheres. The last terms of the Lorenz-Mie series with $n > \alpha$ represent rays reaching the surface of the particle at grazing incidence. They cannot be described in terms of geometrical optics, and a more complex physical and mathematical description is required (Nussenzweig 1979). For the purpose of this chapter, it is worthwhile to remark that these rays originate the surface waves, the contribution of which to scattering, C^{sw}, has to be added to the contributions previously quoted.

Consequently, the angular cross sections of scattering for supermicron-sized spheres illuminated by a light source with wavelengths in the near-UV or visible range are given by the expressions

$$C_{VV}(\theta) = C^{\text{diff}}(\theta) + C_{VV}^{\text{refr}}(\theta) + C_{VV}^{\text{refl}}(\theta) + C_{VV}^{\text{sw}}(\theta) \qquad (9\text{-}16)$$

$$C_{HH}(\theta) = C^{\text{diff}}(\theta) + C_{HH}^{\text{refr}}(\theta) + C_{HH}^{\text{refl}}(\theta) + C_{HH}^{\text{sw}}(\theta) \qquad (9\text{-}17)$$

The cross sections for reflection and refraction are normally much higher than those for surface waves. The latter dominates only in the backscattering region near $\theta = 180°$. However, there is another scattering region near $\theta = 90°$, where surface wave effects are comparable to the effects of geometrical optics. In the side scattering, beyond the limit angle, the contribution of refraction disappears and, furthermore, the horizontally polarized cross section C_{HH}^{refl} is low because the Brewster angle is located near this angular region.

These effects are elucidated in Fig. 9-2, where the dependence of the polarization ratio C_{HH}/C_{VV} at $\theta = 90°$ and $\theta = 60°$ on the size of the particle for a transparent sphere with $m = 1.5$ is given. In the plot the interferences and ripple structures that are predicted by theory for a truly monodisperse scatterer have been averaged in order

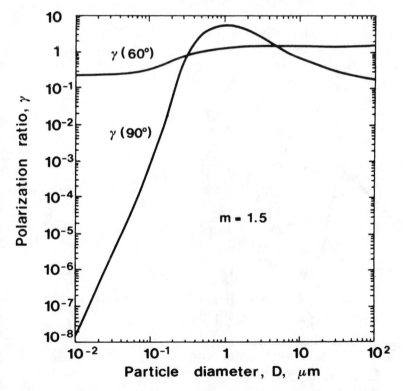

Figure 9-2 Polarization ratio at $\theta = 90°$ and $\theta = 60°$ for transparent spheres ($m = 1.5$) versus the diameter.

to discuss the main features of scattering. At $\theta = 90°$ the fast increase of the polarization ratio in the small particle range indicates a deviation from the dipole behavior, which results in a polarization ratio equal to zero. For larger particles the polarization ratio passes through a peak for particles with diameters of about 1 μm. For even larger particles it declines toward the asymptotic value predicted by the Fresnel equations for external reflection. This limit is obtained for spheres larger than 100 μm. Scattering due to surface waves, which is preferentially horizontally polarized, is initially prevailing or comparable to the scattering due to external reflection. In this way, it contributes significantly to the overall polarization of the scattered radiation. The subsequent decrease of the polarization ratio for larger particles can be explained theoretically. Both C_{HH}^{sw} and C_{VV}^{sw} are approximately proportional to the perimeter of the particle, whereas the contributions of geometrical optics are proportional to their surface area. The polarization ratio at $\theta = 60°$, also reported in Fig. 9-2, has a smooth transition from its value in the Rayleigh regime to the asymptotic value in the geometrical optics regime. This value was attained for particles as small as 1 μm (Glantschnig and Chen 1981; Beretta, Cavaliere, and D'Alessio 1984) because at this angle the surface wave cross sections are much lower than the contributions due to the geometrical optics.

A more complete insight into the dependence of the polarization ratio, $\gamma = C_{HH}/C_{VV}$, on the size and optical properties of supermicron spheres can be gained by analyzing the whole angular scattering pattern. Figure 9-3 shows that γ does not depend upon size in the forward scattering region, which is controlled by geometrical optics. On the contrary, it varies with particle size in the side scattering and backscattering regions due to the different dependencies of the surface waves and reflection cross sections on the size.

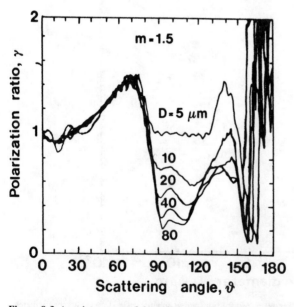

Figure 9-3 Angular pattern of the polarization ratio for transparent droplets with different sizes.

The dependence of the polarization ratio upon the imaginary part of the refractive index is shown in Fig. 9-4 for a 5-μm particle. The polarization ratio decreases with increasing k at all scattering angles except in the rainbow region for $k > 0.1$. For this exception, γ amounts to that of a purely reflecting sphere. This behavior is due to the attenuation of the refracted light passing through the sphere, which increases progressively with increasing imaginary part of the refractive index.

It is worthwhile to remark that the decrease of γ in the forward scattering region depends upon the optical path length kD. Therefore an increase in size for a droplet with constant absorption will produce a similar decrease of γ. A more quantitative discussion of these effects and their relevance to diagnostics is presented in one of the following sections.

All scattering characteristics discussed up to now deal with two components of the scattered electric field. These components can be written using different notation, e.g., the Stokes representation (Beretta, Cavaliere, and D'Alessio 1983a; Shurcliff 1962), but they can be used only on the basis of two independent measurements. There is another characteristic that can be used for the same purpose, namely, the phase of

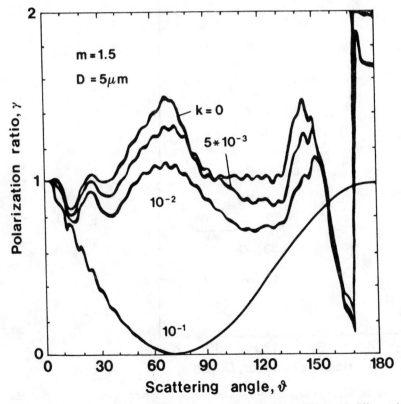

Figure 9-4 Angular pattern of the polarization ratio for absorbing droplets with different absorption coefficients.

the scattered wave (δ), which is related to the amplitude functions S_1 and S_2 by the following expression:

$$tg\,\delta = \frac{\text{Re}\,(S_1)\,\text{Im}\,(S_2) - \text{Re}\,(S_2)\,\text{Im}\,(S_1)}{\text{Re}\,(S_1)\,\text{Im}\,(S_2) + \text{Re}\,(S_1)\,\text{Im}\,(S_2)} \tag{9-18}$$

The same information can also be assumed for all cases where the interference between two scattered waves can be determined.

Angular oscillations of the scattering intensities originate for large particles from interference between rays scattered at the same angle along different optical paths. The phase shifts between the rays are in most cases linearly proportional to the size parameter. They depend on trigonometric functions of the scattering angle and on the refractive index. The number of angular oscillations is approximately equal to α. In other words, the number of oscillations per unit scattering angle is linearly proportional to particle diameter. This is shown in Fig. 9-5 for two different refractive indices

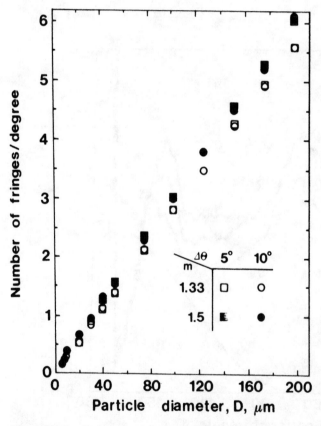

Figure 9-5 Number of fringes per degree versus droplet diameter for two refraction indices as predicted by Lorenz-Mie theory around 90°. The solid symbols give results for a refractive index $m = 1.33$, and the open symbols for $m = 1.5$.

(m = 1.33 and 1.5). This diagrammatic explanation of the angular oscillations is, of course, consistent with the framework of geometrical optics. This scheme is not applicable for smaller particles or for angular regions where the refracted rays are condensed in focal lines, even though the characteristics of interference effects are preserved. In fact, for small particles ($\alpha < 50$) the number of oscillations scales with the dimensionless diameter as well.

In some applications the angular oscillations can be obtained from the interference between scattered waves excited by two different incident fields following optical paths that differ for a fraction of a wavelength (see Chapter 7; Bauckhage 1988; Bohren and Huffmann 1983).

The theory of scattering and extinction by homogeneous spherical particles is a first approximation for the study of condensed phases in combustion. In many real cases, particles with different sizes, optical properties, and shapes are present in the measuring volume. The theoretical treatment of scattering by irregular particles is very difficult (Bohren and Huffmann 1983; Heller 1977; Holland and Cagne 1970; Shuerman 1980; Zerull, Giese, and Weiss 1971). Shape effects are often modeled with a distribution of equivalent spheres or spheroids, which is a reasonable approximation (D'Alessio 1981; Bohren and Huffmann 1983) for scattering in the forward region. Heterogeneous particles exhibit an internal variation of the optical properties, and thus their scattering cannot be predicted exactly. Scattering from stratified spheres furnishes a first-order approximation model for comparison with the experimental data. Examples of both cases will be discussed in more detail later.

For very dilute clouds of particles, it is still possible to measure some scattering properties of the individual particles and to obtain statistically the properties of the cloud. Although this method is discussed in detail elsewhere, a short section of this paper will be devoted to this approach in order to discuss the correspondence between measured and experimental scattering cross sections of droplets in the side scattering region.

For dense particle clouds, e.g., for particle clouds containing a considerable amount of submicronic particles, single-particle detection is not feasible, and ensemble scattering methods have to be used. The scattering and extinction coefficients from all particles present in the scattering volume are simultaneously measured. In the case of spherical particles in a single-scattering regime, scattering and extinction coefficients are related to the cross sections by the equations

$$Q_{VV}(\lambda, \theta) = \int_0^\infty NC_{VV}(D, m, \lambda, \theta)F(D)\,dD \tag{9-19}$$

$$Q_{HH}(\lambda, \theta) = \int_0^\infty NC_{HH}(D, m, \lambda, \theta)F(D)\,dD \tag{9-20}$$

$$k_{ext}(\lambda) = \int_0^\infty NC_{ext}(D, m, \lambda, \theta)F(D)\,dD \tag{9-21}$$

where N is the number density of the particles and $F(D)$ is the size distribution function. For different scattering regimes the simplifying assumption of bimodal distributions allows separation of the contributions of submicron (mainly soot in our cases) and supermicron particles (mainly droplets in our cases). Hence

$$Q_{ii} = Q_{ii}^s + Q_{ii}^d = N^s \int C_{ii}^s F^s(D) \, dD + N^d \int C_{ii}^d F^d(D) \, dD \qquad (9\text{-}22)$$

$$Q_{ext} = Q_{ext}^s + Q_{ext}^d = N^s \int C_{ext}^s F^s(D) \, dD + N^d \int C_{ext}^d F^d(D) \, dD \qquad (9\text{-}23)$$

where the superscripts s and d stand for soot and droplets, respectively.

The following sections will be devoted to experimental measurements of the monochromatic scattering and extinction coefficients in combustion systems and to their theoretical interpretation. The treatise is given stepwise. Thus, the first terms of the right sides of Eqs. (9-22) and (9-23) will be examined by studying sooting flames. In these systems, only soot particles are present. The second step consists of the study of the cross sections and scattering coefficients of droplets of different chemical composition at room temperature. Finally, both terms of the equations will be considered in the interpretation of the behavior of sprays in a combustion environment.

9-3 OPTICAL CHARACTERIZATION OF SOOT CLOUDS

A first characterization of the optical properties of soot particles in flames is derived from measurements of the angular pattern of the scattering coefficients. These measurements may be carried out at different heights above the burner in rich premixed flames. Figure 9-6 reports some typical angular distributions of $Q_{VV}(\theta)$, $Q_{HH}(\theta)$, and the depolarized component $Q_{HV}(\theta)$ both in the soot-forming region and in the soot particle agglomeration zone of CH_4/O_2 rich flames. In the soot-forming region the vertical polarized scattering coefficient Q_{VV} does not depend upon the scattering angle and Q_{HH} has its minimum at $\theta = 90°$. This is in agreement with the Rayleigh limit of the Lorenz-Mie theory. The presence of a depolarization component demonstrates a slight disagreement with the results predicted for homogeneous small spheres by Eq. (9-10). Measurements at other wavelengths have shown that part of the depolarization is due to emission of broadband fluorescence from PAHs. However, even in flame regions or under conditions where this contribution can be neglected, the scattering itself contains a depolarized component, which is not angularly dependent and equals 2% or 3% of the vertical component at $\theta = 90°$. The experimental scattering pattern in the soot growth region is in agreement with the model of a polydispersion of Rayleigh anisotropic spheres (Nussenzweig 1979):

$$Q_{VV} = N \int_0^\infty \frac{\pi D^2}{4} \left(\frac{\pi D}{\lambda}\right)^4 \left|\frac{m^2 - 1}{m^2 + 2}\right|^2 F(D) \, dD \qquad (9\text{-}24)$$

$$Q_{HH} = Q_{VV} \cos^2 \theta + Q_{HV} \sin^2 \theta \qquad (9\text{-}25)$$

$$Q_{HV} = Q_{VH} = \rho_V Q_{VV} \qquad (9\text{-}26)$$

where $\rho_V = Q_{HV}/Q_{VV}$ is the depolarization ratio. The extinction coefficient is given by the expression

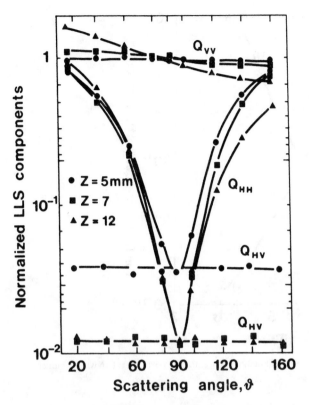

Figure 9-6 Angular distributions of the normalized scattering coefficients Q_{VV} and Q_{HH} at different heights above the burner for a methane-oxygen flame.

$$k_{ext} = N \int_0^\infty \frac{\pi D^2}{4} \left(\frac{\pi D}{\lambda} \right) \text{Im} \left(\frac{m^2 - 1}{m^2 + 2} \right) F(D) \, dD \qquad (9\text{-}27)$$

Thus the scattering and extinction coefficients are proportional to the sixth and third moment of the size distribution function, respectively. Combined measurements of both coefficients give the total number concentration and the $D_{6.3}$ average size of the soot particles present in the cloud for known complex refractive index. Examples of measurements of D and N in a premixed CH_4/O_2 flame with a C/O ratio equal to 0.55 is reported in Fig. 9-7. These types of measurements have furnished data for kinetic studies of soot inception, surface growth, and agglomeration.

Scattering and extinction measurements at different wavelengths shed some light on the structural properties of soot particles. Figure 9-8 shows that the scattering and extinction coefficients measured inside the soot-forming region of a premixed C_2H_4/O_2 flame (Menna and D'Alessio 1982) do not follow the λ^{-4} and λ^{-1} laws but present relevant peaks near 300 nm. This behavior results from the dispersion of the optical properties of soot particles that exhibit a resonance in the near UV.

The complex refractive index of soot can be quantitatively evaluated in this spectral

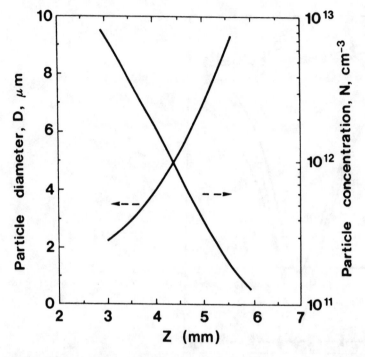

Figure 9-7 Particle diameter (μm $\times 100$) and number concentration versus height above the burner for a methane-oxygen flame with C/O = 0.55.

range from near UV to visible from these measurements, and Eqs. (9-24) to (9-27) provide the average size and number concentration of the soot particles or the optical properties at a reference wavelength from independent measurements. Detailed discussion of the procedures and assumptions are given elsewhere (Vaglieco, Beretta, and D'Alessio 1990) and only the final results are shown in Fig. 9-9. It appears that the optical properties of soot particles are very different from those of graphite and other types of carbon films obtained by deposition under nonequilibrium conditions.

Although the interpretation of the optical properties on the basis of the structural properties of soot particles is beyond the scope of this chapter, it is worthwhile to point out that the behavior of the real and imaginary parts of the refractive index are consistent with a Lorenz-Drude model and that the resonance is due to a transition of π electrons in an excited band.

Interpretation of the angular pattern of the scattering coefficients and their wavelength dependence in the soot agglomeration region is more difficult. Electron microscopy shows that soot particle agglomerates have irregular shapes.

From a scattering point of view the agglomerates may be regarded as spherical multiple-center scatterers. In normal pressure flames their size falls into a range where Lorenz-Mie theory predicts particle size dependent angular and spectral distributions. Examples of evaluation of the average size or the parameters of a two-parameter size

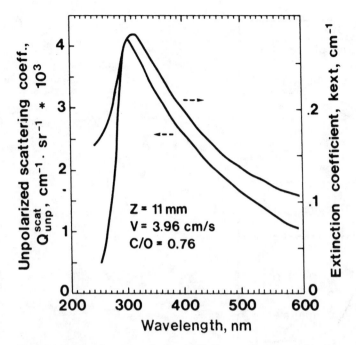

Figure 9-8 Scattering and extinction coefficients versus the wavelength measured at 11 mm above the burner in a C_2H_4-O_2 flame.

distribution function from these measurements are given in the literature (Bockhorn et al. 1981).

9-4 DETERMINATION OF THE SCATTERING CROSS SECTIONS OF SINGLE DROPLETS

In order to verify experimentally the scattering model of droplets based on the Lorenz-Mie theory outlined in Section 9-2, an optical apparatus as shown in Fig. 9-10 has been designed.

A vertically polarized argon-ion laser beam with a radially uniform intensity profile passes through a beam splitter to obtain two beams of equal intensity. One of them passes through a $\lambda/2$ plate to rotate the polarization plane. A polarizing beam splitter then recombines the beams. After passing the combined beam through a $\lambda/4$ plate, two clockwise and anticlockwise circularly polarized beams are produced. The two beams are crossed in the measuring volume. In this way, no light intensity variations occur in the measuring volume, but the polarization changes, generating a pattern of "plane polarization fringe." Single calibrated droplets, produced by a Berglund-Liu atomizer, generate a burst of scattered light when passing through the fringe pattern. The modulated amplitude of the burst is related to the polarization ratio

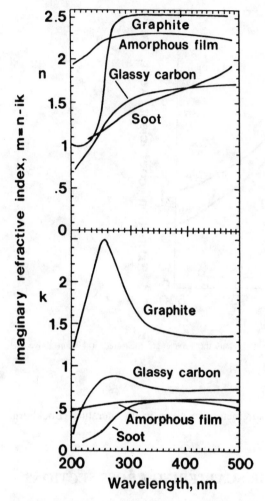

Figure 9-9 Real and imaginary part of the refractive index of soot, graphite, glassy, and amorphous carbon versus wavelength.

$\gamma(\theta) = I_{HH}/I_{VV} = C_{HH}/C_{VV}$, and its frequency is proportional to the velocity of the droplets. An example of such measurements is reported in Fig. 9-11, where the angular patterns of the polarization ratio for monodisperse droplets, 42 and 21 μm, of water ($m = 1.33$) are compared with those computed with Lorenz-Mie theory. The nearly perfect agreement between theory and experiments indicates that measurements of the polarization of the scattered light can be employed to give reliable results in the diagnostics of sprays.

On this basis, a more quantitative experimental and theoretical analysis of the contribution of the surface waves to the scattering at $\theta = 90°$ was executed; it was found that the part of the cross section due to surface waves is proportional to the

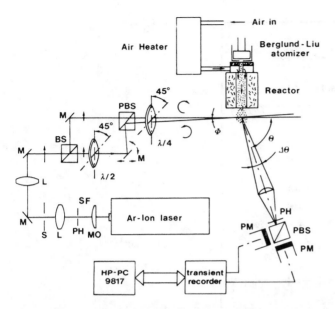

Figure 9-10 Setup for single-droplet experiments. BS, beamsplitter; L, lens; M, mirror; MO, microscope objective; PBS, polarizing beamsplitter; PH, pin hole; PM, photomultiplier; S, stop.

diameter of the droplet, while that due to external reflection is proportional to the surface area. This results in

$$C_{VV}^{d}(90°) = C_{VV}^{d,sw}(90°) + C_{VV}^{d,refl}(90°) = A_1 D + A_2 D^2 \qquad (9\text{-}28)$$

$$C_{HH}^{d}(90°) = C_{HH}^{d,sw}(90°) + C_{HH}^{d,refl}(90°) = A_3 D + A_4 D^2 \qquad (9\text{-}29)$$

Since in the micron range, $C_{VV}^{d,sw} << C_{VV}^{d,refl}$, the measured cross section for the component polarized in the vertical plane is approximately equal to that due to reflection. Hence the polarization ratio is given by the expression

$$\gamma(90°) = \frac{C_{HH}}{C_{VV}} = \frac{A_3}{A_2}\left(\frac{1}{D}\right) + \frac{A_4}{A_2} \qquad (9\text{-}30)$$

Thus for small droplets, $\gamma(90°)$ is inversely proportional to the droplet diameter. It reaches the asymptotic value due to reflection only for droplets larger than 80 μm. Experimental measurements and numerical analysis show that approximately the same results are obtained for droplets of hydrocarbons with a refractive index between 1.5 and 1.6.

The experimental apparatus is based on an interferometric principle that allows simultaneous determination of the joint distribution of size and velocity.

The same apparatus and technique have been used for measuring the scattering cross sections and the polarization ratio of partially absorbing droplets near the Brewster angle ($60 \leq \theta \leq 70°$). In this angular regime, geometrical optics are adequate ap-

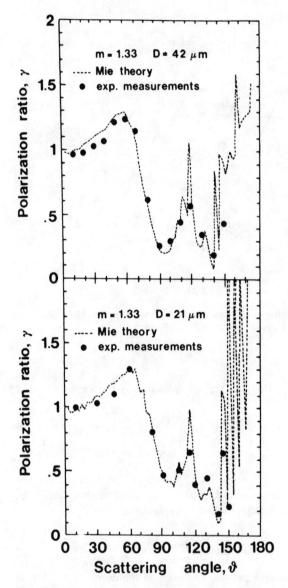

Figure 9-11 Comparison of experimental angular distributions of the polarization ratio with computed values for two sizes of water droplets.

proximations, since surface wave effects may be neglected in comparison with reflection and refraction. However, the refraction cross section is exponentially reduced from its transparent value C_{VV}^{refl} due to the partial absorption of the ray inside the droplet.

$$C_{VV}(\theta) = C_{VV}^{refl}(\theta) + C_{VV}^{refr} \exp - [(4\pi k/\lambda)L] \tag{9-31}$$

$$C_{HH}(\theta) = C_{HH}^{refl}(\theta) + C_{HH}^{refr} \exp - [(4\pi k/\lambda)L] \tag{9-32}$$

At the Brewster angle the reflective scattering cross section for horizontally polarized light is small, and thus the damping effects on the refractive contribution are more evident in this region. Figure 9-12 shows the dependence of the polarization ratio upon the optical depth. Theoretical predictions are compared with experimental results of isopropyl alcohol droplets with calibrated size and variations of the absorption coefficient, obtained with increasing concentration of a red dye. This effect may be exploited for estimating the size of the droplets when the imaginary part of the refractive index is known. More interestingly, this effect can be used to infer variations of the

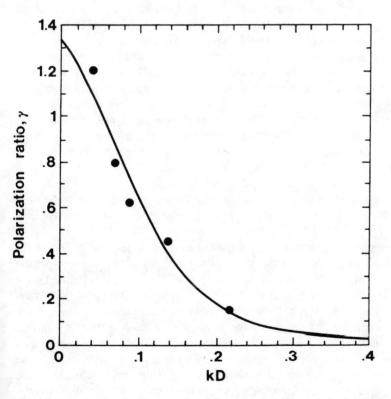

Figure 9-12 Comparison between experimentally measured and theoretical values of the polarization ratio dependent on the optical depth for isopropyl alcohol 40-μm droplets with different concentrations of a red dye.

optical properties, e.g., as a consequence of a chemical reaction, for droplets whose size is known from other measurements, e.g., from scattering measurements.

This effect is currently employed for studying liquid phase pyrolysis and oxidation effects for droplets introduced into high-temperature furnaces. Some examples of this method will also be reported in the discussion of the optical properties of sprays of hydrocarbons in isothermal and burning conditions.

9-5 OPTICAL CHARACTERIZATION OF SPRAYS

Sprays of Liquid Hydrocarbons in the Isothermal Case

The effects outlined in the previous section may also be applied for the optical characterization of sprays with polydisperse clouds of droplets. A systematic study of sprays using single-droplet measurements and statistical treatment are presented in Chapter 7. Hence ensemble measurements of the scattering coefficients are illustrated in this section.

Practical liquid fuels are a mixture of many hydrocarbons, and their refractive indices, $m = n - ik$, change according to their composition. The real part n can be easily estimated from dependence on the composition, considering that the molar refraction is nearly an additive property of the bond refractivity. Large saturated hydrocarbons have n values between 1.4 and 1.5 and aromatics between 1.5 and 1.6.

The variation of the imaginary part of the refractive index k is much more conspicuous, as is shown in Fig. 9-13. In this figure, values of k for a light oil and a heavy oil are represented in the wavelength range between 600 and 200 nm. Only aromatic or large unsaturated molecules that contain delocalized π electrons absorb light in this energy range; thus the high values of k in the ultraviolet are mainly attributed to the aromatic fraction of the fuels, constituted of compounds with single or double rings with side chains.

Heavy oils also contain polycyclic aromatic hydrocarbons with heterocyclic rings containing O, N, and S atoms. They are responsible for absorption in the visible in this case.

These chemical differences are also manifest in the angular distribution of the scattering coefficients obtained in the visible from sprays of light and heavy fuels atomized under similar conditions (compare Fig. 9-14). The droplets of light oil show an angular distribution of γ typical for transparent spheres. The polarization ratio attains a high value in the forward scattering region and a minimum in the rainbow in the backscattering region. This is similar to the results reported for water in Fig. 9-11. On the contrary, the pattern of the polarization ratio of heavy oil exhibits a decrease in the forward scattering region and does not show any minimum in the backscattering region because the refractive component is strongly attenuated.

Q_{VV} and Q_{HH} in the side scattering region, where surface wave effects are significant, are almost proportional to the first two moments of the size distribution function

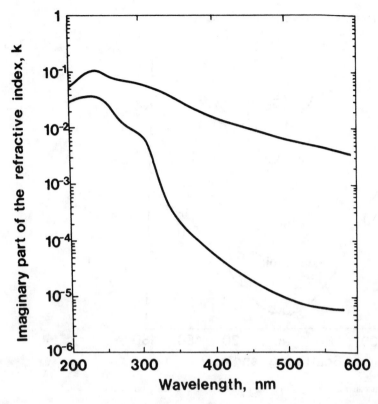

Figure 9-13 Dependence of imaginary part of the refractive index k on wavelength for light and heavy oils.

$$Q_{HH} = N \int_0^\infty C_{HH}F(D) \, dD \propto \int_0^\infty DF(D) \, dD \qquad (9\text{-}33)$$

$$Q_{VV} = N \int_0^\infty C_{VV}F(D) \, dD \propto \int_0^\infty D^2F(D) \, dD \qquad (9\text{-}34)$$

Thus from the values of the two scattering coefficients Q_{VV} and Q_{HH} it is possible to obtain the D_{21} average diameter of the droplets and their number concentration without determining the size distribution function. This method has been extensively employed in studying the atomization properties of pressure and air-assisted atomizers (Beretta, Cavaliere, and D'Alessio 1983*a*, *b*, 1984) and, more recently, those of ultrasonic atomizers (Tamai 1987).

It is worthwhile to point out that this method results in average sizes of about 3–4 μm for efficient atomization. These values are significantly smaller than those estimated by photographic or holographic techniques. The reason for this discrepancy is that ensemble scattering takes appropriately into account the smaller droplets that are not detectable with photographic methods.

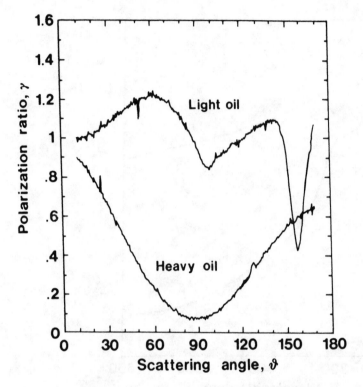

Figure 9-14 Angular patterns of the polarization ratio for light and heavy oil sprays.

A discussion of the relevance of these results for a more detailed understanding of atomization is outside the scope of this chapter. However, it is obvious from the measurements that the momentum exchange between the liquid and gas phase is locally more efficient than considered before. This momentum exchange may be controlled by the small-scale structure of the turbulent flow field.

In principle, it is also possible to obtain the size distribution function by solving the integral equations given by the angular distribution of the scattering coefficients. This is the standard procedure for measurements in the forward diffraction lobe (Bayvel and Jones 1981).

The same results can be achieved by measuring the scattering coefficients at the Brewster angle at different wavelengths across the threshold of an absorption band of the liquid. In this case, the horizontally polarized cross section decreases due to the internal absorption of the refracted contribution, as discussed before. The integral equation takes the form

$$Q_{HH}(\lambda) \propto A \int \exp\left[-\frac{4\pi k(\lambda)}{\lambda} D\right] D^2 F(D) \, dD \qquad (9\text{-}35)$$

which can be solved by an iterative nonlinear numerical procedure to determine the size distribution function (Beretta et al. 1986).

An example of the method is given in Fig. 9-15. This figure gives the normalized scattering coefficients for sprays of light oils generated by pressure and air-assisted atomizers for wavelengths between 300 and 420 nm. The scattering coefficients fall off near 360 nm, just where the absorption of the hydrocarbon mixture becomes significant.

Figure 9-16 reports the size distribution function obtained for the two atomization conditions. This type of measurement again indicates rather small mean droplet sizes and a narrow size distribution.

Sprays of Coal Water Slurries

The optical study of solid-liquid slurries is more complex than that of liquids for which the model of homogeneous spherical droplets is excellent. The following discussion of the atomization of coal water slurries is a good illustration of this point (Beretta et al. 1985).

Figure 9-17 reports the experimental angular distribution of the polarization ratio $\gamma = Q_{HH}(\theta)/Q_{VV}(\theta)$ measured for water and coal water sprays (CWS) under different

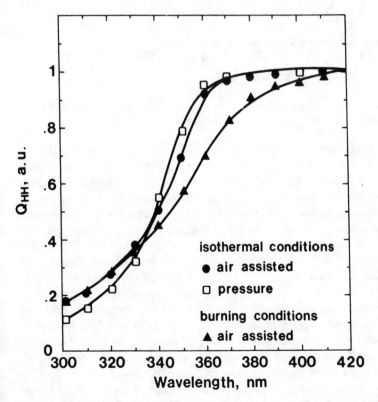

Figure 9-15 Dependence of the horizontally polarized scattering coefficients for light oil spray flames from pressure nozzles and air-assisted atomizers.

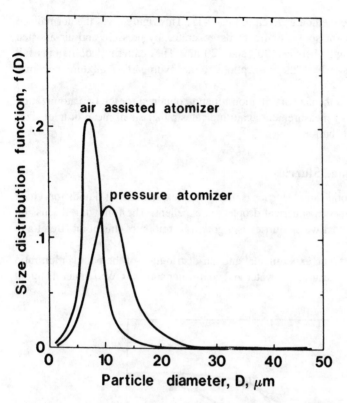

Figure 9-16 Size distribution functions measured with the spectral ensemble light scattering (SELS) technique for sprays from pressure nozzles and air-assisted atomizers.

conditions. The upper curve shows the familiar behavior of water. The lowest curve has been obtained for a coal water spray introduced in high-temperature environments, thus evaporating all the water. This dried CWS is representative of the optical properties of coal particles. The intermediate curves refer to CWS atomized by an air-assisted atomizer with different airflow rates. Quantitative interpretation of the intermediate curves requires a scattering model for these two component sprays. Simpler scattering models consider the two phases to be totally segregated or assume that all the water covers the coal particles. In the latter case the appropriate scattering model is the Lorenz-Mie theory for coated spheres (Toon and Ackerman 1981). Results of computations according to this model are given in Fig. 9-18. The polarization ratio is given as a function of the ratio $\phi = D_c/D$, where D_c is the diameter of the inner core constituted by an absorbing spherical particle ($m_c = 1.8 - i0.1$), and D is the external diameter of the sphere, the external layer consisting of water ($m_w = 1.33$). The optical properties at $\theta = 60°$ and $\theta = 90°$ are coincident with those of water for $\phi \leq 0.7$, while for larger ϕ they are similar to those of coal. The coated sphere model is equivalent to the model of totally segregated phases, where the water phase is constituted by water droplets with a very small coal core ($\phi < 0.7$) and the coal phase consists of coal particles with a very thin layer of water ($\phi > 0.7$).

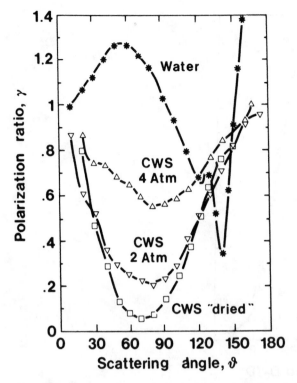

Figure 9-17 Measured angular distribution of the polarization ratio for water and CWS sprays under different conditions.

For this model the scattering coefficients are the superposition of the scattering coefficients of the two phases

$$Q_{HH}^{cws}(\theta) = Q_{HH}^{w}(\theta) + Q_{HH}^{c}(\theta) \qquad (9\text{-}36)$$

and

$$Q_{VV}^{cws}(\theta) = Q_{VV}^{w}(\theta) + Q_{VV}^{c}(\theta) \qquad (9\text{-}37)$$

In the forward region at $\theta = 60°$ near the Brewster angle, all four cross sections are proportional to the surface area. By consequence, the scattering coefficients are proportional to the total surface area per unit volume, S_j. Measurements of the polarization ratio at this angle give directly the ratio of the total surface area of the two phases $R = S_c/S_w$. At $\theta = 90°$ the scattering coefficient Q_{HH}^{w} is proportional to the averaged perimeter per unit volume of the water phase. The other three coefficients are proportional to the specific surface area. By consequence, the polarization ratio at this angle depends both on the average diameter D_{21} of the water phase and on the surface ratio R.

These considerations are elucidated in Fig. 9-19, where the measured polarization ratios at $\theta = 60°$ and $\theta = 90°$ for CWS atomized with increasing flow rates of the

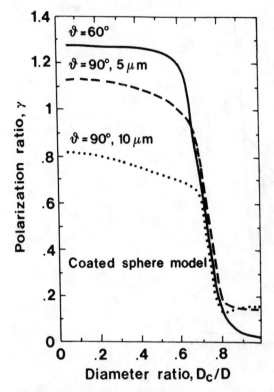

Figure 9-18 Computed polarization ratio for the coated sphere model as a function of the inner/outer diameter ratio at $\theta = 60°$ and $90°$ for two droplet sizes.

atomization air are given. The average diameter of the water droplets decreases down to values smaller than 5 μm with more efficient atomization, and the surface area ratio R increases by an order of magnitude.

The effect of improved atomization can also be observed for other coal-liquid slurries such as coal–light oil. It may be responsible for the ignition delays of CWS or, on the contrary, for a precombustion of the liquid in the case of a combustible liquid phase (J.M. Béer, personal communication, 1987).

Sprays in Combustion Systems

In a burning spray the droplets not only vaporize, but may also undergo chemical reactions in the liquid phase. In addition, soot particles are formed in the gas phase in regions with high fuel-air ratios. The scattering coefficients in burning sprays consist of micronic particles, whose optical properties are changing, and submicronic particles. The scattering coefficients can be described by Eq. (9-22) as discussed in Section 9-2.

In the initial part of the flame, where the spray has a high momentum and is not yet dispersed completely into the air, it is possible to follow separately the different

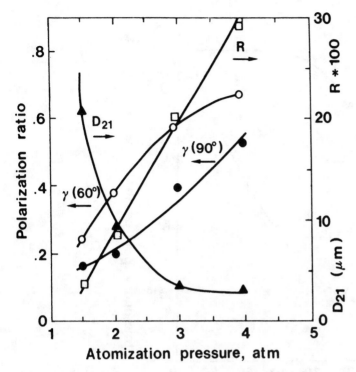

Figure 9-19 Measured profiles of $\gamma(60°)$, $\gamma(90°)$, D_{21}, and R (surface ratio) as a function of the atomization pressure.

effects. This situation prevails in the center of a spray surrounded by a flame. Figure 9-15 gives the spectral distribution of the Q_{HH} coefficients at $\theta = 67°$ for this case. Comparison with the corresponding values obtained in isothermal conditions shows a decrease of the scattering coefficients at longer wavelengths. This effect cannot be related to the size of the droplets but is due to a strong increase of the imaginary part of the refractive index. In the visible, k increases from 10^{-5} to values higher than 10^{-3} (Beretta et al. 1986). This blackening of the droplets has been confirmed independently by sampling the flames in the corresponding region and spectrophotometric measurements of the samples (Prati 1988).

The darkening of the initially transparent droplets implies that the droplet size may be derived from scattering measurements in the forward region in addition to measurements in the side scattering region as discussed above. Combined measurements in both scattering regions give simultaneously the average size and the imaginary part of the refractive index of the droplets (Beretta, Cavaliere, and D'Alessio 1982). Results of this type of measurements are given in Fig. 9-20, where the axial profiles of both quantities are reported for a light oil flame. The increase of the average size during vaporization is easily explained, considering that in a polydisperse spray the smaller droplets disappear at a faster rate than the larger ones.

The interpretation of the increase of k requires complex considerations of a chem-

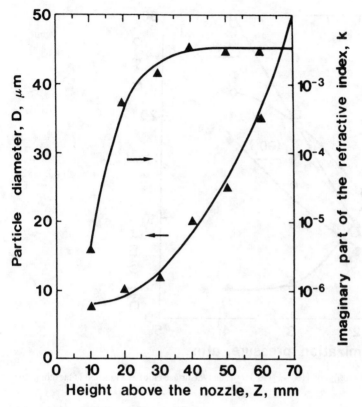

Figure 9-20 Axial profiles of the droplet model diameter and of the imaginary part of the refractive index for a light oil flame.

ical and spectroscopic nature, whose discussion is outside the scope and limit of this chapter. It is sufficient to point out that this effect is an indicator of the oxidation and/ or polymerization of the aromatic components of the fuel, which at room temperature have their absorption bands in the ultraviolet. Carrying out the scattering measurements, also in spatial regions where combustion takes place, submicron soot particles provide a more substantial contribution to the overall scattering.

An illustration of these combined effects is given in Fig. 9-21 in terms of the polarization pattern measured at different heights above the burner on the axis of a light oil spray flame. At $z = 15$ mm above the atomization nozzle, the angular distribution of the polarization ratio is similar to that predicted by the theory for slightly absorbing micron spheres (see Fig. 9-14). Further downstream, the polarization ratio decreases. A minimum in the rainbow region is still detectable, and at $z = 200$ mm the polarization ratio follows closely the theoretical relation for absorbing submicron particles $\gamma(\theta) = \cos^2 \theta$. These measurements demonstrate the transition from the original fuel droplets via partially absorbing and evaporating ones to the formation of submicronic soot particles from the fuel vapor. The values of the polarization ratio at $\theta = 90°$ can be assigned to specific properties of the scatterers. For $\gamma(90°) = 2 \times$

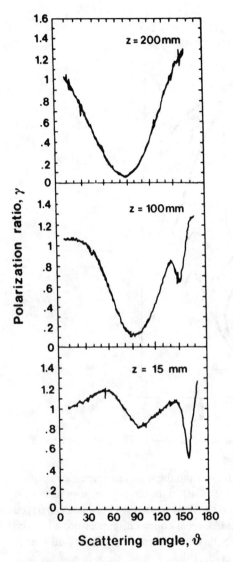

Figure 9-21 Angular distribution of the polarization ratio measured at different heights above the burner in a light oil flame.

10^{-2}, only soot particles are present in the scattering volume, while the exclusive presence of droplets is assured as far as $\gamma(90°) \geq 10^{-1}$ (Beretta, Cavaliere, and D'Alessio 1982). The structure of spray flames can be characterized in terms of the scattering coefficients and polarization ratio. Figures 9-22 and 9-23 give maps of the scattering coefficients $Q_{VV}(90°)$ and the polarization ratios $\gamma(90°)$ of a swirled light oil flame. The central part of the spray is initially occupied by droplets, since $\gamma(90°)$ varies from 0.7 to 0.1. The vaporization of the droplets and the entrainment of the

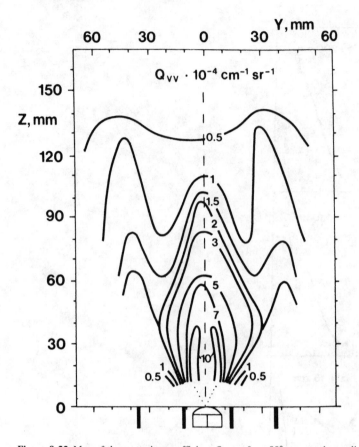

Figure 9-22 Map of the scattering coefficient Q_{VV} at $\theta = 90°$ measured on a light oil spray flame.

swirled air can be observed. The final upper region and the initial external recirculation zone are occupied by soot particles, since $\gamma \leq 2 \times 10^{-2}$. In this region, soot formation and oxidation can be studied. With this method the flame structures have been analyzed for burning sprays of light oils with different chemical composition pressure atomized (Beretta, D'Alessio, and Noviello 1986; Barbella et al., in press) and heavy oil flames atomized with air-assisted nozzles (Beretta et al. 1984).

Polarization ratio measurements allow clear discrimination between soot and droplets. A more quantitative interpretation of the results arises from additional radially resolved measurements of the scattering and extinction coefficients. From these additional measurements the average size and the volume fraction of soot particles are obtained by applying the scattering/extinction method discussed in Section 9-3 (Beretta et al. 1981), while the average size and volume fraction of the droplets are obtained from the polarization ratio measurements at $\theta = 90°$.

An example is reported in Fig. 9-24. This figure shows the distribution of soot in the high-temperature external recirculation region that surrounds the fuel spray in a moderately swirled light oil spray flame similar to gaseous diffusion flames. The

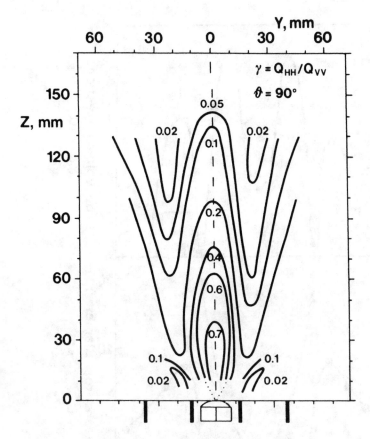

Figure 9-23 Map of the polarization ratio measured on a light oil spray flame.

central region is still occupied by droplets whose sizes and volume fraction show the typical distributions produced by a hollow-cone spray atomizer.

Transient Dense Spray

The analysis of transient sprays, e.g., in diesel combustion, requires simultaneously adequate time and space resolution. These requirements can be met with quantitative imaging techniques, which are the object of this section.

The term imaging refers to the capability of sampling a large number of points in a two-dimensional space field (Browse 1987; Dario and De Rossi 1985). The dimensions of the sampling volume are comparable to the distance between sampling volumes. The imaging is said to be quantitative when the signal from each sampling point is a measure in terms of accuracy and repeatability. In principle, it is possible to scan a two-dimensional field by shifting a single detector from point to point with high spatial resolution. It is evident that the scanning time has to be shorter than the characteristic time of the field evolution. The maps of the scattering coefficients and

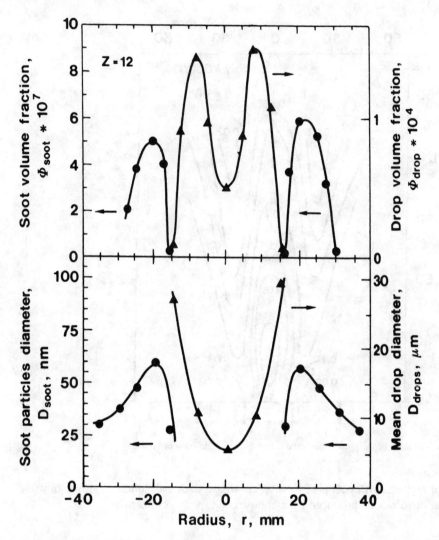

Figure 9-24 Radial distribution of the sizes and volume fractions of soot and droplets in a light oil flame at $z = 12$ mm.

the polarization ratio given in Figs. 9-22 and 9-23 can be regarded approximately as quantitative imaging. Even though they have been obtained from single-point measurements, spatial interpolation can satisfactorily simulate the high-density information from a larger number of points on an array detector. In steady state phenomena the use of array detectors is suggested more for convenience than on principle. On the other hand, it is a compulsory choice in the analysis of, e.g., diesel sprays. This is an unsteady phenomenon in which the spatial distribution of the condensed phase can be obtained only from multipoint measurements within very short characteristic times (less than 1 μs).

At any rate, the application to diesel sprays reported here can be considered as a guideline to other studies on transient processes, e.g., turbulent flows (Winter, Lam, and Long 1987). The optical setup described below can be a reference for quantitative imaging of fluorescence (Suntz et al. 1988), Raman scattering (Long et al. 1983), and emission measurements.

Figure 9-25 shows schematically a technical solution adopted for the study of the formation of diesel sprays (Ragucci et al. 1989; Cavaliere et al. 1990a). The light source is a double-pulsed neodymium: yttrium/aluminum/garnet (Nd: YAG) laser tuned to its second harmonic. The laser beam is shaped into a thin light sheet through two confocal cylindrical lenses. The collecting optic is placed along an axis perpendicular to the laser beam and focuses the scattering images with magnification ratios up to 10 by a set of macro-objective lenses. These images are subsequently split with a 50-mm cube and focused on two microchannel plate intensifiers (MCP) gated by the laser pulse. The photoelectrons coming from the MCP are finally directed through bundles of fiber optics to a 576 × 384 interlaced CCD camera with negligible lateral charge diffusion. The video signals of each camera are sent to analog-to-digital (A/D) converters sampling 512 × 512 pixels and stored into a buffer memory. Selected digital images are processed by a host computer, where image enhancement, noise reduction, stray light subtraction, intensity calibration, and comparison between the images are performed numerically.

The simultaneous use of two arrays allows two-dimensional "frozen" light-scattering measurements at two different polarization states, or different scattering angles, or different wavelengths. In this way, all the effects presented in the previous sections for single-point measurements can be used for the two-dimensional case. By operating the laser in a double-pulse mode, instantaneous temporal variations of the measured quantities can be followed. In principle, it is possible to determine the two-dimensional velocity field of the droplets or spatial correlations of properties in burning sprays.

The mathematical correlation of two quantitative images recorded on two different cameras can be performed only on picture elements (pixels) corresponding to the same

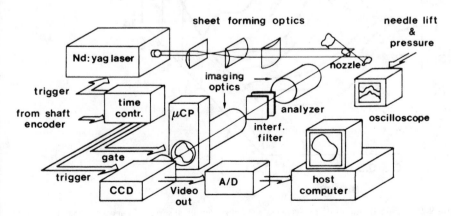

Figure 9-25 Optical setup for two-dimensional laser light scattering.

objective point. Possible shift and magnification differences between the two images can be corrected through numerical codes within the measurable uncertainty quantified in terms of pixels. Therefore mathematical operations must be performed on signals averaged over a group of pixels larger than this uncertainty (grouping technique). Within these limits, single-point and laser light scattering techniques are equivalent. For these kinds of applications, the discussion in the previous sections gives satisfactory references.

On the other hand, quantitative single images deserve some additional comments. The very potential of the imaging consists of the possible local correlation of properties in two-dimensional space. This is relevant for the analysis of flame structure but also, from the diagnostic point of view, for rejecting multiple scattering in dense sprays. In fact, digital images of atomized liquids in dense sprays contain signals from multiple scattering. Multiple scattering in dense sprays smears out the smaller structures and produces a nearly continuous background. However, the imaging technique and the connected numerical procedures allow the background to be easily circumscribed and subtracted from the signal. In this way, the light scattered from the prominent shapes is statistically elaborated.

Measurements reported here refer to an imaging study of the light-scattered sub-

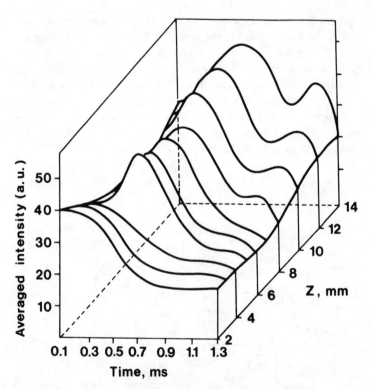

Figure 9-26 Temporal profiles of the averaged scattering intensity at several distances from the nozzle for a spray injected at atmospheric pressure and room temperature.

sequent cross sections of diesel sprays (1–15 mm from the nozzle) at times after the beginning of the injection ranging from 0.1 to 1.3 ms. For these measurements a shot-by-shot averaging procedure (Cavaliere et al. 1990a, b) has been used.

At first the absolute value of the spatially averaged scattered intensity was computed because this quantity in a single-scattering regime is proportional to the specific total surface area of the liquid phase. The temporal evolution of the averaged data, defined as the atomization parameter, is given in Fig. 9-26 for different distances from the nozzle. The second type of information derived from the digital images is the number of the nonconnected areas in the maps of the scattering intensities, which (for fixed values of the atomization parameter) gives a measure of the liquid dispersion. The temporal evolution of this parameter is reported in Fig. 9-27 for the same set of images as in Fig. 9-26. It appears that the atomization parameter is high at short times (<0.5 ms) near the nozzle ($z < 4$ mm), after about half of the injection period at distances higher than 4 mm, and during the needle shutdown period at large distances from the nozzle. The dispersion parameter is low in the early spray period and increases at large distances and large times.

Without entering into a physical discussion of diesel spray formation, given in detail elsewhere (Cavaliere et al. 1988, 1990b), it is worthwhile to point out some qualitative arguments. Time-varying fuel injections generate sprays that cannot be

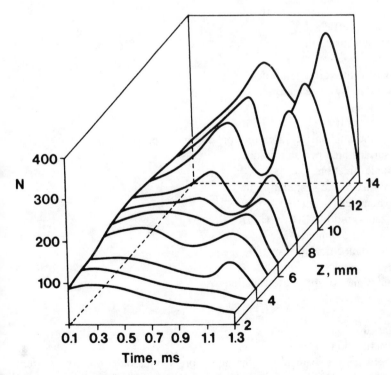

Figure 9-27 Temporal profiles of the number of the nonconnected structures at several distances from the nozzle for a spray injected at atmospheric pressure and room temperature.

analyzed in terms of characteristics observed under stationary conditions. Different regimes of atomization have to be considered in order to justify the recorded two-dimensional scattering pattern. The different regimes occur in length and time scales comparable to those of the combustion processes. They have to be taken properly into account for predicting the performance and pollutant emission from diesel engines.

9-6 NOMENCLATURE

a_n	scattering coefficient
b_n	scattering coefficient
C_{abs}	absorption cross section
C_{ext}	extinction cross section
C_{scatt}	scattering cross section
D	particle diameter
D_c	diameter of the inner core
$\mathbf{E}_0$	electric field intensity
$F(D)$	size distribution function
I_i	incident energy flux
J_1	Bessel function of first order
k	imaginary part of the refractive index
k_{ext}	extinction coefficient
L	optical path inside the droplet
m	complex refractive index
n	real part of the refraction index
N	number density of scatterers
$\mathbf{P}$	dipole moment
Q	scattering coefficient
R	ratio of the total surface area for coal and water phase ($= S_c/S_w$)
S	total surface area per unit volume
S_1, S_2	amplitude function
W_s	energy flux of the scattered light
α	size parameter ($= \pi D/\lambda$)
γ	polarization ratio ($= (C_{HH}/C_{VV})$
δ	phase of the scattered wave
ϵ	complex dielectric function
ϵ'	real part of dielectric function
ϵ''	imaginary part of dielectric function
θ	scattering angle
λ	wavelength
π_n	angular function
τ_n	angular function
ϕ	ratio between the diameters of the inner core and the outer one for a coated sphere ($= D_c/D$)

Subscripts and Superscripts

H   light polarized parallel to the scattering plane
V   light polarized perpendicular to the scattering plane
c   coal
d   droplet
s   soot
w   water

REFERENCES

Barbella, R., Beretta, F., Ciajolo, A., D'Alessio, A., Prati, M. V., and Tamai, R. 1990. *Twenty-second symposium (international) on combustion*, p. 1983. Pittsburgh, Pa.: The Combustion Institute.

Bauckhage, K. 1988. Applications of the phase-Doppler-method to spray analysis. In *Proceedings of the 4th ICLASS 88*, p. 279. Japan: Sundai.

Bayvel, L. P., and Jones, A. R. 1981. *Electromagnetic scattering and its applications*. London: Applied Science Publishers.

Beretta, F., Cavaliere, A., and D'Alessio, A. 1982. *Nineteenth symposium (international) on combustion*, p. 1359. Pittsburgh, Pa.: The Combustion Institute.

Beretta, F., Cavaliere, A., and D'Alessio, A. 1983a. Experimental and theoretical analysis of the angular pattern distribution and polarization state of the light scattered by isothermal sprays and oil flames. *Combust. Flame* 49:183.

Beretta, F., Cavaliere, A., and D'Alessio, A. 1983b. *Sixth symposium on air breathing engines*, p. 186. New York: American Institute of Aeronautics and Astronautics.

Beretta, F., Cavaliere, A., and D'Alessio, A. 1984. Drop size and concentration in a spray by sideward laser light scattering measurements. *Combust. Sci. Technol.* 36:19.

Beretta, F., D'Alessio, A., and Noviello, C. 1986. *Twenty-first symposium (international) on combustion*, p. 1133. Pittsburgh, Pa.: The Combustion Institute.

Beretta, F., Cavaliere, A., Ciajolo, A., D'Alessio, A., Di Lorenzo, A., Langella, C., and Noviello, C. 1981. *Eighteenth symposium (international) on combustion*, p. 1091. Pittsburgh, Pa.: The Combustion Institute.

Beretta, F., Cavaliere, A., D'Alessio, A., De Michele, G., and Pasini, S. 1984. *Rev. Gen. Therm.* 276:827.

Beretta, F., Cavaliere, A., D'Alessio, A., and Ragucci, R. 1985. Ensemble laser light scattering technique for the analysis of atomization on coal-water slurry. In *Proceedings of the 3rd ICLASS 85*.

Beretta, F., Cavaliere, A., D'Alessio, A., Massoli, P., and Ragucci, R. 1986. *Twenty-first symposium (international) on combustion*, p. 675. Pittsburgh, Pa.: The Combustion Institute.

Birks, J. B. 1970. *Photophysics of aromatic molecules*. New York: Wiley-Interscience.

Bockhorn, H., Fetting, F., Meyer, U., Reck, R., and Wannemacher, G. 1981. *Eighteenth symposium (international) on combustion*, p. 1137. Pittsburgh, Pa.: The Combustion Institute.

Bohren, C. F., and Huffmann, D. R. 1983. *Absorption and scattering of light by small particles*. New York: John Wiley.

Born, M., and Wolf, E. 1970. *Principles of optics*. New York: Pergamon.

Browse, R. A. 1987. Feature-based tactile object recognition. *IEEE Trans. Patterns Anal. Mach. Intel.* PAMI9(6):779.

Cavaliere, A., Ragucci, R., D'Alessio, A., and Noviello, C. 1988. Analysis of dense sprays through two-dimensional laser light scattering. In *Proceedings of the conference on spray combustion*. (ILAS Europe Meeting) Rouen, France.

Cavaliere, A., Ragucci, R., D'Alessio, A., and Noviello, C. 1990a. *Proceedings of the conference on heat and mass transfer in gasoline and diesel engine*, ed. D. B. Spalding. Washington, D.C.: Hemisphere.

Cavaliere, A., Ragucci, R., D'Alessio, A., and Noviello, C. 1990b. *Twenty-second symposium (international) on combustion*, p. 1973. Pittsburgh, Pa.: The Combustion Institute.

Chen, S. H. 1981. Interaction of thermal neutrons and photons with matter. In *Scattering techniques applied to supramolecular and nonequilibrium systems*, eds. S. H. Chen, B. Chu, and R. Mossal, p. 3. New York: Plenum.

D'Alessio, A. 1981. Laser light scattering and fluorescence diagnostics of rich flames produced by gaseous and liquid fuels. In *Particulate carbon: formation during combustion*, eds. D. C. Siegla and G. W. Smith, p. 207. New York: Plenum.

D'Alessio, A., Cavaliere, A., and Menna, P. 1983. Theoretical models for the interpretation of light scattering by particles present in combustion systems. In *Soot in combustion systems and its toxic properties*, eds. J. Lahaye and G. Prado, p. 327. New York: Plenum.

D'Alessio, A., Beretta, F., Cavaliere, A., and Menna, P. 1983. Laser light scattering and fluorescence in fuel rich flames: techniques and selected results. In *Soot in combustion systems and its toxic properties*, eds. J. Lahaye and G. Prado, p. 355. New York: Plenum.

Dario, P., and De Rossi, D. 1985. Tactile sensors and the gripping challenge. *IEEE Spectrum* 18:46.

Glantschnig, W. J., and Chen, S. 1981. Light scattering from water droplets in the geometrical optics approximation. *Appl. Opt.* 20:2499.

Heller, W. 1977. Fast precise size distribution from light scattering extrema. In *Optical polarimetry*, SPIE vol. 112, p. 158. New York: Society of Photo-Optical Instrumentation Engineers.

Holland, A. C., and Cagne, G. 1970. The scattering of polarized light by polydisperse systems of irregular particles. *Appl. Opt.* 9:1113.

Kerker, M. 1969. *The scattering of light and other electromagnetic radiation*. San Francisco, Calif.: Academic.

Logan, N. A. 1965. Survey of some early studies of the scattering of plane waves by a sphere. *Proc. IEEE* 16:773.

Long, M. B., Fourguette, D. C., Escoda, M. C., and Layne, C. B. 1983. Instantaneous Ramanography of a turbulent diffusion flame. *Opt. Lett.* 8:244.

Massoli, P., Beretta, F., and D'Alessio, A. 1989. *Appl. Opt.* 28(6):1200.

Menna, P., and D'Alessio, A. 1982. *Nineteenth symposium (international) on combustion*, p. 1421. Pittsburgh, Pa.: The Combustion Institute.

Nussenzweig, H. M. 1979. *J. Opt. Soc. Am.* 69:1068.

Prati, M. V. 1988. Tesi di Laurea in Ingegneria Chimica, Università di Napoli, October 1988.

Ragucci, R., Cavaliere, A., D'Alessio, A., and Massoli, P. 1989. Quantitative measurements from 2-D elastic scattering patterns in the analysis of two phase flows. Paper read at the 4th International Conference, Computational Methods and Experimental Measurements, CMEM'89, May 1989, Capri, Italy.

Shuerman, D. W., ed. 1980. *Light scattering by irregularly shaped particles*. New York: Plenum.

Shurcliff, W. A. 1962. *Polarized light*. Cambridge, Mass.: Harvard University Press.

Suntz, R., Becker, H., Monkhouse, P., and Wolfrum, J. 1988. Two-dimensional visualization of the flame front in an internal combustion engine by laser-induced fluorescence of OH radicals. *Appl. Phys. B* 47:287.

Tamai, R. 1987. Tesi di Laurea in Ingegneria Meccanica, Università di Napoli, October 1987.

Toon, B., and Ackerman, T. P. 1981. *Appl. Opt.* 20:2499.

Vaglieco, B. M., Beretta, F., and D'Alessio, A. 1990. In situ evaluation of the soot refractive index in the U.V.-visible from the measurements of the scattering and extinction coefficients in rich flames. *Combust. Flame.* 79:3.

van de Hulst, H. 1957. *Light scattering from small particles*. New York: John Wiley.

Winter, M., Lam, J. K., and Long, M. B. 1987. Techniques for high-speed digital imaging of gas concentrations in turbulent flows. *Exp. Fluids* 5:177.

Zerull, R. M., Giese, R. H., and Weiss, K. 1971. Scattering functions of nonspherical dielectric and absorbing particles vs. Mie theory. *Appl. Opt.* 16:777.

TEN

GENERALIZED LORENZ-MIE THEORY AND APPLICATIONS TO OPTICAL SIZING

G. Gouesbet, G. Gréhan, and B. Maheu

10-1 INTRODUCTION

In a large number of combustion systems the fuel is produced from a vaporizing spray or from solid particles (in coal-pulverized flames). For instance, a very good review by Faeth (1977) distinguishes prevaporizing systems, afterburners, lean combustors, carburetors, ramjets, liquid rocket engines, gas turbine combustors, industrial furnaces, and diesel engines. A great effort is devoted to the control and understanding of these systems, requiring more elaborate skills to design sophisticated computer codes, which must be validated by comparisons with experimental results. The design of such codes, which involves physical data concerning complex phenomena (vaporization and combustion in more or less steady, premixed situations, possibly including spray processes in which particle interactions are very significant), is far from having reached a fully satisfactory state of development (Faeth 1977; Sirignano 1983). Often we have to be content with experimental correlations. In all cases, extensive and precise measurements of space- and time-dependent particle size spectra are required, thus creating the necessity of designing optical sizing instrumentation.

The development of optical sizing techniques is not only relevant to the control, understanding, and design of specific spray and particle combustion systems but also, more generally, to the study of particle two-phase flow systems (as for instance, the laden flows encountered by chemical and mechanical engineers in pipes and conduits or in chemical engineering standard processes such as fluidization, sedimentation, and pneumatic transports).

339

Consequently, simultaneous measurements of sizes and velocities (and possibly concentrations) of discrete particles transported by flows are of increasing interest both for the researcher, wishing to understand and describe the laws of nature, who is inevitably faced with multiphase problems, and for the engineer, who has to design and control various plants and processes involving particle heat and mass transfer phenomena. Emphasis must be put on nonintrusive optical techniques, which in principle, do not disturb the medium under study and provide us with versatile opportunities.

Several laser-based techniques have been designed that are now in current use, among them the pedestal, visibility, phase Doppler, or top-hat beam techniques. For details, the reader can refer to Gouesbet and Gréhan (1988), who provide an up-to-date account.

When optically measuring the size of spray droplets (possibly simultaneously with velocity and concentration), a laser source is usually required. Under some circumstances, the laser beam is expanded, and/or the particles are so small that the laser source can be safely considered a plane wave. To design instruments and to interpret data, one can then rely, more or less completely, on the basic theory concerning the scattering of a plane wave by a homogeneous, isotropic, spherical particle (here referred as a Mie scatter center), namely, the Lorenz-Mie theory (LMT).

However, very often the situation is not so ideal. The in-going laser beams are usually focused, and we may question the validity of the LMT when the particles are not small with respect to the Gaussian laser probe diameter or to the width of the plateau of a top-hat beam (Allano et al. 1984; Gouesbet, Gréhan, and Kleine 1984; Gréhan and Gouesbet 1986). Then the use of LMT in designing instruments and/or processing data can be misleading. We might have to rely on a more general theory, permitting us to compute the properties of the light scattered by a Mie scatter center illuminated by a nonplane wave (generalized Lorenz-Mie theory, or GLMT). Such a theory now exists. The aim of this chapter is to present this theory by gathering the information we published on the topic in the last 10 years or so.

Section 10-2 will present the GLMT in its more general formulation (for a Mie scatter center arbitrarily located in an arbitrarily shaped beam). Section 10-3 is devoted to the special case when the incident beam is a laser Gaussian beam, including the condition when the scatter center is located on the axis of the beam. Numerical problems to solve for practical applications are considered in Sections 10-4 and 10-5, while Section 10-6 is devoted to optical sizing applications of the GLMT.

10-2 GENERAL FORMULATION OF THE GLMT

This section is a synthetic account of Gouesbet, Maheu, and Gréhan (1988*a*, *b*), Maheu, Gréhan, and Gouesbet (1987*b*), and Maheu, Gouesbet, and Gréhan (1988). Less general formulations have been discussed by Kim and Lee (1983), Morita et al. (1968), Tam and Corriveau (1978), Tsai and Pogorzelski (1975), Yeh, Colak, and Barber (1982), and also by our team tracing back to 1980, whose references will be cited later.

Bromwich Formulation

We shall have to solve the Maxwell equations accounting for the boundary conditions defined by the scattering problem under consideration. The Bromwich formulation (Bromwich 1919; Poincelot 1963) enables us to obtain solutions satisfying the boundary conditions in special curvilinear coordinate systems (base e_1, e_2, e_3) defined by the following properties: (1) the system is orthogonal, (2) $e_1 = 1$, and (3) the ratio e_2/e_3 does not depend on the first coordinate. Such is the case for the spherical coordinate system (r, θ, φ) used in Fig. 10-1.

We consider an electromagnetic sinusoidal wave varying in time like $e^{i\omega t}$, in which ω is the angular frequency. The components of a vector $\mathbf{V}$ are designated by V_r, V_θ, V_φ in a (r, θ, φ) coordinate system at point P. The letter V stands for E (electric field) or H (magnetic field).

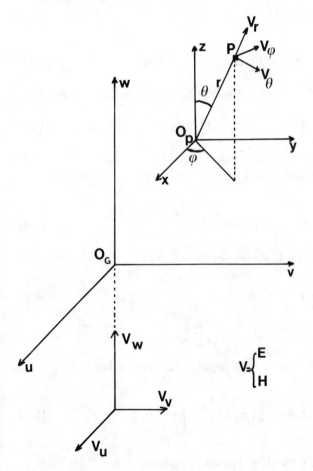

Figure 10-1 Coordinate systems for the beam description (beam center O_G) and for the scattered fields (scatter center O_P; scattered field properties computed at point P).

In the Bromwich formulation the solution of the Maxwell equations is written as the sum of two special solutions with the proviso that the boundary conditions must be satisfied. These special solutions are the transversal magnetic (TM) wave, for which $H_r = 0$, and the transversal electric (TE) wave, for which $E_r = 0$. The TM and TE fields are deduced from Bromwich scalar potentials (BSP), U_{TM} and U_{TE}, respectively. Any BSP, U, complies with the following equation:

$$\frac{\partial^2 U}{\partial r^2} + k^2 U + \frac{1}{r^2 \sin \theta} \frac{\partial}{\partial \theta} \left(\sin \theta \frac{\partial U}{\partial \theta} \right) + \frac{1}{r^2 \sin^2 \theta} \frac{\partial^2 U}{\partial \varphi^2} = 0 \qquad (10\text{-}1)$$

in which k is the wave number,

$$k = \omega \sqrt{\mu \epsilon} = M \frac{\omega}{c} \qquad (10\text{-}2)$$

in which μ and ϵ are the permeability and the permittivity of the medium, respectively, M its complex refractive index, and c the speed of light.

The following BSP,

$$U = r \, \Psi_n^1(kr) \, P_n^m (\cos \theta) \begin{Bmatrix} \sin \\ \cos \end{Bmatrix} (m\varphi) \qquad (10\text{-}3)$$

is a solution of Eq. (10-1); $\Psi_n^1(kr)$ is the spherical Bessel function defined by

$$\Psi_n^1(kr) = \left(\frac{\pi}{2kr} \right)^{1/2} J_{n+1/2}(kr) \qquad (10\text{-}4)$$

in which $J_{n+1/2}(kr)$ is a Bessel function of half-integer order.

In the present paper, the associated Legendre polynomial $P_n^m (\cos \theta)$ is defined as

$$P_n^m (\cos \theta) = (-1)^m (\sin \theta)^m \frac{d^m P_n(\cos \theta)}{(d \cos \theta)^m} \qquad (10\text{-}5)$$

in which $P_n (\cos \theta)$ is the order n Legendre polynomial.

We introduce the Ricatti-Bessel functions:

$$\Psi_n(kr) = kr \Psi_n^1(kr) \qquad (10\text{-}6)$$

$$\xi_n(kr) = \Psi_n(kr) + i(-1)^n \left(\frac{\pi kr}{2} \right)^{1/2} J_{-n-1/2}(kr) \qquad (10\text{-}7)$$

in which $J_{-n-1/2}(kr)$ is a Bessel function of negative half-integer order.

The following BSP,

$$U = \xi_n(kr) P_n^m(\cos \theta) \begin{Bmatrix} \sin \\ \cos \end{Bmatrix} (m\varphi) \qquad (10\text{-}8)$$

is also a solution of Eq. (10-1).

The use of either Ψ_n or ξ_n depends on boundary conditions, since $\Psi_n(kr)$ remains finite for $r = 0$, while $\xi_n(kr)$ tends to a spherical wave description for $r \to \infty$.

From the definition of the TM and TE waves,

$$H_{r,\text{TM}} = E_{r,\text{TE}} = 0 \tag{10-9}$$

When the BSPs are determined, the other field components are obtained from the following relations:

$$E_{r,\text{TM}} = \frac{\partial^2 U_{\text{TM}}}{\partial r^2} + k^2 U_{\text{TM}} \tag{10-10}$$

$$E_{\theta,\text{TM}} = \frac{1}{r} \frac{\partial^2 U_{\text{TM}}}{\partial r \partial \theta} \tag{10-11}$$

$$E_{\varphi,\text{TM}} = \frac{1}{r \sin \theta} \frac{\partial^2 U_{\text{TM}}}{\partial r \partial \varphi} \tag{10-12}$$

$$H_{\theta,\text{TM}} = \frac{i\omega\epsilon}{r \sin \theta} \frac{\partial U_{\text{TM}}}{\partial \varphi} \tag{10-13}$$

$$H_{\varphi,\text{TM}} = (-) \frac{i\omega\epsilon}{r} \frac{\partial U_{\text{TM}}}{\partial \theta} \tag{10-14}$$

$$E_{\theta,\text{TE}} = (-) \frac{i\omega\mu}{r \sin \theta} \frac{\partial U_{\text{TE}}}{\partial \varphi} \tag{10-15}$$

$$E_{\varphi,\text{TE}} = \frac{i\omega\mu}{r} \frac{\partial U_{\text{TE}}}{\partial \theta} \tag{10-16}$$

$$H_{r,\text{TE}} = \frac{\partial^2 U_{\text{TE}}}{\partial r^2} + k^2 U_{\text{TE}} \tag{10-17}$$

$$H_{\theta,\text{TE}} = \frac{1}{r} \frac{\partial^2 U_{\text{TE}}}{\partial r \partial \theta} \tag{10-18}$$

$$H_{\varphi,\text{TE}} = \frac{1}{r \sin \theta} \frac{\partial^2 U_{\text{TE}}}{\partial r \partial \varphi} \tag{10-19}$$

Scattering Problem

The center of the scatterer (diameter d, complex refractive index M relative to the surrounding nonabsorbing medium) is located at point O_P of a Cartesian coordinate system $O_\text{P}xyz$ (Fig. 10-1). The incident wave propagates from the negative z to the positive z. In that section, we only assume that we know the expressions for the radial electric and magnetic components, E_r and H_r, respectively, in the (r, θ, φ) system. The other elements of Fig. 10-1 will be introduced later.

The aim is to compute the properties of the scattered light observed at a point $P(r, \theta, \varphi)$ and some integral associated quantities. The following steps must be carried out.

1. Write out the BSPs, U_{TM} and U_{TE}, for the incident wave (superscript i), the scattered (also named external) wave (superscript e), and the internal wave inside the sphere (superscript sp). We shall then be faced with six BSPs: U^i_{TM}, U^i_{TE}, U^e_{TM}, U^e_{TE}, U^{sp}_{TM}, U^{sp}_{TE}.

2. We are particularly interested in the scattered wave defined by U^e_{TM} and U^e_{TE}. These functions will contain an infinite number of unknown coefficients, which will be determined using boundary conditions at the surface of the sphere.

3. The scattered field components will then be determined from the external BSPs with the aid of Eqs. (10-10)–(10-19).

4. The formulation will then be completed by deriving formulas for the scattered intensities, phase angle, cross sections, and radiation pressure force components.

BSPs for the Incident Wave

From the definition of the TE wave, $E_{r,TE} = 0$. Then, using relation Eq. (10-10), the electric field is simply

$$E_r = E_{r,TM} = \frac{\partial^2 U^i_{TM}}{\partial r^2} + k^2 U^i_{TM} \tag{10-20}$$

in which U^i_{TM} is the TM-BSP for the incident wave.

The equations being linear and E_r being in general a sum of terms, U^i_{TM} can be researched as a sum of corresponding terms. That procedure is more pedagogic and has been used by Gouesbet, Maheu, and Gréhan (1988a) in the case of Gaussian beams. In the present paper we shall be more concise and therefore state directly,

$$U^i_{TM} = E_0 \sum_{n=1}^{\infty} \sum_{m=-n}^{+n} C^{pw}_n g^m_{n,TM} \, r\Psi^1_n(kr) \, P^{|m|}_n (\cos \theta) \exp (im\varphi) \tag{10-21}$$

in which coefficients C^{pw}_n (pw is plane wave) appear in the Bromwich formulation of the LMT (Gouesbet and Gréhan 1982), and are given by

$$C^{pw}_n = \frac{1}{k} i^{n-1} (-1)^n \frac{2n+1}{n(n+1)} \tag{10-22}$$

In Eq. (10-21), polynomials P^0_n identify with the Legendre polynomials P_n.

To determine the unknown coefficients $g^m_{n,TM}$, we substitute for Eq. (10-21) on the right-hand side of Eq. (10-20) and identify the left-hand side of Eq. (10-20) with the E_r value for the beam, producing an equation for the unknown coefficients. We also use the differential equation for spherical Bessel functions (Poincelot 1963),

$$\left[\frac{d^2}{dr^2} + k^2 \right] (r\Psi^1_n(kr)) = \frac{n(n+1)}{r} \Psi^1_n(kr) \tag{10-23}$$

To isolate the unknown coefficients in the resulting equation, we successively invoke orthogonality relations for the exponentials and for the P^m_n,

$$\int_0^{2\pi} \exp\left[i(m - m')\varphi\right]d\varphi = 2\pi\delta_{mm'} \tag{10-24}$$

$$\int_0^{\pi} P_n^m (\cos\theta) P_l^m (\cos\theta) \sin\theta d\theta = \frac{2}{2n + 1} \frac{(n + m)!}{(n - m)!} \delta_{nl} \tag{10-25}$$

Then we can obtain a better expression by means of the orthogonality relations for the Ψ_n^1,

$$\int_0^{\infty} \Psi_n^1(kr)\Psi_m^1(kr)d(kr) = \frac{\pi}{2(2n + 1)}\delta_{nm} \tag{10-26}$$

Multiplying the equation to solve by the adequate integral operators to take advantage of Eqs. (10-24)–(10-26), we obtain

$$g_{n,\text{TM}}^m = \frac{(2n + 1)^2}{2\pi^2 n(n + 1)C_n^{\text{pw}}} \frac{(n - |m|)!}{(n + |m|)!} \int_0^{\pi} \int_0^{2\pi} \int_0^{\infty} \frac{E_r(r, \theta, \varphi)}{E_0}$$

$$\cdot r\Psi_n^1(kr)P_n^{|m|}(\cos\theta) \exp(-im\varphi) \sin\theta \, d\theta \, d\varphi \, d(kr) \tag{10-27}$$

Similarly, the BSP U_{TE}^i is set to be

$$U_{\text{TE}}^i = H_0 \sum_{n=1}^{\infty} \sum_{m=-n}^{+n} C_n^{\text{pw}} g_{n,\text{TE}}^m r\Psi_n^1(kr)P_n^{|m|}(\cos\theta) \exp(im\varphi) \tag{10-28}$$

Then, working with the radial magnetic component H_r instead of radial electric component E_r, we obtain

$$g_{n,\text{TE}}^m = \frac{(2n + 1)^2}{2\pi^2 n(n + 1)C_n^{\text{pw}}} \frac{(n - |m|)!}{(n + |m|)!} \int_0^{\pi} \int_0^{2\pi} \int_0^{\infty} \frac{H_r(r, \theta, \varphi)}{H_0}$$

$$\cdot r\Psi_n^1(kr)P_n^{|m|}(\cos\theta) \exp(-im\varphi) \sin\theta \, d\theta \, d\varphi \, d(kr) \tag{10-29}$$

For a plane wave, all the coefficients g_n^m are zero, but for $|m| = 1$ (see section below on more special cases).

Note that expressions for U_{TM}^i, U_{TM}^e, $g_{n,\text{TM}}^m$, and $g_{n,\text{TE}}^m$ can be expressed in terms of the Ricatti-Bessel functions $\Psi_n(kr)$ instead of the spherical Bessel functions $\Psi_n^1(kr)$ (relation (10-6)). Also, we might introduce the spherical surface harmonics $Y_n^m(\theta, \varphi)$ given by

$$Y_n^m(\theta, \varphi) = \left[\frac{2n + 1}{4\pi} \frac{(n - m)!}{(n + m)!}\right]^{1/2} P_n^m(\cos\theta) \exp(im\varphi) \tag{10-30}$$

A different but equivalent derivation of the expressions (10-27) and (10-29) is given by Maheu, Gouesbet, and Gréhan (1988).

BSPs for the External and Sphere Waves

We define an "external wave" as the wave scattered by the particle and a "sphere wave" as the wave inside the particle. The BSPs for the external (U_{TM}^e, U_{TE}^e) and sphere ($U_{\text{TM}}^{\text{sp}}$, $U_{\text{TE}}^{\text{sp}}$) waves are set to be

$$U_{TM}^e = \frac{-E_0}{k} \sum_{n=1}^{\infty} \sum_{m=-n}^{+n} C_n^{pw} A_n^m \xi_n(kr) P_n^{|m|}(\cos\theta) \exp(im\varphi) \qquad (10\text{-}31)$$

$$U_{TE}^e = \frac{-H_0}{k} \sum_{n=1}^{\infty} \sum_{m=-n}^{+n} C_n^{pw} B_n^m \xi_n(kr) P_n^{|m|}(\cos\theta) \exp(im\varphi) \qquad (10\text{-}32)$$

$$U_{TM}^{sp} = \frac{kE_0}{k_{sp}^2} \sum_{n=1}^{\infty} \sum_{m=-n}^{+n} C_n^{pw} E_n^m \Psi_n(k_{sp}r) P_n^{|m|}(\cos\theta) \exp(im\varphi) \qquad (10\text{-}33)$$

$$U_{TE}^{sp} = \frac{kH_0}{k_{sp}^2} \sum_{n=1}^{\infty} \sum_{m=-n}^{+n} C_n^{pw} D_n^m \Psi_n(k_{sp}r) P_n^{|m|}(\cos\theta) \exp(im\varphi) \qquad (10\text{-}34)$$

in which k_{sp} is the wave number in the sphere material. The functions $\xi_n(kr)$ are used in Eqs. (10-31)–(10-32) to satisfy later the boundary conditions in the limit $r \to \infty$, i.e., to produce the components of a spherical wave in this limit.

Determination of the External Wave

The scattering coefficients A_n^m and B_n^m of the external wave are determined by writing the tangential continuity of the electric and magnetic fields at the surface of the sphere ($r = d/2$), the field components being obtained from the set Eqs. (10-10)–(10-19), using Eqs. (10-21), (10-28), and (10-31)–(10-34). These are four boundary conditions that lead to a four-equation system. Assuming a nonmagnetic particle ($\mu_{sp}/\mu = 1$) and an accessory relation valid for a nonmagnetic particle,

$$M = \frac{k_{sp}}{k} = \left(\frac{\epsilon_{sp}}{\epsilon}\right)^{1/2} \qquad (10\text{-}35)$$

we readily obtain

$$A_n^m = a_n g_{n,TM}^m \qquad (10\text{-}36)$$

$$B_n^m = b_n g_{n,TE}^m \qquad (10\text{-}37)$$

in which a_n and b_n are the usual scattering coefficients of the LMT:

$$a_n = \frac{\Psi_n(\alpha)\Psi_n'(\beta) - M\Psi_n'(\alpha)\Psi_n(\beta)}{\xi_n(\alpha)\Psi_n'(\beta) - M\xi_n'(\alpha)\Psi_n(\beta)} \qquad (10\text{-}38)$$

$$b_n = \frac{M\Psi_n(\alpha)\Psi_n'(\beta) - \Psi_n'(\alpha)\Psi_n(\beta)}{M\xi_n(\alpha)\Psi_n'(\beta) - \xi_n'(\alpha)\Psi_n(\beta)} \qquad (10\text{-}39)$$

in which α is the size parameter $\pi d/\lambda$ (λ being the wavelength in the surrounding medium) and β is $M\alpha$. The prime symbol means the derivative of the function for the value of the argument indicated between parentheses.

Scattered Field Components

From the external BSPs U^e_{TM}, U^e_{TE}, the field components of the scattered wave are obtained using Eqs. (10-10)–(10-19) and relations (10-38)–(10-39). We obtain

$$E_r = -kE_0 \sum_{n=1}^{\infty} \sum_{m=-n}^{+n} C_n^{pw} a_n g_{n,TM}^m$$

$$\cdot \, [\xi_n''(kr) + \xi_n(kr)] P_n^{|m|}(\cos \theta) \exp(im\varphi) \qquad (10\text{-}40)$$

$$E_\theta = -\frac{E_0}{r} \sum_{n=1}^{\infty} \sum_{m=-n}^{+n} C_n^{pw} [a_n g_{n,TM}^m \xi_n'(kr) \tau_n^{|m|}(\cos \theta)$$

$$+ \, m b_n g_{n,TE}^m \xi_n(kr) \Pi_n^{|m|}(\cos \theta)] \exp(im\varphi) \qquad (10\text{-}41)$$

$$E_\varphi = -\frac{iE_0}{r} \sum_{n=1}^{\infty} \sum_{m=-n}^{+n} C_n^{pw} [m a_n g_{n,TM}^m \xi_n'(kr) \Pi_n^{|m|}(\cos \theta)$$

$$+ \, b_n g_{n,TE}^m \xi_n(kr) \tau_n^{|m|}(\cos \theta)] \exp(im\varphi) \qquad (10\text{-}42)$$

$$H_r = -kH_0 \sum_{n=1}^{\infty} \sum_{m=-n}^{+n} C_n^{pw} b_n g_{n,TE}^m [\xi_n''(kr)$$

$$+ \, \xi_n(kr)] P_n^{|m|}(\cos \theta) \exp(im\varphi) \qquad (10\text{-}43)$$

$$H_\theta = \frac{H_0}{r} \sum_{n=1}^{\infty} \sum_{m=-n}^{+n} C_n^{pw} [m a_n g_{n,TM}^m \xi_n(kr) \Pi_n^{|m|}(\cos \theta)$$

$$- \, b_n g_{n,TE}^m \xi_n'(kr) \tau_n^{|m|}(\cos \theta)] \exp(im\varphi) \qquad (10\text{-}44)$$

$$H_\varphi = \frac{iH_0}{r} \sum_{n=1}^{\infty} \sum_{m=-n}^{+n} C_n^{pw} [a_n g_{n,TM}^m \xi_n(kr) \tau_n^{|m|}(\cos \theta)$$

$$- \, m b_n g_{n,TE}^m \xi_n'(kr) \Pi_n^{|m|}(\cos \theta)] \exp(im\varphi) \qquad (10\text{-}45)$$

in which we have used

$$k = \omega\mu \frac{H_0}{E_0} = \omega\epsilon \frac{E_0}{H_0} \qquad (10\text{-}46)$$

and defined generalized Legendre functions according to

$$\tau_n^k(\cos \theta) = \frac{d}{d\theta} P_n^k(\cos \theta) \qquad (10\text{-}47)$$

$$\Pi_n^k(\cos \theta) = \frac{P_n^k(\cos \theta)}{\sin \theta} \qquad (10\text{-}48)$$

For $k = 1$, τ_n^k and Π_n^k are the usual Legendre functions appearing in the LMT, namely, τ_n and Π_n. Note the following correspondence to classical notations: $\tau_n^1 = \tau_n$, $\Pi_n^1 = \Pi_n$, but $P_n^0 = P_n$.

Scattered Field Components in the Far Field

In most cases [but not always; see Slimani et al. (1984)], interest is limited to the far field defined by the inequality $r >> \lambda$, leading to an asymptotic expression for the function ξ_n (Kerker 1969):

$$\xi_n(kr) \rightarrow i^{n+1} \exp(-ikr) \tag{10-49}$$

Then, Eqs. (10-40)–(10-45) reduce to

$$E_r = H_r = 0 \tag{10-50}$$

$$E_\theta = \frac{iE_0}{kr} \exp(-ikr) \sum_{n=1}^{\infty} \sum_{m=-n}^{+n} \frac{2n+1}{n(n+1)} [a_n g^m_{n.\text{TM}} \tau^{|m|}_n (\cos\theta)$$
$$+ imb_n g^m_{n.\text{TE}} \Pi^{|m|}_n (\cos\theta)] \exp(im\varphi) \tag{10-51}$$

$$E_\varphi = \frac{-E_0}{kr} \exp(-ikr) \sum_{n=1}^{\infty} \sum_{m=-n}^{+n} \frac{2n+1}{n(n+1)} [ma_n g^m_{n.\text{TM}} \Pi^{|m|}_n (\cos\theta)$$
$$+ ib_n g^m_{n.\text{TE}} \tau^{|m|}_n (\cos\theta)] \exp(im\varphi) \tag{10-52}$$

$$H_\varphi = \frac{H_0}{E_0} E_\theta \tag{10-53}$$

$$H_\theta = (-)\frac{H_0}{E_0} E_\varphi \tag{10-54}$$

The scattered wave has become a transverse wave.

Scattered Intensities

The scattered intensities are computed with the aid of the Poynting theorem:

$$S^+ = \frac{1}{2}\text{Re}\{E_\theta H^*_\varphi - E_\varphi H^*_\theta\} \tag{10-55}$$

in which S^+ is a dimensionless quantity produced by using the normalizing condition

$$\frac{1}{2}\left(\frac{\epsilon}{\mu}\right)^{1/2} E^2_0 = 1 \tag{10-56}$$

From now on, the condition (10-56) will always be used, although we shall not mention it anymore.

Injecting Eqs. (10-51)–(10-54) into (10-55), and separating the contributions I^+_θ and I^+_φ to S^+, we obtain the scattered intensities:

$$\begin{bmatrix} I^+_\theta \\ I^+_\varphi \end{bmatrix} = \frac{\lambda^2}{4\pi^2 r^2} \begin{bmatrix} |\mathscr{S}_2|^2 \\ |\mathscr{S}_1|^2 \end{bmatrix} \tag{10-57}$$

in which $\mathscr{S}_1$ and $\mathscr{S}_2$ are defined by

$$E_\theta = \frac{iE_0}{kr} \exp\left(-ikr\right) \mathcal{S}_2 \qquad (10\text{-}58)$$

$$E_\varphi = (-)\frac{E_0}{kr} \exp\left(-ikr\right) \mathcal{S}_1 \qquad (10\text{-}59)$$

Phase Angle

Phase angle δ between components E_θ and E_φ characterizes the state of polarization of the scattered wave. It is an exercise to obtain

$$\tan \delta = \frac{\mathrm{Re}\,(\mathcal{S}_1)\,\mathrm{Re}\,(\mathcal{S}_2) + \mathrm{Im}\,(\mathcal{S}_1)\,\mathrm{Im}\,(\mathcal{S}_2)}{\mathrm{Im}\,(\mathcal{S}_1)\,\mathrm{Re}\,(\mathcal{S}_2) - \mathrm{Re}\,(\mathcal{S}_1)\,\mathrm{Im}\,(\mathcal{S}_2)} \qquad (10\text{-}60)$$

Radiative Balance

A radiative balance is carried out on a sphere of radius $r \gg \lambda$, surrounding the scatterer. The component $S_\perp^+$ of the Poynting vector $\mathbf{S}^+$, perpendicular to the surface of the sphere, gives the energy flux per unit area and per unit of time. The involved field components are total field components written as the sum of the corresponding incident components (superscript i) and scattered components (superscript s). The energy balance is expressed by the integration of $S_\perp^+$ on the surface of the sphere and leads to

$$-C_{\mathrm{abs}} = +\int_{(S)} S_\perp^+ \, dS \qquad (10\text{-}61)$$

in which C_{abs} is the absorption cross section of the scatterer.

The right-hand side of Eq. (10-61) is then found to be the sum of three contributions:

$$\mathcal{I}^i = \int_0^\pi \int_0^{2\pi} \frac{1}{2}\,\mathrm{Re}\,(E_\theta^i H_\varphi^{i*} - E_\varphi^i H_\theta^{i*})r^2 \sin\theta\, d\theta\, d\varphi \qquad (10\text{-}62)$$

$$\mathcal{I}^s = \int_0^\pi \int_0^{2\pi} \frac{1}{2}\,\mathrm{Re}\,(E_\theta^s H_\varphi^{s*} - E_\varphi^s H_\theta^{s*})r^2 \sin\theta\, d\theta\, d\varphi \qquad (10\text{-}63)$$

$$\mathcal{I}^{is} = \int_0^\pi \int_0^{2\pi} \frac{1}{2}\,\mathrm{Re}\,(E_\theta^i H_\varphi^{s*} + E_\theta^s H_\varphi^{i*}$$
$$- E_\varphi^i H_\theta^{s*} - E_\varphi^s H_\theta^{i*})r^2 \sin\theta\, d\theta\, d\varphi \qquad (10\text{-}64)$$

The first integral $\mathcal{I}^i$ corresponds to an energy balance of the nonperturbed incident fields. The surrounding medium, being nonabsorbing, $\mathcal{I}^i$ must be zero (see the following section). The second integral $\mathcal{I}^s$ corresponds to an energy balance of the scattered fields and thus, is equal to the scattering cross section C_{sca} of the particle. The third integral is then found to be minus the extinction cross section,

$$\mathcal{I}^{is} = +\int_{(S)} S_\perp^+ \, dS - \mathcal{I}^i - \mathcal{I}^s = -C_{\mathrm{abs}} - C_{\mathrm{sca}} = -C_{\mathrm{ext}} \qquad (10\text{-}65)$$

Incident Field Balance

From the BSPs for the incident fields, U_{TM}^{i} and U_{TE}^{i}, Eqs. (10-21)–(10-28), and using Eqs. (10-10)–(10-19), the incident field components E_θ^{i}, E_φ^{i}, H_θ^{i}, and H_φ^{i} are found to be

$$E_\theta^{\text{i}} = \frac{E_0}{r} \sum_{n=1}^{\infty} \sum_{m=-n}^{+n} C_n^{\text{pw}} (g_{n.\text{TM}}^m \Psi_n' \tau_n^{|m|} + m g_{n.\text{TE}}^m \Psi_n \Pi_n^{|m|}) \exp(im\varphi) \qquad (10\text{-}66)$$

$$E_\varphi^{\text{i}} = \frac{E_0}{r} \sum_{n=1}^{\infty} \sum_{m=-n}^{+n} C_n^{\text{pw}} (im g_{n.\text{TM}}^m \Psi_n' \Pi_n^{|m|} + i g_{n.\text{TE}}^m \Psi_n \tau_n^{|m|}) \exp(im\varphi) \qquad (10\text{-}67)$$

$$H_\theta^{\text{i}} = \frac{H_0}{r} \sum_{n=1}^{\infty} \sum_{m=-n}^{+n} C_n^{\text{pw}} (-m g_{n.\text{TM}}^m \Psi_n \Pi_n^{|m|} + g_{n.\text{TE}}^m \Psi_n' \tau_n^{|m|}) \exp(im\varphi) \qquad (10\text{-}68)$$

$$H_\varphi^{\text{i}} = \frac{H_0}{r} \sum_{n=1}^{\infty} \sum_{m=-n}^{+n} C_n^{\text{pw}} (-i g_{n.\text{TM}}^m \Psi_n \tau_n^{|m|} + im g_{n.\text{TE}}^m \Psi_n' \Pi_n^{|m|}) \exp(im\varphi) \qquad (10\text{-}69)$$

in which the arguments (kr) and $(\cos \theta)$ are now omitted for convenience.

We inject Eqs. (10-66)–(10-69) in relation (10-62), integrate successively with respect to φ using relation (10-24), then to θ (see appendix), to obtain (with A_{np} real numbers)

$$\mathcal{J}^{\text{i}} = \text{Im} \left(\sum_{p=-\infty}^{+\infty} \sum_{n=|p| \neq 0}^{\infty} A_{np} |C_n^{\text{pw}}|^2 (|g_{n.\text{TM}}^p|^2 - |g_{n.\text{TE}}^p|^2) \Psi_n \Psi_n' \right) = 0 \qquad (10\text{-}70)$$

which is equal to zero, as expected (since all terms in the summations are real).

Scattering Cross Section

The scattering cross section can be computed from $\mathcal{J}^{\text{s}}$ in Eq. (10-63) or, equivalently, from

$$C_{\text{sca}} = \int_0^\pi \int_0^{2\pi} (I_\theta^+ + I_\varphi^+) r^2 \sin \theta \, d\theta \, d\varphi \qquad (10\text{-}71)$$

Integrating over φ (Eq. (10-24)), then over θ (appendix), we obtain

$$C_{\text{sca}} = \frac{\lambda^2}{\pi} \sum_{n=1}^{\infty} \sum_{m=-n}^{+n} \frac{2n+1}{n(n+1)} \frac{(n+|m|)!}{(n-|m|)!}$$

$$\cdot (|a_n|^2 |g_{n.\text{TM}}^m|^2 + |b_n|^2 |g_{n.\text{TE}}^m|^2) \qquad (10\text{-}72)$$

Extinction Cross Section

C_{ext} is obtained from Eqs. (10-64)–(10-65). The incident components (superscript i) are given by Eqs. (10-66)–(10-69), and the scattered components by the set Eqs. (10-51)–(10-54). Integrating over φ (Eq. (10-24)), then over θ (appendix), and rearranging, we obtain

$$C_{ext} = \frac{4\pi}{k^2} \operatorname{Re}\left(\sum_{n=1}^{\infty} \sum_{m=-n}^{+n} \frac{2n+1}{n(n+1)} \frac{(n+|m|)!}{(n-|m|)!} \right.$$

$$\cdot \{(-i)^n \exp(ikr)[a_n^* \Psi_n' |g_{n,TM}^m|^2 - ib_n^* \Psi_n |g_{n,TE}^m|^2]$$

$$\left. + i^n \exp(-ikr)[ia_n \Psi_n |g_{n,TM}^m|^2 + b_n \Psi_n' |g_{n,TE}^m|^2]\} \right) \qquad (10\text{-}73)$$

The radius of the sphere surrounding the particle being arbitrary, let us take it in the limit $r \to \infty$. Then we can use the limit expression (Brillouin 1949)

$$\Psi_n(kr) \to \frac{1}{2}[(-i)^{n+1} \exp(ikr) + i^{n+1} \exp(-ikr)] \qquad (10\text{-}74)$$

leading to the final result,

$$C_{ext} = \frac{\lambda^2}{\pi} \operatorname{Re}\left(\sum_{n=1}^{\infty} \sum_{m=-n}^{+n} \frac{2n+1}{n(n+1)} \frac{(n+|m|)!}{(n-|m|)!} \right.$$

$$\left. \cdot (a_n |g_{n,TM}^m|^2 + b_n |g_{n,TE}^m|^2) \right) \qquad (10\text{-}75)$$

Radiation Pressure

The reduced momentum of the wave is equal to the ratio of the reduced Poynting vector S^+ over the speed c of light. The radiation pressure force exerted by the beam on the scatterer is proportional to the net momentum removed from the incident beam. The x, y, z components of the reduced radiation pressure force $\mathbf{F}^+$ can be characterized by three pressure cross sections $C_{pr,x}$, $C_{pr,y}$, and $C_{pr,z}$, which will be examined separately.

In the z direction, the pressure cross section $C_{pr,z}$ can be defined as [generalizing van de Hulst (1957)]

$$C_{pr,z} = cF_z^+ = \overline{\cos\theta}\, C_{ext} - \overline{\cos\theta}\, C_{sca} \qquad (10\text{-}76)$$

in which the overbar mean value designates an integration of the reduced Poynting vector weighted by $\cos\theta$.

The first term is the forward momentum removed from the beam, and the second is minus the forward momentum given by the scatterer to the scattered wave. These terms identify with the expressions $(-\mathscr{J}^{is} \overline{\cos\theta})$ and $(-\mathscr{J}^s \overline{\cos\theta})$, respectively, from Eqs. (10-63) and (10-64). There is no radiation pressure from the integral $\mathscr{J}^i$, since the unperturbed field does not leave any momentum to the scatterer. Gouesbet, Gréhan, and Maheu (1985) set the first term equal to C_{ext}, an approximation justified by the fact that only a waist location of the particle was considered. It means that we assumed that the wave front on the scatterer was nearly a plane. Here, this assumption must be relaxed.

The second term on the right-hand side of Eq. (10-76) can also be written as

$$\overline{\cos \theta} \, C_{\text{sca}} = \int_0^\pi \int_0^{2\pi} [I_\theta^+ + I_\varphi^+] r^2 \sin \theta \cos \theta \, d\theta \, d\varphi \qquad (10\text{-}77)$$

Integrating over φ (Eq. (10-24)), then over θ (appendix), we obtain

$$\overline{\cos \theta} \, C_{\text{sca}} = -\frac{2\lambda^2}{\pi} \sum_{n=1}^\infty \sum_{p=-n}^{+n} p \, \frac{2n+1}{n^2(n+1)^2} \frac{(n+|p|)!}{(n-|p|)!}$$

$$\cdot \text{Re} \, (ia_n b_n^* g_{n,\text{TM}}^p g_{n,\text{TE}}^{p*}) - \frac{1}{(n+1)^2} \frac{(n+1+|p|)!}{(n-|p|)!}$$

$$\cdot \text{Re} \, (a_n a_{n+1}^* g_{n,\text{TM}}^p g_{n+1,\text{TM}}^{p*} + b_n b_{n+1}^* g_{n,\text{TE}}^p g_{n+1,\text{TE}}^{p*})$$

$$(10\text{-}78)$$

The expression for the first term on the right-hand side of Eq. (10-76) is somewhat different:

$$\overline{\cos \theta} \, C_{\text{ext}} = \int_0^\pi \int_0^{2\pi} \frac{1}{2} \, \text{Re} \, (E_\varphi^i H_\theta^{s*} + E_\varphi^s H_\theta^{i*} - E_\theta^i H_\varphi^{s*} - E_\theta^s H_\varphi^{i*})$$

$$\cdot r^2 \sin \theta \cos \theta \, d\theta \, d\varphi \qquad (10\text{-}79)$$

Equation (10-79) is evaluated by performing the φ and θ integrations (Eq. (10-24) and appendix). Then, assuming a very large radius of the spherical surface used for integration, we can replace the Ricatti-Bessel functions by their asymptotic expressions (Eq. (10-74)), leading to

$$\overline{\cos \theta} \, C_{\text{ext}} = \frac{\lambda^2}{\pi} \sum_{n=1}^\infty \sum_{p=-n}^{+n} \left(\frac{1}{(n+1)^2} \frac{(n+1+|p|)!}{(n-|p|)!} \right.$$

$$\cdot \text{Re} \, [(a_n + a_{n+1}^*) g_{n,\text{TM}}^p g_{n+1,\text{TM}}^{p*}$$

$$+ (b_n + b_{n+1}^*) g_{n,\text{TE}}^p g_{n+1,\text{TE}}^{p*}]$$

$$\left. - p \frac{2n+1}{n^2(n+1)^2} \frac{(n+|p|)!}{(n-|p|)!} \, \text{Re} \, [i(a_n + b_n^*) g_{n,\text{TM}}^p g_{n,\text{TE}}^{p*}] \right) \qquad (10\text{-}80)$$

hence

$$C_{\text{pr},z} = \frac{\lambda^2}{\pi} \sum_{n=1}^\infty \sum_{p=-n}^{+n} \left(\frac{1}{(n+1)^2} \frac{(n+1+|p|)!}{(n-|p|)!} \right.$$

$$\cdot \text{Re} \, [(a_n + a_{n+1}^* - 2a_n a_{n+1}^*) g_{n,\text{TM}}^p g_{n+1,\text{TM}}^{p*}$$

$$+ (b_n + b_{n+1}^* - 2b_n b_{n+1}^*) g_{n,\text{TE}}^p g_{n+1,\text{TE}}^{p*}]$$

$$+ p \frac{2n+1}{n^2(n+1)^2} \frac{(n+|p|)!}{(n-|p|)!}$$

$$\left. \cdot \text{Re} \, [i(2a_n b_n^* - a_n - b_n^*) g_{n,\text{TM}}^p g_{n,\text{TE}}^{p*}] \right) \qquad (10\text{-}81)$$

The pressure cross section in the x direction is defined similarly to Eq. (10-76) with the weighting coefficient $\sin \theta \cos \varphi$ instead of $\cos \theta$.

$$C_{\text{pr},x} = cF_x^+ = \overline{\sin \theta \cos \varphi} \, C_{\text{ext}} - \overline{\sin \theta \cos \varphi} \, C_{\text{sca}} \tag{10-82}$$

The second term on the right-hand side is written

$$\overline{\sin \theta \cos \varphi} \, C_{\text{sca}} = \int_0^\pi \int_0^{2\pi} (I_\theta^+ + I_\varphi^+) r^2 \sin^2 \theta \cos \varphi d\theta d\varphi \tag{10-83}$$

Again, integrating over φ, then over θ (appendix), and rearranging,

$$\overline{\sin \theta \cos \varphi} \, C_{\text{sca}} = \frac{\lambda^2}{\pi} \sum_{p=1}^\infty \sum_{n=p}^\infty \sum_{m=p-1 \neq 0}^\infty \frac{(n+p)!}{(n-p)!}$$

$$\cdot \left[[\text{Re} \, (U_{mn}^{p-1} + U_{nm}^{-p})] \left(\frac{1}{m^2} \delta_{m,n+1} - \frac{1}{n^2} \delta_{n,m+1} \right) \right.$$

$$\left. + \frac{2n+1}{n^2(n+1)^2} \delta_{nm} [\text{Re} \, (V_{mn}^{p-1} - V_{nm}^{-p})] \right] \tag{10-84}$$

in which

$$U_{nm}^p = a_n a_m^* g_{n,\text{TM}}^p g_{m,\text{TM}}^{p+1*} + b_n b_m^* g_{n,\text{TE}}^p g_{m,\text{TE}}^{p+1*} \tag{10-85}$$

$$V_{nm}^p = ib_n a_m^* g_{n,\text{TE}}^p g_{m,\text{TM}}^{p+1*} - ia_n b_m^* g_{n,\text{TM}}^p g_{m,\text{TE}}^{p+1*} \tag{10-86}$$

The first term on the right-hand side of Eq. (10-82) is Eq. (10-79) with $\cos \theta$ replaced by $\sin \theta \cos \varphi$. The procedure is similar to that applied to Eq. (10-79), leading finally to

$$C_{\text{pr},x} = cF_x^+ = \frac{\lambda^2}{2\pi} \sum_{p=1}^\infty \sum_{n=p}^\infty \sum_{m=p-1 \neq 0}^\infty \frac{(n+p)!}{(n-p)!}$$

$$\cdot \left[[\text{Re} \, (S_{mn}^{p-1} + S_{nm}^{-p} - 2U_{mn}^{p-1} - 2U_{nm}^{-p})] \right.$$

$$\cdot \left(\frac{1}{m^2} \delta_{m,n+1} - \frac{1}{n^2} \delta_{n,m+1} \right)$$

$$\left. + \frac{2n+1}{n^2(n+1)^2} \delta_{nm} [\text{Re} \, (T_{mn}^{p-1} - T_{nm}^{-p} - 2V_{mn}^{p-1} + 2V_{nm}^{-p})] \right] \tag{10-87}$$

with

$$S_{nm}^p = (a_n + a_m^*) g_{n,\text{TM}}^p g_{m,\text{TM}}^{p+1*} + (b_n + b_m^*) g_{n,\text{TE}}^p g_{m,\text{TE}}^{p+1*} \tag{10-88}$$

$$T_{nm}^p = -i(a_n + b_m^*) g_{n,\text{TM}}^p g_{m,\text{TE}}^{p+1*} + i(b_n + a_m^*) g_{n,\text{TE}}^p g_{m,\text{TM}}^{p+1*} \tag{10-89}$$

Finally, we have, similar to Eq. (10-82),

$$C_{\text{pr},y} = cF_y^+ = \overline{\sin \theta \sin \varphi} \, C_{\text{ext}} - \overline{\sin \theta \sin \varphi} \, C_{\text{sca}} \tag{10-90}$$

which is evaluated following the same procedure, leading to

$$C_{pr,y} = cF_y^+ = \frac{\lambda^2}{2\pi} \sum_{p=1}^{\infty} \sum_{n=p}^{\infty} \sum_{m=p-1\neq0}^{\infty} \frac{(n+p)!}{(n-p)!}$$

$$\cdot \left[[\text{Im } (S_{mn}^{p-1} + S_{nm}^{-p} - 2U_{mn}^{p-1} - 2U_{nm}^{-p})] \right.$$

$$\cdot \left(\frac{1}{m^2} \delta_{m,n+1} - \frac{1}{n^2} \delta_{n,m+1} \right)$$

$$\left. + \frac{2n+1}{n^2(n+1)^2} \delta_{nm} [\text{Im } (T_{mn}^{p-1} - T_{nm}^{-p} - 2V_{mn}^{p-1} + 2V_{nm}^{-p})] \right] \quad (10\text{-}91)$$

which is identical to Eq. (10-87) but with Re replaced by Im.

For a beam axis location in a Gaussian beam, $C_{pr,x}$ and $C_{pr,y}$ are zero [see Eq. (10-143)]. Obviously, they are also zero for a plane wave. More generally, they must be nonzero only for a noncircularly symmetric situation.

10-3 GAUSSIAN BEAMS

The general formulation is now specified for the case of Gaussian beams (laser beam in its fundamental basic TEM_{00} mode). This requires a description of the beam, permitting the explicit expression of components E_r and H_r appearing in Eqs. (10-27) and (10-29). This case of arbitrary location of the scatterer center in Gaussian beams has been discussed by Gouesbet, Maheu, and Gréhan (1988a, b) and Maheu, Gréhan, and Gouesbet (1987b).

Cartesian Description of the Incident Wave in the (u, v, w) System of Coordinates (Fig. 10-1)

We consider a Gaussian beam whose waist middle is O_G. Another Cartesian coordinate system $(O_G uvw)$ is used, with $O_G u$ parallel to $O_P x$, and similarly for the other axes. Axis $O_G w$ is the axis of the beam. The nonzero Cartesian field components are E_u, H_v, E_w, and H_w. We set

$$\mathbf{O_P O_G} = (x_0, y_0, z_0) \quad (10\text{-}92)$$

Following Lax, Louisell, and McKnight (1975) and Davis (1979), we introduce a small parameter s (which can be called a beam parameter),

$$s = w_0/l = 1/(kw_0) \quad (10\text{-}93)$$

in which w_0 is the beam waist radius and l the spreading length. According to Davis (1979), the vector potential may be expanded in power series of s. Neglecting in Davis' formulation all terms with powers of s greater than 1, we define what we call the order L (lowest order) of approximation to the description of the Gaussian beam. The field components are then given by

$$E_v = H_u = 0 \tag{10-94}$$

$$E_u = E_0 \Psi_0 \exp(-ikw) \tag{10-95}$$

$$E_w = -\frac{2Qu}{l} E_u \tag{10-96}$$

$$H_v = H_0 \Psi_0 \exp(-ikw) \tag{10-97}$$

$$H_w = -\frac{2Qv}{l} H_v \tag{10-98}$$

in which E_0 and H_0 are linked by

$$E_0/H_0 = \sqrt{\mu/\epsilon} \tag{10-99}$$

The lowest order function Ψ_0 is the well-known fundamental mode solution given by

$$\Psi_0 = iQ \exp(-iQh_+^2) \tag{10-100}$$

in which

$$h_+^2 = u_+^2 + v_+^2 \tag{10-101}$$

$$Q = \frac{1}{i + 2w_+} \tag{10-102}$$

Here u_+, v_+, w_+ are reduced quantities defined by

$$u_+ = u/w_0 \qquad v_+ = v/w_0 \tag{10-103}$$

$$w_+ = w/l \tag{10-104}$$

The precise status of the order L of approximation is extensively discussed by Gouesbet, Maheu, and Gréhan (1985). The point of concern is that Eqs. (10-94)–(10-98) do not fully satisfy the Maxwell equations. However, when this set is introduced in the Maxwell equations, we observe directly that the approximation is perfect for the whole space in the limit $s \to 0$, the relative introduced errors being $O(s^2)$. As stated in the introduction, s is a small parameter with typically $s^2 \simeq 10^{-6}$. In other terms the introduced formal inconsistency is of no practical relevance for numerical results, in most cases.

We can also introduce an order L^- of approximation in which we furthermore neglect the fields E_w and H_w. In general, this degree of approximation is also good, as evidenced by a formal discussion (Gouesbet, Maheu, and Gréhan 1985) and numerical results (sections 10-4 and 10-5), but should be restricted to cases when the scatterer is not located too far from the beam axis. Alternatively, if necessary, more precise descriptions of the beam can be designed following Davis' (1979) procedure, adding higher powers of s terms in Eqs. (10-94)–(10-98).

Cartesian Description of the Incident Wave in the $(O_P\ xyz)$ System

We readily obtain

$$E_y = H_x = 0 \tag{10-105}$$

$$E_x = E_0 \Psi_0 \exp\left[-ik(z - z_0)\right] \tag{10-106}$$

$$E_z = -\frac{2Q}{l}(x - x_0)E_x \tag{10-107}$$

$$H_y = H_0 \Psi_0 \exp\left[-ik(z - z_0)\right] \tag{10-108}$$

$$H_z = -\frac{2Q}{l}(y - y_0)H_y \tag{10-109}$$

in which Ψ_0 is given by relation (10-100) with the following complementary relations:

$$h_+^2 = \frac{1}{w_0^2}\left[(x - x_0)^2 + (y - y_0)^2\right] \tag{10-110}$$

$$Q = \frac{1}{i + 2(\zeta - \zeta_0)} \tag{10-111}$$

$$\zeta = \frac{z}{l} \qquad \zeta_0 = \frac{z_0}{l} \tag{10-112}$$

Components E_r and H_r

From Eqs. (10-105)–(10-109), we readily obtain, by mere projections, the field components E_r and H_r in the $(r,\ \theta,\ \varphi)$ system:

$$E_r = E_0\Psi_0\left[\cos\varphi\sin\theta\left(1 - \frac{2Q}{l}r\cos\theta\right) + \frac{2Q}{l}x_0\cos\theta\right]\exp(K) \tag{10-113}$$

$$H_r = H_0\Psi_0\left[\sin\varphi\sin\theta\left(1 - \frac{2Q}{l}r\cos\theta\right) + \frac{2Q}{l}y_0\cos\theta\right]\exp(K) \tag{10-114}$$

with

$$K = -ik(r\cos\theta - z_0) \tag{10-115}$$

Relations Eqs. (10-113) and (10-114) enable us to compute coefficients $g_{n,\mathrm{TM}}^m$ and $g_{n,\mathrm{TE}}^m$ [relations Eqs. (10-27) and (10-29)], then closing the GLMT for Gaussian beams. At the order L^- of approximation, the Q terms in Eqs. (10-113)–(10-114) must be deleted.

It is, however, convenient to identify where the argument φ appears and express this dependence on φ in terms of the trigonometric functions sin and cos involved in the general expressions for the BSPs.

The argument φ appears explicitly in Eqs. (10-113)–(10-114) and is also contained in the function Ψ_0, which can be rewritten as

$$\Psi_0 = \Psi_0^0 \Psi_0^\varphi \tag{10-116}$$

$$\Psi_0^0 = iQ \exp\left(-iQ \frac{r^2 \sin^2 \theta}{w_0^2}\right) \exp\left(-iQ \frac{x_0^2 + y_0^2}{w_0^2}\right) \tag{10-117}$$

$$\Psi_0^\varphi = \exp\left(\frac{2iQ}{w_0^2} r \sin \theta(x_0 \cos \varphi + y_0 \sin \varphi)\right) \tag{10-118}$$

in which Ψ_0^0 does not depend on φ.

We modify Ψ_0^φ by replacing the functions sin and cos with exponentials of imaginary arguments; then we expand the resulting exponentials and finally expand terms of the form $(a + b)^j$ to obtain

$$\Psi_0^\varphi = \sum^{jp} \Psi_{jp} \exp\left[i\varphi(j - 2p)\right] \tag{10-119}$$

with

$$\sum^{jp} = \sum_{j=0}^{\infty} \sum_{p=0}^{j} \tag{10-120}$$

$$\Psi_{jp} = \left(\frac{iQr \sin \theta}{w_0^2}\right)^j \frac{(x_0 - iy_0)^{j-p}(x_0 + iy_0)^p}{(j - p)!(p)!} \tag{10-121}$$

When O_P is located on the beam axis ($x_0 = y_0 = 0$), then all the Ψ_{jp} appearing in Eq. (10-119) are zero, except

$$\Psi_{00} = 1 \tag{10-122}$$

The formulation given by Gouesbet, Maheu, and Gréhan (1988a) to express Ψ_0^φ is equivalent but more complicated than necessary.

The radial components E_r and H_r then become

$$E_r = E_0 \frac{F}{2} \left(\sum^{jp} \Psi_{jp} \exp(ij_+\varphi) + \sum^{jp} \Psi_{jp} \exp(ij_-\varphi)\right)$$

$$+ E_0 x_0 G \sum^{jp} \Psi_{jp} \exp(ij_0\varphi) \tag{10-123}$$

$$H_r = H_0 \frac{F}{2i} \left(\sum^{jp} \Psi_{jp} \exp(ij_+\varphi) - \sum^{jp} \Psi_{jp} \exp(ij_-\varphi)\right)$$

$$+ H_0 y_0 G \sum^{jp} \Psi_{jp} \exp(ij_0\varphi) \tag{10-124}$$

in which

$$F = \Psi_0^0 \sin \theta \left(1 - \frac{2Q}{l} r \cos \theta\right) \exp(K) \tag{10-125}$$

$$G = \Psi_0^0 \frac{2Q}{l} \cos \theta \exp (K) \qquad (10\text{-}126)$$

and

$$j_+ = j + 1 - 2p = j_0 + 1 \qquad (10\text{-}127)$$

$$j_- = j - 1 - 2p = j_0 - 1 \qquad (10\text{-}128)$$

Coefficients g_n^m

Expressions (10-123)–(10-124) are then introduced in relations Eqs. (10-27) and (10-29), and the φ integration is performed, leading to the final expression for coefficients g_n^m:

$$g_{n,\text{TM}}^m = \frac{1}{C_n^{\text{pw}}} \frac{(2n+1)^2}{\pi n(n+1)} \frac{(n-|m|)!}{(n+|m|)!} \int_0^\pi \int_0^\infty$$

$$\cdot \left[\frac{F}{2} \left(\sum_{j_+ = m}^{jp} \Psi_{jp} + \sum_{j_- = m}^{jp} \Psi_{jp} \right) + x_0 G \sum_{j_0 = m}^{jp} \Psi_{jp} \right]$$

$$\cdot r \Psi_n^1(kr) P_n^{|m|} (\cos \theta) \sin \theta \, d\theta \, d(kr) \qquad (10\text{-}129)$$

$$g_{n,\text{TE}}^m = \frac{1}{C_n^{\text{pw}}} \frac{(2n+1)^2}{\pi n(n+1)} \frac{(n-|m|)!}{(n+|m|)!} \int_0^\pi \int_0^\infty$$

$$\cdot \left[\frac{F}{2} \left(-i \sum_{j_+ = m}^{jp} \Psi_{jp} + i \sum_{j_- = m}^{jp} \Psi_{jp} \right) + y_0 G \sum_{j_0 = m}^{jp} \Psi_{jp} \right]$$

$$\cdot r \Psi_n^1(kr) P_n^{|m|} (\cos \theta) \sin \theta \, d\theta \, d(kr) \qquad (10\text{-}130)$$

Symbol Σ_c^{jp} designates sum Σ^{jp} restricted to condition c.

Beam Axis Location

When O_P is located on the beam axis, the theory is simplified significantly because, as we observed previously, all Ψ_{jp} are equal to zero, but $\Psi_{00} = 1$, Eq. (10-122). More explicitly, we may write

$$\Psi_{jp}(x_0 = y_0 = 0) = \delta_j^0 \qquad (10\text{-}131)$$

Consequently, all the coefficients $g_{n,\text{TM}}^m$ and $g_{n,\text{TE}}^m$ become equal to zero, except for $|m| = 1$. For $|m| = 1$, we find

$$g_{n,\text{TM}}^1 = g_{n,\text{TM}}^{-1} = \frac{1}{2} g_n \qquad (10\text{-}132)$$

$$g_{n,\text{TE}}^1 = -g_{n,\text{TE}}^{-1} = -\frac{i}{2} g_n \qquad (10\text{-}133)$$

in which the coefficients g_n are given by

$$g_n = \frac{k(2n + 1)}{i^{n-1}(-1)^n \pi n(n + 1)} \int_0^\pi \int_0^\infty Fr\Psi_n^1(kr)P_n^1(\cos \theta)$$

$$\cdot \sin \theta \, d\theta \, d(kr) \tag{10-134}$$

Equations (10-132)–(10-133) are to be compared with Eqs. (10-149)–(10-150) for the plane wave case ($g_n = 1$).

These coefficients appeared in the works by Gouesbet and Gréhan (1982) and Gouesbet, Gréhan, and Maheu (1985). They also appear in the works by Gréhan, Gouesbet, and Rabasse (1980) and Gouesbet (1985) in some GLMT expressions given without any demonstration for the non-off-axis special case, following the references Gouesbet and Gréhan (1980a, b), which were not released in the open literature. They can be more explicitly rewritten as

$$g_n = \frac{2n + 1}{\pi n(n + 1)} \frac{1}{(-1)^n i^{n-1}} \int_0^\pi \int_0^\infty iQ \exp\left(-iQ \frac{r^2 \sin^2 \theta}{w_0^2}\right) \exp(ikz_0)$$

$$\cdot \exp(-ikr \cos \theta)\left(1 - \frac{2Q}{l}r \cos \theta\right)$$

$$\cdot \Psi_n(kr)P_n^1(\cos \theta) \sin^2 \theta \, d\theta \, d(kr) \tag{10-135}$$

All the expressions given in the general formulation (Section 10-2) also simplify. The expressions for the scattered intensities become

$$\begin{pmatrix} I_\theta^+ \\ I_\varphi^+ \end{pmatrix} = \frac{\lambda^2}{4\pi^2 r^2} \begin{pmatrix} i_2 \cos^2 \varphi \\ i_1 \sin^2 \varphi \end{pmatrix} \tag{10-136}$$

in which

$$i_j = |S_j|^2 \tag{10-137}$$

in which the amplitude functions S_j are

$$S_1 = \sum_{n=1}^\infty \frac{2n + 1}{n(n + 1)} g_n[a_n \Pi_n(\cos \theta) + b_n \tau_n(\cos \theta)] \tag{10-138}$$

$$S_2 = \sum_{n=1}^\infty \frac{2n + 1}{n(n + 1)} g_n[a_n \tau_n(\cos \theta) + b_n \Pi_n(\cos \theta)] \tag{10-139}$$

For the phase angle, we obtain

$$\tan \delta = \frac{\mathrm{Re}\,(S_1)\,\mathrm{Im}\,(S_2) - \mathrm{Re}\,(S_2)\,\mathrm{Im}\,(S_1)}{\mathrm{Re}\,(S_1)\,\mathrm{Re}\,(S_2) + \mathrm{Im}\,(S_1)\,\mathrm{Im}\,(S_2)} \tag{10-140}$$

The cross sections reduce to

$$C_{sca} = \frac{\lambda^2}{2\pi} \sum_{n=1}^\infty (2n + 1)|g_n|^2 (|a_n|^2 + |b_n|^2) \tag{10-141}$$

$$C_{\text{ext}} = \frac{\lambda^2}{2\pi} \, \text{Re} \left(\sum_{n=1}^{\infty} (2n + 1)|g_n|^2 \, (a_n + b_n) \right) \tag{10-142}$$

Finally, for the radiation pressure force components, we have

$$F_x^+ = F_y^+ = 0 \tag{10-143}$$

and

$$C_{\text{pr},z} = \frac{\lambda^2}{2\pi} \sum_{n=1}^{\infty} \frac{2n + 1}{n(n + 1)} |g_n|^2 \, \text{Re} \, [(a_n + b_n - 2a_n b_n^*)] + \frac{n(n + 2)}{n + 1}$$

$$\cdot \text{Re} \, [g_n g_{n+1}^* (a_n + b_n + a_{n+1}^* + b_{n+1}^* - 2a_n a_{n+1}^* - 2b_n b_{n+1}^*)] \tag{10-144}$$

If we exclude pressure cross section $C_{\text{pr},z}$, the above formulation with $x_0 = y_0 = 0$ is identical to that of the LMT with only the addition of the g_n coefficients. The expression for the pressure cross section differs from that for the LMT due to the wave front curvature, which has been discussed in the text following Eq. (10-76), and neglected for this term by Gouesbet, Gréhan, and Maheu (1985).

More Special Cases

If, furthermore, $z_0 = 0$, as assumed by Gouesbet and Gréhan (1982) and Gouesbet, Gréhan, and Maheu (1985), the g_n reduce to

$$g_n = \frac{1}{i^{n-1}(-1)^n \pi} \frac{2n + 1}{n(n + 1)} \int_0^{\pi} \int_0^{\infty} \sin^2 \theta f \exp \, (-ikr \cos \theta)$$

$$\cdot \Psi_n(kr) P_n^1(\cos \theta) \, d\theta \, d(kr) \tag{10-145}$$

in which the radial basic function f is given by

$$f = iQ \exp \left(-iQ \frac{r^2 \sin^2 \theta}{w_0^2} \right) \left(1 - \frac{2Q}{l} r \cos \theta \right) \tag{10-146}$$

in agreement with the results obtained by Gouesbet, Gréhan, and Maheu (1985) except for the pressure cross section $C_{\text{pr},z}$, which was given with an approximation.

At the order of approximation L^- discussed by Gouesbet and Gréhan (1982), the radial basic function simplifies to

$$f = iQ \exp \left(-iQ \frac{r^2 \sin^2 \theta}{w_0^2} \right) \tag{10-147}$$

Finally, if $w_0 \to \infty$, the incident beam becomes a plane wave. Then we find that all Ψ_{jp} and g_n^m are zero except

$$\Psi_{00} = 1 \tag{10-148}$$

$$g_{n,\text{TM}}^1 = g_{n,\text{TM}}^{-1} = \frac{1}{2} \tag{10-149}$$

$$g_{n,\text{TE}}^{1} = (-)g_{n,\text{TE}}^{-1} = (-)\frac{i}{2} \qquad (10\text{-}150)$$

Using Eqs. (10-148)–(10-150) [which are equivalent to setting $g_n = 1$ in Eqs. (10-132) and (10-133)], it is an exercise to show that we recover the usual expressions for the LMT, although some algebra is required for $C_{\text{pr},z}$. The LMT has become a special case of our GLMT, as it should for correctness.

10-4 NUMERICAL COMPUTATIONS OF THE COEFFICIENTS g_n

Generalities

Starting from the formulas of the GLMT or from those of its special cases, the computation of numerical results, required for concrete applications, can be divided into three tasks: (1) computations of coefficients $g_{n,\text{TM}}^{m}$ and $g_{n,\text{TE}}^{m}$ (general case) or of coefficients g_n (axis location), (2) computations of other mathematical functions involved in the theory, and (3) assembling the terms and series to compute the required scattered properties.

Points (2) and (3) are not discussed in the present chapter to focus our interest on the new coefficients (g_n^m, g_n), which are mainly the new quantities introduced in the GLMT.

Historically, the theory was first developed for the axis location case (or even the beam waist center case) before being fully generalized. Thus our knowledge concerning numerical computations is much more extended for the g_n than for the g_n^m.[*] In this section, we discuss coefficients g_n, discussion of the g_n^m being postponed to Section 10-5. Furthermore, as we observed, in the case of on-axis location, the GLMT formulas basically own the same structure as those of the LMT, but for the appearance of the new coefficients g_n. Consequently, the computation of the scattered quantities reduces to a simple adaptation of the existing codes for Mie theory, such as the code SUPERMIDI (Gréhan and Gouesbet 1979a, b; Gréhan, Gouesbet, and Rabasse 1980).

At present, we know three methods to compute coefficients g_n. These methods are presented below, ordered from the slowest to the fastest running.

Quadrature Computations

For Gaussian beam scattering with axis location, the expressions for the g_n originally appeared in the theory as quadratures like relations Eqs. (10-134)–(10-135). For axisymmetric non-Gaussian beams and axis location of point O_P, we might expect a reduction of coefficients g_n^m to coefficients g_n, starting from relations Eqs. (10-27), (10-29), with disappearance of the φ integration. For nonaxisymmetric non-Gaussian beams, such a reduction to a set of g_n is not expected. These last two statements are more or less intuitive and need to be verified.

*New steps have been achieved since this chapter was written (1988).

The evaluation of g_n through quadratures is not an easy task because the functions $\Psi_n^1(kr)$ and $P_n^1(\cos\theta)$ in the integrand have oscillatory behaviors. It consequently requires large computing times even on mainframe computers. One hour of CPU time is typical for a single g_n with an IBM-3090 of the Centre InterRégional de Calcul Electronique (CIRCE; Orsay, France). The accuracy of the results may be limited, depending on the quality of the numerical integration routine. For non-Gaussian beams, if the reduction to g_n is possible, the quadrature method only requires modification of the integrand.

Finite Series

It is possible to consider the quadrature expressions for the g_n as resulting from the expansion of E_r and H_r in spherical harmonics or, more precisely, in surface harmonics of the first kind (Maheu, Gouesbet, and Gréhan 1988), using an expansion theorem that is valid for any $f(\theta, \varphi)$ function provided that it as well as its first two derivatives are continuous (Robin 1957–59) and also using the fact that the r variable can be separated from the θ and φ in the E_r and H_r expressions.

It is also possible to use Neumann's expansion, i.e., Bessel function expansion (Watson 1962, p. 524), under the following form (Maheu, Gouesbet, and Gréhan 1988):

$$x^{1/2}g(x) = \sum_{n=0}^{\infty} c_n J_{n+1/2}(x) \qquad (10\text{-}151)$$

in which coefficients c_n can be computed from the McLaurin expansion of $g(x)$.

In order to produce the above standard equation in the GLMT, the radial component expressions (E_r and H_r) are rearranged in the following manner. First, the φ dependence is eliminated using an integral operator, then the θ dependence is also avoided by specifying the E_r and $\partial E_r/\partial \cos\theta$ expressions for $\theta = \pi/2$, and finally, the spherical Bessel functions are expressed in terms of half-integer order Bessel functions. Details are given by Gouesbet, Gréhan, and Maheu (1988a, b).

For Gaussian beams (with axis location), we obtain (with $p \geq 0$)

$$g_{2p+1} = Z_0 \exp(ikz_0) \sum_{j=0}^{p} \frac{p!}{j!(p-j)!} \frac{\Gamma(p+j+3/2)}{\Gamma(p+3/2)} (-4Z_0 s^2)^j \qquad (10\text{-}152)$$

$$g_{2p+2} = \frac{1}{k} \exp(ikz_0) \sum_{j=0}^{p} \frac{p!}{j!(p-j)!} \frac{\Gamma(p+j+5/2)}{\Gamma(p+5/2)}$$

$$\cdot \left(A - \frac{jB}{Z_0}\right)(-4Z_0 s^2)^j \qquad (10\text{-}153)$$

$$A = kZ_0 + kZ_0'' - \frac{2}{l}\epsilon_L Z_0^2 \qquad (10\text{-}154)$$

$$B = -kZ_0 Z_0'' \qquad (10\text{-}155)$$

$$Z_0 = iQ(\theta = \pi/2) = \frac{1}{1 + (2iz_0/l)} \tag{10-156}$$

$$Z_0'' = \frac{1}{kr} \left. \frac{\partial Q}{\partial \theta} \right|_{\theta = \pi/2} = \frac{2}{kl[i - (2z_0/l)]^2} \tag{10-157}$$

Here ϵ_L is equal to 1 or zero, depending on whether the Gaussian beam is described at the order L or L^-, respectively. Note that the expressions for the g_n, with n odd, do not depend on the order of approximation (L or L^-).

Several further assumptions can be introduced by steps in Eqs. (10-152) and (10-153). Let us, for instance, consider the case when $O_P \equiv O_G$ ($z_0 = 0$), assume an order L^- of approximation ($\epsilon_L = 0$), and finally, neglect s^2, which is small with respect to $O(1)$. Then we obtain

$$g_{2p+1} = \sum_{j=0}^{p} \frac{p!}{j!(p-j)!} \frac{\Gamma(p+j+3/2)}{\Gamma(p+3/2)} (-4s^2)^j \tag{10-158}$$

$$g_{2p+2} = \sum_{j=0}^{p} \frac{p!}{j!(p-j)!} \frac{\Gamma(p+j+5/2)}{\Gamma(p+5/2)} (-4s^2)^j \tag{10-159}$$

Relations Eqs. (10-158) and (10-159) have been obtained by Ferguson and Currie (1986).

The g_n computations using finite series provide good numerical results, with a much better accuracy then even quadratures in most cases. More important, they are carried out much faster and at a much lower cost because they only need PCs, with a typical CPU time of 1 s per g_n. However, the method requires extra algebraic work when the description of the beam is modified. Also, in exotic situations [for rays passing far from the axis of the beam, that is, for great values of n (see the following section) the meaning of "great" depending on the value of the beam waist], numerical problems arise because the series starts behaving erratically. These problems can be overcome by programing computations in the Mulisp language (available from Soft Warehouse, Honolulu, Hawaii) in order to cut off truncature errors (the Mulisp language permits "infinite" accuracy computations for rational numbers, the actual accuracy being only limited by the storage capability of the computers).

Localized Interpretation

The two methods discussed above are rigorous and, consequently, formally equivalent. The finite series method is much more advantageous in terms of cost and required time. The third method is even faster, since g_n computations can be carried out "instantaneously" on pocket computers. The price to pay is loss of accuracy, which is neglectable in most cases. At the beginning of its development, the method relies on a physical interpretation of the coefficients g_n that we call the localized interpretation. The localized interpretation is designed at the order L^- of approximation.

We first consider a Gaussian beam at the order L^-, with $O_P \equiv O_G$.

In the study of plane wave scattering by spheres having a large diameter with respect to the light wavelength λ, a principle of localization was formulated by van de Hulst (1957), according to which "a term of order n corresponds to a ray passing the origin at a distance $(n + 1/2)(\lambda/2\pi)$." Here, the term of order n is the term of a Lorenz-Mie series containing Bessel functions and spherical harmonics, the integer n ranging from 1 to ∞. We have extended the localization principle to the case of Gaussian beams. At the order L^- of description of the Gaussian beam, the amplitude E_x/E_0 of the beam has been assumed to be described by the leading term $\exp[-(\rho/w_0)^2]$ at a distance $\rho = r \sin \theta$ from the axis of the beam [from Eqs. (10-106) and (10-116)–(10-118)] specified for the case under study. Consequently, the term of order n in Mie series [Eqs. (10-138), (10-139)] corresponds to a ray passing a distance $\rho_n = [(n + 1/2)\lambda/2\pi]$, and the amplitude E_x/E_0 is identified with coefficient g_n of the GLMT and reads as follows:

$$g_n = \exp\left[-\left(\frac{(n + 1/2)\lambda}{2\pi w_0}\right)^2\right] \tag{10-160}$$

The plane wave case can be obtained in the limit $w_0 \to \infty$, leading to $g_n = 1$, a result that we know is true from Gouesbet and Gréhan (1982) and Gouesbet, Gréhan, and Maheu (1985). The original van de Hulst formulation is then recovered.

When Eq. (10-160) is introduced in the GLMT for an axis location, we obtain what we call a localized approximation to the GLMT. The g_n computed from Eq. (10-160) are called the localized approximations to the g_n (in the framework of the localized interpretation).

The validity of this approach has been firmly evidenced on the basis of different kinds of results: (1) examination of Rayleigh-Gans scattering with Gaussian illumination of the particle, (2) comparisons of scattering diagrams in Gaussian scattering with those produced by other authors, in some special cases, (3) comparisons with optical levitation experimental data, and (4) direct comparisons between g_n values computed using Eq. (10-160) as well as the two previous methods (quadratures and finite series). These results are discussed by Gréhan, Maheu, and Gouesbet (1986a, b) and Maheu, Gréhan, and Gouesbet (1987a, b, 1988). See also Gréhan and Gouesbet (1980).

Examples will be given later in a more general context. Furthermore, the localized character of the coefficients g_n is directly discussed and evidenced by Maheu, Gréhan, and Gouesbet (1987a, 1989).

However, Eq. (10-160) refers to a case too special to permit full use of the GLMT. We have now to generalize it. In this section, we shall discuss the generalization to the case ($x_0 = y_0 = 0$, $z_0 \neq 0$: axis location, but not necessarily at the beam waist center). This generalization will be based on guesses and conjectures and will eventually be proved to be efficient and successful. To succeed in this guesswork, we need to give a more formal derivation to the localized interpretation.

For the special case $O_P \equiv O_G$, at the order L^-, the radial component E_r reads [specifying Eq. (10-113), and remarking that the $Q \cos \theta$ terms pertain to the description at the order L]:

$$E_r = \{E_0 \exp(-ikz)\cos\varphi\sin\theta\}\, iQ \exp\left(-iQ\frac{r^2\sin^2\theta}{w_0^2}\right) \qquad (10\text{-}161)$$

in which the term in braces is the plane wave term and Q stands for $Q(z_0 = 0) = 1/(i + 2z/l)$.

Furthermore, from Eq. (10-21) specified for the special case under study, we have

$$U_{\mathrm{TM}}^i = \cos\varphi \sum_{n=1}^{\infty} g_{n,\mathrm{TM}}\, \mathscr{A}_n \qquad (10\text{-}162)$$

in which $\partial\mathscr{A}_n/\partial\varphi = 0$ and the subscript TM is added to g_n. In Eqs. (10-161) and (10-162), we observe that the dependence on φ involves the same function $\cos\varphi$.

Let us consider a function $h(z, r\sin\theta)$ and introduce the transformed set of functions $\overline{h}_n(z, r\sin\theta)$ given by

$$\overline{h}_n(z, r\sin\theta) = h(0, \rho_n) \qquad (10\text{-}163)$$

Now, to the subscript n in the U_{TM}^i series, we associate a $g_{n,\mathrm{TM}}$ coefficient given by

$$g_{n,\mathrm{TM}} = \overline{iQ \exp\left(-iQ\frac{r^2\sin^2\theta}{w_0^2}\right)} \qquad (10\text{-}164)$$

In other words, $g_{n,\mathrm{TM}}$ is obtained from the nonplane wave factor in Eq. (10-161), to which we apply the overbar transformation.

Similarly, we have

$$H_r = [H_0 \exp(-ikz)\sin\varphi\sin\theta]\, iQ \exp\left(-iQ\frac{r^2\sin^2\theta}{w_0^2}\right) \qquad (10\text{-}165)$$

$$U_{\mathrm{TE}}^i = \sin\varphi \sum_{n=1}^{\infty} g_{n,\mathrm{TE}}\, \mathscr{B}_n \qquad (10\text{-}166)$$

and, applying the same procedure as above,

$$g_{n,\mathrm{TE}} = \overline{iQ \exp\left(-iQ\frac{r^2\sin^2\theta}{w_0^2}\right)} \qquad (10\text{-}167)$$

Hence, noting that $\overline{Q}(z_0 = 0) = 1/i$, we have

$$g_{n,\mathrm{TM}} = g_{n,\mathrm{TE}} = g_n = \exp\left[-\left(\frac{\rho_n}{w_0}\right)^2\right] \qquad (10\text{-}168)$$

which is identical to Eq. (10-160).

We now conjecture that the same procedure can be generalized to design a localized approximation for the case $(x_0 = y_0 = 0, z_0 \neq 0)$.

At the order L^- of approximation, we then have

$$\begin{pmatrix} E_r \\ H_r \end{pmatrix} = \left[\begin{pmatrix} E_0\cos\varphi \\ H_0\sin\varphi \end{pmatrix} \exp(-ikz)\sin\theta \right] \qquad \text{(continued)}$$

$$\cdot \, iQ \, \exp \left(-iQ \, \frac{r^2 \sin^2 \theta}{w_0^2} \right) \exp (ikz_0) \qquad (10\text{-}169)$$

in which Q is now given by its general expression Eq. (10-111). Furthermore, U_{TM}^i and U_{TE}^i remain proportional to $\cos \varphi$ and $\sin \varphi$, respectively.

Then, we obtain

$$g_{n,TM} = g_{n,TE} = g_n = iQ \, \exp \left(-iQ \, \frac{r^2 \sin^2 \theta}{w_0^2} \right) \exp (ikz_0) \qquad (10\text{-}170)$$

that is,

$$g_n = i\overline{Q} \, \exp \left[-i\overline{Q} \left(\frac{\rho_n}{w_0} \right)^2 \right] \exp (ikz_0) \qquad (10\text{-}171)$$

in which now,

$$\overline{Q} = Q(z = 0) = \frac{1}{i - (2z_0/l)} \qquad (10\text{-}172)$$

Numerical Results and the Validity of our Generalization of the Localized Interpretation

Recent numerical results for the g_n computations, and comparisons between the three methods (quadratures, limited series, localized approximation), are discussed by Gouesbet, Gréhan, and Maheu (1988b). In this chapter, we only reproduce a table from this last reference, concerning the case ($x_0 = y_0 = 0$, $z_0 \neq 0$). Then the g_n are computed using Eq. (10-135) for the quadratures, Eqs. (10-152) and (10-153) for the finite series, and Eq. (10-171) for the localized interpretation.

Table 10-1 Comparison of g_n coefficients for a localization along the beam axis

	Quadrature computation Eq. (10-135) A	Series computation Eqs. (10-152), (10-153), order L D	Series computation Eqs. (10-152), (10-153), order L^- E	Localized interpretation computation Eq. (10-171) G
g_1	(0.277300, 0.345338)	(0.280006, 0.348707)	(0.280006, 0.348707)	(0.279888, 0.348741)
g_2	(0.274657, 0.342330)	(0.279808, 0.348750)	(0.279907, 0.348728)	(0.279684, 0.348784)
g_3	—	(0.279511, 0.348814)	(0.279511, 0.348814)	(0.279387, 0.348848)
g_4	—	(0.279115, 0.348899)	(0.279214, 0.348878)	(0.278990, 0.348933)
g_{30}	—	(0.234722, 0.356328)	(0.234818, 0.356317)	(0.234610, 0.356326)
g_{45}	—	(0.181297, 0.359650)	(0.181297, 0.359649)	(0.181207, 0.359610)
g_{60}	(0.099474, 0.320927)	(0.111850, 0.354374)	(0.111933, 0.354387)	(0.111799, 0.354300)
g_{80}	—	(0.006395, 0.322089)	(0.004661, 0.322121)	(0.006420, 0.321998)
g_{100}	—	(−0.091237, 0.252183)	(−0.091195, 0.252228)	(−0.091136, 0.252132)

The beam diameter is $2w_0 = 20\lambda$, and the waist ordinate, $z_0 = \pi 10^{-4}$. The wavelength is 0.5 μm.

Only a few results have been produced by the quadrature method due to the computing cost. Hence the most important part of this section concerns the comparison between results obtained using finite series and those obtained from the localized interpretation.

Table 10-1 is for a laser beam with a beam waist diameter equal to $2w_0 = 20\lambda = 10 \ \mu m$, and $z_0 = 10^{-4}\pi$. The agreement between results in columns A, D, and G is very good. Results in columns A and D should be, in principle, identical due to the mathematical equivalence of the corresponding expressions used to compute them. The observed differences are attributed to the limited numerical accuracy, mainly in computing the quadratures. The good agreement between columns D and G provides us with a validation of our generalization of the localized interpretation.

The practical net consequence of these results is that GLMT computations can be efficiently carried out on microcomputers (Corbin et al. 1988).

10-5 NUMERICAL COMPUTATIONS OF THE COEFFICIENTS g_n^m

Rigorous Methods

As discussed in Section 10-4 for the g_n coefficients, there are also two rigorous methods to compute the coefficients $g_{n,\text{TM}}^m$ and $g_{n,\text{TE}}^m$.

The quadrature method relies on the evaluation of Eqs. (10-27) and (10-29) in the general formulation, becoming relations Eqs. (10-129) and (10-130) for Gaussian beams.

These expressions can again be obtained in terms of surface harmonics of the first kind expansion as discussed in Section 10-4 under finite series (Maheu, Gouesbet, and Gréhan 1988). Due to the expected increasing cost and time-consuming character of these computations, we never used this method.

The Neuman expansion technique (see section above on finite series) can again be used, providing us with finite series expressions that can be programed using PCs, leading to cheap and fast running computations. The algebraic expressions are, however, too lengthy to be given here. Furthermore, they depend on the parity of n and m. Also, for $m = 2k$ (even), we must distinguish the cases $k = 0$, $k > 0$, and $k < 0$, and for $m = 2k + 1$ (odd), we must distinguish the cases $k \geq 0$ and $k < 0$. For a given n and m, expressions are also different whether we consider the coefficients $g_{n,\text{TM}}^m$ or $g_{n,\text{TE}}^m$. These expressions are given by Gouesbet, Gréhan, and Maheu (1988*a*) for a Gaussian beam (at orders L and L^-). For a non-Gaussian beam, or a different description of a Gaussian beam, extra algebraic work is required to derive new expressions. However, even if the obtained expressions are lengthy, they are easy to program. Consequently, the finite series method is efficient (and accurate) for handling nonaxis locations. Results will be given later.

Localized Approximation

We have succeeded in generalizing the localized interpretation from the case ($x_0 = y_0 = z_0 = 0$) to the case ($x_0 = y_0 = 0$, $z_0 \neq 0$). We now show that it is possible to achieve a final generalization to the arbitrary location case, providing the cheapest and fastest way of computing the coefficients $g_{n,\text{TM}}^m$ and $g_{n,\text{TE}}^m$.

We are now faced with a subscript n and a superscript m. For subscript n, we assume that the reduction procedure explained in the section above on localized interpretation remains valid. However, we need some clue concerning the status of superscript m.

The n corresponds to the fact that BSP U_{TM}^i and U_{TE}^i are expanded in series of the $\Sigma_{n=1}^{\infty}$ kind. These series induce, through the localization principle, a discretization of the space conveyed by the introduction of discrete radial locations ρ_n. Due to this discretization, the radial wave character of the fields is lost, permitting us to discuss the problem in terms of geometry, that is, rays.

We might similarly think that the m is associated with an angular discretization in space, on angle φ. Associated with a bundle of rays at distance ρ_n, we might then introduce a further discretization φ_m. Since m ranges from $(-n)$ to $(+n)$, we should then be faced with $(2n + 1)$ bundle of rays for each n bundle of rays. However, this is not correct because there is a fundamental difference between an n discretization and an m discretization. The n ranges on an infinite domain (from 1 to infinity), reflecting the fact that r also ranges on an infinite domain (from zero to infinity). Conversely, the m ranges on a finite domain ($-n$ to $+n$), reflecting the fact that φ also ranges on a finite domain (zero to $+2\pi$). Then, if we introduce an angle φ_{-n} to $m = -n$, we shall have to introduce an angle φ_{+n} to $m = +n$ with, for instance, $\varphi_{-n} \simeq 0$ and $\varphi_{+n} \simeq 2\pi$. For a continuous modification of the angle φ, modulo 2π at $\varphi = 0 = 2\pi \pmod{2\pi}$ would then correspond to a discontinuous modification of $m = -n$ to $m = +n$.

To avoid this discontinuous modification, the domains φ_{-n} and φ_{+n} must be matched. Such a matching reintroduces a wave character with the condition that the φ wave be continuous at $\varphi = 0$. In other words, to the m, we prefer to associate a wave mode. When n increases, the number of φ wave modes increases because the circumference $2\pi\rho_n$ is larger, permitting us to introduce a larger number of different wavelengths. Formally, these wave modes are associated with the terms $P_n^{|m|}(\cos \theta)$ $\exp(im\varphi)$ in the BSPs. As it stands now, the argument is rather intuitive and heuristic and might be given a more formal basis in the future. However, our present hands-on representation (discretization in the r domain corresponding to n bundle of rays, and wave character in the angular domain corresponding to m wave modes) is enough to successfully design the generalization of the localization approximation. We actually attempted to discretize angle φ also, but without success.

At the order L^- of approximation (for which the localized interpretation has been designed up to now), the field component E_r reduces to [from Eq. (10-113)]

$$E_r = [E_0 \exp(-ikz) \cos \varphi \sin \theta] \exp(ikz_0) \, iQ \exp\left(-iQ \frac{r^2 \sin^2 \theta}{w_0^2}\right)$$

$$\cdot \exp\left(-iQ \frac{x_0^2 + y_0^2}{w_0^2}\right) \sum^{jp} \Psi_{jp} \exp[i\varphi(j - 2p)] \tag{10-173}$$

in which the φ dependence is explicit. Equation (10-173) can be rewritten as

$$E_r = [E_0 \exp(-ikz) \sin\theta] \exp(ikz_0)\, iQ$$

$$\cdot \exp\left(-iQ\,\frac{r^2 \sin^2\theta}{w_0^2}\right) \exp\left(-iQ\,\frac{x_0^2 + y_0^2}{w_0^2}\right)\frac{1}{2}$$

$$\cdot \left(\sum^{jp} \Psi_{jp} \exp(ij_+\,\varphi) + \sum^{jp} \Psi_{jp} \exp(ij_-\,\varphi)\right) \qquad (10\text{-}174)$$

in which we recall

$$j_+ = j - 2p + 1 \qquad (10\text{-}175)$$

$$j_- = j - 2p - 1 \qquad (10\text{-}176)$$

We now process m by isolating a contribution E_r^m to E_r, proportional to $\exp(im\varphi)$:

$$E_r^m = \{E_0 \exp(-ikz)\exp(im\varphi)\sin\theta\}\exp(ikz_0)\,iQ$$

$$\cdot \exp\left(-iQ\,\frac{r^2 \sin^2\theta}{w_0^2}\right)\exp\left(-iQ\,\frac{x_0^2 + y_0^2}{w_0^2}\right)\frac{1}{2}$$

$$\cdot \left(\sum_{j_+=m}^{jp}\Psi_{jp} + \sum_{j_-=m}^{jp}\Psi_{jp}\right) \qquad (10\text{-}177)$$

We recall that

$$U_{\mathrm{TM}}^i = \sum_{n=1}^{\infty}\sum_{m=-n}^{+n} g_{n,\mathrm{TM}}^m \mathscr{A}_n^m \exp(im\varphi) \qquad (10\text{-}178)$$

in which $\partial\mathscr{A}_n^m/\partial\varphi = 0$.

The block in braces in Eq. (10-177) is similar to that in Eq. (10-161). It contains a plane wave contribution $E_0 \exp(-ikz)\cdot\sin\theta$ multiplied by the term $\exp(im\varphi)$, which expresses the φ dependence in U_{TM}^i [in Eqs. (10-161) and (10-162), this φ dependence simply involved $\cos\varphi$]. We now conjecture that the second part on the right-hand side of Eq. (10-177) can be n discretized using the overbar transformation, leading to

$$g_{n,\mathrm{TM}}^m = \exp(ikz_0)\,i\overline{Q}\exp\left[-i\overline{Q}\left(\frac{\rho_n}{w_0}\right)^2\right]\exp\left(-i\overline{Q}\,\frac{x_0^2 + y_0^2}{w_0^2}\right)$$

$$\cdot\frac{1}{2}\left(\sum_{j_+=m}^{jp}\overline{\Psi}_{jp} + \sum_{j_-=m}^{jp}\overline{\Psi}_{jp}\right) \qquad (10\text{-}179)$$

Specifying separately the cases $m = 2l$ and $m = 2l + 1$, the double summation $\Sigma_{j_+=m}^{jp}$ and $\Sigma_{j_-=m}^{jp}$ can be reduced to simple summations. After some algebraic work, we obtain

$$g^{2l}_{n,\text{TM}} = \exp\left(ikz_0\right) i\overline{Q} \exp\left[-i\overline{Q}\left(\frac{\rho_n}{w_0}\right)^2\right]$$

$$\cdot \exp\left(-i\overline{Q}\,\frac{x_0^2 + y_0^2}{w_0^2}\right) H^{2l}_{\text{TM}}(\overline{\Psi}_{jp}) \qquad (10\text{-}180)$$

$$g^{2l+1}_{n,\text{TM}} = \exp\left(ikz_0\right) i\overline{Q} \exp\left[-i\overline{Q}\left(\frac{\rho_n}{w_0}\right)^2\right]$$

$$\cdot \exp\left(-i\overline{Q}\,\frac{x_0^2 + y_0^2}{w_0^2}\right) H^{2l+1}_{\text{TM}}(\overline{\Psi}_{jp}) \qquad (10\text{-}181)$$

in which

$$H^{2l}_{\text{TM}}(\Psi_{jp}) = \frac{1}{2} \sum_{j=0}^{\infty} \left(\Psi_{2j+1,j+1} + \Psi_{2j+1,j}\right) \qquad l = 0$$

$$H^{2l}_{\text{TM}}(\Psi_{jp}) = \frac{1}{2} \left(\sum_{j=l-1}^{\infty} \Psi_{2j+1,j+1-l} + \sum_{j=l}^{\infty} \Psi_{2j+1,j-l}\right) \qquad l > 0 \qquad (10\text{-}182)$$

$$H^{2l}_{\text{TM}}(\Psi_{jp}) = \frac{1}{2} \left(\sum_{j=|l|}^{\infty} \Psi_{2j+1,j+1-l} + \sum_{j=|l|-1}^{\infty} \Psi_{2j+1,j-l}\right) \qquad l < 0$$

and

$$H^{2l+1}_{\text{TM}}(\Psi_{jp}) = \frac{1}{2} \left(\sum_{j=l}^{\infty} \Psi_{2j,j-l} + \sum_{j=l+1}^{\infty} \Psi_{2j,j-l-1}\right) \qquad l \geq 0$$

$$H^{2l+1}_{\text{TM}}(\Psi_{jp}) = \frac{1}{2} \left(\sum_{j=|l|}^{\infty} \Psi_{2j,j-l} + \sum_{j=|l|-1}^{\infty} \Psi_{2j,j-l-1}\right) \qquad l < 0 \qquad (10\text{-}183)$$

For the coefficients $g^m_{n,\text{TE}}$ the generalization proceeds similarly using H_r and U^i_{TE} instead of E_r and U^i_{TM}. We obtain

$$g^m_{n,\text{TE}} = \exp\left(ikz_0\right) i\overline{Q} \exp\left[-i\overline{Q}\left(\frac{\rho_n}{w_0}\right)^2\right]$$

$$\cdot \exp\left(-i\overline{Q}\,\frac{x_0^2 + y_0^2}{w_0^2}\right) \frac{1}{2i} \left(\sum_{j_+ = m}^{jp} \overline{\Psi}_{jp} - \sum_{j_- = m}^{jp} \overline{\Psi}_{jp}\right) \qquad (10\text{-}184)$$

which becomes

$$g^{2l}_{n,\text{TE}} = \exp\left(ikz_0\right) i\overline{Q} \exp\left[-i\overline{Q}\left(\frac{\rho_n}{w_0}\right)^2\right]$$

$$\cdot \exp\left(-i\overline{Q}\,\frac{x_0^2 + y_0^2}{w_0^2}\right) H^{2l}_{\text{TE}}(\overline{\Psi}_{jp}) \qquad (10\text{-}185)$$

$$g^{2l+1}_{n,\text{TE}} = \exp\left(ikz_0\right) i\overline{Q} \exp\left[-i\overline{Q}\left(\frac{\rho_n}{w_0}\right)^2\right]$$

$$\cdot \exp\left(-i\overline{Q}\,\frac{x_0^2 + y_0^2}{w_0^2}\right) H_{\mathrm{TE}}^{2l+1}\,(\overline{\Psi}_{jp}) \qquad (10\text{-}186)$$

in which

$$H_{\mathrm{TE}}^{2l}(\Psi_{jp}) = \frac{1}{2i}\sum_{j=0}^{\infty}(\Psi_{2j+1,j+1} - \Psi_{2j+1,j}) \qquad l = 0$$

$$H_{\mathrm{TE}}^{2l}(\Psi_{jp}) = \frac{1}{2i}\left(\sum_{j=l-1}^{\infty}\Psi_{2j+1,j+1-l} - \sum_{j=l}^{\infty}\Psi_{2j+1,j-l}\right) \qquad l > 0 \qquad (10\text{-}187)$$

$$H_{\mathrm{TE}}^{2l}(\Psi_{jp}) = \frac{1}{2i}\left(\sum_{j=|l|}^{\infty}\Psi_{2j+1,j+1-l} - \sum_{j=|l|-1}^{\infty}\Psi_{2j+1,j-l}\right) \qquad l < 0$$

$$H_{\mathrm{TE}}^{2l+1}(\Psi_{jp}) = \frac{1}{2i}\left(\sum_{j=l}^{\infty}\Psi_{2j,j-l} - \sum_{j=l+1}^{\infty}\Psi_{2j,j-l-1}\right) \qquad l \geq 0$$

$$\qquad\qquad\qquad\qquad\qquad\qquad\qquad\qquad\qquad\qquad\qquad (10\text{-}188)$$

$$H_{\mathrm{TE}}^{2l+1}(\Psi_{jp}) = \frac{1}{2i}\left(\sum_{j=|l|}^{\infty}\Psi_{2j,j-l} - \sum_{j=|l|-1}^{\infty}\Psi_{2j,j-l-1}\right) \qquad l < 0$$

These expressions can be very simply programed and evaluated on PCs. More details are given by Gouesbet (1987). If we apply the same procedure at the order L of approximation, we find $g_{n,\mathrm{TM}}^m(L) = g_{n,\mathrm{TM}}^m(L^-)$ and $g_{n,\mathrm{TE}}^m(L) = g_{n,\mathrm{TE}}^m(L^-)$.

Validity of Generalization of the Localized Approximation

Table 10-2 presents a comparison between values of coefficients g_n^m, computed using finite series expressions at the order L and localized approximation expressions [Eqs. (10-180), (10-181), (10-185), (10-186)]. The agreement between both methods is impressive and constitutes a strong validation of the generalization of the localized

Table 10-2 Comparison of g_n^m coefficients for different off-axis locations

$X_0 = Y_0$	Exact	Localized
0 μm	$g_1^{-1} = (0.49796, -0.31703 \times 10^{-1})$ $g_1^1 = (0.49798, -0.31703 \times 10^{-1})$	$g_1^{-1} = (0.49770, -0.31667 \times 10^{-1})$ $g_1^1 = (0.49770, -0.31702 \times 10^{-1})$
2 μm	$g_1^{-1} = (0.36247, -0.15700 \times 10^{-1})$ $g_1^1 = (0.36247, -0.15700 \times 10^{-1})$	$g_1^{-1} = (0.36227, -0.15645 \times 10^{-1})$ $g_1^1 = (0.36226, -0.15737 \times 10^{-1})$
5 μm	$g_1^{-1} = (0.06794, 0.43020 \times 10^{-2})$ $g_1^1 = (0.06794, 0.43020 \times 10^{-2})$	$g_1^{-1} = (0.06791, 0.43405 \times 10^{-2})$ $g_1^1 = (0.06790, 0.43585 \times 10^{-2})$
10 μm	$g_1^{-1} = (0.00015, 0.74212 \times 10^{-4})$ $g_1^1 = (0.00015, 0.74212 \times 10^{-4})$	$g_1^{-1} = (0.00015, 0.74548 \times 10^{-4})$ $g_1^1 = (0.00015, 0.73791 \times 10^{-4})$

Here, $\omega_0 = 5$ μm, $\lambda = 0.5$ μm, and $z_0 = 10$ μm. The wavelength is 0.5 μm. The "exact" column has been computed with formulas given by Gouesbet, Gréhan, and Maheu (1988a). The "localized" column has been computed with Eqs. (10-181) and (10-182).

approximation (and of the conjectures leading to it), at least for $|m| = 1$. For $|m| \neq$ 1 the localized approximation must be completed by introducing simple normative coefficients on the right-hand side of Eqs. (10-179) and (10-184). Of course, for $|m| = 1$ these coefficients are equal to 1. Details are not given here because they concern current work that is not fully completed but that is expected to appear shortly in the open literature. However, the localized approximation is fully completed, leading to results similar to those in Table 10-2, whatever the value of n and m.

Actually, starting from a localization principle and generalizing our localized approximation procedure by steps, using a guess and conjecture game with a drastic economy of mathematical tools, we manage to reach a good agreement in Table 10-2. These results are not mere coincidence, and we expect that future work will reveal the deeper significance, at present hidden to us but grasped intuitively to some extent.

10-6 APPLICATIONS OF THE GLMT

Efficiently computing coefficients g_n^m has been made possible only recently. Consequently, applications of the GLMT in its more general formulation (nonaxis location) have not yet been produced. This section discusses applications of the GLMT in the restricted case of axis location, in which we only need to compute coefficients g_n and to implement them in existing Mie codes. More examples of computations and applications in this restricted framework can be found in the works by Gréhan, Maheu, and Gouesbet (1986b, 1988), Maheu, Gréhan, and Gouesbet (1987b), Corbin et al. (1988).

Comparisons with Optical Levitation Data

The optical levitation technique introduced by Ashkin (1970), then developed by Roosen, Delaunay, and Imbert (1977) is well suited to experimental study of the interaction between a laser beam and a particle. The first experimental scattering diagram produced by this technique is apparently due to Gréhan and Gouesbet (1980). The technique is now currently in use in our laboratory. Figure 10-2 shows an example of recent data.

The experimental setup is basically the same as that of Gréhan and Gouesbet (1980) but for the introduction of an automatic device to obtain and process the data. The scattered light from the levitating scatter center is collected by a 200-μm-diameter optical fiber, located at 20 cm from the particle and fed to a detector. The accuracy of angular displacements is $0.01°$. When the scattering diagram is recorded, the particle is collected and its diameter measured by optical microscopy. The refractive index of the material is determined using the Becke method.

In Fig. 10-2 the particle diameter equals 21.69 μm, the refractive index is $M = 1.552 \pm 0.15$, and scattering angle θ ranges from 0 to $20°$. The experimental scattering diagram is compared with results from the LMT (curve marked by asterisks) and from the GLMT (Gaussian beam, solid curve). The improvement in the comparisons is significant. The Gaussian character of the beam produces an efficient smoothing of

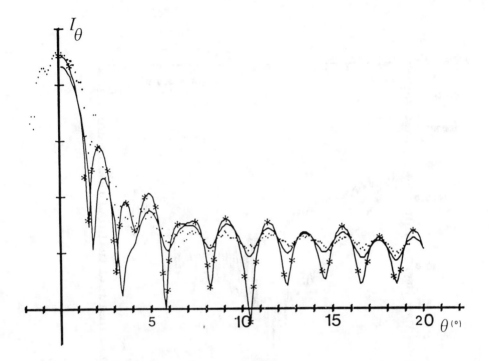

Alpha= 132.4.

Dpart.= 21.69 um Lambda= .5145 um

Figure 10-2 Experimental scattering diagram. Decimal logarithm of the intensity function $|S_2|^2$ versus θ. Dotted curve, measured values; solid curve, plane wave computation; and curve with asterisks, Gaussian wave computation.

the lobe structure, leading to a better agreement with the experiment than with the LMT.

Theoretical Scattering Diagrams

Figures 10-3 and 10-4 show scattering diagrams computed in the LMT and GLMT frameworks, for a size parameter $\alpha = \pi d/\lambda = 50$. The Gaussian beam has a dimensionless diameter $b_0 = 2\pi w_0/\lambda = 25$. Figure 10-3 shows the intensity function $|S_2|^2$ for a transparent particle ($M = 1.5$), and Fig. 10-4 shows the other intensity function $|S_1|^2$ for an absorbing particle ($M = 1.5-1.0i$). For the transparent particle, we observe a loss of visibility except for θ around 80° and an overall decrease of $|S_2|^2$, which reaches 3 orders of magnitude near $\theta \simeq 80°$. For the absorbing particle the lobe structure is poor for plane wave as well as for Gaussian beams. However, there is a loss of visibility near the forward direction. In contrast with the transparent particle case, the backward intensity is not affected, a fact that might be useful for optical

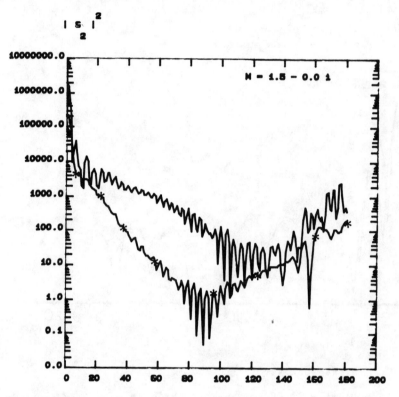

Figure 10-3 Computed scattering diagrams. Decimal logarithm of intensity function $|S_2|^2$ versus θ. Parameters are $\alpha = 50$, $M = 1.5 - 0.0i$. Solid curve, plane wave; and curve with asterisks, Gaussian beam ($2\pi w_0/\lambda = 25$).

sizing with backward collection of the scattered light. Scattering diagrams $|S_1|^2$ (for the transparent case) and $|S_2|^2$ (for the absorbing case) would lead to similar results and comments.

Collected Scattered Powers

We consider the scattered powers collected by a lens with a diameter of 3.6 cm located 30 cm from the scatter center illuminated by a Gaussian beam with a diameter $2w_0 = 10$ μm, versus the radius of the particle. Figures 10-5 to 10-7 show the decimal logarithm of the collected power versus decimal logarithm of the scatterer radius for a transparent particle ($M = 1.5$) and for scattering angles equal to 0, 90°, and 180°, respectively, and $\varphi = 0°$. Figure 10-8 is for an absorbing particle ($M = 1.5-1.0i$) in backward collection ($\theta = 180°$).

When the particle radius $r = d/2$ is smaller than the beam radius w_0, the LMT correctly describes the phenomenon. However, when the diameter increases, we observe significant differences between LMT and GLMT computations, which can reach

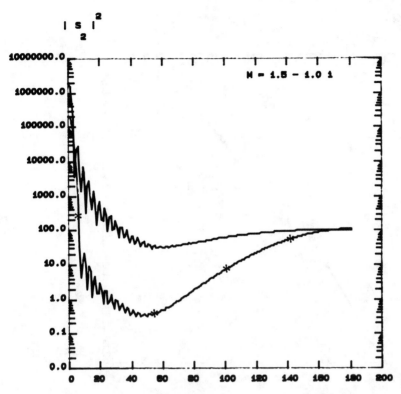

Figure 10-4 Computed scattering diagrams. Decimal logarithm of intensity function $|S_1|^2$ versus θ. Parameters are $\alpha = 50$, $M = 1.5 - 1.0i$. Solid curve, plane wave; and curve with asterisks, Gaussian beam ($2\pi w_0/\lambda = 25$).

several orders of magnitude. For an absorbing particle in backward collection, however (Fig. 10-8), LMT and GLMT lead to nearly the same results, as noted in the previous section (Fig. 10-4). Differences would only appear using linear scales.

Perspectives

The GLMT is fully completed from a formal point of view. For axis location, when the only new coefficients to compute are the g_n, any computer program for LMT can be easily adapted to handle GLMT computations. This requires a subroutine to compute the g_n by finite series, or even a single statement [Eq. (10-171)] in the main program when the localized interpretation is used.

Extensive results can now be produced at will. Examples of applications to laser Doppler anemometry (LDA) can be found in the work by Ferguson and Currie (1986).

For arbitrary location of the particle, it is now possible to efficiently and accurately compute the coefficients g_n^m.

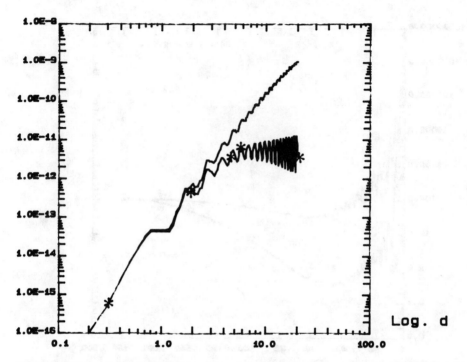

Figure 10-5 Collected power. Decimal logarithm of the collected power versus decimal logarithm of the particle radius. Forward collection $\theta = 0°$. Parameters are $M = 1.5-0.0i$, lens diameter 3.6 cm, located 30 cm from the particle center, $w_0 = 5$ μm. Solid curve, plane wave; and curve with asterisks, Gaussian beam.

In the near future, the GLMT will permit the production of extensive results for the scattering of Gaussian beam by a sphere. For two crossing incident Gaussian beams, as in LDA, it will permit accurate predictions of Doppler signals for particles larger than the beam diameter. Such computations and predictions will enable researchers to produce a rigorous theory for some existing optical sizing methods such as the visibility and the pedestal techniques, or the phase Doppler technique.

For the top-hat technique (Allano et al. 1984; Gouesbet, Gréhan, and Kleine 1984; Gréhan and Gouesbet 1986), results have been obtained using the localized approximation to compute the coefficients g_n associated with the incident top-hat beam. More general computations are now feasible, using the coefficients g_n^m.

These new opportunities will permit improvement of the design of instruments and data processing software packages, and also possible development of new instruments. Consequently, we expect the GLMT to become a new theoretical tool useful for two-phase flow studies when large discrete particles are transported by flows, including spray combustion systems.

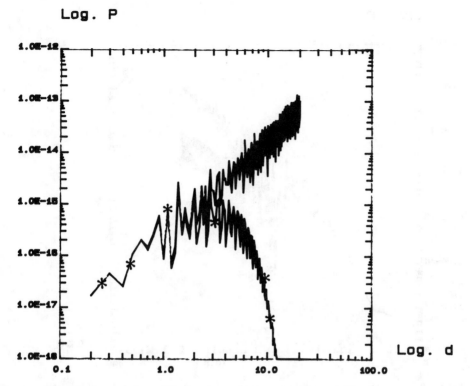

Figure 10-6 Collected power. Decimal logarithm of the collected power versus decimal logarithm of the particle radius. Side collection $\theta = 90°$. Parameters are $M = 1.5–0.0i$, lens diameter 3.6 cm, located 30 cm from the particle center, $w_0 = 5$ μm. Solid curve, plane wave; and curve with asterisks, Gaussian beam.

10-7 CONCLUSION

This chapter presents the state of the art in the development of the generalized Lorenz-Mie theory. The formulation is complete for the scattering by one sphere arbitrarily located in an arbitrary incident beam. Special cases involve Gaussian beam scattering, axis location of the scatter center, and classical Lorenz-Mie theory, which is now a special case of a much more general formulation. It involves the appearance of new sets of coefficients (g_n^m or/and g_n), which contain information about the incident beam.

Three methods are known to compute the new coefficients g_n^m (g_n). Two are rigorous (quadratures, finite series), while the third (localized approximation) emerges from a generalization of the van de Hulst principle of localization. The finite series and the localization approximation methods are accurate and fast running, permitting extensive production of results.

The GLMT is a new theoretical tool to design optical sizing instruments and to process data. It is expected to become a standard tool, long needed because most light

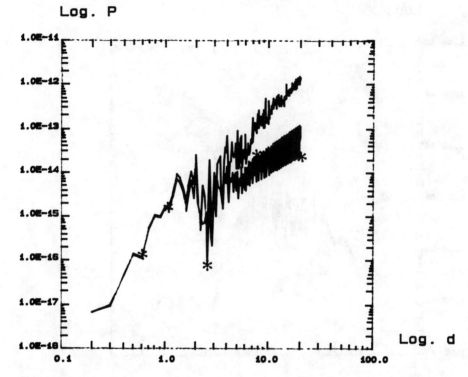

Figure 10-7 Collected power. Decimal logarithm of the collected power versus decimal logarithm of the particle radius. Backward collection $\theta = 180°$. Parameters are $M = 1.5-0.0i$, lens diameter 3.6 cm, located 30 cm from the particle center, $w_0 = 5$ μm. Solid curve, plane wave; and curve with asterisks, Gaussian beam.

sources in optical sizing systems are Gaussian beams rather than plane waves. It will permit advances in two-phase flow studies when large discrete particles are transported by flows, including spray combustion systems.

10-8 APPENDIX: USEFUL INTEGRAL RELATIONS

The integral relations given below are required to carry out integrations in the main text. Some are trivial to establish; others are more difficult. In any case, in order to establish them, one might go back to basic treatises such as those by Robin (1957–59) or Arfken (1968). To avoid misleading computations, it is necessary to observe that two different definitions of the polynomials $P_n^m(\cos \theta)$ exist. Our definition is given by Eq. (10-5). For more details, see Gouesbet, Maheu, and Gréhan (1988b). We have

$$\int_0^{2\pi} \cos \varphi \exp (ik\varphi) \exp (-ik'\varphi) \, d\varphi = \pi(\delta_{k',k+1} + \delta_{k,k'+1}) \quad (10\text{-A1})$$

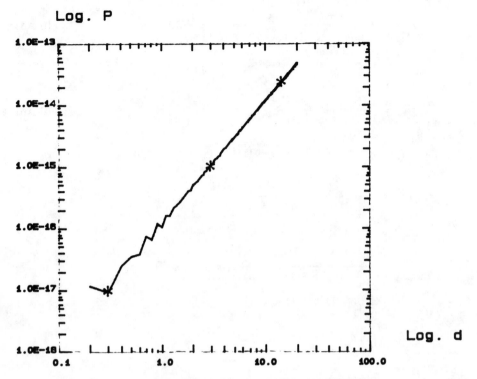

Figure 10-8 Collected power. Decimal logarithm of the collected power versus decimal logarithm of the particle radius. Backward collection $\theta = 180°$. Parameters are $M = 1.5-1.0i$, lens diameter 3.6 cm, located 30 cm from the particle center, $w_0 = 5$ μm. Solid curve, plane wave; and curve with asterisks, Gaussian beam.

$$\int_0^{2\pi} \sin \varphi \, \exp \, (ik\varphi) \, \exp \, (-ik'\varphi) \, d\varphi = i\pi(\delta_{k,k'+1} - \delta_{k',k+1}) \quad \text{(10-A2)}$$

$$I_1 = \int_0^{\pi} (\tau_n^{|p|}\tau_m^{|p|} + p^2 \Pi_n^{|p|}\Pi_m^{|p|}) \sin \theta \, d\theta$$

$$= \frac{2n(n + 1)}{2n + 1} \frac{(n + |p|)!}{(n - |p|)!} \delta_{nm} \quad \text{(10-A3)}$$

$$I_2 = \int_0^{\pi} (\Pi_n^{|p|}\tau_m^{|p|} + \Pi_m^{|p|}\tau_n^{|p|}) \sin \theta \, d\theta = 0 \qquad p \neq 0 \quad \text{(10-A4)}$$

$$I_3 = \int_0^{\pi} (\tau_n^{|p|}\tau_m^{|p|} + p^2 \Pi_n^{|p|}\Pi_m^{|p|}) \cos \theta \sin \theta \, d\theta$$

$$= \frac{2(n - 1)(n + 1)}{(2n - 1)(2n + 1)} \frac{(n + |p|)!}{(n - 1 - |p|)!} \delta_{m,n-1} \qquad \text{(continued)}$$

$$+ \frac{2(m-1)(m+1)}{(2m-1)(2m+1)} \frac{(m+|p|)!}{(m-1-|p|)} \delta_{n,m-1} \qquad \text{(10-A5)}$$

$$I_4 = \int_0^\pi (\tau_n^{|p|} \Pi_m^{|p|} + \tau_m^{|p|} \Pi_n^{|p|}) \cos\theta \sin\theta\, d\theta$$

$$= \frac{2}{2n+1} \frac{(n+|p|)!}{(n-|p|)!} \delta_{nm} \qquad \text{(10-A6)}$$

$$I_5 = \int_0^\pi (\tau_n^{|p|} \tau_m^{|p+1|} + p(p+1) \Pi_n^{|p|} \Pi_m^{|p+1|}) \sin^2\theta\, d\theta$$

$$= \frac{2}{(2n+1)(2m+1)} \frac{(m+p+1)!}{(m-p-1)!} [(n-1)(n+1)\delta_{n,m+1}$$

$$- (m-1)(m+1)\delta_{m,n+1}] \qquad p \geq 0$$

$$= \frac{2}{(2n+1)(2m+1)} \frac{(n-p)!}{(n+p)!} [(m-1)(m+1)\delta_{m,n+1}$$

$$- (n-1)(n+1)\delta_{n,m+1}] \qquad p < 0 \qquad \text{(10-A7)}$$

$$I_6 = \int_0^\pi [p\Pi_n^{|p|} \tau_m^{|p+1|} + (p+1)\Pi_m^{|p+1|} \tau_n^{|p|}] \sin^2\theta\, d\theta$$

$$= \frac{2}{2n+1} \frac{(n+p+1)!}{(n-p-1)!} \delta_{nm} \qquad p \geq 0$$

$$= \frac{-2}{2n+1} \frac{(n-p)!}{(n+p)!} \delta_{nm} \qquad p < 0 \qquad \text{(10-A8)}$$

10-9 NOMENCLATURE

a_n	scattering coefficient, Eq. (10-38)
A	defined in Eq. (10-154)
A_n^m	$= a_n g_{n,\text{TM}}^m$, Eq. (10-36)
A_{np}	defined in Eq. (10-70)
$\mathscr{A}_n$	defined in Eq. (10-162)
b_0	beam waist parameter
b_n	scattering coefficient, Eq. (10-39)
B	defined in Eq. (10-155)
B_n^m	$= b_n g_{n,\text{TM}}^m$, Eq. (10-37)
c	speed of light, Eq. (10-2)
c_n	McLaurin expansion coefficient, Eq. (10-151)
C_{abs}	absorption cross section, Eq. (10-61)
C_{ext}	extinction cross section, Eq. (10-65)
C_{sca}	scattering cross section, Eq. (10-65)

$C_{\mathrm{pr},x}$	pressure cross section, x direction
$C_{\mathrm{pr},y}$	pressure cross section, y direction
$C_{\mathrm{pr},z}$	pressure cross section, z direction, Eq. (10-76)
C_n^{pw}	plane wave coefficient, Eq. (10-22)
E	electric field
f	radial basic function, Eq. (10-146)
F	function, Eq. (10-125)
$\mathbf{F}^+$	reduced radiation pressure force
g_n	finite beam coefficient, on-axis case, Eq. (10-134)
$g_{n,\mathrm{TM}}^m$	finite beam TM coefficient, Eqs. (10-21) and (10-27)
$g_{n,\mathrm{TE}}^m$	finite beam TE coefficient, Eqs. (10-28) and (10-29)
G	function, Eq. (10-126)
h_+	reduced beam axis distance, Eq. (10-101)
H	magnetic field
Im	imaginary part
I_θ^+, I_φ^+	scattered intensities, Eq. (10-57)
j_+	defined in Eq. (10-127)
j_-	defined in Eq. (10-128)
$J_{n+1/2}(\quad)$	Bessel function of half-integer order, Eq. (10-4)
$\mathscr{J}^i$	defined in Eq. (10-62)
$\mathscr{J}^s$	defined in Eq. (10-63)
$\mathscr{J}^{is}$	defined in Eq. (10-64)
k	wave number, Eq. (10-2)
K	defined in Eq. (10-115)
l	spreading length, Eq. (10-93)
L	approximation order of the beam description
L^-	approximation order of the beam description
M	complex refractive index, Eq. (10-2)
$P_n(\cos\theta)$	order n Legendre polynomial, Eq. (10-5)
$P_n^m(\cos\theta)$	associated Legendre polynomial, Eq. (10-5)
Q	defined in Eq. (10-102)
Re	real part
s	inverse beam waist parameter, Eq. (10-93)
S^+	Poynting vector, Eq. (10-55)
S_1, S_2	amplitude function, Eqs. (10-138) and (10-139)
$\mathscr{S}_1$, $\mathscr{S}_2$	amplitude function, Eqs. (10-58) and (10-59)
S_{nm}^p	defined in Eq. (10-88)
T_{nm}^p	defined in Eq. (10-89)
u, u_+	Cartesian coordinate, reduced, Eq. (10-103)
U	Bromwich scalar potential, Eq. (10-1)
U_{nm}^p	defined in Eq. (10-85)
v, v_+	Cartesian coordinate, reduced, Eq. (10-103)
$\mathbf{V}$	vector E or H
V_{nm}^p	defined in Eq. (10-86)
w_0	beam waist radius, Eq. (10-93)

w, w_+	Cartesian coordinate, reduced, Eq. (10-104)
$Y_n^m(\theta, \varphi)$	spherical surface harmonic, Eq. (10-30)
Z_0	defined in Eq. (10-156)
Z_0''	defined in Eq. (10-157)
α	size parameter of the scatterer
β	second size parameter of the scatterer
δ	phase angle, Eq. (10-60)
δ_{nm}	Kronecker symbol
ϵ	complex permittivity, Eq. (10-2)
ζ, ζ_0	reduced coordinates, Eq. (10-112)
λ	wavelength
μ	complex permeability, Eq. (10-2)
ξ_n	Ricatti-Bessel function, Eq. (10-7)
$\Pi_n(\cos\theta)$	Legendre function
$\Pi_n^k(\cos\theta)$	generalized Legendre function, Eq. (10-48)
ρ_n	discretized distance from the axis
$\tau_n(\cos\theta)$	Legendre function
$\tau_n^k(\cos\theta)$	generalized Legendre function, Eq. (10-47)
Ψ_{jp}	function, Eq. (10-121)
$\Psi_n(\ )$	Ricatti-Bessel function, Eq. (10-6)
Ψ_0	function, Eq. (10-116)
Ψ_0^φ	function, Eq. (10-118)
Ψ_0^0	function, Eq. (10-117)
$\Psi_n^1(\ )$	spherical Bessel Function, Eq. (10-4)
ω	angular frequency
$\Sigma_{\ }^{jp}$	defined in Eq. (10-120)

Superscript

‾

transformation, Eq. (10-163)

REFERENCES

Allano, D., Gouesbet G., Gréhan, G., and Lisiecki, D. 1984. Droplet sizing using a ''top-hat'' laser beam technique. *J. Phys. D. Appl. Phys.* 17(1):43–58.

Arfken, G. 1968. *Mathematical methods for physicists.* San Diego, Calif.: Academic.

Ashkin, A. 1970. Acceleration and trapping of particles by radiation pressure. *Phys. Rev. Lett.* 24(4):156–59.

Brillouin, L. 1949. The scattering cross-sections of spheres for electromagnetic scattering. *J. Appl. Phys.* 20:1110–25.

Bromwich, T. J. 1919. Electromagnetic waves. *Philos. Mag.* S6(38):143–64.

Corbin, F., Gréhan, G., Gouesbet, G., and Maheu, B. 1988. Interaction between a sphere and a Gaussian beam: computations on a microcomputer. *J. Particle Characterization* 5:103–8.

Davis, L. W. 1979. Theory of electromagnetic beams. *Phys. Rev. A* 19(3):1177–79.

Faeth, G. M. 1977. Current status of droplet and liquid combustion. *Progr. Energy Combust. Sci.* 3:191–224.

Ferguson, D. C., and Currie, L. G. 1986. Theoretical evaluation of LDA techniques for two-phase flow measurements. Paper 18-1 read at Third International Symposium on Applications of Laser Anemometry to Fluid Mechanics, 7–9 July 1986, Lisbon, Portugal.

Gouesbet, G. 1985. Optical sizing, with emphasis on simultaneous measurements of velocities and sizes of particles embedded in flows. In *Measurement techniques in heat and mass transfer*, ed. R. I. Soloukhin and N. H. Afgan, pp. 27–50. New York: Springer-Verlag.

Gouesbet, G. 1987. A discussion of the localized approximation to the GLMT. Case of an arbitrary location of the scatter center. Laboratoire d'Energétique des Systèmes et Procédés internal report ESP/G/3/87/ I. Rouen, France.

Gouesbet, G., and Gréhan, G. 1980a. A formalism to compute the scattered intensities from an isotropic, homogeneous, spherical, non-magnetic particle located on the axis of a Gaussian beam, using Bromwich functions. Internal report TTI/GG/1/80/I.

Gouesbet, G., and Gréhan, G. 1980b. A formalism to compute the scattered intensities from an isotropic, homogeneous, spherical, non-magnetic particle located on the axis of an axisymmetric incident light profile, using Bromwich functions. Laboratoire d'Energétique des Systèmes et Procédés internal report TTI/GG/80/06/IV. Rouen, France.

Gouesbet, G., and Gréhan, G. 1982. Sur la généralisation de la théorie de Lorenz-Mie. *J. Opt.* 13(2):97–103.

Gouesbet, G., and Gréhan, G., eds. 1988. *Proceedings of the international symposium; optical particle sizing: theory and practice.* New York: Plenum.

Gouesbet, G., Gréhan, G., and Kleine, R. 1984. Simultaneous optical measurement of velocity and size of individual particles in flows. Paper read at Second International Symposium on Applications of Laser Anemometry to Fluid Mechanics. 2–4 July 1984, Lisbon, Portugal.

Gouesbet, G., Gréhan, G., and Maheu, B. 1985. Scattering of a Gaussian beam by a Mie scatter center, using a Bromwich formulation. *J. Opt.* 16(2):83–93. (Republished in *Selected papers on light scattering*, SPIE vol. 951, ed. M. Kerker, pp. 361–71. New York: Society of Photo-Optical Instrumentation Engineers.)

Gouesbet, G., Gréhan, G., and Maheu, B. 1988a. Expressions to compute the coefficients g_n^m in the generalized Lorenz-Mie theory, using finite series. *J. Opt.* 19(1):35–48.

Gouesbet, G., Gréhan, G., and Maheu, B. 1988b. Computations of the coefficients g in the generalized Lorenz-Mie theory using three different methods. *Appl. Opt.* 27(23):4874–83.

Gouesbet, G., Maheu, B., and Gréhan, G. 1985. The order of approximation in a theory of the scattering of a Gaussian beam by a Mie scatter center. *J. Opt.* 16(5):239–47. (Republished in *Selected papers on light scattering*, SPIE vol. 951, ed. M. Kerker, pp. 352–60. New York: Society of Photo-Optical Instrumentation Engineers.)

Gouesbet, G., Maheu, B., and Gréhan, G. 1988a. Scattering of a Gaussian beam by a sphere using a Bromwich formulation: case of an arbitrary location. *J. Particle Characterization* 1:1–8. (Also in Gouesbet and Gréhan, 1988, pp. 27–42.)

Gouesbet, G., Maheu, B., and Gréhan, G. 1988b. Light scattering from a sphere arbitrarily located in a Gaussian beam, using a Bromwich formulation. *J. Opt. Soc. Am. A* 5:1427–43.

Gréhan, G., and Gouesbet, G. 1979a. Mie theory calculations: new progress, with emphasis on particle sizing. *Appl. Opt.* 18(20):3489–93.

Gréhan, G., and Gouesbet, G. 1979b. The computer program SUPERMIDI for Mie theory calculations, without "practical" size or refractive index limitations. Laboratoire d'Energétique des Systèmes et Procédés internal report TTI/GG/79/03/20. Rouen, France.

Gréhan, G., and Gouesbet, G. 1980. Optical levitation of a single particle to study the theory of quasi-elastic scattering of light. *Appl. Opt.* 19(15):2485–87.

Gréhan, G., and Gouesbet, G. 1986. Simultaneous measurements of velocities and sizes of particles in flows using a combined system incorporating a top-hat beam technique. *Appl. Opt.* 25(19):3527–38.

Gréhan, G., Gouesbet, G., and Rabasse, C. 1980. The computer program SUPERMIDI for Lorenz-Mie theory and research of one-to-one relationships for particle sizing. Paper read at Symposium on Long Range and Short Range Optical Velocity Measurements. 15–18 September 1980, at Institut Franco-Allemand de Saint-Louis.

Gréhan, G., Maheu, B., and Gouesbet, G. 1986a. Scattering of laser beams by Mie scatter centers: numerical results using a localized approximation. *Appl. Opt.* 25(19):3539–48.

Gréhan, G., Maheu, B., and Gouesbet, G. 1986b. Localized approximation to the generalized Lorenz-Mie theory and its application to optical sizing. Paper read at Fifth International Congress on Applications of Lasers and Electro-Optics: The Changing Frontiers of Flow and Particle Diagnostics, 10–13 November 1986, Arlington, Virginia.

Gréhan, G., Maheu, B., and Gouesbet, G. 1988. Diffusion de la lumière par une sphère dans le cas d'un faisceau d'extension finie, 2 partie, théorie de Lorenz-Mie généralisée, applications à la granulométrie optique. *J. Aerosol Sci.* 19(1):55–64.

Kerker, M. 1969. *The scattering of light and other electromagnetic radiation.* San Diego, Calif.: Academic.

Kim, L. S., and Lee, S. S. 1983. Scattering of laser beams and the optical potential well for a homogeneous sphere. *J. Opt. Soc. Am.* 73(3):303–12.

Lax, W., Louisell, W. H., and McKnight, W. B. 1975. From Maxwell to paraxial optics. *Phys. Rev. A* 11(4):1365–70.

Maheu, B., Gréhan, G., and Gouesbet, G. 1987a. Generalized Lorenz-Mie theory: first exact values and comparisons with the localized approximation. *Appl. Opt.* 26(1):23–26.

Maheu, B., Gréhan, G., and Gouesbet, G. 1987b. Laser beam scattering by individual spherical particles: theoretical progress and applications to optical sizing. *J. Particle Characterization* 4:141–46. (Also in Gouesbet and Gréhan, 1988, pp. 77–88.)

Maheu, B., Gréhan, G., and Gouesbet, G. 1988. Diffusion de la lumière par une sphère dans le cas d'un faisceau d'extension finie, 1 partie, théorie de Lorenz-Mie généralisée, les coefficients g et leur calcul numérique. *J. Aerosol Sci.* 19(1):47–53.

Maheu, B., Gréhan, G., and Gouesbet, G. 1989. Ray localization in Gaussian beams. *Opt. Commun.* 70(4):259–62.

Maheu, B., Gouesbet, G., and Gréhan, G. 1988. A concise presentation of the generalized Lorenz-Mie theory for arbitrary location of the scatterer in an arbitrary incident profile. *J. Opt.* 19(2):59–67.

Morita, N., Tanaka, T., Yamasaki, T., and Nakanishi, Y. 1968. Scattering of a beam wave by a spherical object. *IEEE Trans. Antennas Propag.* AP16(6):724–27.

Poincelot, P. 1963. *Précis d'électromagnétisme théorique.* Paris: Masson.

Robin, L. 1957–59. *Fonctions sphériques de Legendre et fonctions sphéroïdales,* eds. 1, 2, 3. Paris: Gauthier-Villars Edition.

Roosen, G., Delaunay, B., and Imbert, C. 1977. Etude de la pression de radiation exercée par un faisceau lumineux sur une sphère réfringente. *J. Opt. Paris* 8(3):181–87.

Sirignano, W. A. 1983. Fuel droplet vaporization and spray combustion theory. *Progr. Energy Combust. Sci.* 9:291–322.

Slimani, F., Gréhan, G., Gouesbet, G., and Allano, D. 1984. Near-field Lorenz-Mie theory and its application to microholography. *Appl. Opt.* 23(22):4140–48.

Tam, W. G., and Corriveau, R. 1978. Scattering of electromagnetic beams by spherical objects. *J. Opt. Soc. Am.* 68(6):763–67.

Tsai, W. C., and Pogorzelski, R. J. 1975. Eigenfunction solution of the scattering of beam radiation fields by spherical objects. *J. Opt. Soc. Am.* 65(12):1457–63.

Van de Hulst, H. C. 1957. *Light scattering by small particles.* New York: John Wiley.

Watson, G. N. 1962. *A treatise of the theory of Bessel functions,* 2nd ed. New York: Cambridge University Press.

Yeh, C. W., Colak, S., and Barber, S. 1982. Scattering of sharply focused beams by arbitrarily shaped dielectric particles: an exact solution. *Appl. Opt.* 21(24):4426–33.

ELEVEN

FOURIER TRANSFORM INFRARED EMISSION/ TRANSMISSION SPECTROSCOPY IN FLAMES

P. R. Solomon and P. E. Best

11-1 INTRODUCTION

A better understanding of the combustion process would enable improved combustion systems to be designed and developed. In investigating the behavior of combusting systems, it is desirable to have nonintrusive techniques to monitor the composition, physical properties, and temperatures of the various phases (gas, soot particles, droplets, aerosols, coal, char, and fly ash) present. Techniques to measure these phenomena must be capable of following the behavior of the solid particles as well as the gas species. Many of the techniques for gas analysis have been recently reviewed (Penner, Wang, and Bahadori 1984a, b; Hardesty 1984). For particle temperature, two- or more-color pyrometry has been used extensively (Timothy, Sarofim, and Beer 1982; Mitchell and McLean 1982; Cashdollar and Hertzberg 1983; Mackowski et al. 1983; Altenkirch et al. 1984; Seeker et al. 1981). Advances in optical emission techniques to measure simultaneously size, velocity, and temperature of single particles have also been reported (Macek and Bulik 1984; Tichenor et al. 1984).

Recently, Fourier transform infrared (FT-IR) spectroscopy has shown promise as a versatile technique. FT-IR transmission spectroscopy has been used as an in situ

P. E. Best's permanent address is Physics Department and Institute of Materials Science, University of Connecticut, Storrs, Connecticut 06268.

This work was supported under the U.S. Department of Energy contracts DE-AC21-85MC22050 and DE-AC21-86MC23075 and NSF grant CBT-8420911. The two-dimensional tomography graphics were prepared by Michael Danchak and Irena Ilovici of the Hartford Graduate Center. The authors are grateful for their collaboration on this phase of the work. The authors wish to thank Judith Wornat of CSIRO for reviewing the chapter and making helpful comments.

diagnostic technique to determine both gas concentrations (Solomon, Hamblen, and Carangelo 1982, 1984) and temperatures (Solomon et al. 1982; Solomon and Hamblen 1985; Ottesen and Stephenson 1982; Ottesen and Thorne 1983). An FT-IR emission and transmission (E/T) spectroscopic technique has been applied to particles, gases, and soot in flames (Best, Carangelo, and Solomon 1986; Solomon et al. 1986a, b, 1987a, b, c, d, e). The E/T technique has been used previously for gases and soot (Mackowski et al. 1983; Altenkirch et al. 1984; Tourin 1962, 1966; Limbaugh 1985; Vervisch and Coppalle 1983).

Advantages of the FT-IR E/T technique include (1) the capability to determine separate temperatures and concentrations for individual gas species as well as for solid particles (including soot) by employing different regions of the infrared spectrum, (2) the capability to determine temperatures as low as 100°C and, consequently, the ability to follow particle or droplet temperatures prior to ignition, (3) the ability to make measurements in densely loaded streams to study cloud effects [in this respect, the technique is a good complement to the single-particle measurements (Macek and Bulik 1984; Tichenor et al. 1984)], (4) the capability to separate the radiative contribution from soot and from char particles, and (5) the capability to measure particle sizes.

The disadvantage of FT-IR E/T spectroscopy is that the measurements are for an ensemble of particles over a line of sight, and tomographic techniques must be used to obtain spatially resolved data. In addition, spectra acquisition times of the order of 0.1 s are longer than time scales associated with transient features that are present in turbulent flows. For such flows the technique measures time-averaged properties. The tomography technique can be used successfully for steady flames.

This chapter describes the application of the FT-IR E/T technique to study combustion. FT-IR measurements of temperature using the relative intensities of the rotational bands (Solomon et al. 1982; Solomon and Hamblen 1985; Ottesen and Stephenson 1982; Ottesen and Thorne 1983) will not be considered here. The chapter considers experiments with coal, ethylene, hexane spray flames, and coal water fuel (CWF) spray flames. The theory of the FT-IR E/T analysis is presented, and the data reduced according to the theory are compared with the results obtained by other methods. Recent results employing tomography to obtain spatially resolved data from the line-of-sight measurements (Chien et al. 1988) are also discussed.

11-2 EXPERIMENT

FT-IR Spectrometer

The measurement of electromagnetic spectra by the Fourier transform technique has a number of advantages over wavelength dispersive methods, as discussed in texts and monographs (Griffiths 1975, 1978; Bell 1972). First, the Fourier transform technique processes all wavelengths of a spectrum simultaneously. For this reason it can be used to measure spectral properties of particulate streams, which are notoriously difficult to maintain at a constant flow rate. The technique is extremely rapid; a low noise emission or transmission spectrum at low resolution (4 cm^{-1}) can be recorded in less than a second.

Second, radiation through the interferometer is amplitude modulated, and only such radiation is detected. Because of its unmodulated nature, the particle emission passing directly to the detector does not interfere with the measurements of scattering or transmission. However, there can be indirect interference from sample emission, which is reflected back from the interferometer through the sample. This component can be recognized by its out-of-phase relationship to the radiation from the source, and correction for this effect is straightforward.

The FT-IR instruments we have used include a Nicolet model 7199, a Nicolet model 20SX, and a Bomem Michelson 110, equipped with globar sources for transmission measurements, and utilizing mercury-cadmium-telluride detectors for both emission and transmission measurements. An indium antimonide detector has also been employed for measuring transmission spectra.

Reactors

Co-annular laminar diffusion flame. To allow the results of the FT-IR technique applied to soot particles to be compared with the results of other methods, a series of measurements were made on a coannular laminar diffusion flame burning ethylene in air. A photograph of the flame is shown in Fig. 11-1. This flame has been previously studied by Santoro and coworkers, and its soot particle distribution is well characterized (Santoro, Yeh, and Semerjian 1985; Santoro and Semerjian 1984; Santoro, Semerjian, and Dobbins 1983). Similar flames have been studied by Kent and Wagner (1984*a*,

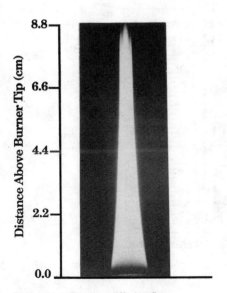

Figure 11-1 Ethylene diffusion flame.

b), Boedeker and Dobbs (1986), Solomon et al. (1987*b*), Chien et al. (1988), and Markstein and deRis (1984). The burner geometry that has been previously described (Santoro, Semerjian, and Dobbins 1983) consists of two concentric brass tubes 11.1 mm and 101.6 mm ID, with fuel flowing through the central tube and air through the outer passage. For the experiments described in this chapter, the fuel flow rate Q was set so that the luminous tip was at 88 mm from the tube end. For these conditions, the ethylene flow rate was 2.85 cm³/s.

Transparent wall reactor for coal combustion. A transparent wall reactor (TWR) has been used to study coal combustion in a flame where particle radiance can be measured without scattering from hot walls (Solomon et al. 1987*e*, 1988). The facility shown in Fig. 11-2 consists of an electrically heated furnace and a heat exchanger. Dry air is passed through the heat exchanger and exits the top of the furnace through a screen to smooth the flow. The preheated gas stream, which is 10 cm in diameter,

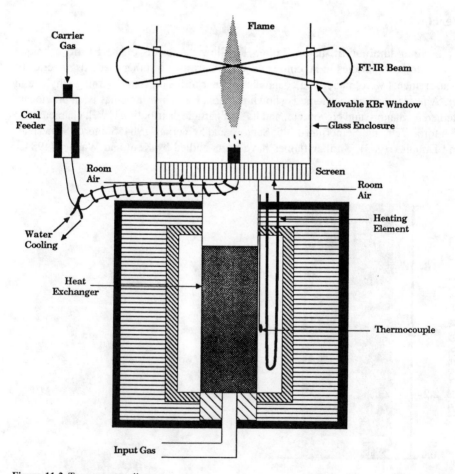

Figure 11-2 Transparent wall reactor.

remains relatively hot and undisturbed for approximately 20 cm above the screen. Coal entrained in a cold carrier gas (dry air) is injected through a coaxial 4-mm-diameter tube.

The feed rates for the carrier gas and the hot gas were 225 mL/min and 173 L/min, respectively. The hot gas exits the screen at a temperature of 850°C and a velocity of 2.8 m/s. Short-exposure photographs show particle tracks that indicate that the coal particles are moving at about 2.5 m/s near the injector and 4.8 m/s in the flame.

An octagonal glass enclosure shields the reacting stream from room air currents. The enclosure has movable KBr windows to allow access to the flame by the FT-IR spectrometer (a modified Nicolet 20SX). The spatial resolution is approximately 5 mm. Figure 11-3 shows photographs of three flames: a lignite that ignites close to the nozzle, a low volatile bituminous coal that ignites (apparently homogeneously in the gas phase) higher up, and a close-up of a high volatile bituminous coal flame showing volatile flamelets surrounding the particles, which are much larger than the particles.

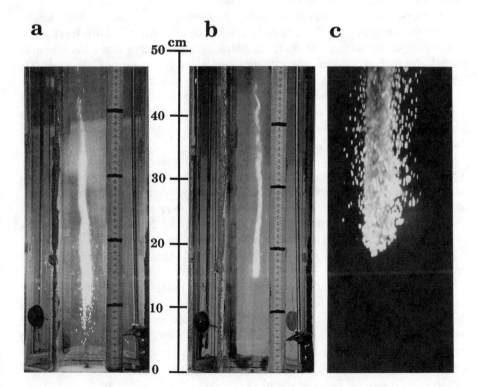

Figure 11-3 Photographs of (*a*) Zap North Dakota lignite flame and (*b*) a Pocahontas bituminous coal flame. The scale in the center is distance above the injector nozzle. (*c*) Closeup of a Pittsburgh seam bituminous coal flame.

Such volatile flames have been observed previously (McLean, Hardesty, and Pohl 1981; Timothy et al. 1987).

The samples used in these experiments were sieved fractions of a lignite, a sub-bituminous coal, and several bituminous coals. Samples have been demineralized by the Bishop and Ward (1958) technique. Chars were prepared from some of the coals in nitrogen in a previously described entrained flow reactor (Solomon et al. 1982; Solomon and Hamblen 1985; Serio et al. 1987). The characteristics of the coals have been published previously (Serio et al. 1987; Vorres 1987; Best et al. 1987; Solomon, Serio, and Heninger 1986). Table 11-1 summarizes the coals, preparation procedures, and references to the previously published data. The samples were also characterized in a thermogravimetric analyzer (TGA) to determine their weight loss at constant heating rate in nitrogen (to determine the volatile content) and in air (to provide a relative measure of the samples' reactivity (Best et al. 1987; Solomon, Serio, and Heninger 1986) and the ash concentration).

Spray combustion facility. A spray facility for testing the combustion characteristics of liquid fuels is shown in Fig. 11-4. Both hydrocarbon liquid and CWF sprays have been burned in this system (Solomon et al. 1987*d*). The system allowed testing of the FT-IR diagnostics on a reasonably large flame.

The spray-down geometry shown in Fig. 11-4 was chosen so that the high-density residues do not reenter the flame under the influence of gravity. A down-spray flame into still air is not suitable for study, of course, because such a flame has a turning point to become an upward going flame, due to buoyancy forces. We avoid this transition by embedding the nozzle-flame system in a relatively high velocity, downward flow of air. The furnace is in an "air-tight" enclosure, which has the floor and rear walls lined with steel sheet. The high-velocity downward flow of air about the flame results from the location of the opening into the enclosure being directly above the suction vent.

The FT-IR spectrometer is also located within the furnace room. This spectrometer is on a movable hoist table, so that its 0.4-cm-diameter beam can be positioned to probe any line-of-sight across the flame. The IR probe beam is taken from the interferometer to the flame region by mirrors, which are mounted on the end of beams. The spectrometer used in this work was a Nicolet model 20SX, which employed 4 cm^{-1} resolution. Most spectra were measured with a total of 128 scans, which take 1.5 min to record. Present work employs a Bomem Michelson M110 FT-IR spectrometer.

The facility employed an air assist Parker Hannifin model EDL 6850661mL nozzle. This nozzle was modified for the CWF work to narrow the spray cone. All air and fuel supplies, as well as the FT-IR spectrometer controls, reside outside of the furnace room. The hexane flow through the nozzle was induced by constant pressure over the liquid reservoir. A Robbins-Myers Ramoy 61011 slurry pump was used to pump the CWF.

Coupling of Furnaces to the FT-IR Spectrometer

The coupling of the FT-IR spectrometer to the various furnaces is illustrated in Fig. 11-5. The optical focus of the IR beam can be steered using several mirrors to the

Table 11-1 Sample characteristics

Sample	Sample abbreviation in figures	Classification or preparation conditions	Nominal particle size, μm	TGA analysis				Sample reference
				Ash, %*	DAF volatile matter, %†	T_{cr}, °C‡	$T_{10\%}$ weight loss, °C*	
Zap	Z	North Dakota lignite, dry	45–75	7.3	43.5	434	360	Serio et al. (1987)
Zap 900°C char	ZC 900	EFR pyrolysis, 1 m/s 900°C, 66 cm, in N$_2$	45–75	12.7	22.0	418	388	Serio et al. (1987)
Zap 1500°C char	ZC 1500	EFR pyrolysis, 1 m/s 1500°C, 66 cm, in N$_2$	45–75	14.3	8.8	494	458	Serio et al. (1987)
Argonne Zap	AZ	North Dakota lignite, dry	45–75	8.2	40.5	418	335	Vorres (1987)
Zap demineralized	DZ	HCl, HF washings, dry	45–75	0.5	44.7	550	383	Best et al. (1987)
Rosebud	R	Subbituminous coal, dry	45–75	15.1	42.8	485	400	Serio et al. (1987)
Rosebud 900°C char	RC 900	EFR pyrolysis, 1 m/s 900°C, 66 cm, in N$_2$	45–75	23.5	19.1	443	440	Serio et al. (1987)

(Table continues on next page)

EFR: entrained flow reactor; DAF: dry, ash-free.

*TGA analysis in air at 30 K/min to 900°C.

†TGA analysis in N$_2$ at 30 K/min to 900°C.

‡TGA analysis in N$_2$ at 30 K/min to 900°C followed by in air at 30 K/min to 900°C; critical temperature defined by Best et al. (1987) and Solomon, Serio, and Heninger (1986).

Table 11-1 Sample characteristics (Continued)

Sample	Sample abbreviation in figures	Classification or preparation conditions	Nominal particle size, μm	Ash, %*	DAF volatile matter, %†	T_{cr}, °C‡	$T_{10\%}$ weight loss, °C*	Sample reference
					TGA analysis			
Rosebud 1500°C char	RC 1500	EFR pyrolysis, 1 m/s 1500°C, 66 cm, in N_2	45–75	23.4	2.6	545	575	Serio et al. (1987)
Rosebud demineralized	DR	HCl, HF washings, dry	45–75	3.7	39.1	500	432	Best et al. (1987)
Argonne Pittsburgh	Pit	Bituminous coal (HVAB)	45–75	3.2	37.2	600	488	Vorres (1987)
Argonne Pocahontas	P	Bituminous coal (LVB)	45–75	5.1	20.0	628	512	Vorres (1987)

EFR: entrained flow reactor; DAF: dry, ash-free.
*TGA analysis in air at 30 K/min to 900°C.
†TGA analysis in N_2 at 30 K/min to 900°C.
‡TGA analysis in N_2 at 30 K/min to 900°C followed by in air at 30 K/min to 900°C; critical temperature defined by Best et al. (1987) and Solomon, Serio, and Heninger (1986).

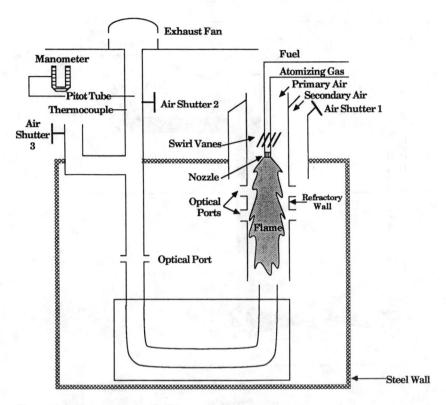

Figure 11-4 Spray combustion facility.

sample stream, at distances of up to a meter from the interferometer. Both transmission and emission spectra can be measured with this arrangement. Emission measurements are made with the movable mirror placed in the beam to divert the beam to the emission detector as shown in Fig. 11-5(a). Transmission measurements are made with the movable mirror out of the beam as shown in Fig. 11-5(b).

In an ideal emission-transmission experiment, both measurements are made on the identical sample. Up to now, the emission and transmission measurements were made sequentially along the same optical path, for a sample in a nominal steady condition. Another method has been used, in which the two measurements are made simultaneously through adjacent beam paths. Both approaches have their difficulties. In the measurements we report, some of the nonideal results are due to time variations in the flow. In this regard we had more difficulty maintaining a uniform flow of solid particles, compared with that for gaseous or liquid samples.

11-3 ANALYSIS

FT-IR E/T Method

For multiphase reacting systems, measurements are made of the transmittance and the radiance, and from these a quantity called the normalized radiance is calculated. The

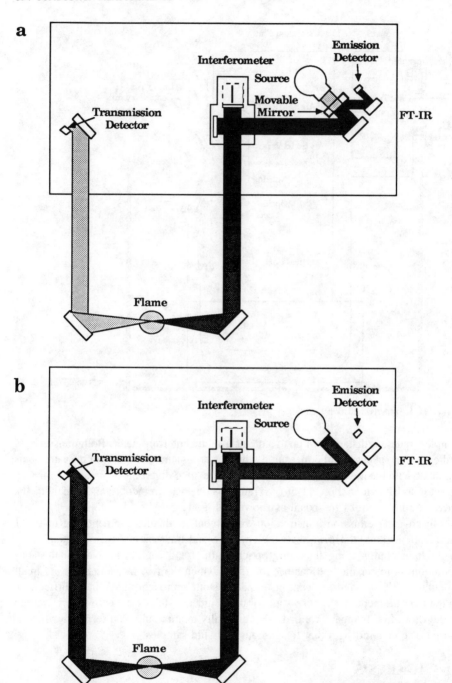

Figure 11-5 Spectrometer and transfer optics. (*a*) Emission measurement and (*b*) transmission measurement. Reproduced from Solomon et al. (1987*c*), with permission.

analysis, which follows Siegel and Howell (1972), has been presented previously (Best, Carangelo, and Solomon 1986). The relevant equations for a homogeneous medium are presented below.

It is recognized that most flames are not spatially homogeneous along the line of sight. In this case, application of the analysis yields the average properties along a line of sight. Alternatively, tomographic methods can be employed to obtain spatially resolved data (Chien et al. 1988).

Transmittance. The transmittance τ_ν is defined in the usual manner,

$$\tau_\nu = I_\nu/I_{0\nu} \tag{11-1}$$

where $I_{0\nu}$ is the intensity transmitted in the absence of a sample, while I_ν is that transmitted with the sample stream in place. The terms are defined in the nomenclature.

For a medium containing gases and soot with absorption coefficients α_ν^g and α_ν^s and particles of geometrical cross-sectional area A at a density of N particles cm^{-3},

$$\tau_\nu = \exp\left[-(\alpha_\nu^s + \alpha_\nu^g + NAF_\nu^t)L\right] \tag{11-2}$$

where F_ν^t is the ratio of the total cross section (extinction) to geometric cross section and L is the path length; τ_ν is sometimes plotted as a percent.

Concentrations can be determined from measured values of τ_ν using Eq. (11-2). Gas concentrations can be determined from selected bands by determining the amplitude above the broad background. The relationship between the gas absorption coefficient α_ν^g and its concentration must be known (e.g., by calibration), as well as the path length L.

In the absence of particles, the soot volume fraction can be determined using the amplitude of transmission at any position in the spectrum away from interference from gas lines. The absorption coefficient for soot α_ν^s may be calculated from its particle diameter and complex index of refraction (Solomon et al. 1987b, c).

In the absence of soot, both particle area A and number density N may be determined from τ_ν (Solomon et al. 1987c). As described in the example, the area is determined by comparing the measured shape of τ_ν with values calculated using Mie theory (Bohren and Huffman 1983), employing the particle's complex index of refraction. The number density is then determined by the amplitude of τ_ν.

When both soot and particles are present, the two contributions may be separated if the spectral shapes are known, as discussed in the results section.

Radiance. To determine the sample radiance R_ν, the radiative power emitted and scattered by the sample with background subtracted, S_ν, is measured and converted to the sample radiance,

$$R_\nu = S_\nu/W_\nu \tag{11-3}$$

where W_ν is the instrument response function measured using a blackbody cavity of known temperature. The radiance measurement detects both radiation emitted, as well as radiation scattered or refracted by the particles.

R_ν is given by (Best, Carangelo, and Solomon 1986; Siegel and Howell 1972)

$$R_\nu = \{[\alpha_\nu^s R_\nu^b(T_s) + \alpha_\nu^g R_\nu^b(T_g) + NA\epsilon_\nu R_\nu^b(T_p) + NAF_\nu^{s'} R_\nu^b T(_w)]$$

$$\times [1 - \exp[-(\alpha_\nu^s + \alpha_\nu^g + NAF_\nu^t) L]]\} (\alpha_\nu^s + \alpha_\nu^g + NAF_\nu^t)^{-1} \quad (11\text{-}4)$$

where $R_\nu^b(T_g)$, $R_\nu^b(T_s)$, $R_\nu^b(T_p)$, and $R_\nu^b(T_w)$ are the blackbody emission spectra at the temperatures of the gas, soot, particle, and wall, respectively; ϵ_ν is the particle's spectral emittance, and $F_\nu^{s'}$ is the particle's cross section for scattering radiation into the spectrometer. In deriving this expression, we have included the scattering terms $F_\nu^{s'}$ and F_ν^t for particles but assume no scattering for soot particles in the IR region.

Normalized radiance. The normalized radiance R_ν^n is defined as

$$R_\nu^n = R_\nu/(1 - \tau_\nu) \quad (11\text{-}5)$$

From Eqs. (11-2) and (11-4), the normalized radiance is

$$R_\nu^n = \frac{\alpha_\nu^s R_\nu^b(T_s) + \alpha_\nu^g R_\nu^b(T_g) + NA\epsilon_\nu R_\nu^b(T_p) + NAF_\nu^{s'} R_\nu^b(T_w)}{\alpha_\nu^s + \alpha_\nu^g + NAF_\nu^t} \quad (11\text{-}6)$$

For the geometry of the three reactors considered in this chapter, scattering from particles, $F_\nu^{s'}$, may be neglected, since T_w is room temperature.

The measurement of temperature is made by determining the values of T_s, T_g, T_p, and T_w required in Eq. (11-6) to make the calculated values of R_ν^n agree with the measured value. The values of α_ν^s, α_ν^g, and NAF_ν^t on the right-hand side of Eq. (11-6) may be determined from transmission measurements. Values of ϵ_ν and $F_\nu^{s'}$ must be determined by independent measurements.

In the simplest case where the sample is a single component such as soot, Eq. (11-6) reduces to

$$R_\nu^n = R_\nu^b(T_s) \quad (11\text{-}7)$$

The temperature is determined by varying a trial value of T_s and comparing the Planck function $R_\nu^b(T_s)$ to the measured normalized radiance R_ν^n, either visually or by using an automated least squares fitting routine. More complicated procedures are used for multiphase samples, as will be discussed in the remainder of the chapter.

Deviations from Simple Analysis

There are three cautions to be taken concerning the interpretation of transmission or emission data. These can decrease the accuracy of the method, limit its range of applicability, or require corrections to the simple equations.

Stimulated emission. This effect occurs in cases where the thermal energy kT is large in comparison with the photon energy $h\nu$, where k is the Boltzmann constant and h is Planck's constant. In this limit, apparent absorption coefficients, α, can be different from the true coefficients, α'. The two coefficients are related in the following manner:

$$\alpha_\nu = [1 - \exp(-h\nu/kT)]\alpha_\nu' \quad (11\text{-}8)$$

The effect is due to emission, which is stimulated by the transmitted beam (Siegel and Howell 1972). The exponential term is negligible for large values of $h\nu/kT$ but becomes an important factor at combinations of small frequencies and high temperatures. For example, there is a difference of 18% between α and α' when measurements are made at the 2350 cm^{-1} band of CO_2 at a temperature of 2000 K. Species concentrations are calculated from true absorption coefficients integrated across the absorption band (Siegel and Howell 1972).

For a thin gas sample in a high-temperature enclosure, energy balance analysis shows that the true absorptivity $(1 - \tau_\nu)$ is equal to the total emissivity.

$$1 - \exp{(-\alpha_\nu'^g L)} \approx \alpha_\nu'^g L = \epsilon_\nu + (\alpha_\nu'^g - \alpha_\nu^g)L \qquad (11\text{-}9)$$

where ϵ_ν is the emissivity due to spontaneous emission and the second term on the right is due to stimulated emission. On rearranging,

$$\alpha_\nu^g L = \epsilon_\nu \qquad (11\text{-}10)$$

For this high-temperature gas in a room temperature enclosure, it is the apparent absorptivity, $\alpha_\nu^g L$, and the spontaneous radiance, $\epsilon_\nu R_\nu^b$, that are measured. From Eqs. (11-5) and (11-10), it can be seen that Eq. (11-4) still holds in the presence of stimulated emission.

Non-Beer's law behavior. A deviation from the simple analysis occurs in regions of a spectrum where the instrumental width is greater than the width of an absorption feature. This can happen at high concentrations of small gas molecules such as CO, CO_2, CH_4, and H_2O. In this case, measured absorption coefficients may not be directly proportional to gas concentrations (Siegel and Howell 1972). Beer's law is not applicable. To extract information about gases in such a spectral region, calibration measurements must be made in which transmittance is related to gas concentration and temperature. As an alternative, particularly in cases where a direct calibration cannot be made, a band model can be used to simulate the situation (Limbaugh 1985; Siegel and Howell 1972).

The straightforward use of the E/T method to determine temperature, in the manner of Eq. (11-4), only applies to samples that obey Beer's law. Band models must be used to determine temperatures when Beer's law does not apply. Alternatively, the analysis may be applied to regions where Beer's law does hold, such as the wings of a band or other bands where the absorbance is lower.

Nonhomogeneous samples. For nonhomogeneous samples, the analysis of the transmission spectra gives the average of the absorption coefficient over the line of sight. The interpretation of temperature is more complicated. In the case of negligible self-absorption, the measurement of temperature will be the average temperature weighted by the species concentration along the line of sight. We have generally interpreted the data by the simplest assumption that there are two or three populations at separate temperatures. This assumption is a gross simplification of what is surely a much more complex situation, and the solutions derived are certainly not unique. For nonnegligible self-absorption, interpretation of the data is far more complicated, but some useful information can be obtained, as will be discussed later in the chapter.

Alternatively, the best way to handle nonhomogeneous cases is to perform a tomographic reconstruction to obtain point spectra. The E/T analysis for the homogeneous case is then applicable.

Tomography

Transmission. Tomography refers to the construction of a three-dimensional image by stacking up two-dimensional slices of the image. The two-dimensional images can be mathematically "reconstructed" from measured projections, which are derived from line-of-sight measurements across the object. The term image is applied loosely here, referring to fields of the variables concerned, whether presented in tabular, graphical, or image format. In transmission tomography the projections are of absorbance.

Both the reconstruction and the required measurements are simplified for objects of axial symmetry. The most common reconstruction techniques in such cases use radial inversion based on Abels equations (Tourin 1966; Pearce 1958; Hughey and Santavicca 1982), sometimes referred to as onion peeling because the determination of the two-dimensional image begins at the exterior shell, proceeding inward shell by shell as in a peeling process (Ludwig et al. 1973). More recent reconstruction techniques can produce two-dimensional slices of an image for objects of arbitrary shape (Ray and Semerjian 1983; Santoro et al. 1981; Semerjian et al. 1981; Shepp and Logan 1974).

The spatial dependencies of both gas concentrations and temperatures have previously been obtained from tomography applied to *monochromatic* measurements of a steady state flame (Elder, Jerrick, and Birkeland 1965). In spectral regions for which Beer's law applies, the two-dimensional image reconstructed from transmittance leads to the determination of the spatial dependence of absorbances, and hence concentrations [Eq. (11-2)].

All of the above results apply to gas samples without a continuum phase such as soot. The simultaneous presence of continuum and band features reduces the accuracy of monochromatic gas band measurements. The presence of soot cannot be corrected for when only one frequency is employed. This limitation is removed when the image reconstruction techniques described above are applied wave number by wave number to build up local IR absorbance spectra. Then accurate local concentrations can be determined for soot and for all of the gases present in the flame. To date, only the case of a low-absorbance sample (percent transmittance $>80\%$) has been studied in this manner (Chien et al. 1988; Best et al. 1990).

The spectral tomography reviewed here is for the ethylene diffusion flame described previously. In these measurements, 21 parallel scans were made at each of 10 different heights in the flame. The beam diameter was reduced using 3.5-mm-diameter apertures after the focusing mirrors (see Fig. 11-5). The resulting beam diameter was less than 1 mm over the region of the flame. The slight beam divergence has negligible affect on the tomography results. For each height, the transmission scans were converted to absorbance ($= -\log_{10} \tau$), as it is the projections of this quantity across the flame that are the appropriate inputs to image reconstruction algorithms.

A Fourier reconstruction of the two-dimensional images from the projections was performed using the program published by Shepp and Logan (1974). Other reconstruction methods are, of course, possible (Hughey and Santavicca 1982; Ludwig et al. 1973; Shepp and Logan 1974). The results of the reconstruction were consistent with the experimental data, in that transmittance across the flame diameter calculated from the image properties agreed with the measured transmittance.

Emission. A straightforward application of reconstruction techniques to scans of radiance is not possible because of self-absorption in the sample. There are a number of cases for which the self-absorbance has been corrected to obtain local species concentrations and temperatures from line-of-sight emission and transmission. In the case of small absorbance (percent transmission >80%), an emission measurement can be directly corrected by an absorption measurement made along the same path. A radial inversion can be applied directly to the emission thus corrected, to obtain local radiances (Freeman and Katz 1960).

For samples with larger absorbance (80% > percent transmission > 20%) an iterative procedure in conjunction with radial inversion has been used to give local radiances (Elder, Jerrick, and Birkeland 1965). In addition, for those cases for which there is not a linear relationship between absorbance and concentration (Beer's law does not apply) a band model in conjunction with radial inversion has led to the determination of local gas species concentrations and temperatures (Limbaugh 1985; Brewer and Limbaugh 1972).

In the work described in this chapter, measurements of radiance and transmission were made for 21 parallel scans at each of 10 heights in the ethylene diffusion flame. The flame had high transmission, and therefore the radiance could be corrected for self-absorption using the direct method (Freeman and Katz 1960). The Fourier reconstruction technique was applied to these corrected radiance spectra to give local radiances. These reconstruction procedures were applied to each 16 cm^{-1} wave number interval from 500 to 6500 cm^{-1}, to build up local emission and transmission IR spectra. The analysis of these local spectra was performed in the same manner as for thin homogeneous samples.

Examples

In this section, several simple examples are considered that show the characteristic spectra for gases, soot, and particles. Application of the analysis to derive concentrations, temperatures, and particle sizes is also illustrated.

Gases. The simplest case to consider is that for gases alone, where Eq. (11-2) becomes

$$\tau_\nu = \exp\left(-\alpha_\nu^g L\right) \qquad (11\text{-}11)$$

For these studies, an electrically heated tube, the heated tube reactor (HTR), was used to heat a mixture of acetylene, CO_2, and butane with helium to a known temperature. Figure 11-6(a) presents the transmittance spectrum. The wavelength-dependent absorption coefficient α_ν^g is a function of the concentration, and so the peak heights can be used to monitor the change in species concentration when reactions occur.

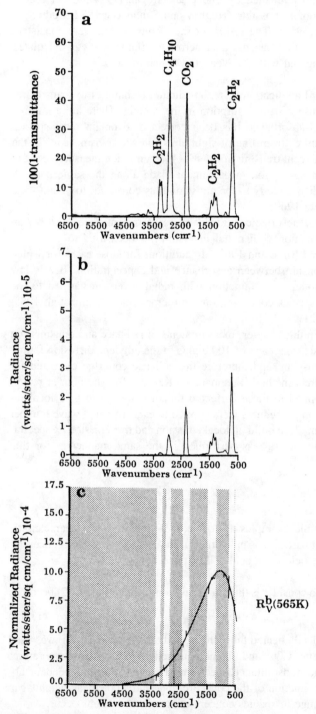

Figure 11-6 Spectra for acetylene, CO_2, and butane in helium at 555 K in the heated tube reactor: (a) 100(1-transmittance), (b) radiance, and (c) normalized radiance. The blackbody spectrum at 565 K is fit, ignoring the regions away from the absorption bands that are shaded.

Figure 11-6(b) shows the radiance R_ν obtained under the same conditions as Fig. 11-6(a). As can be seen, the gas mixture emits in the same bands where it absorbs.

Figure 11-6(c) presents the normalized radiance R_ν^n. For gases alone, Eq. (11-7) applies; that is, in regions where α_ν^g is nonzero, the normalized radiance, R_ν^n is at the theoretical blackbody radiance, corresponding to the temperature of the gas T_g. Figure 11-6(c) shows a theoretical blackbody curve for $T_g = 565$ K, which is the "best fit" to the measured spectrum in both spectral shape and amplitude in the regions where there are emission bands. In regions of low emission ($\alpha_\nu^g \approx 0$), which are shaded, R_ν^n is noisy due to the division by small values of $(1 - \tau_\nu)$ and has not been plotted. The best fits were determined visually over the whole spectrum by overlaying calculated and observed spectra. An automated least squares fitting routine has now been employed, giving similar results. The blackbody temperature of 565 K is in good agreement with the average thermocouple temperature of 555 K.

These and other measurements show the gas temperatures determined by this method in a homogeneous gas sample to be within ± 10 K for temperatures as high as 2000 K.

Soot. This section considers the slightly more complicated case of mixtures of soot and gas. In the simplest case, soot was formed by the pyrolysis of butane in nitrogen in an entrained flow reactor (Solomon et al. 1982; Solomon and Hamblen 1985; Serio et al. 1987) at 1573 K. Figure 11-7 shows transmittance, radiance, and normalized radiance spectra. The transmittance spectrum (Fig. 11-7(a)) shows a sloping continuum, typical of soot with the absorption bands for gases, mainly CO_2, acetylene, methane, ethylene, and polycyclic aromatic hydrocarbons (PAHs) superimposed. The magnitude of 100(1-transmittance) for the broad continuum is proportional to the volume fraction of soot f_v integrated along the line of sight. Here f_v can be calculated from this amplitude if the complex index of refraction is known (Solomon et al. 1987b).

The radiance spectrum in Fig. 11-7(b) shows emission bands for the gases and a continuum for soot. The continuum is nongraybody due to the frequency-dependent absorption.

In going from the radiance to the normalized radiance spectrum, Fig. 11-7(c), the nonblackbody effects are removed, so that one can use a blackbody fit to get the soot temperatures. In regions where gas contributions are zero, the normalized radiance is equal to the theoretical blackbody at the temperature of the soot. A theoretical blackbody curve for $T_s = 1280$ K yields the best match to the experimental curve in both spectral shape and amplitude in the regions away from the gas bands. The uncertainty in determining the radiance is typically within $\pm 5\%$. At 1280 K this results in $\pm 30°C$ uncertainty in temperature. The E/T value of 1280 K is in good agreement with a suction pyrometer measurement (1295 K).

When a gas is at the same temperature as the soot, then the normalized radiance in the gas absorption regions will be at the blackbody intensity for the common temperature, as shown by setting $T_s = T_g$ and $N = 0$ in Eq. (11-6). Then Eq. (11-6) reduces to

$$R_\nu^n = \frac{\alpha_\nu^s R_\nu^b(T_s) + \alpha_\nu^g R_\nu^b(T_s)}{\alpha_\nu^s + \alpha_\nu^g} = R_\nu^b(T_s) \qquad (11\text{-}12)$$

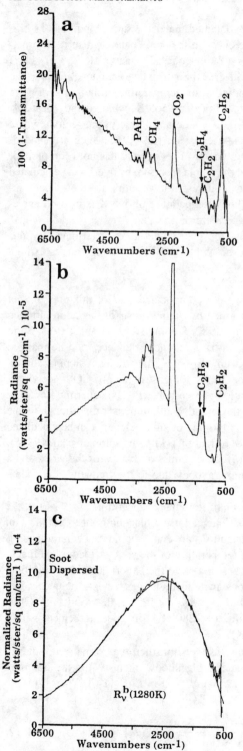

Figure 11-7 Butane and CO_2 in the entrained flow reactor at 1300°C, 66 cm in nitrogen: (a) 100(1-transmittance), (b) radiance, and (c) normalized radiance. Reproduced from Solomon et al. (1987b), with permission.

Coal and char emittance. To determine the temperature of a particle by the E/T technique or by pyrometry, it is necessary to know its spectral emittance. Such measurements were made in a heated tube reactor (HTR), which provides hot particles such as coal or char surrounded by cold walls ($T_p \gg T_w$) (Best et al. 1986; Solomon et al. 1986a). Then, according to Eq. (11-6), in the absence of gas and soot,

$$R_\nu^n = \frac{\epsilon_\nu R_\nu^b(T_p)}{F_\nu^t} \tag{11-13}$$

If T_p is known, then ϵ_ν can be determined. For particles large in comparison with the wavelength, $F_\nu^t \approx 1$.

To achieve the condition where the particle temperature is known, the particles reach thermal equilibrium with the HTR reactor and then exit the reactor and immediately pass the FT-IR aperture. In Fig. 11-8 we show the radiance (R_ν), 100(1-transmission), 100($1 - \tau_\nu$), and the normalized radiance (R_ν^n) from char emerging from the HTR at a temperature of 983 K. The ($1 - \tau_\nu$) is a measure of the intersected surface area of particles, i.e., 3.7% of the IR beam is blocked. The normalized radiance was compared with a number of graybody curves at different values of constant ϵ. The $\epsilon = 0.7$ curve at 1000 K gives the best fit with the experimental data. The temperature is in good agreement with the thermocouple measurement. The value of ϵ is consistent with values previously determined for graphite (Foster and Howarth 1968).

The second example is for lignite particles held at a temperature of 782 K for a short time (~0.1 s). This time is sufficiently short that at 782 K, appreciable pyrolysis has not occurred. The normalized radiance spectrum in Fig. 11-9(a) is compared with a graybody ($\epsilon = 0.90$) at the average thermocouple temperature of 782 K. While the previously formed char shows a graybody shape close to the average temperature, the lignite does not have a graybody shape. The graybody and experimental curves are close only in the range 1600–1000 cm^{-1}, where the emittance [Fig. 11-9(b)], calculated from Eq. (11-13) (assuming $F_\nu^t = 1$) is close to 0.9. For the size of coal particles used here, only specific bands between 1600 and 1000 cm^{-1} (corresponding to the strongest absorption bands in coal) have a spectral emittance near 0.9.

The results can be understood by considering that a particle's emittance is related to its absorbance. Particles will emit only where they have sufficiently strong absorption bands. As discussed by Hottel and Sarofim (1967), for large particles (where diffraction can be neglected) the absorption can be calculated from geometrical optics considering all possible rays through the particles. The absorption within the particle can be calculated using absorbance values measured by the KBr pellet method (Solomon, Hamblen, and Carangelo 1982; Solomon and Carangelo 1982, 1988; Reisser et al. 1984). Figure 11-9(c) shows the absorbance measured by the KBr pellet method. The correspondence between the high absorbance bands and the regions of high emissivity is apparent. The most significant difference between the spectra of Figs. 11-9(a) and 11-9(c) is the presence of a steeply sloping background going toward large wave numbers in the pellet spectrum [Fig. 11-9(c)] and its absence in the emittance spectrum [Fig. 11-9(b)]. The pellet spectrum is from a transmission measurement that does not distinguish between the absorption and scattered components of the total extinction cross section (Solomon et al. 1987b, c).

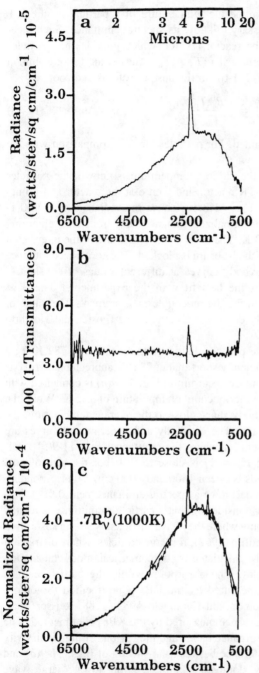

Figure 11-8 (*a*) Radiance, (*b*) 100(1-transmittance), and (*c*) normalized radiance $R/(1 - \tau)$ from char formed at 1300°C in a previous test and subsequently measured at an asymptotic tube temperature of 1073 K (measured by thermocouples in the center of the flow and at the wall), 115-cm reaction distance. Reproduced from Solomon et al. (1987*c*), with permission.

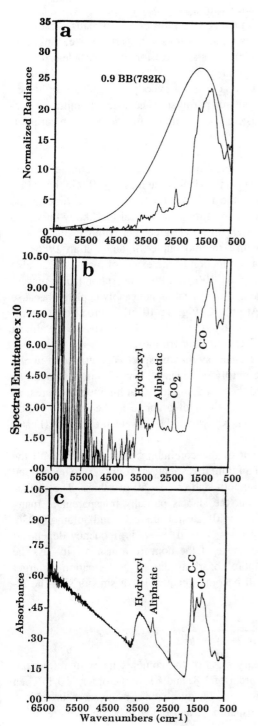

Figure 11-9 Spectra for 500 mesh size fraction of Zap North Dakota lignite. (*a*) Normalized radiance, (*b*) spectral emittance, and (*c*) KBr pellet spectrum. Reproduced from Solomon et al. (1987*c*), with permission.

By comparing the normalized radiance with a graybody for particles at the same temperature, it can be shown that the emittance (the ratio of the normalized radiance to a blackbody curve at the particle temperatures) varies with particle size, rank, and extent of pyrolysis (Solomon et al. 1987c). R_v^n is smallest for small particle sizes and approaches the blackbody curve for large size particles. The emittance of lignite is least gray, while the emittance of anthracite is close to graybody. The spectral emittance also approaches a graybody at high extents of pyrolysis. The spectral emittance variation with rank and extent of pyrolysis is consistent with the functional group variation with these variables.

Particle temperatures. The FT-IR E/T technique was used to determine particle temperatures for nonisothermal conditions in the HTR, assuming $\epsilon \approx 0.9$ in the 1600–1000 cm^{-1} region. Figure 11-10 shows the E/T spectra for increasing time in the reactor. The FT-IR temperatures, obtained by fitting theoretical blackbody curves to the experimental normalized radiance, were in good agreement with calculated temperatures and temperatures measured with a thermocouple (Solomon et al. 1986a).

Changes in emittance of particles as a function of the extent of pyrolysis can also be observed in Fig. 11-10. At 883 K [Fig. 11-10(a)], all the absorption bands present in raw coal can be seen, except that the hydroxyl peak is noticeably depleted because of a reduction in hydrogen bonding. At 963 K [Fig. 11-10(b)], a broad continuum, characteristic of char, is beginning to grow. By 1050 K [Fig. 11-10(d)], we see almost a graybody continuum with CO_2 and H_2O peaks superimposed.

From the spectra discussed so far, we can say that the spectral emittance of lignite of this size increases continuously with pyrolysis, reaching a maximum of about 0.9, when primary pyrolysis is complete and then reducing to 0.7 at high temperatures, as the char becomes more graphitic. The emissivity of graphite in the infrared region between 2 and 10 μm (5000–1000 cm^{-1}) is reported to be between 0.6 and 0.8 (Foster and Howarth 1968).

The relationship of the development of the continuum char spectrum with the extent of pyrolysis will be the subject of a future investigation. However, to understand this variation, qualitatively consider a smooth slab of coal 50 μm thick. The surface reflectivity ρ_v is of the order of 6%, but the slab is partially transparent in many spectral regions (i.e., the transmissivity, $\tau_v \neq 0$) (Best, Carangelo, and Solomon 1986; Solomon et al. 1987a). The emissivity $\epsilon_v = 1 - \rho_v - \tau_v$ is frequency dependent and in many regions is less than 0.9 because of the nonzero transmissivity. As the coal pyrolyzes, the char will become more absorbing due to the formation of larger aromatic ring clusters, and the slab will become opaque. For a smooth slab the reflectance is

$$\rho = \frac{(n - 1)^2 + k^2}{(n + 1) + k^2}$$

where n and k are the real and imaginary parts of the complex index of refraction. For a low-temperature char, n is of the order of 1.8, and k is of the order of 0.1. Then $\rho = 0.08$ and, since $\tau = 0$, $\epsilon = 0.92$. So, on pyrolyzing, ϵ increases in regions where it was previously low due to transparency, and the char becomes more like a

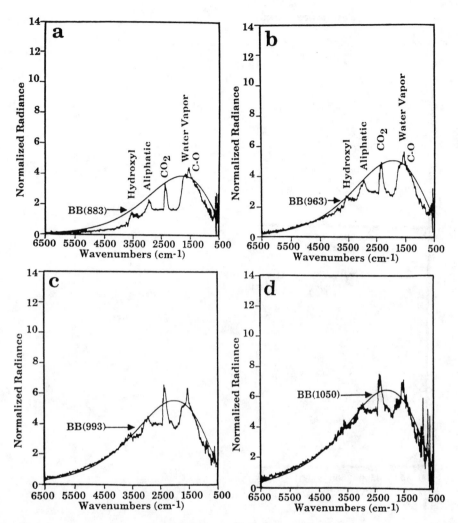

Figure 11-10 Comparison of normalized radiance with theoretical graybody curves ($\epsilon = 0.9$) for chars at increasing extents of pyrolysis. Reproduced from Solomon et al. (1987c), with permission.

graybody. The higher the temperature, the larger k becomes, and the more graphite-like the char becomes. For graphite, n is of the order of 2.5 and k is of the order of 1.2; then $\rho = 0.27$ and $\epsilon = 0.73$. The increase in k with pyrolysis leads first to an increase in ϵ because of reduced transmission, then to a decrease due to increased reflection.

Particle size. Size information can be obtained from the $(1 - \tau_\nu)$ spectra for particles with $D < 80$ μm. For example, $(1 - \tau_\nu)$ in Fig. 11-11(c) increases at low wave numbers (long wavelengths) due to diffraction effects. The shape of $100(1 - \tau_\nu)$ may

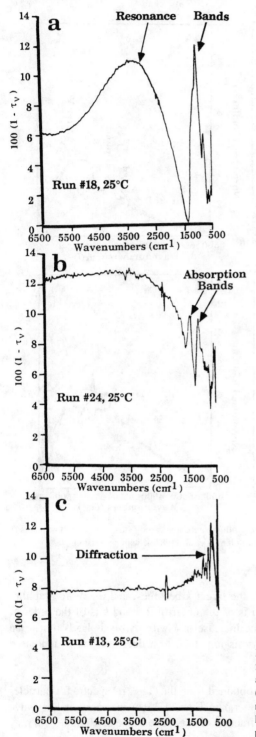

Figure 11-11 Infrared extinction $100(1 - \tau_\nu)$ for particles exiting the heated tube reactor at an asymptotic tube temperature of 25°C. (*a*) Silica (1.4 g/min feed rate), (*b*) fly ash (3.4 g/min feed rate), and (*c*) Ohio 6 coal (2.0 g/min feed rate). Reproduced from Solomon et al. (1987*c*), with permission.

be calculated using Rayleigh theory for large particles (Gumprecht and Sliepcevich 1957) or Mie theory for small particles (Bohren and Huffman 1983).

The effect illustrated in Fig. 11-11(c) is for particles near 80-μm diameter. The diffraction effect increases as the particles decrease in size and is very sensitive to size in the 1- to 30-μm range. Figures 11-11(a) and 11-11(b) illustrate results for fine (<20-μm diameter) mineral particles. The spectra show the characteristic absorption bands of the minerals (peaks in the 500–1500 cm^{-1} region). The spectra also illustrate the reduced attenuation at large wavelengths (i.e., low wave numbers) characteristic of fine particles. The silica particles [Fig. 11-11(a)], which have a very tight size distribution, show an increase in attenuation at 3500 cm^{-1} (~3 μm). This is a resonance effect, where the wavelength of the radiation matches the diameter of the particle. In this case, because the silica is very porous, its effective diameter is smaller than the observed diameter of 11 μm. The resonance is not so apparent for the ash [Fig. 11-11(b)] because it has a very broad size distribution.

Figure 11-11(c) shows the results for a sample of coal with large size particles (i.e., >40 μm diameter). The characteristics of such large particles are that they scatter or absorb almost all of the light incident on them. Any radiation hitting the particle that is not absorbed will be scattered. Consequently, the attenuation is proportional to the area of the particle, and no absorption effects can be seen. Although there is noise in the spectra at low wave numbers, it can be seen that there is an increase in attenuation due to diffraction effects. Lower noise at low wave numbers can be obtained using an FT-IR, which extends into the far IR.

Figure 11-12 illustrates the use of theoretical reference spectra to interpret the observed spectra. The reference spectra were generated using Mie theory. The theory and computer code were published by Bohren and Huffman (1983), and so no further discussion of the theory will be given here. Mie theory is a general solution of Maxwell's equations for an isotropic, homogeneous sphere (diameter D) with a complex index of refraction $m = n + ik$ in a medium of different index of refraction. The complex index of refraction contains all of the electromagnetic properties of nonmagnetic materials. The complex index of refraction has been determined using a transmission method discussed by Solomon et al. (1987c) for the samples of silica, ash, and Ohio 6 coal (for which spectra are shown in Fig. 11-11). The absorbance ($-\log_{10} \tau$) is used in Fig. 11-12.

When Mie theory is used to calculate the extinction properties of monodisperse spheres, it is found that surface resonances and other interference effects dominate the spectra. Such features are not observed in our data, and the scattering spectra should be compared with those calculated for scattering by a distribution of particle sizes, as done by Bohren and Huffman (1983). That is the approach we have used. Use of a size distribution also makes the application of a theory for spheres to represent irregularly shaped particles a reasonable approximation.

The particle size distribution we have chosen is a two-parameter (b and q) distribution, where the number of particles P having diameters between D and $D + dD$ is given by

$$P = \frac{6bq}{\pi} D^{q-4} \exp(-bD^q) \, dD \qquad (11\text{-}14)$$

which is the Rosin-Rammler function (Chigier 1981).

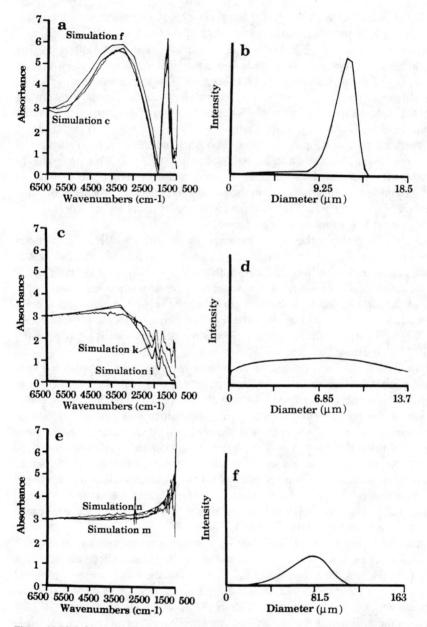

Figure 11-12 Infrared absorbance for particles. The spectra are arbitrarily scaled to give the same amplitude at 6500 wave numbers. (*a*) Comparison of measured and computed extinction efficiency for silica. Theory curve c is for $D = 5$ μm, solid particles; f is for $D = 12$ μm with a pore volume of 65%. (*b*) Particle size distribution for simulation f. (*c*) Comparison of measured and computed extinction efficiency for ash. Theory curve i is for $D = 3$ μm, solid particles; k is for $D = 7$ μm with a pore volume of 57%. (*d*) Particle size distribution for simulation k. (*e*) Comparison of measured and computed extinction efficiency for coal particles of two size cuts (200 × 325 and 324 × 400). (*f*) Particle size distribution for simulation m. Reproduced from Solomon et al. (1987*c*), with permission.

Calculated extinction spectra for several characteristic size distributions are shown in Figs. 11-12(*a*), 11-12(*c*), and 11-12(*e*). The size distribution parameters [*b*, *q*, the diameter of maximum intensity, *D* (peak), and width ΔD at half the intensity)] are presented in Table 11-2. The spectra are normalized to give the same amplitude at 6500 cm^{-1}, and the value of F_{6500}^t is given in Table 11-2. The observed spectra normalized to the same amplitude at 6500 cm^{-1} can then be compared with the reference spectra.

Figure 11-12(*a*) shows the results for silica. The reference spectra are for a tight size distribution as measured for this sample by scanning electron microscope (SEM). The best fit between the spectrum and the reference spectrum is for an average diameter of 5 μm (simulation c). This is substantially smaller than the observed diameter of 11 μm. The discrepancy is caused by the material's porosity (~65% void fraction). This changes the effective index of refraction. Calculations were made using an index of refraction based on an average between the silica and air (the voids). This average employs the Garnett relation for inhomogeneous materials given by Bohren and Huffman (1983). The resulting calculation for 12-μm-diameter spheres (58% voids) is now in excellent agreement with the observed spectrum.

Figure 11-12(*c*) presents the results for ash. Here a wide distribution was used in agreement with the SEM results. The same problem of void fraction exists here, as the powder has a density of 0.75 g/cm^3, which suggests a void fraction of 57%. While the best fit for solid material is for a diameter of 3 μm, the fit for a 57% void fraction is 7 μm, in good agreement with the observed distribution.

Results for coal are presented in Fig. 11-12(*e*). For particles larger than 20-μm diameter, the spectra are reasonably smooth with a slight upward slope going toward low wave numbers and with a sharper rise approaching 500 cm^{-1}. The variation with particle size is not drastic, so the sensitivity of the fitting procedure is much lower than for the small particles. Nevertheless, the observed spectra for two size fractions of coal (200 $\times$ 325 and 325 $\times$ 400) do show the expected difference, and can be used to infer relative size. Figure 11-12(*e*) compares the measured spectra for the two

Table 11-2 Particle distribution parameters

	b	*q*	ΔD, μm	*D*(peak), μm	Porosity (void fraction)	F_{6500}^t
Silica						
Simulation c	9.36×10^{-11}	14	1.3	5	0	1.96
Simulation f	4.45×10^{-16}	14	2.4	12	0.58	1.53
Ash						
Simulation i	8.23×10^{-4}	5	3.1	3	0	2.64
Simulation k	1.35×10^{-5}	4.2	13.5	7	0.57	1.95
Coal						
Simulation m	2.04×10^{-14}	7	40	80	—	1.12
Simulation n	1.53×10^{-13}	7	30	60	—	1.10

size fractions of Ohio 6. These are expected to have an average difference of 18 μm in diameter. Because of the large amount of noise at low wave numbers, the theoretical fits in Fig. 11-12(c) are done above 1000 cm^{-1}. The reference spectra that best fit are for 60- and 80-μm diameters. The simulated spectra do show the right trends but were for a larger average radius than expected.

11-4 RESULTS

Laminar Diffusion Flame (Line-of-Sight Average)

The FT-IR E/T diagnostics were applied to a coannular laminar ethylene diffusion flame described earlier (see Fig. 11-1) (Solomon et al. 1987b). Measurements of the soot particle concentration and temperature were obtained for several axial positions above the burner lip with the FT-IR beam passing through the center of the flame. The spectra obtained at 10- and 50-mm axial positions are shown in Fig. 11-13. The transmittance spectra in Fig. 11-13(a) show a small amount of soot present at 10 mm, but a substantial amount of unburned hydrocarbons. In contrast, the 50-mm spectrum shows a sloping continuum, typical of soot, with the absorption bands for gases mainly CO_2 and H_2O, but no hydrocarbon species.

The normalized radiance spectra in Fig. 11-13(b) also show interesting differences. The 50-mm spectrum matches a theoretical blackbody curve for 1620 K over the whole spectrum. This is expected when the same average temperature exists for both the soot and gas.

The 10-mm spectrum shows a temperature of about 1470 K for the soot but a higher temperature for the CO_2 and a lower temperature for the hydrocarbon gases. These represent the different temperatures of these species, which are concentrated at different radial positions in the flame.

The results are summarized in Fig. 11-14. The concentration for soot, and 100(1 $-$ τ_ν) values for CO_2 and ethylene are presented in Fig. 11-14(a). To allow comparison with the results of Santoro et al. (Santoro, Yeh, and Semerjian 1985; Santoro and Semerjian 1984; Santoro, Semerjian, and Dobbins 1983) and Kent and Wagner (1984a, b), the data are presented on a nondimensional scale $Z = \mathscr{D}h/Q$, where Q is the fuel flow rate in cm^3/s, h is the height above the burner, and $\mathscr{D}$ is the diffusion coefficient for ethylene in nitrogen (0.156 cm^2/s).

Since the FT-IR technique gives a line-of-sight measurement, the soot concentration presented in Fig. 11-14(a) is the volume fraction f_ν, integrated along a line passing through the flame center as discussed by Solomon et al. (1987b).

These results were compared with the data of Kent and Wagner (1984a) (for $Q = 4.6$ cm^3/s) and with values of f_ν recalculated from the data of Santoro, Semerjian, and Dobbins (1983), using Mie theory, and integrated along the line of site through the center of the flame. The latter data are plotted as open squares and referred to in Fig. 11-14 as the NBS (National Bureau of Standards) data. The shape of the FT-IR soot concentration is in good agreement with the NBS data and data of Kent and Wagner (1984a). The magnitudes are not identical but are within the uncertainty in calculating the soot volume fraction from the IR transmittance.

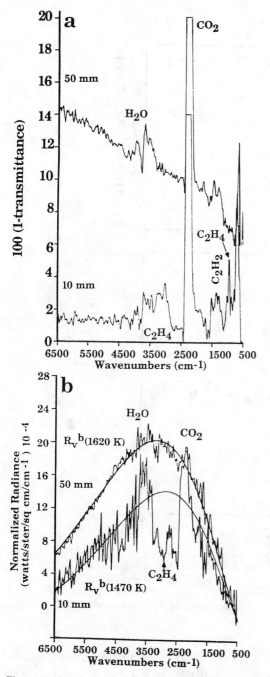

Figure 11-13 Spectra for combustion of ethylene in a coannular diffusion flame at 10 and 50 mm above the burner lip: (*a*) 100(1-transmittance) (50-mm spectrum is displaced vertically for clarity), and (*b*) normalized radiance along a line of sight passing through the axis. Reprinted from Solomon et al. (1987*b*), with permission.

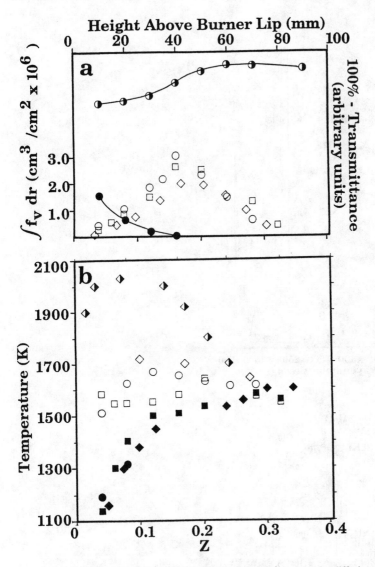

Figure 11-14 (*a*) Species concentration and (*b*) temperature for coannular diffusion flame as a function of height above the burner lip. FT-IR measurements: open circles, soot; semi-open circles, CO_2; solid circles, ethylene. NBS data: (Solomon et al. 1987*b*): open squares, soot; solid squares, centerline. Data of Kent and Wagner (1984*a*): open diamonds, soot; solid diamonds, centerline; semi-open diamonds, peak. Reprinted from Solomon et al. (1987*b*), with permission.

For ethylene and CO_2, values of $100(1 - \tau_\nu)$, which are related to concentrations, are presented to show that these data are available from the spectra as well. The ethylene concentration decreases as the fuel is pyrolyzed and burned. The CO_2 concentration is high even early in the flame and then increases further as the soot is burned.

Figure 11-14(*b*) presents the average temperatures along a line of sight passing through the flame axis determined for the soot (open circles). These are compared with the temperatures measured on an identical burner at NBS using the rapid insertion thermocouple technique previously described (Santoro, Yeh, and Semerjian 1985). In order to compare the thermocouple measurements with the FT-IR line-of-sight measurements, the thermocouple temperature measurements made at positions along the diameter were weighted by the measured soot volume fraction at each position to yield an average value in the region of the flame where soot is present (open squares) (Solomon et al. 1987*b*).

Data of Kent and Wagner (1984*a*) (using the Kurlbaum method for soot) are also shown (open diamonds). Since radial temperature variations are likely, the FT-IR soot temperatures are obtained using the best fits to spectral shape to obtain an average temperature along the line of site. This is equivalent to the Kurlbaum method. The fits to the amplitude and shape are between 10 and 50 K lower than fits to shape alone. The FT-IR temperatures are in good agreement with those of Kent and Wagner (1984*a*) but are up to 100 K higher than those measured with the thermocouple, especially in the region of high soot concentrations. It is likely that soot coats the thermocouple, giving somewhat low values due to both the thermal resistance and the high emittance of the soot.

Temperature data are also presented for the ethylene gas at 10 and 20 mm (solid circles). Since the ethylene is confined to the center of the flame, its temperature is a good measure of the centerline gas temperature. The data up to 20 mm, above which the ethylene concentration is too small to use, agree with the centerline temperature measured at NBS on an identical burner (solid squares) and by Kent and Wagner (solid diamonds), both using a thermocouple rapidly inserted to avoid coating with soot.

Tomography in Laminar Diffusion Flame

Line-of-sight FT-IR measurements of both transmittance and radiance were obtained for 21 chords at each of 10 heights in a coannular ethylene diffusion flame (Solomon 1987*a*). This small, axially symmetric flame always had a transmittance greater than 0.8, and therefore the transmittance measurement could be applied directly to correct for self-absorption (Freeman and Katz 1960) in the radiance spectra.

From these measurements, local spectra were obtained for transmittance, radiance, and normalized radiance. Examples are presented in Fig. 11-15. An InSb detector was used for the transmittance measurements, and so the spectra are only plotted above 1700 cm^{-1}. From these data, concentrations and temperatures were obtained for the soot, as well as for CO_2, H_2O, alkanes, alkenes, and alkynes. Figure 11-15(*c*) shows temperatures derived for the soot and CO_2.

For H_2O and soot, these concentrations and temperatures for slices at 10 heights above the burner lip are plotted in Figs. 11-16 and 11-17, respectively, as a function of the distance from the centerline. The results of the local soot concentrations obtained in this manner were compared with laser extinction/scattering data obtained by Santoro, Yeh, and Semerjian (1985) for a flame formed by the same burner under nominally identical conditions. As shown in Fig. 11-18, there is very good agreement between the results obtained from these two different techniques.

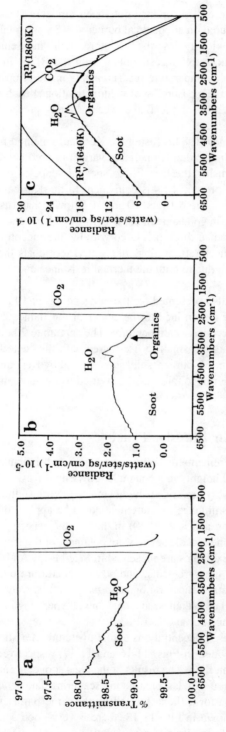

Figure 11-15 Results of Fourier reconstruction of the IR spectra at a radial position 2 mm from the axis and at a height of 4 cm in the ethylene diffusion flame. (*a*) Transmittance showing a considerable amount of soot. (*b*) Radiance reconstructed after the original measurements were corrected for self-absorption. (*c*) Normalized radiance. In a single wavelength E/T measurement a value of 1835 K would be deduced for the CO_2 temperature, 1680 K for H_2O. After correcting for the presence of both species, the temperature of soot was found to be 1860 K (*Best et al. 1990*).

149

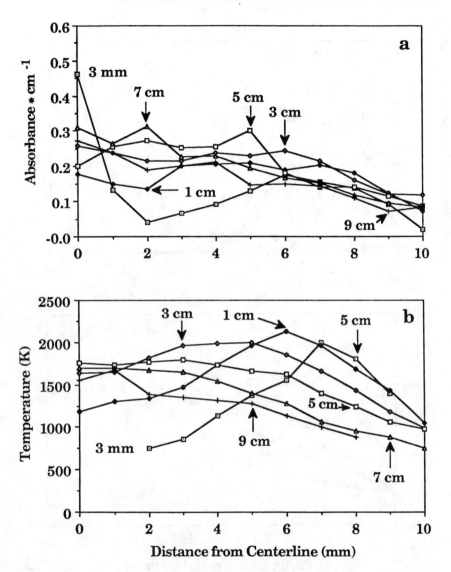

Figure 11-16 Composite graph from the results of the tomographic reconstruction for the ethylene diffusion flame showing the spatial variation of H_2O absorbance and temperature. The concentration is directly proportional to absorbance (*Best et al. 1990*).

Temperature comparisons were made with the thermocouple measurement of gas temperatures (lumped) by Santoro, Yeh, and Semerjian (1985) and the nonintrusive coherent anti-Stokes Raman spectroscopy (CARS) temperature measurement for CO_2 by Boedeker and Dobbs (1986). The temperatures from the FT-IR tomography are about 100°C higher than those from thermocouple measurement with an additional 100°C increment in the soot zone. On the other hand, there is excellent agreement

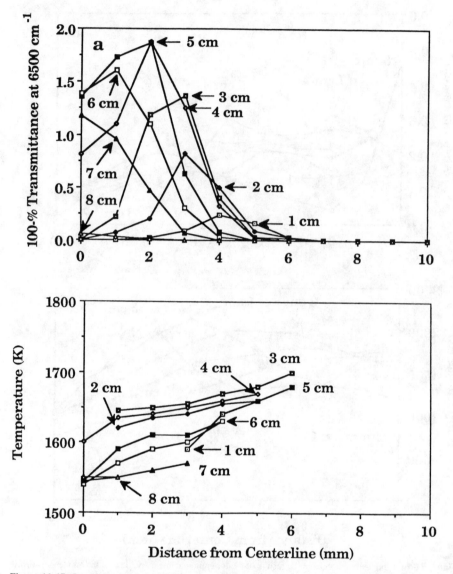

Figure 11-17 Composite graph from the results of the tomographic reconstruction for the ethylene diffusion flame showing the spatial variation of soot absorbance and temperature. The concentration is directly proportional to absorbance (*Best et al. 1990*).

between FT-IR and CARS. It is most probable that radiation heat loss from a soot-covered thermocouple is the cause of low temperatures in the thermocouple measurements.

Images, both direct and derived from the tomographic results, are presented in Figs. 11-19 and 11-20. The reconstructed images are for the plane of the flame containing the axis. In Fig. 11-19, a photograph of the flame is shown alongside

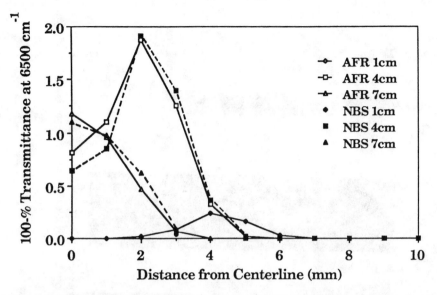

Figure 11-18 Comparison between spatial-dependent measurements that are proportional to soot concentration, normalized at a distance of 2 mm from the centerline. There is good agreement between the two sets of data, obtained from FT-IR tomography (AFR) (*Best et al. 1990*) and laser scattering (NBS) (*Santoro, Semerjian, and Dobbins 1983*).

reconstructed images (gray scale) of soot and CO_2 concentrations, and CO_2 temperatures. In both Figs. 11-19 and 11-20 the scale increases toward white. Three "compressed" central plane flame images in Fig. 11-20 show soot concentration, H_2O temperature gradient, and H_2O concentration gradient.

Examination of the CO_2 temperature image shows the location of the flame front about the surface of highest temperature (Fig. 11-19). Also in that figure it can be seen that maximum CO_2 concentration is attained after soot burnout. The strong correlation between the direct image and that of soot concentration is apparent in the figure.

Use of the gradient fields is described with the help of the images in Fig. 11-20. High spatial gradients are indicated by the blocks of white, whereas the lines are artifacts of the image-handling technique. It can be seen that gas temperature gradients are strongly suppressed in regions of appreciable soot concentrations [Figs. 11-20(*a*) and 11-20(*b*)].

The presentation of Fig. 11-17(*a*) should be compared with that of Fig. 11-20(*b*), to see the impact of the different representations for the soot concentration field. In making the graphical representation of H_2O concentration, field measurements at alternate flame heights only were included so as to avoid a cluttered diagram [Fig. 11-16(*a*)]. The image of the gradient of this field showed two extremely high gradients on axis, one low and one high in the flame [Fig. 11-20(*c*)]. That at low, but not high, flame height showed up in the graphical representation [Fig. 11-16(*a*)]. Examination of the data in tabular form shows that a very high concentration of H_2O appears on-

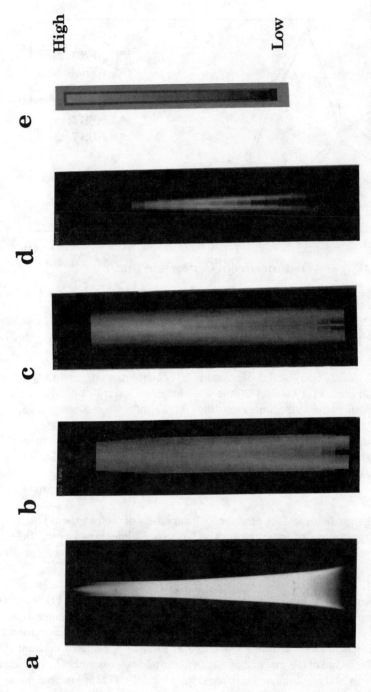

High

Low

a   b   c   d   e

Figure 11-19 (*a*) Photograph of the ethylene diffusion flame. Central plane images reconstructed from FT-IR transmission and radiance spectra. (*b*) CO_2 temperature. (*c*) CO_2 concentration. (*d*) Soot concentration. The flame luminance is almost completely due to the soot. The gray scale for Figs. 11-19(*b*) to 11-19(*d*) is shown in Fig. 11-19(*e*).

a

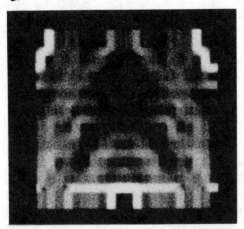

b

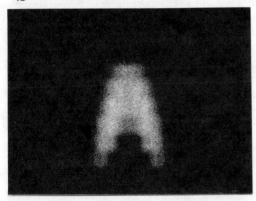

c

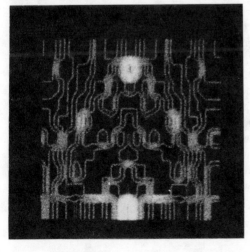

Figure 11-20 Compressed central plane images reconstructed from FT-IR transmission and radiance spectra. (*a*) CO_2 temperature gradient, (*b*) soot concentration, and (*c*) H_2O concentration gradient.

axis at a height of 8 cm in the flame. For the amount of data collected in this tomography study, image construction is the representation of choice for the purpose of recognizing data trends, and correlations between the different fields.

While a complete analysis of these data has not been made, FT-IR tomography has demonstrated the unique capability of measuring local species concentrations and temperatures in mixed phase systems. The measurements are so new that their impact on the combustion field has not been assessed.

Coal Flames

FT-IR E/T measurement. Results on flame properties have been obtained for a series of samples including coal, demineralized coal, and char (Solomon et al. 1988). FT-IR measurements were made along the center of the flame described in Section 11-2 at several positions above the coal injector.

Figure 11-21 presents transmittance and normalized radiance spectra obtained for a Rosebud subbituminous coal flame at several distances above the nozzle: 5 cm above the nozzle is in the region prior to ignition, 9 cm is at the beginning of the ignition region, 11 cm is in the ignition region, 13 and 14 cm are in the brightest part of the flame, and 17, 20, and 25 cm are where burnout occurs.

Figure 11-21(c) presents the transmittance spectra as 100% transmittance. In the optically thin limit, the amplitude of (100% transmittance) is proportional to the intersected surface area of the particles, NAL, times the extinction efficiency, F_ν^t, where $F_\nu^t = 1$ in the absence of diffraction. The attenuation at 5 cm is almost entirely from coal particles. The spectrum slopes due to diffraction effects, and as discussed above, the shape of τ_ν, which is proportional to F_ν^t, may be used to determine particle size.

At 9 cm a few particles have ignited, and attenuation from CO_2 can be seen. At 11 cm the attenuation is flatter due to the appearance of soot. The soot attenuation slopes downward toward low frequencies (0 cm^{-1}), the particle attenuation slopes upward toward low frequencies, and the sum appears flat. The attenuation at 11 cm and at 13 cm contains larger contributions from CO_2 and H_2O, a contribution from soot, as well as a contribution from the particles.

To resolve the spectrum into particle and soot contributions, F_ν^t for the particles is assumed to have the same shape as prior to ignition, and the particle transmittance is assumed to be equal to the measured transmittance extrapolated to 0 cm^{-1} (where the attenuation from soot goes to zero). A straight-line extrapolation is made below 3500 cm^{-1}, excluding the region of the spectrum containing CO_2 and H_2O bands. The uncertainty estimated by considering the possible high and low extrapolation lines is ±10% of the transmittance. The soot contribution is the difference between the particle attenuation and the total, as shown in Fig. 11-21(c) (11 cm). Much larger soot contributions were observed for bituminous coals. Above the ignition region, at 14, 17, 20, and 25 cm, 100% transmittance progressively reduces as the particles burn out. The 17-, 20-, and 25-cm cases show the presence of ash (the dips near 100 cm^{-1} due to the Christiansen effect). Such dips, which are observed near strong absorption

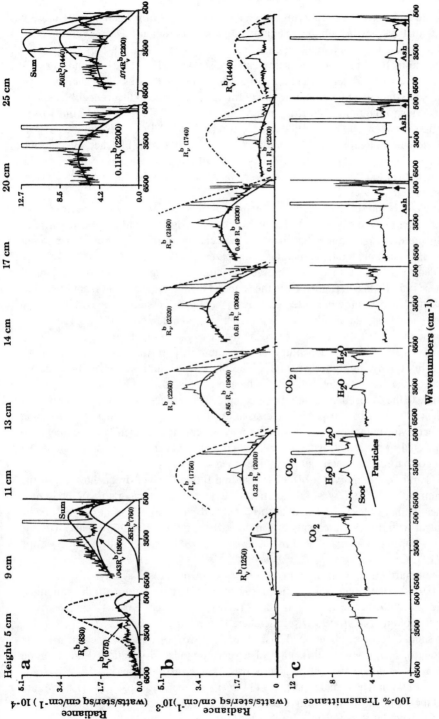

Figure 11-21 Spectra of Rosebud subbituminous coal flame in the TWR for several positions above the coal injection nozzle. (a) Normalized radiance on expanded scales, (b) normalized radiance on a common scale, and (c) 100% transmittance. The dashed lines represent blackbody distributions fit to the CO_2 band. The solid lines represent graybody fits to coal, char, ash, and soot particles. The theoretical blackbody distributions are designated as $R_v^b(T)$, where T is the component temperature in Kelvins.

423

bands, are due to reduced scattering as the material's index of refraction more closely matches that of the surrounding medium. Dips in this region have previously been observed for coal ash and small silica spheres, as shown in Fig. 11-11.

The normalized radiance measurements are shown in Fig. 11-21(b) and Fig. 11-21(a) on an expanded scale for the 5-, 9-, 20-, and 25-cm cases. The average temperatures are obtained by comparing the normalized radiance to theoretical blackbody curves. Prior to ignition (5 cm), a particle temperature of 575 K is determined from the region between 1000 and 1600 cm^{-1}, where for coal [which is nongray (Best, Carangelo, and Solomon 1986; Solomon et al. 1986a, 1987a, c; Fletcher, Baxter, and Ottesen 1987)], $\epsilon \sim 1$. There must be about 2% of the particles ignited based on the radiance above 3000 cm^{-1}, but these were ignored in the analysis at 5 cm^{-1}. They are considered for 9 cm.

At 9 cm the presence of radiation intensity above 4000 cm^{-1} in the spectra shows that a few particles have ignited. The combination of ignited and unignited particles is fit theoretically, assuming that there are two classes of particles: very hot burning char particles and moderate temperature coal particles. The spectra for the two classes are added to provide a fit to the observed spectrum. The temperature of the ignited particles (1950 K) is determined from the shape of the region above 4000 cm^{-1} (where the contribution from the unignited particles is negligible) assuming a graybody ($\epsilon = 0.8$) for char (Best, Carangelo, and Solomon 1986; Solomon et al. 1986a, 1987a, c). The value of $\epsilon_\nu = 0.8$ is a guess based on the observation that the emissivity of char lies between 0.7 and 0.9, depending on the extent of pyrolysis. The spectrum is fit using a blackbody multiplier of 0.043. The blackbody multiplier M is the constant fraction of the theoretical blackbody, which produces the best match in shape and amplitude to the experimental data. For a completely homogeneous sample of graybody particles, M would be the particle's emissivity. For the case considered here, some particles may be unignited, or ash particles may be present at a much lower temperature than the coal particles and have a very low emittance. Then, M is approximately the fraction of particles in each category times their emissivity.

Assuming an emissivity of 0.8 for char and fitting the high-frequency end of the spectrum indicates that 5.4% of the particles are ignited. The temperature of the ignited particles is about 750°C higher than the preheated gas stream. This is in reasonable agreement with the measurements of Timothy, Sarofim, and Beer (1982) and with the diffusion-burning limit calculated by Waters et al. (1986) and Timothy, Sarofim, and Beer (1982).

The contribution from unignited particles is fit by adding a graybody spectrum ($\epsilon = 0.9$) for the remaining particles (94.6%). The temperature of the unignited particles is fit to match the region 1000–1600 cm^{-1}, where coal has an emissivity near 0.9 (Solomon et al. 1987a, c; Fletcher, Baxter, and Ottesen 1987).

It should again be mentioned that assuming two populations, each at a single temperature, is a gross simplification of a very complicated situation. Equally good or better fits to the data might be achieved with other assumptions.

In the region where many particles (11 cm), and then most particles (13 cm), are ignited the shape of the spectrum is consistent with a temperature of 2010 K and 1900 K, respectively. It is not clear why these differences are observed.

A value of M of 0.32 at 11 cm is consistent with only partial ignition of the particles, and of 0.85 at 13 cm with complete ignition, and of $\epsilon = 0.7$–0.9 for char (Best, Carangelo, and Solomon 1986; Solomon et al. 1987a, c). In the upper part of the flame the particle contributions to the radiance become progressively lower as the particles burn out, leaving fly ash. The contributions from unignited particles or ash for these two cases and for 14, 17, and 20 cm have been ignored as being small in comparison with the radiance from ignited particles.

The CO_2 temperatures can be obtained from the normalized radiance, as well. In Fig. 11-21 a rough estimate of the CO_2 temperature was made by fitting a blackbody to the CO_2 amplitude in the normalized radiance. This approximation is reasonably accurate when the CO_2 bands are large in comparison with the soot or particle contribution to the spectrum. For the most accurate determination of gas temperatures by the E/T technique, the gas emission and transmission lines should be corrected for the presence of soot or particles. This was done for the CO_2 temperatures presented in Fig. 11-22. Comparison of Figs. 11-21 and 11-22 indicates that the method of laying blackbody temperature curves over the total spectra, without correcting for particles or soot, contributes errors of the order of 100°C in the present case.

The results of Fig. 11-21 are summarized in Fig. 11-22. Figure 11-22 also includes an estimate of the particles' residence time based on photographic measurements of their velocity. Figure 11-22(b) presents the particle and CO_2 temperatures and M and Fig. 11-22(a) presents the soot, CO_2, and particle contributions to the infrared absorption as a function of distance above the nozzle. The magnitude of these absorptions is related to the respective concentrations. Prior to ignition, the temperatures, CO_2, and soot absorptions and the values of M for ignited particles are low. At ignition the CO_2 and particle temperatures and the CO_2 and soot absorptions go up sharply. In the region where there is soot, the particle temperature is an average of both char and soot temperatures. The influence of the soot temperature on the particle temperature has been analyzed by Grosshandler (1984). In this region, M increases to over 0.8. After ignition, the particle and soot absorptions both fall rapidly as they oxidize. M falls as the ash fraction increases. For the emission at 25 cm [Fig. 11-21(a)] a strong contribution can be seen in the radiance near 1000 cm^{-1} from the ash. As for the 9-cm case, a sum of two particle classes is used to fit the data: ignited char and unignited ash. In this case the ash is assumed to be at the gas temperature (1440 K) and to have zero emissivity above 4000 cm^{-1}. The sum spectrum then determines the emissivity ($\epsilon = 0.55$) near 1000 cm^{-1}. This is in reasonable agreement with previous measurements for coal ash (Solomon et al. 1986c).

A correlation of flame properties with sample characteristics. Several flame characteristics that varied with sample properties were examined to determine what sample properties were the controlling factors and why. A comparison of the CO_2 absorbance profiles, which are by-products of the ignition behavior, is presented in Fig. 11-23(a), for the raw and demineralized Rosebud coal, and Rosebud chars produced at 900°C and 1500°C. Ignition for the coals is accompanied by a more rapid increase in CO_2 absorbance than seen for the chars. This is believed to be due to the rapid release of energy from the combustion of volatiles. The position of the ignition also varies for the four samples.

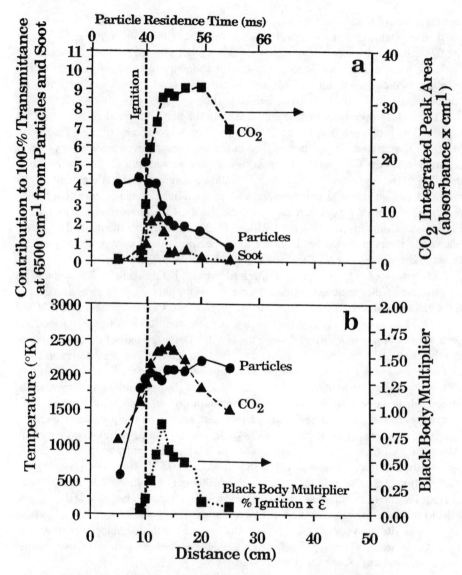

Figure 11-22 Flame properties as a function of distance above the injection nozzle for Montana Rosebud subbituminous coal. (*a*) Attenuation from CO_2, particles, and soot. (*b*) Temperature of particles and CO_2 and fraction of particles ignited times emissivity.

To determine what controls the ignition, measurements have been made of the weight loss of coal in a TGA, both under inert and oxidizing conditions. The objective is to correlate what happens in the TGA at temperatures from 450°–600°C, to what happens during ignition in the TWR at similar particle temperatures. As shown in Fig. 11-23(*b*), the ignition distance correlates well with the temperature for 10% weight loss in air measured at a constant heating rate of 30°C/min in a TGA. The lower the

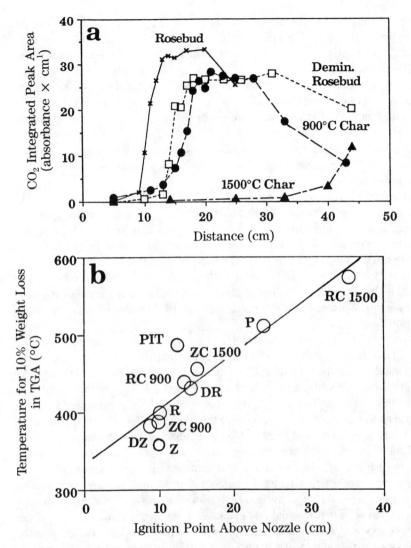

Figure 11-23 Correlation of ignition behavior with coal properties. (*a*) Comparison of CO_2 attenuation in flames of four samples. (*b*) Correlation of ignition point (maximum slope in CO_2 amplitude) with TGA weight loss in air. See Table 11-1 for the meaning of the sample designations.

temperature to achieve 10% weight loss in the TGA, the shorter the ignition distance in the TWR.

To provide additional information on what produces the 10% weight loss, measurements on identical samples were made in the TGA for inert gas (nitrogen) and the weight loss compared with that in air. In the case of the previously formed chars, the TGA weight loss in air is almost exclusively due to char oxidation, since there is no corresponding weight loss in nitrogen and hence the ignition in the TWR must be heterogeneous (i.e., oxidation within the porous solid matrix). For high-rank coals the first 10% weight loss in the TGA under oxidizing conditions is almost the same as in nitrogen and so is mostly due to pyrolysis, consistent with homogeneous ignition in the TWR. For low-rank coals, however, there is a significant early weight loss in the TGA in air due to heterogeneous oxidation, and it appears that there is a significant heterogeneous contribution to the ignition in the TWR. This is consistent with the observation that the demineralized Rosebud coal (which is less reactive to oxidation than the raw coal) is more difficult to ignite [Fig. 11-23(a)]. The results are also consistent with the measurements of Midkiff, Altenkirch, and Peck (1986), who conclude that there is a significant weight loss due to heterogeneous oxidation in combustion for low-rank coals.

A comparison of the soot concentration for three samples is shown in Fig. 11-24(a). The demineralized Rosebud produces about twice the maximum soot yield as the raw sample, and the char produces almost none. As shown in Fig. 11-24(b), the soot production correlates well with the yield of tar as determined in pyrolysis experiments. Pyrolysis was performed in an entrained flow reactor described previously (Solomon et al. 1982; Solomon and Hamblen 1985; Serio et al. 1987). The higher the tar yield, the higher the soot yield. The relationship between tar and soot is consistent with the results of Wornat, Sarofim, and Longwell (1987) and Nenniger, Howard, and Sarofim (1983).

Based on previous infrared measurements of soot in which the complex index of refraction of Dalzell and Sarofim (1969) was used to relate the infrared attenuation to the volume fraction (Solomon et al. 1987b), the mass of soot in the line of sight was calculated and compared with the mass of coal in the same volume. For this calculation the infrared attenuation due to soot determined at 4000 cm^{-1} was expressed in terms of an integral of the soot volume fraction over the beam path length. Thus volume of soot in the beam was determined from measured attenuation. In a similar manner, the attenuation due to particles was expressed in terms of known constants (i.e., extinction coefficient and particle diameter) and the total volume of particles within the beam. Again, measurement of attenuation gave an estimate of the total particle volume. The particle and soot volumes were converted to masses using a value of density for soot that is midway between that for coal and graphite. A maximum value of 1.2% of the Rosebud mass appeared as soot at any line of sight within the flame. Performing the same calculation for the other coals, it appears that no more than ⅓ of the tar appears as soot at any one point. This not unreasonable, considering that soot is simultaneously consumed by oxidation as it is being formed.

Figure 11-25(a) compares the value of M for ignited particles, the raw coal, the demineralized coal, and the 900°C char. M can be less than 1.0 due to unignited particles and/or a low value of emissivity. Prior to ignition, all three samples have low values of M due to unignited particles. At ignition, M for the two coals goes up rapidly as the particles ignite and soot is formed. Above the ignition region, where all the particles are expected to be ignited, M drops rapidly for the raw coal but remains high for the demineralized coal. The source for this effect appears to be ash particles, which are shed from the burning char particles. The ash particles that are shed will increase 100(1-transmittance) without adding significantly to the radiance because of

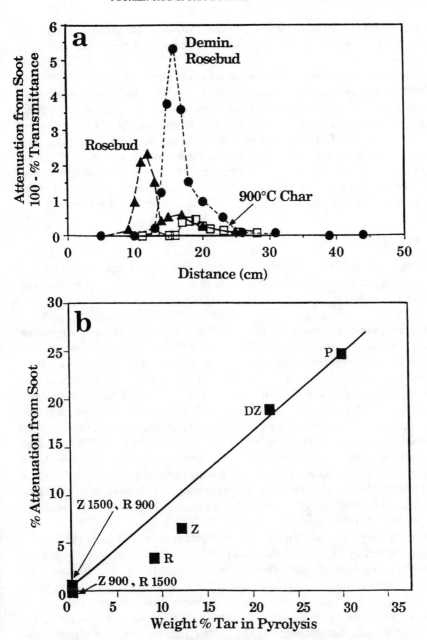

Figure 11-24 Correlation of soot production with tar yield. (*a*) Comparison of soot attenuation for three samples and (*b*) correlation of the maximum soot attenuation with tar yield in pyrolysis. See Table 11-1 for the meaning of the sample designations.

their low temperature and low emissivity. TGA analysis of the captured samples of char early in the flame showed that a significant fraction of the minerals (30%) had already been shed.

An alternative explanation is that the unshed ash particles act as diffuse scatterers, which lowers the emissivity. For example, a layer of sintered transparent particles can have a reflectivity of 50% and hence lower the apparent emissivity. However, a direct measurement of the emissivity of captured char particles using the E/T technique at a lower temperature (800°C) shows the emissivity to be approximately 0.85, suggesting that the reduction of the emissivity due to surface ash is not important.

Figure 11-25(*b*) presents a correlation between M above ignition (where all particles should be ignited and the soot has been consumed) with the ash content of the coal. The higher the ash content, the lower M, in agreement with either hypothesis.

Spray Flames

Experiments were performed on unignited sprays and spray flames of hexane and CWFs in the apparatus shown in Fig. 11-4. These are described in detail by Solomon et al. (1987*d*).

Liquid sprays, noncombusting. Work was performed on sprays without combustion to demonstrate the measurement of droplet size distributions. This was done by comparing the transmittance with theoretically determined reference spectra as described in the analysis section. The shape of the transmittance provides information on the particle size and composition, while the amplitude is proportional to concentration.

The transmittance of the spray measured on-axis at a pressure of 60 psi, is plotted in Fig. 11-26(*a*). Band absorption features due to the hexane can be seen, superimposed on a smooth background (with noise), which increases from 6500 to 1500 cm^{-1}, and subsequently decreases toward 500 cm^{-1}. In the figure a straight line has been drawn at the average intensity below 1500 cm^{-1}, replacing the true spectrum that decreased below 1500 cm^{-1} but was quite noisy. By comparing this spectrum with calculated spectra, it can be shown that the band absorption features in Fig. 11-26(*a*) are due to hexane vapor, rather than liquid.

Mie calculations for each particle size distribution were made with the aid of a program published by Bohren and Huffman (1983). The best fit theoretical spectrum is overlayed on the measured spectrum in Fig. 11-26(*a*). In Fig. 11-26(*b*) we show the droplet size distribution that gave rise to the best fit. This distribution has a maximum at $D_{peak} = 16$ μm, while D_s, the Sauter mean diameter, is equal to 21.5 ± 1.5 μm. If we narrow or broaden the distribution, we see poorer fits in the shape, while the same width distribution moved to smaller diameter or greater diameter also gives a visibly poorer fit in shape. For this size range the extinction measurement is a remarkably sensitive gage of particle size distribution, i.e., slight differences in the particle size distribution show noticeable differences in the FT-IR spectra. This method is an extension of the spectral sizing method used by Hammond (1981) and Perelman and Shifrin (1980).

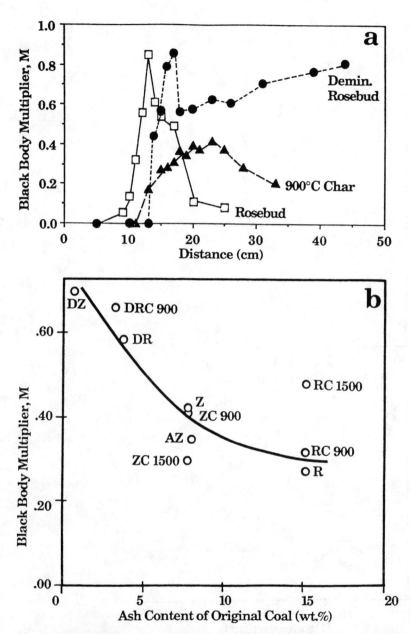

Figure 11-25 Correlation of infrared attenuation with ash content. (*a*) Comparison of blackbody multiplier for three samples and (*b*) correlation of blackbody multiplier in the burnout region with ash content of the coal. See Table 11-1 for the meaning of the sample designations.

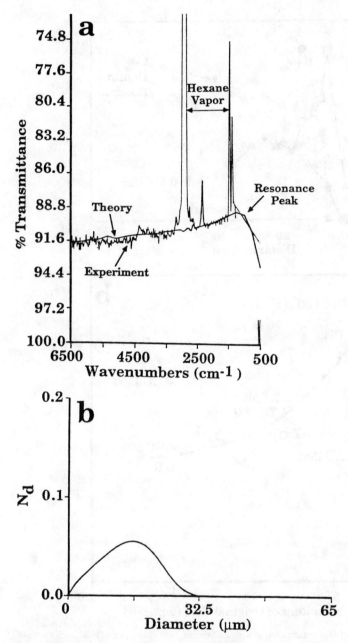

Figure 11-26 Comparison of theoretical and experimental transmittance spectra. (*a*) Spectra. (*b*) Size distribution used for calculation.

Hexane spray combustion. Both emission and transmission measurements were obtained to determine concentrations of hexane pyrolysis and combustion products and their temperatures. The hexane spray flame was ignited by a hydrogen pilot flame, which was then turned off: the flame is of hexane burning in air. Photographs of the flames (Solomon et al. 1987*d*) show them to be about 70 cm in length, with real differences visible due to change in the pressure of atomizing air. The FT-IR E/T diagnostics were applied to these hexane flames.

For an atomizing air pressure of 34 psi, measurements of the combustion species temperatures and concentrations were obtained for a number of axial positions, as well as off-axis positions. Examples of the species that can be detected by FT-IR spectra are shown in transmittance and normalized radiance spectra presented in Figs. 11-27 and 11-28. They include the following:

Unburned hexane in both liquid and vapor phases.
Pyrolyzed hexane as methane, PAH, and soot.
Combustion products, H_2O, CO_2, and CO.

In this spray flame, where the combustion is controlled by mixing, the various components (fuel, soot, CO_2, H_2O, etc.) can be at different temperatures, and these can be seen when compared with theoretical blackbody curves in the normalized spectra, e.g., the low fuel temperature compared with the CO_2 temperature in Fig. 11-28(*a*), and the high methane and CO_2 temperatures compared with the lower soot temperature in Fig. 11-28(*b*). The range of temperatures measured for soot and CO_2 is in agreement with the ranges determined for the laminar ethylene diffusion flame discussed above.

Figures 11-27 and 11-28 show transmission and normalized emission spectra taken through the axis at depths of 3.5 and 30 cm, respectively. The flat, nonzero baseline of the transmission spectrum at a depth of 3.5 cm [Fig. 11-27(*a*)] indicates that hexane droplets exist in the flame at this point. Intermediate spectra between 3.5 and 30 cm show that the flat nonzero baseline disappears at 7 cm, indicating that there are no liquid droplets left in this flame past 7-cm depth. Gas lines can be readily identified on top of the continuum.

The continuum spectrum, in Fig. 11-27(*b*), is made up of a small flat part due to a particle blockage, and a sloping part that intercepts the flat part at 0 cm^{-1}, due to soot. Here we ignore the spectral slope of the droplet extinction spectra due to diffraction effects. The soot concentration is measured by the amplitude of the triangular "wedge" (Fig. 11-27). In Fig. 11-28 the temperatures for soot and gas components are obtained by comparison with the blackbody reference spectra.

The relative concentrations of hexane, methane, soot, and CO_2 that have been derived from transmittance spectra are summarized in Fig. 11-29(*a*) together with temperature information obtained from normalized emission spectra in Fig. 11-29(*b*). A flame photograph for these conditions is presented in Fig. 11-29(*c*). The sequence of hexane combustion along the flame axis is very clearly shown in these data. Close to the nozzle, the combustion process is just beginning (low temperature), and hexane can exist in both the liquid and vapor phase. As the combustion continues, liquid hexane becomes completely vaporized between the depths of 3.5 and 7 cm, while

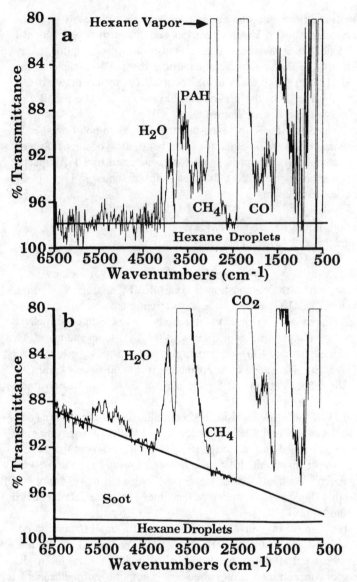

Figure 11-27 Percent transmittance τ_ν for hexane flame (34 psi atomizing air pressure, 0.1L/min fuel flow) at (*a*) 3.5 cm below nozzle and (*b*) 30 cm below nozzle.

methane, a pyrolysis product of hexane, is initially detected in this region, by the appearance of its characteristic emission line at 3002 cm^{-1}, as a shoulder on the hexane peak. The maximum concentration of methane is found to be at a depth of 15 cm. At this point the hexane concentration has decreased to a rather low value. However, PAH, a known soot precursor, indicated by the line at 3080 cm^{-1}, was identified in the spectra at this height. This observations imply that in our flame the pyrolysis of

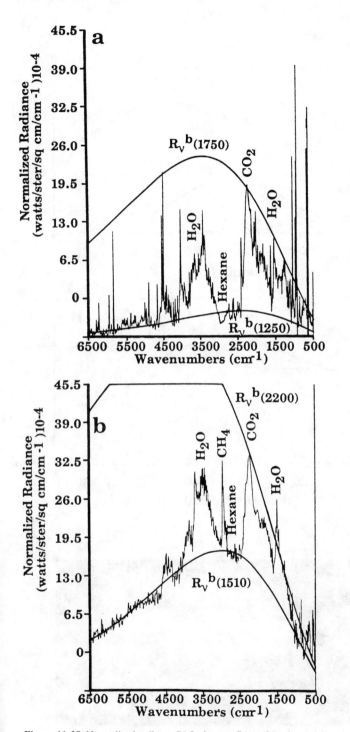

Figure 11-28 Normalized radiance R_ν^n for hexane flame (34 psi atomizing air pressure, 0.1L/min fuel flow) at (*a*) 3.5 cm below nozzle, and (*b*) 30 cm below nozzle.

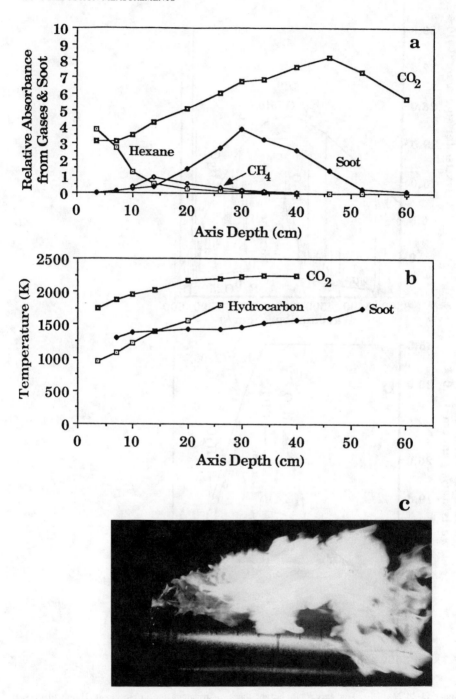

Figure 11-29 Characteristics of the hexane flame (34 psi atomizing air pressure, 0.1 L/min fuel flow). (*a*) Concentrations (arbitrary units). (*b*) Temperatures of species along the flame axis as a function of the depth below the nozzle. (*c*) Photograph of the flame.

hexane is an important step. The PAH emission intensity is seen to increase until a depth of 34 cm is reached. The soot concentration also peaks in this region. Beyond this point, all of the fuel concentrations, including hexane, methane, PAH, and soot, decrease with increasing depth. This observation implies that oxidation reactions predominate in the lower half of the flame. The CO_2 absorbance increases due to increasing CO_2 concentrations, reaching a maximum at 50 cm, at which point all the products have been consumed. Above 50 cm the CO_2 absorbance decreases as the CO_2 is dispersed away from the flame axis.

The fuel, soot, and CO_2 temperatures that were also obtained are shown in Fig. 11-29(b). These temperatures increase as the fuel is consumed.

Combustion of CWF. FT-IR diagnostics were applied to monitor combustion of CWF sprays (Solomon et al. 1987d). Measurements were made along the axis of the spray combustion apparatus described in Section 11-2 as a function of depth below the nozzle. From these measurements, the following quantities can be determined:

Gas concentrations and temperatures for CO_2, H_2O, and CH_4.
Concentrations of particles and soot.
Particle temperatures.
Percent of particles ignited.
Flame radiation intensity from individual components (particles, soot, gases).

To illustrate the measurements, Fig. 11-30 shows the transmittance, radiance, and normalized radiance spectra obtained from the combustion of a CWF at a depth of 25 cm below the nozzle.

The concentration of CO_2 can be estimated by measuring the peak height of the absorbance at 2297 cm^{-1}. The CO_2 temperature is obtained from the normalized radiance spectrum [Fig. 11-30(c)] with the correction for soot and coal particle emissions (as discussed in Section 11-3).

The concentrations of particles and soot are determined from Fig. 11-30(a). The baseline extinction represents the optical blockage of soot and CWF droplets. Since the droplets have a relatively large average size, their blockage in the (transmittance) spectrum is taken to be independent of wave number. The concentration of coal droplets is proportional to the amplitude of the horizontal line that intersects the sloping baseline at 0 cm^{-1}, as shown in Fig. 11-30(a). The amplitude of the sloping baseline above this horizontal line, at 6500 cm^{-1}, is proportional to the concentration of soot.

Particle temperatures were determined from Fig. 11-30(c) by comparison with a theoretical blackbody distribution $R_\nu^b(T)$. This curve is fit in the region where there is no interference from gas lines. While the shape of the experimental curve matches R_ν^b for a temperature of 1980 K, the amplitude of the experimental curve is too low. The explanation for this observation is that a significant fraction of coal particles were not ignited during the combustion process. The measured particle concentrations from the transmittance [Fig. 11-30(a)] contain both hot and cold coal particles. If the blockage of unignited coal particles in the transmittance spectrum is taken out, the normalized particle emission spectrum should fit the theoretical blackbody curve in

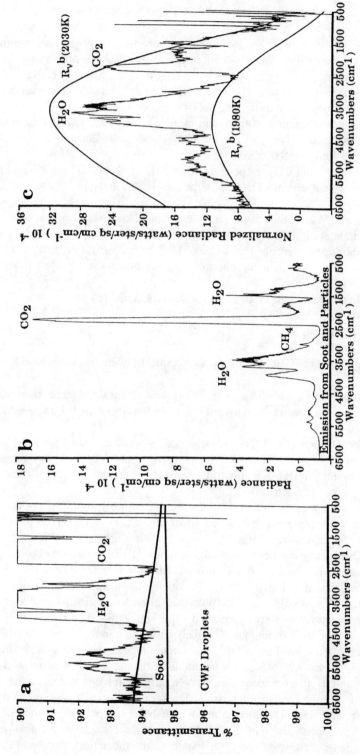

Figure 11-30 Spectra for CWF flame: (*a*) % transmittance, τ_v, (*b*) radiance R_v, and (*c*) normalized radiance R_v^n.

438

both shape and amplitude. By a trial-and-error method, we found the fraction of coal particles that was burning. This is the fraction for which the shape and amplitude of the normalized emission spectrum matched R_ν^b. The emissivity of the ignited char particles was assumed to be 0.7 for this determination.

The use of normalized radiance for particle and soot temperature determinations has considerable advantages over the use of simple radiance. In particular, when simple radiance spectra of nongraybodies (e.g., soot) are interpreted as graybody spectra, significant errors can occur in temperature determination. The soot and hot coal particle temperature, which can be obtained from the fitting of blackbody curves to the normalized radiance spectrum [Fig. 11-30(c)], will however, be correct if the soot and char particles are at the same temperature. If soot and char are at different temperatures, a more complicated fitting procedure is required, as discussed in the previous section.

Spectra such as those in Fig. 11-30 were taken throughout the CWF flame. Data similar to those for the hexane flame were obtained and summarized in Fig. 11-31. The interpretation of these data follows the same logic as described for the coal flame or hexane spray.

11-5 CONCLUSIONS

1. The FT-IR E/T technique is a versatile technique for coal combustion diagnostics, allowing measurements of particle size, concentrations, and temperatures as well as gas compositions, concentrations, and temperatures. FT-IR temperature measurements appear to be accurate within ± 50 K, and concentration measurements to within $\pm 10\%$ when compared with results of other methods.
2. Line-of-sight measurements were made of soot and of gas concentration and temperature in a coannular diffusion flame. The measurements were in good agreement with the results of other investigators.
3. Tomographic methods were successfully applied to the coannular diffusion flame to obtain spatially resolved data from the line-of-sight measurements. These spatially resolved data were also in agreement with the results of other investigators.
4. Line-of-sight measurements were made on a pulverized coal flame to investigate important processes in combustion. A comparison of the ignition of several samples suggests that the rate of ignition correlates with the initial rate of weight loss in air in a TGA experiment at lower temperatures. Comparison of the TGA weight loss in air with that in nitrogen suggests that ignition of chars is heterogeneous, ignition of high-rank coals is homogeneous, but low-rank coals exhibit both homogeneous and heterogeneous contributions to ignition. Soot formation in the pulverized coal flame correlates well with tar yield in pyrolysis, suggesting that tar is the chief precursor of soot.
5. The technique was also successfully applied to a much larger (25-cm diameter, 100 cm long) hexane and a CWF flame. Measurements were made of droplet size and species concentrations and temperatures along lines of sight through the axis of the flame.

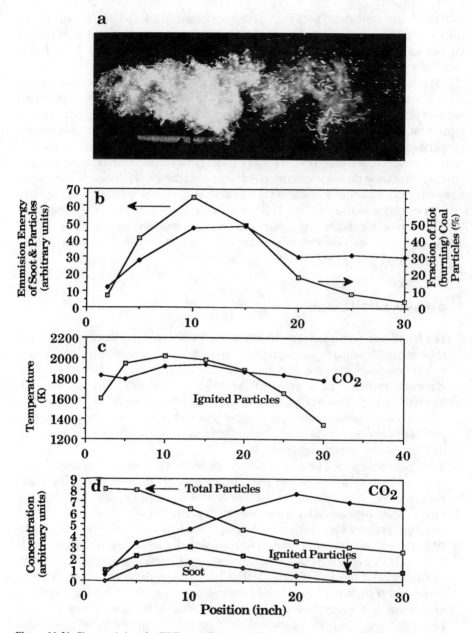

Figure 11-31 Characteristics of a CWF spray flame. (*a*) Flame photograph. (*b*) Emitted energy and fraction of ignited particles. (*c*) Temperature of CO_2 and ignited particles. (*d*) Relative concentration of CO_2, soot, ignited particles, and total particles.

11-6 NOMENCLATURE

A	geometrical cross-sectional area
b, q	parameters of Rosin-Rammler distribution
D	particle diameter
D_s	Sauter mean diameter
$\mathscr{D}$	diffusion coefficient for ethylene in nitrogen
f_v	soot volume fraction
F_v^a	absorption efficiency ($= \epsilon_v$), defined as the absorption cross section divided by the geometric cross section
F_v^t	total efficiency (extinction) for scattering out of the angular acceptance aperture of our instrument plus absorption
F_v^s	scattering efficiency for scattering out of the angular acceptance aperture of our instrument
$F_v^{s'}$	scattering efficiency for scattering wall radiation into the acceptance aperture of our instrument
I_v, I_{0v}	detected intensity
k_v	imaginary part of the index of refraction
L	optical path length through the sample stream
m_v	complex index of refraction ($= n - ik$)
n_v	real part of the index of refraction
N	particle number density
Q	fuel flow rate
R_v	sample radiance
R_v^n	normalized radiance $[= R_v/(1 - \tau_v)]$
$R_v^b(T_w)$	blackbody radiance
S_v	observed emission spectra from sample (corrected for background)
T	temperature
W_v	instrumental response function
α_v^g	absorption coefficient for gases
α_v^s	absorption coefficient for soot
ϵ_v	emittance
v	wave number
τ_v	transmittance ($= I_v/I_{0v}$)

REFERENCES

Altenkirch, R. A., Mackowski, D. W., Peck, R. E., and Tong, T. W. 1984. *Combust. Sci. Technol.* 41:327.

Bell, R. J. 1972. *Introductory Fourier transform spectroscopy.* San Diego, Calif.: Academic.

Best, P. E., Carangelo, R. M., and Solomon, P. R. 1986. *Combust. Flame* 66:47.

Best, P. E., Solomon, P. R., Serio, M. A., Suuberg, E. M., Mott, W. R., Jr., and Bassilakis, R. 1987. *ACS Div. Fuel Chem. Preprints* 32(4):138.

Best, P. E., Chien, P. L., Carangelo, R. M., and Solomon, P. R. 1990. Tomographic reconstruction of FT-IR emission and transmission spectra in a sooting laminar diffusion flame: species concentrations and temperatures. *Combust. Flame* (in press).

Bishop, M., and Ward, D. L. 1958. *Fuel* 37:191.

Boedeker, L. R., and Dobbs, G. M. 1986. *Combust. Sci. Technol.* 46:301.

Bohren, C. F., and Huffman, D. R. 1983. *Absorption and scattering of light by small particles.* New York: John Wiley.

Brewer, L. E., and Limbaugh, C. C. 1972. *Appl. Opt.* 11:1200–4.

Cashdollar, K. L., and Hertzberg, M. 1983. *Combust. Flame* 51:23.

Chien, P. L., Best, P. E., Carangelo, R. M., and Solomon, P. R. 1988. Tomographic reconstruction of Fourier transform infrared (FT-IR) spectra at points within a coannular flame. Poster session presented at 22nd Symposium (International) on Combustion, August 1988, Seattle, Washington.

Chigier, N. 1981. *Energy combustion and environment*, p. 261. New York: McGraw-Hill.

Dalzell, W. H., and Sarofim, A. F. 1969. *J. Heat Transfer* 91:100.

Elder, P., Jerrick, T., and Birkeland, J. W. 1965. *Appl. Opt.* 4(5):589.

Fletcher, T. H., Baxter, L. L., and Ottesen, D. K. 1987. *ACS Div. Fuel Chem. Preprints* 32(3):42.

Foster, P. J., and Howarth, C. R. 1968. *Carbon* 6:719.

Freeman, M. P., and Katz, S. J. 1960. *Opt. Soc. Am.* 50(8):826.

Griffiths, P. R. 1975. *Chemical infra-red Fourier transform spectroscopy.* New York: John Wiley.

Griffiths, P. R. 1978. *Transform techniques in chemistry.* New York: Plenum.

Grosshandler, W. L. 1984. *Combust. Flame* 55:59.

Gumprecht, R. O., and Sliepcevich, C. M. 1953. *J. Phys. Chem.* 57:90.

Hammond, D. C. 1981. *Appl. Opt.* 20:493.

Hardesty, D. R. 1984. An assessment of optical diagnostics for in situ measurements in high temperature coal combustion and conversion flows. Sandia National Laboratory report SAND84-8724.

Hottel, H. C., and Sarofim, A. F. 1967. *Radiative transfer.* New York: McGraw-Hill.

Hughey, B. J., and Santavicca, D. A. 1982. *Combust. Sci. Technol.* 29:167–90.

Kent, J. H., and Wagner, H. G. 1984a. *Combust. Sci. Technol.* 41:245.

Kent, J. H., and Wagner, H. G. 1984b. *Twentieth symposium (international) on combustion*, p. 1007. Pittsburgh, Pa.: The Combustion Institute.

Limbaugh, C. C. 1985. *Infrared methods for gaseous measurement: theory and practice*, ed. J. Wormhoudt. New York: Marcell Dekker.

Ludwig, C. B., Malkums, W., Reardon, J. R., and Thomson, J. A. L. 1973. *Handbook of infrared radiation from combustion gases.* NASA SP-3080, eds. R. Goulard and J. A. L. Thomson. Washington, D. C.: NASA, Scientific and Technical Information Office.

Macek, A., and Bulik, C. 1984. *Twentieth symposium (international) on combustion*, p. 1223. Pittsburgh, Pa.: The Combustion Institute.

Mackowski, D. W., Altenkirch, R. A., Peck, R. E., and Tong, T. W. 1983. *Combust. Sci. Technol.* 31:139.

Markstein, G. H., and deRis, J. 1984. *Twentieth symposium (international) on combustion*, p. 1637. Pittsburgh, Pa.: The Combustion Institute.

McLean, W. J., Hardesty, D. R., and Pohl, J. H. 1981. *Eighteenth symposium (international) on combustion*, p. 1239. Pittsburgh, Pa.: The Combustion Institute.

Midkiff, K. C., Altenkirch, R. A., and Peck, R. E. 1986. *Combust. Flame* 64:253.

Mitchell, R. E., and McLean, W. J. 1982. *Nineteenth symposium (international) on combustion*, p. 1113. Pittsburgh, Pa.: The Combustion Institute.

Nenniger, R. D., Howard, J. B., and Sarofim, A. F. 1983. *Proceedings of the 1983 international conference on coal science*, p. 521. Pittsburgh, Pa.: The Combustion Institute.

Ottesen, D. K., and Thorne, L. R. 1983. Fourier transform infrared (FT-IR) spectroscopic measurements of in situ coal combustion. In *Proceedings of the 1983 international conference on coal science*, p. 621. Pittsburgh, Pa.: The Combustion Institute.

Ottesen, D. K., and Stephenson, D. A. 1982. *Combust. Flame* 46:95.

Pearce, W. J. 1958. *Conference on extremely high temperatures*, pp. 123–35. New York: John Wiley.

Penner, S. S., Wang, C. P., and Bahadori, M. Y. 1984a. *Progr. Energy Combust. Sci.* 10:209.

Penner, S. S., Wang, C. P., and Bahadori, M. Y. 1984b. *Twentieth symposium (international) on combustion*, p. 1149. Pittsburgh, Pa.: The Combustion Institute.

Perelman, A. Y., and Shifrin, K. S. 1980. *Appl. Opt.* 19:1787.
Ray, S., and Semerjian, H. G. 1983. Laser tomography for simultaneous concentration and temperature measurements in reacting flows. Paper read at AIAA 18th Thermophysics Conference, June 1–3, 1983, Montreal, Canada.
Reisser, B. Starsinic, M., Squires, E., Davis, A., and Painter, P. C. 1984. *Fuel* 63:1253.
Santoro, R. J., and Semerjian, H. G. 1984. *Twentieth symposium (international) on combustion*, p. 997. Pittsburgh, Pa.: The Combustion Institute.
Santoro, R. J., Semerjian, H. G., and Dobbins, R. A. 1983. *Combust. Flame* 51:203.
Santoro, R. J., Yeh, T. T., and Semerjian, H. G. 1985. The transport and growth of soot particles in laminary diffusion flames. Paper read at the 23rd ASME/AIChE National Heat Transfer Conference. [Revised manuscript in *Combust. Sci. Technol.* 53:89 (1987).]
Santoro, R. J., Semerjian, H. G., Emmerman, P. J., and Goulard, R. 1981. *Int. J. Heat Mass Transfer* 24:1139.
Seeker, W. R., Samuelsen, G. S., Heap, M. P., and Trolinger, J. D. 1981. *Eighteenth symposium (international) on combustion*, p. 1213. Pittsburgh, Pa.: The Combustion Institute.
Semerjian, H. G., Santoro, R. J., Goulard, R., and Emmerman, P. J. 1981. *Fluid mechanics of combustion systems*, eds. T. Morel, R. P. Lohmann, and J. M. Rackley, p. 119. New York: The American Society of Mechanical Engineers.
Serio, M. A., Hamblen, D. G., Markham, J. R., and Solomon, P. R. 1987. *Energy Fuel* 1:138.
Shepp, L. A., and Logan, B. F. 1974. *IEEE Trans. Nucl. Sci.* NS-21:228–236.
Siegel, R., and Howell, J. R. 1972. *Thermal radiation heat transfer*, section 20-5. New York: McGraw-Hill.
Solomon, P. R., and Carangelo, R. M. 1982. *Fuel* 61:663.
Solomon, P. R., and Carangelo, R. M. 1988. *Fuel* 67:949.
Solomon, P. R., and Hamblen, D. G. 1985. *Chemistry of coal conversion*, ed. R. H. Schlosberg, p. 121. New York: Plenum.
Solomon, P. R., Hamblen, D. G., and Carangelo, R. M. 1982. In *Fuel science*. ACS Symposium Series 205, vol. 4, p. 77. Washington, D. C.: American Chemical Society.
Solomon, P. R., Hamblen, D. G., and Carangelo, R. M. 1984. *Analytical pyrolysis*, ed. K. J. Voorhees, p. 121. London: Butterworths.
Solomon, P. R., Serio, M. A., and Heninger, S. G. 1986. *ACS Div. Fuel Chem. Preprints* 31(2):200.
Solomon, P. R., Hamblen, D. G., Carangelo, R. M., and Krause, J. L. 1982. *Nineteenth symposium (international) on combustion*, p. 1139. Pittsburgh, Pa.: The Combustion Institute.
Solomon, P. R., Serio, M. A., Carangelo, R. M., and Markham, J. R. 1986a. *Fuel* 65:182.
Solomon, P. R., Carangelo, R. M., Hamblen, D. G., and Best, P. E. 1986b. *Appl. Spectrosc.* 40:746.
Solomon, P. R., Hamblen, D. G., Markham, J. R., and Carangelo, R. M., 1986c. Novel diagnostics for particulate properties in combustion. U. S. DOE final report contract DE-AC01-85ER80313.
Solomon, P. R., Carangelo, R. M., Best, P. E., Markham, J. R., and Hamblen, D. G. 1987a. *Twenty-first symposium (international) on combustion*, p. 437. Pittsburgh, Pa.: The Combustion Institute.
Solomon, P. R., Best, P. E., Carangelo, R. M., Markham, J. R., Chien, P. L., Santoro, R. J., and Semerjian, H. G. 1987b. *Twenty-first symposium (international) on combustion*, p. 1763. Pittsburgh, Pa.: The Combustion Institute.
Solomon, P. R., Carangelo, R. M., Best, P. E., Markham, J. R., and Hamblen, D. G. 1987c. *Fuel* 66:897.
Solomon, P. R., Carangelo, R. M., Best, P. E., Markham, J. R., Hamblen, D. G., and Chien, P. L. 1987d. *Fundamentals of physical-chemistry of pulverized coal combustion*. NATO ASI Series 137, eds. J. Lahaye and G. Prado, p. 347. Netherlands: Matrinus Nijhoff Publishers.
Solomon, P. R., Chien, P. L., Carangelo, R. M., Best, P. E., and Markham, J. R. 1987e. Application of FT-IR E/T spectroscopy to study coal combustion phenomena, paper read at the International Conference on Coal Science, September 7–10, 1987, Beijing, China.
Solomon, P. R., Chien, P. L., Carangelo, R. M., Best, P. E., and Markham, J. R. 1988. *Twenty-second symposium (international) on combustion*, p. 211. Pittsburgh, Pa.: The Combustion Institute.
Tichenor, D. A., Mitchell, R. E., Hencken, K. R., and Niksa, S. 1984. *Twentieth symposium (international) on combustion*, p. 1213. Pittsburgh, Pa.: The Combustion Institute.

Timothy, L. D., Sarofim, A. F., and Beer, J. M. 1982. *Nineteenth symposium (international) on combustion*, p. 1123. Pittsburgh, Pa.: The Combustion Institute.

Timothy, L. D., Froelich, D., Sarofim, A. F., and Beer, J. M. 1987. *Twenty-first symposium (international) on combustion*, p. 1141. Pittsburgh, Pa.: The Combustion Institute.

Tourin, R. H. 1962. *Temperature, its measurement and control in science and industry*, vol. 3, part II, chapter 43, ed. C. M. Herzfeld. New York: Reinhold.

Tourin, R. H. 1966. *Spectroscopic gas temperature measurement*. New York: Elsevier.

Vervisch, P., and Coppalle, A. 1983. *Combust. Flame* 52:127.

Vorres, K. S. 1987. *ACS Div. Fuel Chem. Preprints* 32(4):221.

Waters, B. J., Mitchell, R. E., Squires, R. G., and Laurendeau, N. M. 1986. *ACS Div. Fuel Chem. Preprints* 31(1):12.

Wornat, M. J., Sarofim, A. F., and Longwell, J. P. 1987. *Energy Fuels* 1(5):431.

PARTICLE MEASUREMENTS IN
COAL CHAR COMBUSTION

Ranajit Sahu and Richard C. Flagan

12-1 INTRODUCTION

After being introduced into a furnace, coal devolatilizes as it heats, changing both the physical and chemical structure of the carbonaceous matrix. The volatile matter released during this process burns much as do gaseous or vaporized liquid fuels. The carbonaceous residue burns through surface reactions, both on its exterior surface, and within the pores distributed throughout its volume. The rate at which char oxidation proceeds depends both on surface oxidation kinetics and on transport resistances to diffusion of oxidizer into the pores and reaction products from the pores. To understand the combustion of coal, experimental measurements of the physical and chemical properties of coal and the char are essential.

Coal is burned in a wide variety of combustion systems and at a range of temperatures and particle sizes. The chemical and physical transformations experienced by coal are strongly influenced by these environmental conditions. At temperatures below about 1500 K, the char oxidation reactions are slow enough to control the rate of reaction in small (less than 50 μm) particles. Low-temperature combustion is usually carried out in systems that require large fuel particles, i.e., fixed bed (stoker) combustors or fluidized bed combustion systems, so the diffusional resistances are still important in practical low-temperature combustion systems. Fixed and fluidized bed

This work was supported in part by U.S. Department of Energy grant DE-AC22-86PC90751. The authors also wish to thank R. E. Mitchell of Sandia National Laboratories for helpful comments and suggestions.

combustors allow the long residence times required for complete combustion of large fuel particles. The low temperatures are necessary in these systems to minimize the problems of slag and clinker formation from the mineral matter in the coal. High combustion temperatures are usually achieved only in entrained flow combustion systems (pulverized coal). Only particles smaller than about 200 μm can be burned completely within the residence times (few seconds at most) that can be achieved within combustion chambers of practical size.

Coal combustion necessarily involves an inhomogeneous reaction environment, since the burning particles act as localized sinks for reactants and sources of combustion products and intermediates. Studies of combustion kinetics are frequently conducted in dilute systems in order to minimize the interactions between burning char particles, thereby providing a well-known reaction environment. Still, models of the combustion process are needed to infer fundamental reaction mechanisms and rates from such experiments. In any practical combustion system, a range of particle sizes is usually burned. This size variation influences the mass transfer rates and complicates most attempts at fundamental reaction kinetics measurements. This difficulty can be overcome in the laboratory but only through very extreme measures.

In addition to concern over combustion rates and gaseous pollutant formation, the fate of mineral matter in coal combustion is critically important to the development of reliable combustion systems and to the maintenance of environmental quality. Hence another major concern in coal combustion research is the behavior of coal minerals and ash.

This chapter will focus on measurements of those aspects that are unique to coal combustion, namely, the gas-solid reactions and the mineral transformations.

Combustion Rate Measurements

Coal combustion is a complex process. Upon entering a combustor, the solid fuel is heated and undergoes a number of transformations. Coal frequently contains a large quantity of water, which vaporizes early in the heating process. At higher temperatures, the coal devolatilizes, releasing a spectrum of hydrocarbons ranging from low molecular weight vapors to heavy tars. Substantial pressure can accumulate within the coal before the volatiles are vented to the surrounding atmosphere, leading to violent jetting and possible rupturing of the particle (McLean, Hardesty, and Pohl 1981). During devolatilization, some coals become plastic, swelling to many times their original volume and forming hollow particles known as cenospheres (Fig. 12-1). Other coals remain solid and undergo little change in shape or structure during devolatilization. A carbonaceous residue is left after the volatiles are released.

Combustion of the coal involves two classes of oxidation processes: the gas-phase reactions of the volatile matter and the surface oxidation of the char. The volatile matter oxidation is similar to the combustion of the mixtures of hydrocarbon vapors found in other combustion systems, but the char oxidation introduces many new features to the combustion process. Oxygen must diffuse from the surrounding atmosphere to the surface of the char before reaction can take place. The char has a porous microstructure with far more surface area contained in the pores than on the particle exterior.

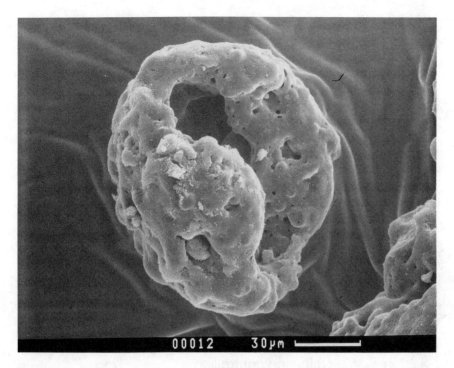

Figure 12-1 Electron micrograph of a char particle produced by pyrolysis of a bituminous coal for 2 s at 1600 K.

Much of the reaction occurs on the internal surface area. The rate of char oxidation is determined by the combined effects of the exterior and interior diffusional resistances, and the intrinsic chemical surface oxidation kinetics. Due to these resistances, the char burns relatively slowly, so long residence times are required to assure complete combustion of coal.

Char oxidation can be separated into four regimes as illustrated in Fig. 12-2 (Mulcahy and Smith 1969). In regime I, chemical reaction rates are slow, and an oxidant molecule that diffuses into the pore structure is likely to diffuse out before reacting. The oxidant concentration in the pores is effectively equal to that in the surrounding atmosphere, and all of the surface area in the char contributes equally to the reaction.

In regime II the reaction rate has become too great for diffusion to maintain the oxidant concentration within the porous microstructure. The oxidant concentration decreases to a minimum at the center of the char particle. The rate of reaction in regime II is controlled by the combined effects of surface oxidation kinetics and pore diffusion. As the rate increases further, diffusion in the region around the particle also limits reaction (regime III). Ultimately, oxygen can only penetrate a thin layer near the particle surface. Still higher rates lead to reaction control by diffusion in the boundary layer (regime IV).

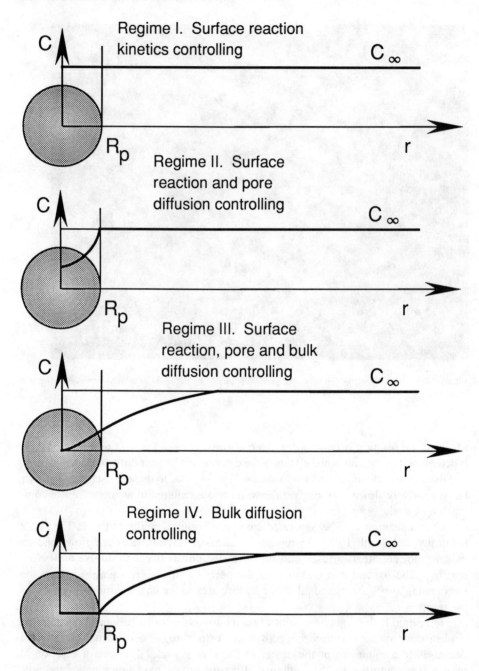

Figure 12-2 Regimes of char particle combustion.

Temperature, particle reactivity, particle size, and oxidizer concentration, all determine which regime applies. Fluidized bed or fixed bed combustion involves combustion of large particles at relatively low temperatures, so regimes I or II apply. On the other hand, pulverized coal combustion uses smaller particles and leads to much higher temperatures and correspondingly faster surface reaction rates, so regimes III or IV apply. Clearly, since the porous microstructure plays a critical role in combustion regimes I–III, knowledge of that structure is essential to the understanding of the combustion process.

In this section we shall briefly review the physical and chemical factors that control the rate at which coal particles burn. The literature on coal combustion is vast. Specific studies are cited primarily for the insights they provide with what measurements need to be made to elucidate the combustion characteristics of a particular coal. Excellent reviews in the general area of pulverized coal combustion are available in the literature. Field et al. (1967) provide a comprehensive description of pulverized coal combustion, including fluid mechanical, thermal, and chemical kinetic effects. They give a succinct review of various topics prior to the mid 1960s. A few years later, Mulcahy and Smith (1969) published a review of the chemical kinetic aspects of pulverized fuel combustion. Laurendeau (1978) describes in great detail much of the progress made in the 1970s in the area of heterogeneous kinetics of coal char combustion. His review gives a comprehensive discussion of the mechanisms and rates of the relevant gas-solid reactions. Mass transfer and diffusion are also examined. Smith (1982) summarizes the field of char kinetics, including reactivity data for a spectrum of carbons over a wide range of temperatures. He proposed a universal intrinsic chemical rate expression for chars. Recently, Smoot and Smith (1985) published a valuable addition to the field of coal combustion. Most of these papers deal with the solid phase in greater detail than the gas phase combustion. Libby and Blake (1979) focused on the reactions in the atmosphere surrounding the burning char particle.

Combustion is often described in terms of the apparent or external rate, i.e., the rate of carbon removal per unit particle external area. Calculation of the apparent rate involves measurements of the particle size, burnout time, and particle temperature. Ideally, particle temperature should be known as a function of its burn time. The apparent rate is a useful measure of reactivity for all regimes of combustion. It does not matter whether there is internal reaction or not. It includes the combined effects of pore diffusion and that of intrinsic chemical reactivity. In many cases, oxidant penetration into the particle is significant, so the effects of internal reaction cannot be neglected. Reaction within the porous microstructure leads to changes in the particle density. The internal morphology of the char can be characterized in terms of the pore volume and pore surface area distributions, and an intrinsic reaction rate. The intrinsic reaction rate is the rate of carbon loss per unit of total particle surface area. Due to the extremely complex pore morphology, diffusion limitations, and energetically non-uniform surface properties that are characteristic of typical coals and chars, the calculation of the intrinsic rate is a challenging task. Experimental measurements of particle size, temperature (as a function of burn time), conversion, pore volume, and pore size distributions are needed. Models capable of predicting particle temperature and conversion as a function of time must also be developed. Validation of a proposed reaction rate or pore diffusion model depends on reliable experimental measurements. There is mounting evidence that some of the species present in mineral matter catalyze carbon oxidation, especially at low temperatures (less than 1500 K). Ash can also physically block pore mouths and prevent access of oxygen for reaction. Neither the catalytic reactions nor physical blocking mechanisms are well enough understood to warrant their inclusion in models at the present time.

In a major study Jenkins, Nandi, and Walker (1973) investigated the role of coal rank, pyrolysis temperature, and mineral matter on the low-temperature reactivity of chars from 21 U.S. coals. Thermogravimetric analysis of the mass loss of chars as they are heated in an oxidizing atmosphere was used to measure the reactivity. After heating a small sample of char to a preset temperature in nitrogen, dry air was admitted, and the mass of the char was continuously recorded. For coals pyrolyzed in N_2 at 600–1000°C, they found that while reactivity increases with the decreasing coal rank, possibly due to the catalytic effects of mineral matter, it also decreases with increasing pyrolysis temperature. They correlated reactivity with the calcium and magnesium contents in the char. The reactivity was not well correlated with other known catalysts like iron, so they speculated that the form and distribution of mineral matter have an important influence on the reactivity.

The role of the internal structure of the chars in determining their reactivities was explored by Dutta and Wen (1977). They found that, in the chemical kinetic regime, char reactivity to oxygen depends more on the degree of gasification and the pore characteristics of the char than it does on the parent coal. However, the char-carbon dioxide reactivity was found to depend more on the characteristics of the parent coal seam than on the char internal structure.

Mahajan, Yarzab, and Walker (1978) proposed a scheme for unifying char gasification reactions with various gases. Observing that the conversion versus time plots for a particular char with different gases (air, water vapor, hydrogen, carbon dioxide, etc.) had similar shapes, they collapsed the various plots into one by normalizing time with that needed to reach 50% conversion, $\tau_{0.5}$. A gradual increase in the oxidation rate was followed by a period of constant oxidation rate and, finally, a long period of slowly decreasing rate as the conversion asymptotically approached unity (Fig. 12-3). While not offering quantitative reasons for such self-similarity, Mahajan, Yarzab, and Walker (1978) identified three major factors governing the shapes of the burnoff curves: (1) the presence of active sites where reaction first occurs, (2) diffusional limitations of the micropores, where many of the active sites responsible for reaction are present, and (3) the amount, form, and degree of dispersion of the mineral matter. More recently, Morgan, Robertson, and Unsworth (1987) showed that low-temperature reactivity depends on the maceral content of the coal. Vitrinites are more reactive than inertinites. They also found that the maceral distribution is a function of coal particle size for conventional grinding systems. Hence char size influences the oxidation rate through both the intrinsic chemical reactivity and the diffusional resistances. Table 12-1 summarizes some of the many low-temperature reactivity studies that have been made for bituminous chars.

Char particle burning rates are generally determined by measurements of the mass loss as a function of time. Thus

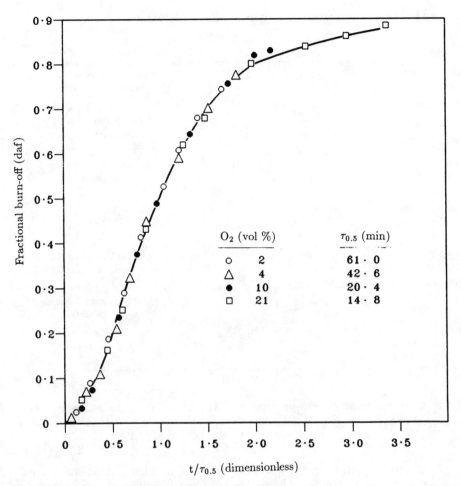

Figure 12-3 Conversion versus dimensionless time in low-temperature oxidation studies (*Mahajan, Yarzab, and Walker 1978*).

$$R = \frac{1}{m}\frac{dm}{dt} \tag{12-1}$$

where R is the rate of mass loss based on instantaneous carbon mass, m. In some cases, m is taken to be the mass of the unburned material, including mineral matter, ash, etc. (Smith and Tyler 1974).

Assuming that the char particles are spheres of uniform diameter D_p and of uniform apparent density ρ_a, the mass rate given in Eq. (12-1) can be converted to an apparent reaction rate per unit of external surface area R_a, given by

$$R_a = \frac{\rho_a D_p}{6} R \tag{12-2}$$

Table 12-1 Low temperature reactivity studies on bituminous chars

Reference	Method	Temperature	Percent O_2	Pressure, atm	Char type
Jenkins, Nandi, and Walker (1973)	TGA	500°C	21	1	from 21 U.S. coals pyrolyzed in N_2 at 600–1000°C
Dutta and Wen (1977)	TGA	424–576°C	21	1	from two U.S. bituminous coals pyrolyzed in N_2 at 1024°C and four process chars
Mahajan, Yarzab, and Walker (1978)	TGA	405°C	21	1	from 16 U.S. coals pyrolyzed in N_2 at 1000°C
Radovic, Walker, and Jenkins (1983)	TGA	550–750 K	21	1	process chars
Tseng and Edgar (1985)	TGA	425–900°C	0.5–100	1	from one U.S. bituminous coal pyrolyzed in N_2 at 1000°C
Knill et al. (1986)	TGA	<500°C	0–100	1	from hydropyrolysis of subbituminous Canadian coals at 600–800°C
Morgan, Robertson, and Unsworth (1986)	TGA	450–650°C	21	1	from hydropyrolysis of a British coal at 1400°C

The apparent reactivity R_a can be expressed in terms of an apparent reaction rate coefficient k_a as

$$R_a = k_a C_s^n \qquad (12\text{-}3)$$

where C_s is the oxygen concentration at the surface of the char particle and n is the apparent reaction order. Similar to the definitions of the apparent reactivity and apparent reaction rate coefficient, the intrinsic reactivity R_i and the intrinsic reaction rate coefficient k_i can be defined as

$$R_i = k_i C_s^m \qquad (12\text{-}4)$$

where m is the true reaction order and R_i is defined as the reaction rate per unit of total accessible surface area. Most studies in the past have assumed that the total area is the same as that measured by gas adsorption measurements, either before combustion or as a function of carbon conversion during combustion. However, recent studies (Sy and Calo 1983; Radovic, Walker, and Jenkins 1983; Wojtowicz, Calo, and Suuberg 1985; Jenkins and Piotrowski 1987; Calo and Perkins 1987) have shown that not all of the accessible surface area, A_T, participates in the reaction. Only a fraction of the surface area designated as the active surface area contributes to the reactivity

$$A_E = f A_T \qquad (12\text{-}5)$$

where A_E and A_T are the active and total surface areas, respectively. It is important to remember that both A_T and f are functions of carbon conversion. Besides, f and therefore A_E will also depend on the reactant gas composition. Generally, however, in the past, intrinsic reaction rate coefficients have been deduced either by assuming A_T constant or at best treating it as a function of conversion but still assuming that $f = 1$. As a further complication, there are differences in the values of A_T obtained by various methods (using different adsorptive gases, for example). Of course, A_E can be directly measured without measuring A_T, but similar questions remain about its absolute value.

The reaction rate coefficients are generally assumed to have an Arrhenius dependence on particle temperature T_p, i.e.,

$$k_1 = A_1 e^{-E_1/RT_p} \qquad l = a, i \qquad (12\text{-}6)$$

where the A are the frequency factors and the E are the activation energies.

Char reactivity studies seek to determine the quantities A_i, E_i, and m from measurements of R and T_p as a function of time. A simple method based on the approach by Smith (1982) will be outlined here to illustrate the data needed for such estimation. Smith's method calculates the intrinsic reaction rate from measurements of the apparent rate, the surface area, and pore properties. It assumes a spherical, isothermal particle. Pore diffusion is considered via an effective diffusivity (Eqs. (12-16) and (12-17) below), and the effectiveness of the particle is related to an appropriate Thiele modulus, these concepts being extended from heterogeneous catalysis.

The quantity R_a can be expressed as

$$R_a = h_m(C_g - C_s) = k_a C_s^n \qquad (12\text{-}7)$$

where C_g is the oxygen concentration in the bulk gas and h_m is a mass transfer coefficient, which for pulverized fuel combustion is given by

$$h_m = \frac{3}{4} \frac{\mathcal{D}}{D_p} \qquad (12\text{-}8)$$

where $\mathcal{D}$ is the bulk gas diffusion coefficient at the mean temperature of the particle boundary layer. Equation (12-8) assumes that there is no velocity slip between the gas and the particle and that the primary product of reaction is carbon monoxide.

Since C_s is generally unknown, it is frequently eliminated from Eq. (12-7) by writing

$$R_a = k_a(1 - \chi)^n C_g^n \qquad (12\text{-}9)$$

where

$$\chi = \frac{R_a}{R_m} \qquad (12\text{-}10)$$

and

$$R_m = h_m C_g \qquad (12\text{-}11)$$

R_m is the maximum possible reaction rate at T_p under diffusion-limited conditions when the surface oxygen concentration becomes zero. The apparent and intrinsic rates can be related by

$$R_a = \eta \rho_a \frac{D_p}{6} a_T k_i C_s^m \qquad (12\text{-}12)$$

where η is an effectiveness factor, the ratio of the actual combustion rate to that attainable if there was no pore diffusion resistance, and a_T is the total specific surface area (per gram of carbon). In terms of the bulk concentration,

$$R_a = \eta \rho_a \frac{D_p}{6} a_T k_i [C_g (1 - \chi)]^m \qquad (12\text{-}13)$$

The effectiveness factor η can be calculated by computing the quantity

$$\Phi = \frac{D_p R_a (m + 1)}{48 \mathcal{D}_e C_g (1 - \chi)} \qquad (12\text{-}14)$$

and using relationships between η and Φ given by Mehta and Aris (1971). Here, Φ is also equal to $\eta \phi^2 (m + 1)/2$, where ϕ, the Thiele modulus, is given by

$$\phi = \frac{D_p}{6} \left(\frac{a_T \rho_a k_i C_s^{m-1}}{\mathcal{D}_e} \right)^{0.5} \qquad (12\text{-}15)$$

The effective pore diffusion coefficient, $\mathcal{D}_e$, in Eqs. (12-14) and (12-15), is calculated by (Satterfield 1970)

$$\mathcal{D}_e = \frac{\mathcal{D}_{pore} \epsilon}{\tau} \qquad (12\text{-}16)$$

where $\mathcal{D}_{pore}$ is the pore diffusion coefficient for diffusion in cylindrical pores, ϵ is the porosity of the solid, and τ is a tortuosity factor that can be as low as 1.5 or as high as 10 but is generally taken as 2. The pore diffusion coefficient depends on the size of the pores. For large pores (greater than 1 μm in radius) it is the same as the molecular bulk diffusion coefficient. For smaller pores, where Knudsen diffusion occurs,

$$\mathcal{D}_{pore} = 9.7 \times 10^3 \bar{r}_p \left(\frac{T_p}{M} \right)^{0.5} \quad cm^2/s \qquad (12\text{-}17)$$

where M is the molecular weight of the diffusing species and $\bar{r}_p$ is the mean pore radius given by

$$\bar{r}_p = \frac{2 \epsilon \tau^{0.5}}{a_T \rho_a} \quad cm \qquad (12\text{-}18)$$

Another method to determine the apparent reactivity of coals and chars is by measuring the particle temperatures of single particles as they burn (Mitchell and

McLean 1982; Timothy, Sarofim, and Beer 1982; Mitchell and Madsen 1986) and then inferring the reactivity from an energy balance

$$\frac{1}{6}D_p\rho_a c_p \frac{dT_p}{dt} = R_a\Delta H - \alpha(T_p - T_\infty) - \sigma\epsilon_p(T_p^4 - T_\infty^4) \qquad (12\text{-}19)$$

where c_p and ϵ_p are the particle specific heat and emissivity, respectively, α is the convective heat transfer coefficient, σ is the Stefan-Boltzmann constant, T_∞ is the temperature far from the particle, and ΔH is the heat of reaction.

Smith (1982) estimated the intrinsic rates of a variety of coal chars and carbons, as shown in Fig. 12-4; the rates varied by about 2 orders of magnitude even at temperatures as high as 1500 K. This variability can be attributed to differences in the chemical structure, the pore structure, and the presence of catalyzing minerals. While the differences narrow considerably when one focuses on a single material, the remaining spread suggests that a more detailed pore model accounting for the distribution of pore sizes and pore growth is needed.

Significant progress in accounting for the effects of pore size distribution and pore growth in particle combustion was achieved with the introduction of the random pore models by Bhatia and Perlmutter (1980) and Gavalas (1980) and the pore-tree model of Simons and Finson (1979). Several recent studies (Gavalas 1981; Lowenberg, Bellan, and Gavalas 1987; Sahu 1988) have used random pore models to include effects of pore growth in particle combustion. Difficulties remain, however, in describing pore connectivity and the effects of large pores, which raise questions about the suitability of the whole class of models that are based on continuum descriptions of transport. These questions are beginning to be addressed by discrete simulations and percolation theory (Sandmann and Zygourakis 1986; Sahimi and Tsotsis 1987; Kerstein and Edwards 1987; Sahu, Flagan, and Gavalas 1989). Thus accurate and careful measurements of pore structure, pore growth, and the effects of mineral matter are essential for model validation.

Table 12-2 summarizes some of the high-temperature reactivity studies done on bituminous chars, while Figs. 12-5 and 12-6 show examples of partially oxidized cenospheric bituminous chars. While the influence of char type on the reactivity decreases with increasing temperature, Smith (1982) found that the intrinsic reactivities of different chars at some temperatures varied by as much as 4 orders of magnitude. His attempt to unify the known reactivities of various carbons over a wide range of temperatures met with limited success. This may be due to the intrinsically different nature of the carbons and their reactivity with oxygen, or due to the different pore structures of the carbons, or both. The chemical effects of ash were not investigated. The measurement of reaction order is also open to question. Mechanisms that have been theoretically proposed for the carbon-oxygen reaction (Nagle and Strickland-Constable 1962; Essenhigh 1981) have not been entirely validated (Laurendeau 1978; Tseng and Edgar 1985). There is, however, increasing evidence that the complex variation of intrinsic reaction rate with temperature observed by Nagle and Strickland-Constable in pyrolytic graphite oxidation also occurs for carbons from widely different

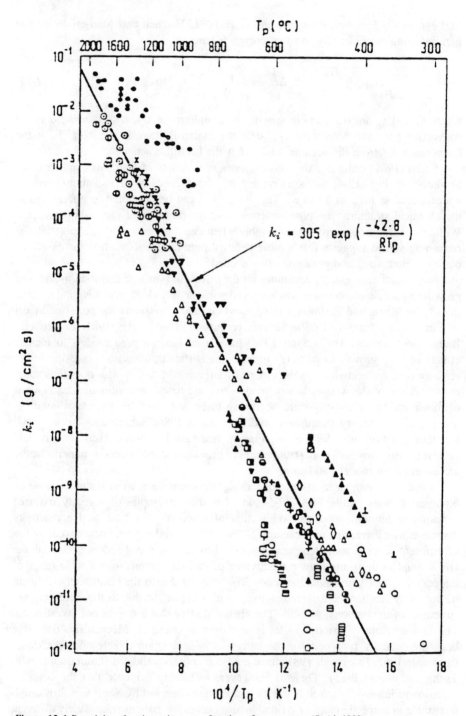

Figure 12-4 Reactivity of various chars as a function of temperature (*Smith 1982*).

Table 12-2 High-temperature reactivity studies on bituminous chars

Reference	Method	Temperature	Percent O_2	Pressure, atm	Char type
Field (1969)	drop tube	1200–1720°C	5, 10	1	low-rank U.K. coal (N_2 at 1600 K)
Smith (1971*a*, *b*)	drop tube	1200–1900 K	21	1	N. Z. bituminous coal (air at 1500 K)
Hamor, Smith, and Tyler (1973)	drop tube	900–2200 K	10, 20	1	Australia brown coal (air at 1600 K)
Smith and Tyler (1974)	drop tube	630–1812 K	10, 20	1	Australia brown coal (air at 1600 K)
Dutta and Wen (1977)	TGA	834–1106°C	0.2–2	1	U.S. bituminous coals (N_2 at 1024°C)
Smith (1978)	see Smith (1978)	see Smith (1978)	see Smith (1978)	1	different porous chars
Young and Smith (1981)	drop tube	1000–1800 K	5–30	1	petroleum coke
Lester, Seeker, and Merklin (1981)	shock tube	1700–2200 K	10–50	5.5–10	U.S. bituminous coals
Wells et al. (1984)	drop tube	1300–1700 K	21	1	process chars
Knill et al. (1986)	drop tube	1100°C	21	1	Canada bituminous coals (600–800°C)
Mitchell (1987)	drop tube	1300–1800 K	0–30	1	bituminous coals/ chars

sources, e.g., carbon black (Park and Appleton 1973) and glassy carbons (Levendis and Flagan 1989).

From this discussion, it should be apparent that experimental studies of the combustion of coal requires detailed measurements of the chemical and physical properties in addition to measurements of the combustion process. Ideally, such properties as the pore size distribution and internal surface area would be measured as a function of time in the combustion process. This requires the development of methods for extraction of partially reacted char particles and quenching of the reactions, so that the properties of the char in the reactor can be determined.

Mineral Matter Transformation and Ash Formation

Coal contains a spectrum of minerals that varies widely from one source to another. The minerals in coal have long been considered important because of their direct

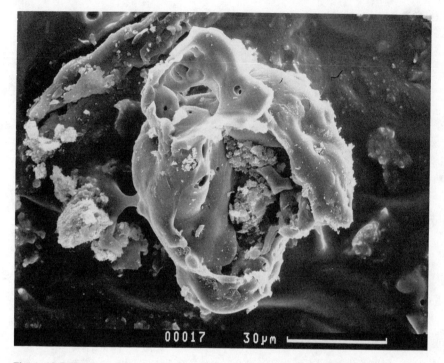

Figure 12-5 Electron micrograph of a bituminous char after 17.4% oxidation.

impacts on combustion system reliability and efficiency (Raask 1985). Particles depositing on heat transfer surfaces reduce heat transfer efficiency and, through chemical reactions between the ash and the tube material, corrode the boiler tubes. The early research into coal ash properties that was motivated by these issues was naturally focused on the particles that account for the bulk of the deposited mass and corrosion.

The mechanism that dominates mass deposition on surfaces is inertial impaction. The efficiency with which particles of diameter D_p will deposit on a collector with a characteristic dimension of L by inertial impaction is a function of the Stokes number,

$$St = \frac{\rho_a U D_p^2}{18 \mu L} \tag{12-20}$$

where ρ_a is the particle density, U is the characteristic fluid velocity, and μ is the fluid viscosity. The Stokes number is essentially the ratio of the distance a particle moving with an initial velocity U would penetrate into a quiescent fluid, i.e., a stopping distance, to the characteristic dimension of the flow. The larger the Stokes number, the more it will deviate from the fluid streamlines when the fluid is deflected by an obstacle, and the more likely it is to reach the surface. Because the Stokes number is proportional to D_p^2, large particles are the most likely to deposit by inertial impaction.

Concern over atmospheric pollution in the early 1970s led to the realization that coal combustion is a major source of heavy metals in the atmosphere, and that those

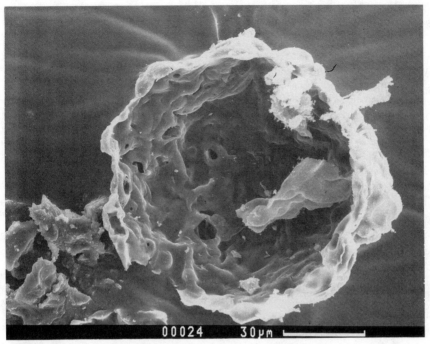

Figure 12-6 Electron micrograph of a bituminous char after 17.4% oxidation.

trace constituents of the coal ash tend to concentrate in the finer particles (Davison et al. 1974). Since small particles are removed from the flue gases less efficiently than are larger particles, and further since they can penetrate deep into the respiratory tract when inhaled, the resulting enrichment of the atmospheric aerosol with toxic trace species catalyzed a large number of studies of the distribution of composition of coal ash with particle size (Davison et al. 1974), and of the mechanisms of ash particle formation and evolution, with emphasis on the smaller particles, particularly those smaller than 1 μm in diameter.

The particulate matter produced when coal is burned thus extends over a broad spectrum of particle sizes, and the composition and other properties of those particles can vary dramatically with particle size. It is generally not sufficient to know the number concentration, particulate mass loading, or mean composition of this aerosol in order to understand its impacts on combustion systems or the environment. Rather, the particle size distribution and distribution of composition with particle size must be determined. The particle size distribution can be expressed in terms of number such that the number of particles in the size interval between D_p and $D_p + dD_p$ is

$$dN = n(D_p)dD_p \qquad (12\text{-}21)$$

or mass

$$dM = m(D_p)dD_p \qquad (12\text{-}22)$$

The variation of composition with particle size can be similarly described, i.e., the mass distribution of species i is

$$dM_i = m_i(D_p)dD_p \tag{12-23}$$

Many early studies of coal ash and other particles presented data in terms of cumulative mass distributions, i.e.,

$$M(D_p) = \int_{D_p=0}^{D_p} m_i(D_p')dD_p' \tag{12-24}$$

This form presupposes that the measurement methods employed detect all smaller particles, a questionable assumption that can lead to erroneous conclusions regarding the importance of particles in a given size range. An intuitive presentation of the size distribution is to plot the differential form on a linear scale, i.e., dM/dD_p versus D_p. Since one is generally concerned with populations that vary over several decades in particle diameter, it is more convenient to plot $dM/d \log D_p$ versus $\log D_p$. When plotted in this form, the size distribution can be visually integrated and gives an undistorted representation of the size regimes that dominate the distribution without biasing the interpretation because of missing or incomplete data.

Flagan and Friedlander (1978) predicted that volatilized coal ash would nucleate and form large numbers of fly ash particles in the submicron size regime. Their calculated number concentrations were orders of magnitude higher than had been observed at that time, and the distribution of particles with respect to size was far narrower than could be measured with the instrumentation that had been used in combustion system measurements up to that time, as illustrated in Fig. 12-7. The predicted particle size at the peak in the number concentration in pulverized coal-fired utility boilers was as small as 0.07 μm. Subsequent field observations reproduced the form of the predicted size distribution, but not the absolute number concentration and mean particle size (G. R. Markowski, personal communication, 1980). Laboratory studies have, however, validated the theoretical predictions (Taylor and Flagan 1981). The reasons for the discrepencies in the field measurements will be addressed later in this paper.

The importance of these fine particles is illustrated by measurements of the submicron particle size distribution made upstream and downstream of an electrostatic precipitator on a coal-fired power plant (Markowski et al. 1980). The submicron aerosol accounted for only 1% of the aerosol measured at the boiler outlet, but 20% of the mass at the outlet to the electrostatic precipitator. This increase in relative importance results from the size-dependent collection efficiency of the electrostatic precipitator.

The technology is considerably less well developed for measurement of particulate properties in combustion effluents than for measurement of gaseous combustion products. Measurement of particles within the combustion zone is far more difficult, and no completely satisfactory methods exist at present. The reasons for these difficulties are tied to the dynamics of the particles and the complex chemical and physical structures of coal ash particles.

Minerals are distributed throughout the volume of the coal as discrete inclusions, but some of the elements of concern here are also found chemically bound into the

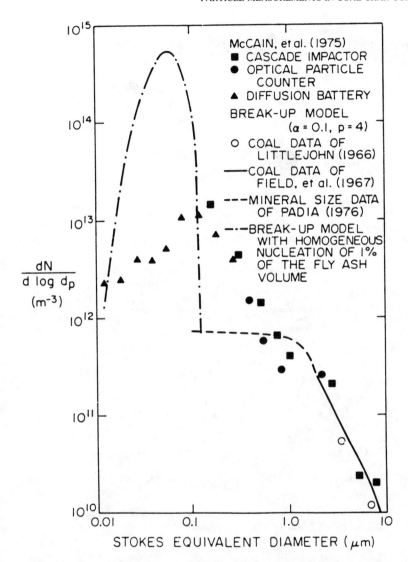

Figure 12-7 Particle size distribution measured in the flue gases of a coal-fired utility boiler and comparison with prediction of an early model (*Flagan and Friedlander 1978*).

carbonaceous structure, e.g., ion-exchanged calcium and alkali metals. The nature of the minerals varies with the origin of the coal and with coal rank. During combustion, these minerals and heteroatoms undergo a number of chemical and physical transformations that ultimately lead to the formation of ash particles.

Figure 12-8 shows the different mechanisms of fly ash particle formation schematically. The bulk of the mineral matter in coal remains as an ash residue when the carbon burns. In high-temperature pulverized coal combustion, the minerals may melt. As the carbon surface recedes due to oxidation, ash droplets adhering to the carbon

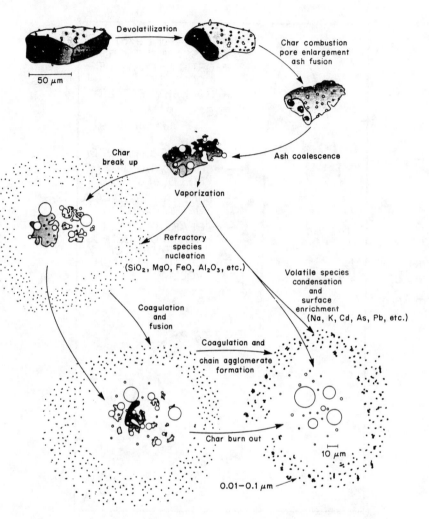

Figure 12-8 Mechanisms of fly ash formation (*Flagan and Seinfeld 1988*).

surface come into contact with one another and may coalesce to form larger droplets. At the same time, oxidation of the porous microstructure leads to loss of the physical integrity of the char structure, and ultimately the formation of a number of fragments from each char particle. Since each of these fragments contains some ash, as they burn out they generate a number of residual ash particles from each parent coal particle.

The mechanism of fragmentation and the distribution of ash particles that results are not yet resolved. A simplistic model that assumes that a few large fragments are formed from each coal particle (of the order of 3–10), and that that number is the same for all particles, reproduced the essential features of the distribution of particulate mass with particle size (Flagan and Friedlander 1978). This result is not surprising, since the mass distribution is dominated by large particles. A more sensitive test of the residual ash formation mechanisms is the number distribution. More recent work makes it clear that the simplistic model grossly underpredicts the number of ash particles produced (Helble and Sarofim 1989; Holve et al. 1981).

Recent theoretical investigations have shown that the fragmentation behavior of burning char particles is directly coupled to the porous microstructure of the char (Kerstein and Niksa 1984; Dunn-Rankin and Kerstein 1986; Helble and Sarofim 1989). Hence an understanding of the behavior of residual coal ash is directly tied to the combustion behavior of coal chars. Fragmentation of the coal char can generate ash residue particles that vary in size from that corresponding to the entire mass of the ash derived from the parent coal particle to the size of the smallest mineral inclusion. Even smaller particles may be produced from the organically bound heteroatoms in the small inclusion-free char fragments. Thus residual ash particles may account for some of the submicron aerosol.

While some submicron ash particles may be derived from the ash residue, and in special cases these may conceivably account for the bulk of the fine ash particles, the variation of the fine ash production with combustion conditions cannot be explained by the residual particle formation mechanism alone. Davison et al. (1974) showed that the enrichment of fine coal combustion derived particles with heavy metals was related to the volatility of the species in question. Flagan and Friedlander (1978) speculated that the volatilized mineral would, in addition to condensing or adsorbing preferentially on the surfaces of fine ash residue particles, form large numbers of very small particles by homogeneous nucleation. These particles would then grow by Brownian coagulation, decreasing the number concentration and increasing the mean size of these ash fume particles, much in the way that soot particles grow in the late stages of a sooting flame.

The evolution of the particle size distribution is described by the general dynamic equation for aerosols, which accounts for Brownian coagulation and heterogeneous condensation, i.e.,

$$\rho \frac{Dn(v)/\rho}{Dt} = \frac{1}{2} \int_0^v K(v - u, u)n(v - u, x, t)n(u, x, t)du$$

$$- n(v, x, t) \int_0^\infty n(u, x, t)du$$

$$+ \frac{\partial[I(v, x, t)n(v, x, t)]}{\partial v} \tag{12-25}$$

where $K(u, v)$ is the coagulation rate coefficient (Flagan and Seinfeld 1988). Small particles coagulate primarily when their Brownian motion brings two particles into contact. For particles larger than the mean free path of the gas molecules, the diffusivity decreases as D_p^{-1}, so Brownian coagulation slows rapidly with increasing particle size. Larger particles can also coagulate due to inertial effects in the turbulent fluid motion (Saffman and Turner 1956).

The size distribution that results from Brownian coagulation of fine particles approaches an asymptotic form known as the self-preserving particle size distribution

(Lai et al. 1972). Condensation onto the fine particles tends to narrow the particle size distribution. Hence for experimental data to be useful in the elucidation of the mechanisms that govern the evolution of the ash fume, size resolution must be sufficient to resolve these sharply peaked particle size distributions.

The theoretical descriptions of aerosol dynamics are based upon the assumption that the particles undergo rapid coalescence upon coagulation. Ash fume particles are frequently observed to be soot-like chain agglomerates, an indication that coalescence is incomplete, at least under some circumstances. The aerodynamic drag of such agglomerates is quite different from that of dense spheres of equal condensed phase volume, as has recently been demonstrated by Schmitt-Ott (1988) using silver agglomerates. Agglomerate particles have received considerable theoretical interest because they appear to be fractal in nature, i.e., the particle mass scales with linear dimension as

$$m \propto r^D \tag{12-26}$$

where D is known as the Hausdorf or fractal dimension (Forrest and Witten 1979; Meakin 1984; Mountain, Mulholland, and Baum 1986). For a geometrically dense particle, $D = 3$, but for a fractal agglomerate, $D < 3$. Thus the particle density decreases with increasing particle size. The collision cross section for agglomerate particles can also be expected to differ dramatically from that of dense spheres. The structure of the agglomerate depends on the mechanism of particle formation. Growth of a cluster by deposition of much smaller particles (diffusion-limited aggregation) forms relatively dense agglomerates $D \approx 2.5$. Growth by coagulation of like-sized clusters (cluster-cluster aggregation) forms agglomerates with much lower density $D \approx 1.8$. The precise fractal dimension depends on the details of the aerosol history. Hence particle structure not only influences the aerosol dynamics, it also contains information about the way the particles evolved. Thus measurements of particle structure are needed to understand the coal ash aerosol.

12-2 COMBUSTION SYSTEMS

Many different types of systems have been used to study coal combustion and related phenomena. Idealized reactor systems have been designed to study the reactivity of coals and chars, while concerns about combustion-generated pollutants and particulate pollutants have prompted research efforts that attempt to simulate practical combustors. This section will briefly describe some of the systems that are currently used in fundamental studies of coal combustion.

Char reactivity experiments seek to relate the rate of carbon loss to experimental conditions. Ultimately, the apparent or intrinsic chemical reaction rate parameters (activation energy and preexponential factor) and the reaction order are sought. Carbon loss can be measured by mass loss or surface regression rate measurements or by combustion gas analysis techniques. Once the carbon loss is known, the reaction rate parameters are found by fitting the experimental data with model predictions and adjusting the rate parameters. This fitting is facilitated if the reactions can be performed in an environment that does not vary with position in the reactor.

The choice of reactor systems for a particular experimental objective depends on the desired

Temperature
Pressure
Energy source/energy transfer mechanism/heating rate
Ambient composition
Flow field
Resolution (spatial and temporal)
Extent of combustion
Fuel form and particle size

Particle temperatures influence reactivity to a great extent. While temperatures in actual pulverized coal combustors can be as high as 2000 K, those in fluidized bed combustors are around 1000 K. Pressurized coal combustion systems are less common than atmospheric systems, but pressurized fluidized bed combustors are increasing in importance, and coal-burning diesel and gas turbine engines are under development. Most laboratory systems are designed to operate at atmospheric pressure, but studies at elevated pressures are needed to support pressurized combustor development.

The choice of the heat source is intimately related to the heating rate required. Electrical heating, radiant heating, flames, plasmas, laser heating, and shock tubes have all been used. While flame heating is convenient, it does vitiate the ambient with the products of combustion, thereby influencing reactivity measurements. The composition of the reaction atmosphere can be changed in a controlled manner to study the reactivities of coal/char with oxygen, hydrogen, carbon dioxide, and water vapor, the gases present in combustors. Plasmas also generate reactive species that may influence combustion reactions and mechanisms if sufficient time is not allowed between plasma generation and particle injection. In spite of these complications, flame and plasma heating are important tools in coal combustion research, since they can produce the range of temperatures encountered in practical coal combustion systems. External electrical resistance heating is generally limited by both reactor construction and heating element material to temperatures below about 2000 K.

For studies of rate processes in entrained flow reactors, the reaction time would ideally be the same for all particles processed. Laminar flow reactors with a dilute stream of particles introduced on the reactor centerline best approximate this ideal. Kinetic studies generally require that such systems be large enough to permit sample extraction from the reaction zone or have optical access to facilitate noninvasive measurements of particle size, particle temperature, and gas composition. Sampling probes introduce a heat load that can alter reactor temperature profiles and possibly introduce sampling biases. Kinetic experiments can examine the evolution of an ensemble of particles, e.g., bulk property measurements made on extracted samples or optical measurements of many particles [either measurements of an entire cloud of particles or single-particle measurements made at a fixed point along a reactor length (Mitchell and McLean 1982)] or measurement of the combustion history of a single particle [for example, temperature histories measured by viewing along the axis of a drop-tube furnace with a two-color pyrometer (Timothy, Sarofim, and Beer 1982; Levendis and Flagan 1987)].

To simulate coal flames in practical combustion systems, large reactors are needed. This introduces the complications of turbulence and, frequently, flow recirculation, but are essential for the study of flame stability and some aspects of pollutant formation. As the temperature is reduced, the reaction rates slow, making entrained flow studies impractical. A variety of approaches for the study of reactions of stationary particles has been developed.

Table 12-3 is a partial summary of the many reactor systems that have been employed in the study of coal combustion. Clearly there is no system that encompasses the complete range of temperature, particle size, and reactor environment. The next three sections will describe some of the systems employed in recent studies.

Table 12-3 Experimental systems used in the study of coal and char combustion

Flow configuration	Reactor type	Applications	Reference
Gas flowing, fuel fixed	fixed bed	oxidation kinetics	Sinford and Eyring, (1956)
			Marsh, O'Hair, and Wynne-Jones (1965)
			Hottel, Williams, and Wu (1966)
	electrically heated rods and fibers	graphite oxidation kinetics	Nagle and Strickland-Constable (1962)
	particle suspended on fiber	oxidation kinetics	Blyhoider and Eyring (1959)
			Tyler, Wouterlood, and Mulcahy (1975)
Quiescent gas, stationary single particle	electrodynamically levitated single particle	oxidation kinetics	Spjut et al., (1986)
			Dudek, Longwell, and Sarofim (1989)
		charignition	Bar Ziv et al. (1989)
Gas flowing, sample fixed	flow tube		Hawtin and Gibson (1966)
	thermogravimetric analyzer	reactivity	Jenkins, Nandi, and Walker (1973)
			Hippo and Walker (1975)
			Dutta and Wen (1977)
			Sahu et al. (1988a)
Gas flowing, sample fixed	single-particle microbalance	reactivity	Walker, Foresti, and Wright (1953)

			Essenhigh, Froberg, and Howard (1965) Strange and Walker (1976)
Entrained flow	isothermal drop tube	pyrolysis, char oxidation, NO_x and ash particle formation, char fragmentation	Field et al. (1976) Sarofim and coworkers Flagan and coworkers
	flat flame burner	char oxidation	Mitchell and coworkers
	plug flow reactor, externally heated	coal combustion and char oxidation	Smith and coworkers Flagan and coworkers Fletcher and coworkers
	plug flow combustor	coal combustion	Field (1969) Howard and Essenhigh (1967)
		NO_x formation, coal ash evolution	Pershing and Wendt (1976)
	pulverized coal combustion simulators	NO_x formation, coal combustion	Smoot and coworkers
Suspended particles	shock tube		Lightman and Street (1968)
		soot oxidation	Park and Appleton (1973)

Thermogravimetric Analysis

The thermogravimetric analyzer (TGA) is used extensively in low-temperature studies of pyrolysis and char oxidation reactions. The TGA consists of a sensitive balance designed to weigh a sample continuously as it is heated and reacts. Figure 12-9 shows typical mass loss data from TGA analysis of char oxidation. A programable temperature controller is used to heat the sample at a prescribed rate. A gas flow continuously supplies the reactants and removes reaction products as samples ranging in mass from 1 to 100 mg are reacted. Temperatures are usually below 1000 K, but some of the newer instruments can be used up to 1700 K. For studies of char oxidation, reaction kinetics become so fast that mass transfer resistances to the bulk sample determine the rate at high temperatures. Hence the TGA is most useful for studies of reactions at temperatures below 1000 K. Even then, rate data frequently need to be corrected to account for diffusional resistances.

Single-Particle Methods

Numerous attempts have been made to reduce the diffusional resistances in thermogravimetric analyses. Since the characteristic time for diffusion is $\tau = l^2/\mathcal{D} \propto m_p^{2/3}$,

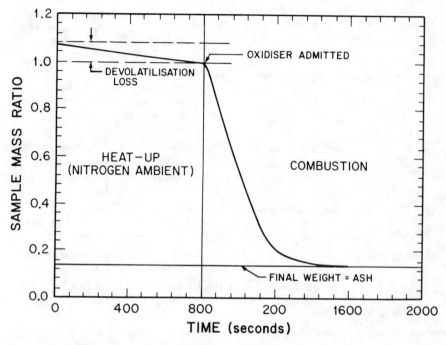

Figure 12-9 Mass as a function of time in thermogravimetric analysis of coal in an oxidizing atmosphere.

where l is a characteristic length scale, $\mathcal{D}$ is the diffusion coefficient, and m_p is the sample mass, the sample mass must be reduced dramatically to significantly accelerate diffusion. One particularly promising approach is the use of the electrodynamic balance (EDB), a modern version of the Millikan oil drop experiment that can suspend and weigh a single charged particle in an electric field (Davies 1987). Particles ranging in size from a fraction of a micron to several hundred microns can be trapped in an EDB, although high-temperature experiments are usually limited to particles larger than about 100 μm.

Figure 12-10 shows the electrode configuration used in the EDB. A charged particle is suspended in the electric field created by a potential across the top and bottom (end-cap) electrodes. An alternating potential on the ring electrode creates a time-varying force on the charged particle. That force increases with displacement from the center of the cell, causing a particle that is not centered to oscillate. Due to the particle's inertia and aerodynamic drag, the motion of the particle lags the field, leading to a time-averaged force that tends to push the particle back toward the center of the chamber. This dynamic focusing makes possible the study of particles undergoing rapid change. The EDB is basically a balance that has an ultimate sensitivity in the range of 10^{-9}–10^{-14} g.

Spjut et al. (1986) developed an electrodynamic thermogravimetric analyzer (EDTGA) in which laser heating is used to raise a particle trapped in an electrodynamic balance to temperatures at which reaction occurs. The particle temperature is measured opti-

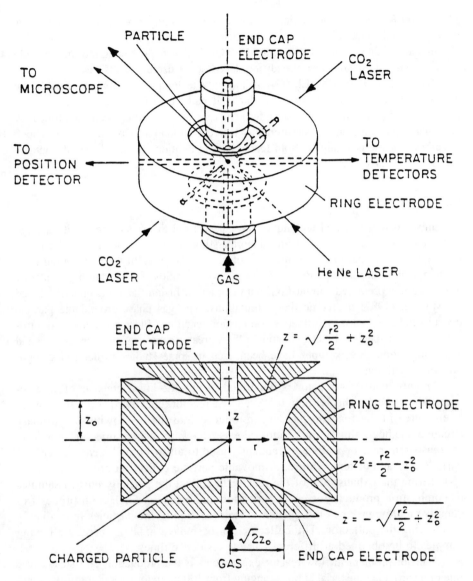

Figure 12-10 Electrode arrangement in an electrodynamic balance (*Bar-Ziv et al. 1989*).

cally, and its mass is determined from the field required to hold it at the null point of the cell. While the EDTGA is a very attractive concept, a number of factors have slowed its development and use. Laser heating of a particle produces photophoretic forces that can push the particle from the view volume unless they are carefully balanced. A hot particle in a cold gas generates a buoyant flow that tends to lift the particle, complicating the measurement of particle mass. Finally, at temperatures above

1400–1500 K, the particle begins to lose charge, making mass loss measurements impossible and frequently causing the particle to be lost from the view volume.

In spite of these complications, the EDTGA is a very promising tool in the study of coal combustion, and has already produced some useful results, as reviewed by Bar-Ziv et al. (1989). The EDTGA is the only technique that makes possible the detailed characterization of the particular coal particle to be burned. Not only can photographs be taken, the porous microstructure of the particle can be probed by gravimetric gas adsorption measurements wherein a monolayer of adsorbed gas such as carbon dioxide is weighed. Both reactivity and ignition characteristics of levitated char particles have been probed.

Flow Reactors

A number of experimental techniques have been used to measure reactivity at high temperatures. Most workers have used laminar-flow drop-tube reactors with heat supplied by a flame, plasma, or some type of electric heating. In vitiated combustion, the temperature depends on the fuel/oxidant mixture ratios. Flame stability limits the temperature in premixed combustion, but the microdiffusion flame systems developed by Mitchell (1987) overcome these limitations. The gas phase around the particle contains radicals that may influence combustion reactions. Plasma heating also introduces reactive species. Electric heating allows greater control of the temperature and gas composition. Shock tubes have been used in some studies to determine high-pressure reactivities.

An entrained flow reactor is illustrated in Fig. 12-11. It consists of an electrically heated air preheater tube followed by a reactor tube. The pulverized sample is entrained in the reactant gas and introduced into the co-flowing hot primary stream at the beginning of the reactor tube using a water-cooled injector. Windows or transparent reactor walls are frequently used to allow optical measurements of particle size and/or temperature. A movable particle collection probe may be inserted from the exhaust end of the reactor to collect the partially burned samples or combustion products for characterization and analysis. Collection probes frequently employ gas or liquid quench to stop reaction of the particles immediately on entering the collector. The collector can be moved axially, thereby allowing samples to be obtained at different particle residence times.

A major problem in flow reactor experiments is the production of a steady feed rate of a particulate material at rates ranging from 100 mg/h to several grams per hour. Two designs that have been used extensively in recent studies are illustrated in Figs. 12-12 and 12-13. The fluidized bed feeder (Fig. 12-12) is well suited for relatively large feed rates and provides direct gravimetric measurements of the amount of material processed. The syringe pump system (Fig. 12-13) works well at smaller feed rates.

12-3 TEMPERATURE MEASUREMENT

The temperatures of particles and the surrounding gas in a combustion environment along with the oxidizer concentration determine the rates of char oxidation, release of

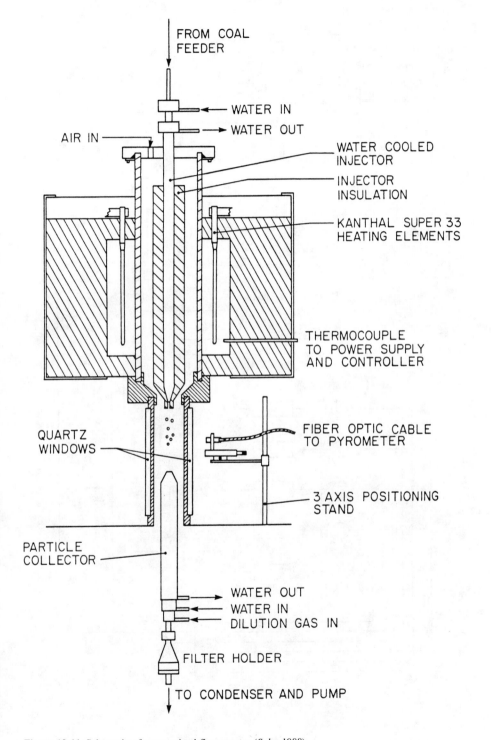

Figure 12-11 Schematic of an entrained flow reactor (*Sahu 1988*).

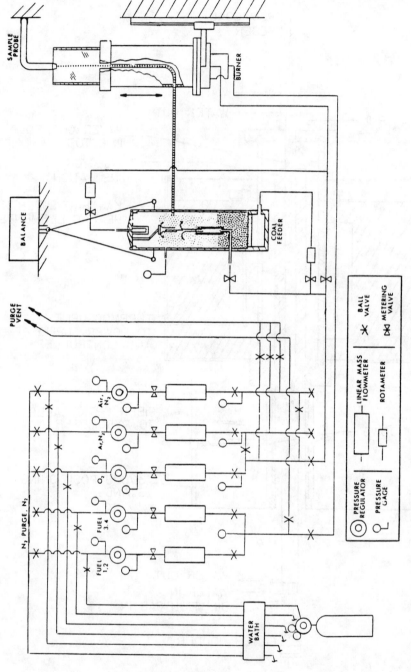

Figure 12-12 Fluidized bed coal/char feeder (*Hardesty, Pohl, and Stark 1978*).

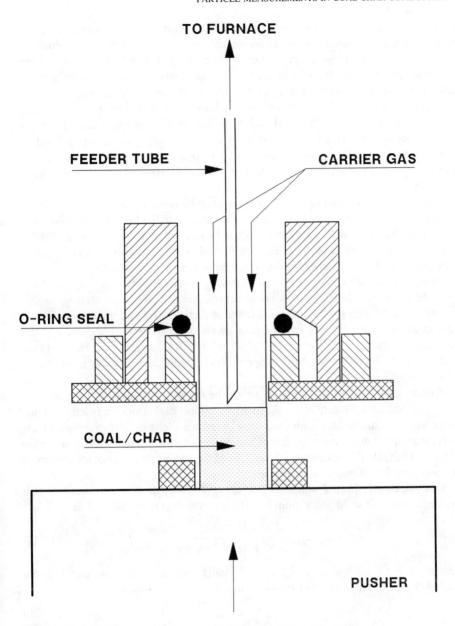

Figure 12-13 Example of a syringe pump coal/char feeder (*Mims et al. 1980*).

heteroatoms in the coal, and the fate of the mineral matter in the combustion environment. Given the temperature of a burning char particle and additional information about the physical and chemical properties of the burning char and the reactions taking place, one can calculate the rate of char oxidation. Thus knowledge of the particle temperature provides important insights into the combustion process and is essential to the interpretation of other measurements related to coal combustion. Flame and particle temperatures were measured earlier by disappearing filament and brightness pyrometry techniques (Gaydon and Wolfhard 1970). These are essentially subjective techniques and produce significant errors due to operator inexperience. Also, it is extremely difficult to measure the temperatures of small moving particles using these techniques.

Recently two-color pyrometry has developed into a versatile tool for single-particle temperature measurement. Temperature-time histories of individual particles have been used to infer reaction kinetic parameters (Levendis and Flagan 1987; Sahu et al. 1988b) and to study their ignition behavior (Levendis et al. 1989a). In this section, we shall review some of the approaches used to determine particle temperatures.

Measurements of the temperature of a cloud or ensemble of burning particles have been used by many investigators in the interpretation of combustion measurements. Since most combustion studies involve a range of particle sizes and, in the combustion of dense suspensions, the gas environment varies dramatically with position in the flame, the uncertainty in the temperature is rather large. Recent studies of char oxidation kinetics have frequently probed the temperatures of individual particles to overcome this weakness.

Particle temperature measurements probe the radiant emissions from the hot particles. One attempts to determine the particle temperature from knowledge of the particle size and the radiant emissions at selected wavelengths. An overview of the general principles and instrument design issues of radiation thermometry is presented by Dixon (1988). In this section we shall focus on the special problems of measurement of temperatures of burning char particles.

The blackbody spectral emittance, or energy emitted from a unit area per unit wavelength per unit solid angle from a surface at absolute temperature T is

$$e_{\lambda b} = \epsilon_\lambda \frac{2C_1}{\lambda^5(e^{C_2/\lambda T} - 1)} \tag{12-27}$$

where C_1 and C_2 are the first and second Planck radiation constants and λ is the wavelength of the radiation. The energy emitted by a real surface is

$$e_\lambda = \epsilon_\lambda e_{\lambda b} \tag{12-28}$$

where ϵ_λ is the spectral emissivity of the surface being probed.

The simplest radiant temperature measurement is brightness pyrometry, in which the absolute emission intensity is measured at a single wavelength. The most common implementation of this technique is the disappearing filament pyrometer, which matches the emission to that from a calibrated filament in the red, at $\lambda = 650$ nm. Newer brightness pyrometers make electronic measurements of the absolute emittance from the particle.

The brightness pyrometer is useful for estimating the temperature of a surface or a dense cloud of particles that fills the field of view of the instrument. It is less useful for measurements of the temperature of a single particle, since the background of the particle is also viewed. For a small particle viewed from a long distance L, compared with its radius R_p, the detector subtends a solid angle $d\Omega$ as viewed from the particle, and the intensity at the detector is

$$I_\lambda = \epsilon_\lambda e_{\lambda b}\pi R_p^2 d\Omega \tag{12-29}$$

Because of the extreme sensitivity of the intensity to temperature, a small uncertainty in the emissivity or particle size can lead to a very large error in the temperature that is estimated from brightness pyrometry.

Much of this uncertainty can be eliminated by comparing intensity measurements at two or more wavelengths, taking advantage of the wavelength dependence of the spectral emittance and assuming that the spectral emissivity does not vary with wavelength over the region of interest. Two-color pyrometers have been used in many studies of coal combustion (Ayling and Smith 1972; McLean, Hardesty, and Pohl 1981; Timothy, Sarofim, and Beer 1982; Tichenor et al. 1984; Jorgensen and Zuiderwyck 1985; Levendis and Flagan 1987). Hoff, Rothman, and Mundy (1985) discuss the calibration of an optical pyrometer against a tungsten strip lamp above 2300°C and show that high accuracies can be achieved. A schematic diagram of a two-color pyrometer is shown in Fig. 12-14.

Spectrally resolved measurements can be made using a monochrometer or bandpass filters to select the wavelength intervals viewed by separate detectors. The radiant emissions can be split for two detection systems using beam splitters (Ayling and Smith 1972; McLean, Hardesty, and Pohl 1981; Jorgenson and Zuiderwyck 1985) or bifurcated fiber optics (Timothy, Sarofim, and Beer 1982; Macek and Bulik 1984; Levendis and Flagan 1987).

The measurements are made over a finite wavelength interval, so the response of detector i is

$$S_i(T) = \int_{\lambda_i}^{\lambda_i + \Delta\lambda_i} f(\lambda)\epsilon_\lambda e_{\lambda b}(\lambda_i, T)\pi R_p^2 d\Omega d\lambda \tag{12-30}$$

where $f(\lambda)$ is the spectral response function of the wavelength selection system in combination with the detector used. For a wavelength interval that is sufficiently narrow, we may assume that the intensity and response are constant over the interval, so the signal may be expressed more simply as

$$S_i(T) \approx f(\lambda_i)\epsilon_{\lambda_i} e_{\lambda b}(\lambda_i, T)\pi R_p^2 d\Omega\Delta\lambda_i \tag{12-31}$$

Taking the ratio of the signals at two different wavelengths, the geometry factors are eliminated, i.e.,

$$R_{12}(T) = \left[f(\lambda_1)\epsilon_{\lambda_1} \frac{2C_1}{\lambda_1^5(e^{C_2/\lambda_1 T} - 1)} \Delta\lambda_1 \right] \left[f(\lambda_2)\epsilon_{\lambda_2} \frac{2C_1}{\lambda_2^5(e^{C_2/\lambda_2 T} - 1)} \Delta\lambda_2 \right]^{-1}$$

$$\tag{12-32}$$

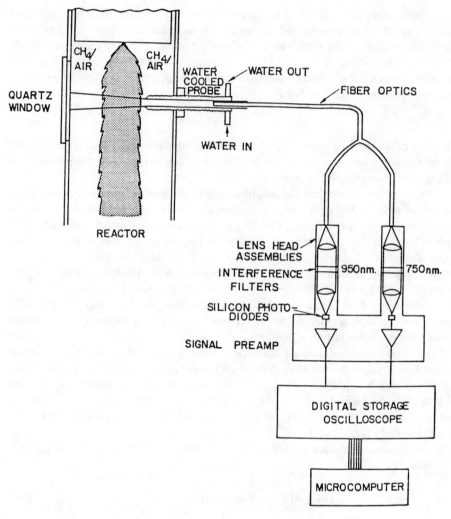

Figure 12-14 Schematic of a two-color pyrometer *(Seeker et al. 1981)*.

A calibration factor for the instrument may be defined as

$$\kappa = \frac{f(\lambda_1)\Delta\lambda_1\lambda_2^5}{f(\lambda_2)\Delta\lambda_2\lambda_1^5} \tag{12-33}$$

The temperature dependence of R_{12} is now

$$R_{12}(T) = \kappa\frac{\epsilon_{\lambda_1}(e^{C_2/\lambda_1 T} - 1)}{\epsilon_{\lambda_2}(e^{C_2/\lambda_2 T} - 1)} \tag{12-34}$$

The emissivity is frequently not known precisely. If the difference between the two wavelengths selected for the measurement is small, the variation in the emissivity is frequently assumed not to be large, at least over the range of wavelengths of the measurement. Making this graybody assumption that $\epsilon_{\lambda_1}(T) = \epsilon_{\lambda_2}(T)$, the ratio of signals further simplifies to

$$R_{12}(T) = \kappa \frac{(e^{C_2/\lambda_1 T} - 1)}{(e^{C_2/\lambda_2 T} - 1)} \tag{12-35}$$

Even if the emissivity does vary with wavelength, the wavelength dependence can be folded into the calibration constant, which can then be determined experimentally if it can be assumed to be independent of temperature. Figure 12-15 shows the intensity measurements at 800 and 1000 nm and the temperature of a single burning bituminous char particle.

In flowing systems, measurements can be made at a particular position by viewing across the flow. This allows measurements to be made of an ensemble of particles that have been processed for the same residence time. For any detector selected, there is a lower limit to the intensity that can be detected. Since the intensity viewed by the pyrometer increases with both particle temperature and size, the minimum temperature that can be determined is a function of particle size. If the particles being probed are distributed over a range of sizes, a biased distribution of temperatures can result.

Tichenor et al. (1984) and Niksa et al. (1984) developed an optical method for simultaneous measurement of particle size, velocity, and temperature in a combustion environment to eliminate this bias. Temperature measurements were made using two-color optical pyrometry. Particle size and velocity were measured by imaging the particle onto a coded aperture as illustrated in Fig. 12-16. Measurement of light scattered from a laser through the trigger slit ensures that the particle will pass through center of the view volumes of the size-encoding slits and will be in proper focus.

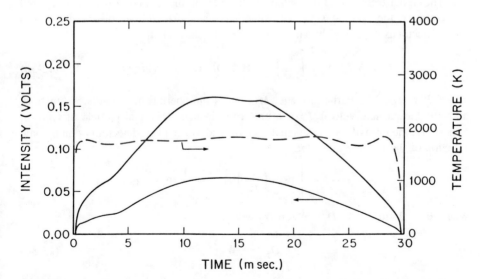

Figure 12-15 Intensity and temperature measurements of a bituminous char particle (*Levendis et al. 1989a*).

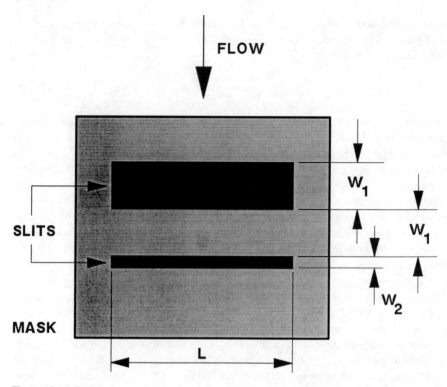

Figure 12-16 Coded aperture for simultaneous temperature and size measurement (*Tichenor and Wang 1982*).

The signal from one of the two photodetectors in the two-color pyrometer is used to determine the particle size. The time-varying signal $S(T, t)$ observed by each photodetector is described by an integral over the view area,

$$S(T, t) = \int_{-\infty}^{\infty} \int_{-\infty}^{\infty} M(x, y)P(x, y - vt)dxdy \qquad (12\text{-}36)$$

where $P(x, y)$ is the particle image, $M(x, y)$ is the aperture transmittance function, and v is the particle velocity. For a uniformly heated, spherical particle of uniform emissivity, the ratio of the peak signals determined by the photodetector is a monotonic function of the particle diameter,

$$\frac{S_1}{S_2} = \frac{4}{\pi} \frac{W_2}{D_p} \left[1 - \left(\frac{W_2}{D_p} \right)^2 \right]^{1/2} + \frac{1}{\pi} (\theta - \sin \theta) \qquad (12\text{-}37)$$

where $\theta = 2 \sin^{-1} (W_2/D_p)$. When $W_2 \gg D_p$,

$$D_p \approx \frac{4}{\pi} W_2 \frac{S_2}{S_1} \qquad (12\text{-}38)$$

For nonspherical particles, the latter value is approximately equal to the dimension of the particle measured in the direction perpendicular to the narrow slit.

While assumptions regarding coal or char particles as graybody emitters are commonly used, the spectral emissivities of coals, chars, and coal minerals are not well known. Solomon et al. (1986) have applied Fourier transform infrared (FT-IR) emission and transmission spectroscopy to measure the spectral emittance of coal and char in the infrared. They found that the spectral emittance of individual particles depends on particle size. Small particles exhibit nongray behavior with high emittance only in the strong infrared absorption bands in the coal. The spectral emittance increases with increasing rank, particle size, and the extent of pyrolysis. Baxter, Fletcher, and Ottesen (1988) have measured the low-temperature (120–200°C) spectral emission characteristics of coal particles in the range 40–115 μm. Nongray behavior was noted at wavelengths between 2.2 and 17 μm. While emissivities of high-rank coals were in the range 0.7–0.98, those for lignites were as low as 0.5. Particles generally exhibited more graybody behavior as particle size, coal rank, and carbon conversion increased. However, partially devolatilized chars of bituminous coals generally behaved as graybodies with emissivities of 0.8 over a wide range of the infrared spectrum. Recent experimental studies by Goodwin and Mitchner (1989) provide additional measurements of emissivities.

Coal devolatilization releases hydrocarbon vapors that can undergo gas phase pyrolysis, leading to the formation of a soot cloud (Grosshandler 1984). The radiant emissions from this cloud during the early stages of combustion can lead to erroneous particle temperature measurements, if not properly accounted for.

Most single-particle pyrometry experiments have been performed in geometries that allow the particle to be observed against a cold, nonluminous background. Macek and Bulik (1984) used a bifurcated fiber optic probe to measure particle temperatures in a fluidized bed. In their experiments, a small number of coal particles were burned in a fluidized bed of inert particles. Fuel particle temperatures were inferred from emission measurements at two wavelengths. Due to the highly emitting background, the data analysis accounted for the radiation from two sources at two different temperatures: the transient emissions from burning fuel particles and the steady state background from the inert bed material.

12-4 PARTICLE MEASUREMENTS

Coal combustion begins with particles of solid fuel and ends with a solid residue of particles that are derived from the mineral matter in the coal. The particulate nature of the fuel allows mass transfer to influence strongly and, in some cases, to control the combustion rate. Much of this chapter has been devoted to measurements of the properties of reacting coal or coal char particles. The mineral matter in coal undergoes many transformations during combustion, as illustrated in Fig. 12-8, leading to the production of ash residue particles that are typically larger than 1 μm in size, and of a fine fume of submicron particles that are formed from volatilized minerals. Partially burned char particles are also found primarily in the supermicron size regime (up to 10 μm), while soot particles formed from hydrocarbon vapors are similar in size to the ash fume. Measurement of the number and mass concentrations of these particles, size distributions, and size-dependent chemical composition are essential in the study of particulate air pollution and of the deposits, erosion, and corrosion that result from coal combustion.

Ideally, one would like to measure the properties of these aerosols with a minimum of perturbations through probing. In situ optical diagnostics provide critical, but limited insight into the properties of the aerosol. Moments of the particle size distribution can be estimated if the particle structure and optical properties are known. Usually the particles are assumed to be spherical, but that assumption is frequently incorrect for combustion aerosols. Extractive sampling allows more direct measurement of particle size distributions but can result in distorted data due to particle losses, phase transformations (solidification of molten materials or condensation of vapors), or coagulation. In this section we shall briefly review both (in situ and extractive) classes of particle measurements, emphasizing those that have been used in the study of coal combustion.

In Situ Measurements

Noninvasive optical measurements are made by light scattering or imaging. Which of the many methods is appropriate to a particular aerosol depends on the dimensionless particle size,

$$\chi = \frac{\pi D_p}{\lambda} \tag{12-39}$$

which is the ratio of the particle circumference (assuming spherical shape) to the wavelength of light employed in the measurement. For $\chi \gg 1$, the particles are large enough that imaging or diffraction methods can be used. For $\chi \ll 1$, the particle structure cannot be resolved, but measurements of light scattering from a cloud of particles can be used to infer moments of the particle size distribution. In the intermediate particle size regime [$\chi = O(1)$], complex resonant structures in the scattered light (Mie 1908) severely complicate the interpretation of optical measurements.

While the number concentrations of large particles may be small enough to permit inspection of individual particles, the fume particles are generally so numerous ($>10^8$–10^{11} cm^{-3}) that a cloud of particles will be contained in all but the smallest view volumes at any time. Hence in situ measurements of fine particles are generally based upon measurements of the light scattered from large numbers of particles. In studies of coal combustion, conditioned sampling may be required to ensure that larger particles are not present in the sample volume when such measurements are made. In situ measurements, either of clouds or of individual particles, are generally limited to small or dilute systems in which the region of interest is optically thin. Extractive sampling must be employed when the density of the aerosol is such that only the perimeter regions can be probed optically.

Light scattering from a cloud of particles. In situ optical probing has been used extensively for the study of soot formation in hydrocarbon flames (Dalzell, Williams, and Hottel 1970). Soot particles are much smaller than the wavelengths of light used to probe the aerosol, so the data have generally been interpreted using models of

Rayleigh scattering ($\pi D_p/\lambda < 0.3$). The absorption coefficient for uniformly sized absorbing particles is

$$k_{abs} = -\frac{\pi^2}{\lambda} \text{Im} \left[\frac{m^2 - 1}{m^2 + 2} \right] ND_p^3 \qquad (12\text{-}40)$$

where m is the complex refractive index of the particulate material. The total scattering coefficient for vertically polarized incident and scattered beams is

$$Q_{vv} = \frac{1}{4} \left[\frac{\pi}{\lambda} \right]^4 \left| \frac{m^2 - 1}{m^2 + 2} \right|^2 ND_p^6 \qquad (12\text{-}41)$$

Since the absorption of light is proportional to the volume of particulate material while the scattering is proportional to the square of the particle volume, the combination of absorption and scattering data can provide information on the mass loading and mean particle size. One important limitation of light-scattering methods is that the optical properties of the particles must be known before the data can be used to infer particle size distributions and concentrations. Dobbins and Mulholland (1984) show that highly distorted profiles of N versus t can result from simplistic analysis of such data.

Angular scattering patterns can also be used for the determination of moments of the particle size distribution (Dalzell, Williams, and Hottel 1970; Heintzenberg and Welch 1982). This method has the advantage of allowing the use of extremely small view volumes, leading to high spatial resolution. For very small particles ($\chi << 1$), the angular distribution of scattered light is independent of particle size. To determine the sizes of such small particles, it is necessary to combine scattering and extinction measurements as described above. Light scattering at fixed angle but multiple wavelengths is an alternate approach to measuring particle size at intermediate values of χ (Ariessohn, Self, and Eustis 1980).

Small particles undergo rapid Brownian motion that can be detected using quasi-elastic light-scattering methods (Clark, Lunacek, and Benedek 1970; Hinds and Reist 1973a, b; Flower 1983; Chowdhury et al. 1984). Two versions of this method are employed in studies of aerosols and colloids: photon correlation spectroscopy and diffusion broadening spectroscopy. The latter method has been used extensively in the study of soot aerosols in flames. The Brownian motion of small particles results in spectral broadening of the scattered light. The resulting line width is directly related to the sizes of the diffusing particles. When monodisperse particles with no mean motion scatter monochromatic light, a Lorentzian line shape results:

$$P(\omega) = \frac{I^2 K^2 \mathscr{D}_p/\pi}{(K^2 \mathscr{D}_p)^2 + (\omega - \omega_0)^2} \qquad (12\text{-}42)$$

Here I is the intensity of the scattered light, $\mathscr{D}$ is the particle diffusivity, ω and ω_0 are the frequencies of the scattered and incident light, respectively, and

$$K = \frac{4\pi}{\lambda_0} \sin \left(\frac{\theta}{2} \right) \qquad (12\text{-}43)$$

is the magnitude of the scattered wave vector, where λ_0 is the wavelength of the incident light and θ is the scattering angle, which is measured from the forward direction, as illustrated in Fig. 12-17. The power spectrum of the scattered light is measured with a spectrum analyzer. Since the particle diffusivity depends on temperature, the temperature at the point where the scattering measurements are made must also be determined. Flower (1983) has used line-of-sight, two-color pyrometer to measure the temperature from the thermal emission of the particles in a sooting flame.

Fraunhofer diffraction has been used to study sprays and other assemblages of particles larger than the wavelength of light. The technique is based upon the Fraunhofer diffraction of a parallel beam of monochromatic light as illustrated in Fig. 12-18 (Swithenbank et al. 1976). A Fourier transform lens is used to focus a stationary light pattern into series of focused rings of light. Light diffracted by the particles at an angle θ, is refracted by the receiving lens, producing a ring on the focal plane at radius, $r = f\theta$, where f is the focal length of the lens. A monodisperse aerosol of spherical particles that are large in comparison with the wavelength of light would produce an intensity distribution,

$$I(\theta) = I_0 \frac{\pi^2 D_p^4}{16\lambda^2} \left[\frac{2J_1(\pi D_p\theta/\lambda)}{(\pi D_p\theta/\lambda)} \right]^2 \tag{12-44}$$

In nonmonodisperse aerosol systems, the diffraction pattern is determined by the particle size distribution,

$$I(\theta) = I_0 \int_0^\infty \frac{\pi^2 D_p^4}{16\lambda^2} \left[\frac{2J_1(\pi D_p\theta/\lambda)}{(\pi D_p\theta/\lambda)} \right]^2 \frac{dN}{dD_p} dD_p \tag{12-45}$$

By inverting the measured intensity distribution as a function of radial position from the focal point to find the particle size distribution corresponding to the measured angular intensity distribution, the particle size distribution can be inferred from the diffraction data. The diffraction pattern is independent of the position of the particles in the original parallel beam of light, so measurements can be made in flowing systems without complications due to particle motion.

For the study of coal ash fumes, measurements are complicated by the presence of particles produced by several different mechanisms. Large residual ash particles differ substantially in their scattering properties from the smaller fume particles. As

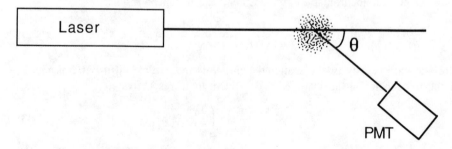

Figure 12-17 Geometry for scattering angle experiments.

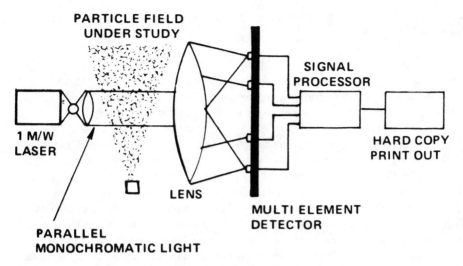

Figure 12-18 Laser diffraction particle size distribution analyzer (*Swithenbank et al. 1976*).

will be shown below, the form of the particle size distribution must be assumed to interpret cloud-scattering data, so the existence of multiple modes in the particle size distribution severely complicates the data interpretation. Since the large particles are much less abundant than the fume particles, conditional sampling only when large particles are not present in the view volume might prove useful.

Particle structure measurements. Fume particles produced by condensation of refractory vapors are generally agglomerates of small spherules. The remarkably similar structures of these pyrogenous fumes have received considerable attention (Forrest and Witten 1979; Witten and Sander 1981; Schaefer 1988). These aggregate particles have been characterized as fractal structures, in which the particle mass scales with radius as

$$m \propto R_p^D \tag{12-46}$$

where D is the fractal dimension. Geometrically dense particles would have $D = 3$, but for the aggregate particles, D ranges from about 1.8 to 2.5. The value of D conveys information about the mechanisms of particle formation and growth and is therefore of considerable interest. Moreover, the dynamics of the aggregation process depends on the aerodynamics of the particles, which, in turn, depends on the particle structure (Mountain, Mulholland, and Baum 1986).

Small-angle scattering methods have been used to measure the mass correlation functions in situ (Schaefer 1987). Scattering at a particular angle can be associated with fluctuations on a length scale l, determined by Bragg's law. Scattering patterns are correlated with the spatial Fourier frequency or momentum transfer

$$Q = \frac{2\pi}{l} = \frac{4\pi}{\lambda} \sin\left(\frac{\theta}{2}\right) \tag{12-47}$$

where λ is the wavelength of the radiation and θ is the scattering angle. For a mass fractal object,

$$I(Q) \propto Q^{-D} \tag{12-48}$$

Small-angle X ray scattering (SAXS) and small-angle neutron scattering (SANS) provide information over dimensions of approximately 1–100 nm (Schaefer 1987). Larger scales can be probed by small-angle light scattering (Hurd and Flower 1988).

Single-particle light scattering. In situ optical measurements of the sizes of individual particles have been demonstrated by several workers (Farmer 1972; Hirleman and Wittig 1976; Holve 1980; Bachalo 1980; Holve et al. 1981; Pettit and Peterson 1983; Holve and Davis 1985). The various instruments provide measurements over the range of sizes from 0.2 to 1000 μm, and for number concentrations as high as 10^{12} m^{-3}. Theoretical investigations of such measurement systems suggest that the limits could be pushed to particle sizes as small as 0.03 μm and concentrations as high as 10^{14} m^{-3} (Holve et al. 1981; Pettit and Peterson 1984). Until such advances can be made, in situ single-particle sizing systems are limited to the measurement of the residues derived from the fuel particles, either partially burned char or ash. Number concentrations of nuclei produced from vapors are too high and sizes are too small for current single-particle counters to be effective for such particles. Nonetheless, single-particle instruments can play an important role in the study of coal combustion, helping to unravel such questions as the extent of char fragmentation and its role in generating fine ash particles.

At the upper end of the single-particle measurement size range, imaging methods can be very effective. Microscopy, photography (either film or video), and holography have all been used to study sprays and, to a lesser extent, coal combustion. On-line size analysis has been achieved by imaging particles onto a mask or an array detector. Trolinger (1976) describes the use of holographic methods for analysis of a field of particles.

Extractive Sampling

Many of the limitations of in situ optical measurements can be overcome by extracting the aerosol from the experimental system and taking advantage of the major advances in aerosol instrumentation that have been made over the past few decades of research into atmospheric air pollution. A new set of problems is introduced, however, as one attempts to ensure that a representative sample is obtained and that the aerosol is not perturbed by the act of sample extraction. Key issues in sample extraction include sampling biases introduced by probe and particle aerodynamics, deposition of particles within the probe, particularly in regions where a hot particle-laden gas is exposed to cool surfaces, and changes in aerosol properties that can take place within the sampling system, especially where fine particles are concerned.

A case in point is the measurement of the fine particles generated by homogeneous nucleation of volatilized ash in coal combustion. Particle concentrations in the combustion zone are initially very high, but the number concentration decreases rapidly due to coagulation in the few seconds' residence time in a typical furnace. Coagulation can be minimized by diluting the sample, taking advantage of the second-order coagulation kinetics. The characteristic time for coagulation is

$$\tau_{\text{coag}} = \frac{1}{KN_0} \qquad (12\text{-}49)$$

where K is the coagulation coefficient and N_0 is the initial number concentration. By reducing the number concentration by an order of magnitude, this time scale can be increased by a corresponding amount.

Some samplers used in early attempts to measure the fine particles in the flue gases had dilution systems installed at the probe exit (Markowski et al. 1980). Because of flow rate limits of the instruments used for aerosol analysis, a sampling probe used to measure particles in the flue gases could easily have a residence time as long or longer than the residence time in the furnace itself, leading to the possibility of significant coagulation after sample extraction. Hence measured number concentrations would be smaller and the mean size larger than those present in the flue gases at the sampling point. By installing a dilution system at the entrance to the sampling probe, coagulation can be quenched immediately, eliminating this sampling bias.

The apparent number concentration is also reduced by particle deposition in the probe. Particle deposition rates can be significantly higher than predicted considering convective diffusion alone. The temperature gradient established between a hot, particle-laden gas and the cool surface of a probe can lead to greatly enhanced particle losses through a process known as thermophoresis (Talbot et al. 1980). The thermophoretic migration velocity of a particle away from a high-temperature region toward a cooler region may be correlated with the thermal dimensionless group,

$$\text{Th} = -\frac{v_T \rho T}{\mu (dT/dx)} \qquad (12\text{-}50)$$

in which dT/dx is the temperature gradient in the direction of motion, μ is the gas viscosity, ρ is the gas density, and v_T is the migration velocity. The dimensionless group Th is essentially constant for particles that are much smaller than the mean free path of the gas molecules. Theoretical predictions suggest that $0.42 < \text{Th} < 1.5$, and experimental data suggest that $\text{Th} \approx 0.5$. Thus for a 0.1-μm particle in air at 573 K, we estimate $v_T = 4.17 \times 10^{-4}(dT/dx)$ cm s^{-1}. Thus for $dT/dx = 10^4$ K cm^{-1}, $v_T = 4.17$ cm s^{-1} (Flagan and Seinfeld 1988). One way to overcome this large velocity is to use a transpiration-cooled sampling probe as illustrated in Fig. 12-19. Mitchell and coworkers have developed an alternate design to quench reactions in studies of char oxidation, illustrated in Fig. 12-20. This design has been duplicated in several laboratories both for their original purpose and for sampling of fine particles.

Aerodynamic biases are most frequently addressed by isokinetic sampling, i.e., by matching the velocity of the gas entering the probe to the fluid velocity at the measurement point. Isokinetic sampling attempts to avoid biases caused by the inability of particles to follow the fluid streamlines at the probe entrance, as illustrated in Fig. 12-21. While isokinetic sampling is critically important for measurements of large particles, it may not be necessary in all cases.

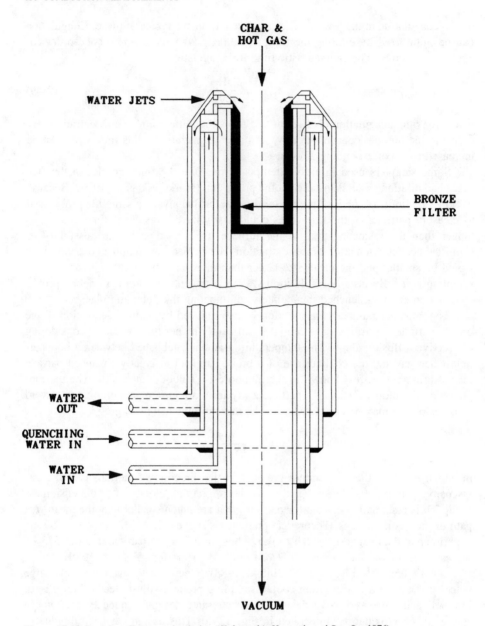

Figure 12-19 Particle collector probe design (*Kobayashi, Howard, and Sarofim 1976*).

The parameter that characterizes the ability of particles to follow fluid motions is the Stokes number, a ratio of a characteristic stopping distance for a particle to the characteristic dimension of the flow, i.e.,

$$\text{St} = \frac{\rho_p U D_p^2}{18 \mu L} C(\text{Kn}) \tag{12-51}$$

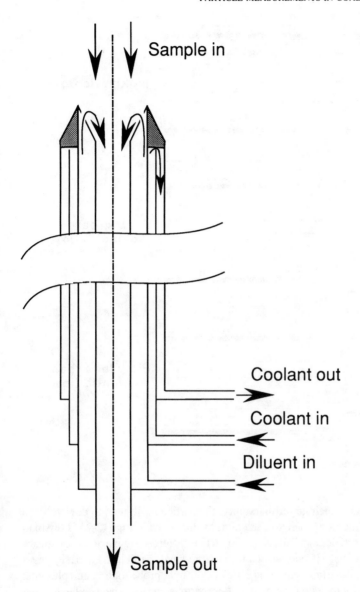

Figure 12-20 Particle collector probe design.

For particles that are small for comparison to the mean free path of the gas molecules λ, i.e., for particles with Knudsen number $(Kn = 2\lambda/D_p) \geq 1$, the Stokes number must be modified as noted above by the addition of a slip correction factor,

$$C(Kn) = 1 + Kn[\alpha + \beta \exp(-\gamma/Kn)] \qquad (12\text{-}52)$$

where the empirically determined constants determined from the Millikan oil drop experiments are $\alpha = 1.257$, $\beta = 0.40$, and $\gamma = 1.10$ (Flagan and Seinfeld 1988).

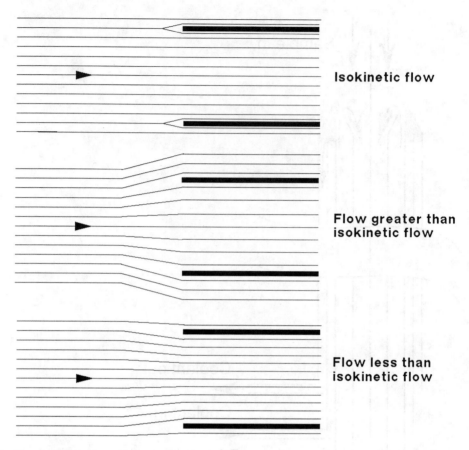

Figure 12-21 Isokinetic sampling.

For Stokes numbers much larger than unity, the particles will tend to deviate from the streamlines when the flow changes direction, as illustrated in Fig. 12-21. For small Stokes numbers, the particles can follow the flow. This difference in particle trajectories can be used advantageously if one wishes to measure small particles. By designing a probe such that the large-particle sampling efficiency is low, probe fouling and plugging can be minimized without jeopardizing the fine particle measurements (Taylor and Flagan 1981). Such a probe is illustrated in Fig. 12-22.

Particles can also be lost to walls of sampling systems by convective diffusion, by inertial impaction at bends in the flow, by sedimentation, and by electrostatic and thermophoretic deposition. Since most aerosols carry a distribution of charges, electric fields can enhance deposition rates of the charged particles. If metal sampling probes and lines are used, the electrostatic forces are limited to image forces in the region near the wall. Many plastics create localized electric fields that result in significant increases in losses of the charged particles over the losses that are observed with metal tubing.

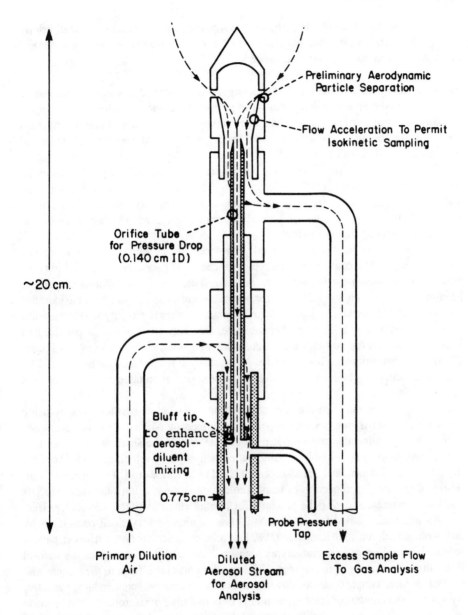

Figure 12-22 Combustor exhaust gas probe (*Taylor and Flagan 1981*).

Aerosol Instrumentation

Instrumentation for the measurement of aerosol particles in extracted sample streams has been developed primarily to analyze aerosols in the atmosphere, so the instruments are generally designed to operate at ambient temperature and pressure, and at relatively low particle concentrations. Pui and Liu (1988) provide an excellent review of the state of the art for atmospheric aerosol measurements. In this section we shall briefly review the technologies that have seen use or have particular promise for aerosol measurements in coal combustion streams.

Aerodynamic measurements. Particles larger than about 0.3 μm in diameter can be classified aerodynamically using inertial impactors. Cyclone separators are useful for measurements of particles larger than a few microns in diameter. These instruments are useful for chemical characterization of the aerosol, since the particles are collected in size-classified samples.

Figure 12-23 illustrates a multistage impactor. In the impactor a particle-laden jet impinges on a substrate. Particles larger than a critical size impact on the substrate and, ideally, stick. The size of particles that can be collected on a particular stage scales with the Stokes number. A typical impactor is a cascade of several stages that collect successively smaller particles.

The impaction substrate is generally treated with an adhesive layer to reduce particle bounce. The coating can interfere with chemical measurements, and limits the temperature at which the impactor can be used. The need for an adhesive layer can be eliminated by using a virtual impactor or dichotomous sampler in which the particles are concentrated aerodynamically into a stagnant or slowly flowing region for subsequent deposition on a filter or other substrate. One version of this is the particle trap impactor in which the particles are concentrated in a small cavity (Biswas and Flagan 1988). This instrument has been operated successfully at temperatures as high as 500°C, and can conceivably be operated at higher temperatures if suitable materials are used for the construction.

The operating range of the impactor can be extended to smaller sizes by using small jets, as in the microorifice impactor developed by Marpel, Liu, and Kuhlmey (1981), by operating the impactor at reduced pressure so the small particles begin to slip relative to the gas, possibly in combination with high jet velocities, up to the sonic limit (Hering, Flagan, and Friedlander 1978). De la Mora et al. (1984) have demonstrated the inertial impaction of heavy molecules in a gas using supersonic gas jets and low pressures, showing the potential for inertial classification to molecular sizes.

An alternate method for aerodynamic classification of very small particles is by diffusion. Lundgren and Rangaraj (1982) developed a screen type diffusion battery for stack sampling at temperatures as high as 260°C. In this instrument the aerosol flows through a series of fine screens. The smaller particles diffuse to the screens and are lost. A filter sample collected downstream of the screens will contain large particles. By varying the number of screens, a number of cumulative measurements can be made in parallel. Again, with suitable materials of construction, this measurement method could conceivably be operated at relatively high temperatures.

Optical measurements. Optical particle counters measure particle size by light scattering from individual particles. The aerosol is commonly aerodynamically focused so that the particles pass through the center of the view volume. This eliminates much of the uncertainty encountered in in situ optical counting at the intensity to which the particles are exposed. To extend the size range to very small sizes, active scattering

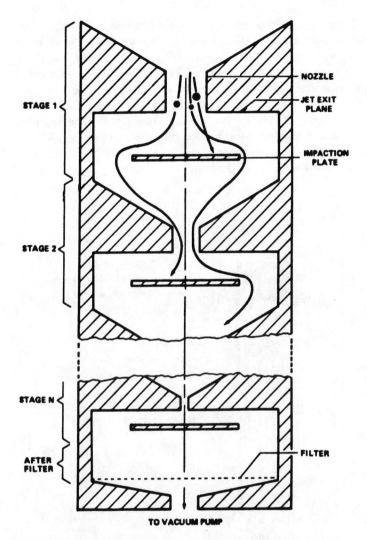

Figure 12-23 Schematic diagram of a cascade impactor (*Pui and Liu 1988*).

instruments have been developed. In these instruments the particles are passed through the resonant cavity of a helium-neon laser. At small sizes, the optical particle counters are capable of quite high resolution, provided the optical properties of the particles are known, but in the range of the Mie structure resonances the scattered intensity is a multiple valued function of the particle size.

Electrical mobility measurements. The measurement of the size distributions of particles smaller than 100-nm diameter is most commonly performed using electrical mobility analysis in which the particles are given an electrical charge and then classified in an electrical field. The electrical aerosol analyzer (EAA), originally developed by

Whitby and Clark (1966), has been widely used for particle size distribution measurements. Figure 12-24 is a schematic of the EAA. The aerosol is introduced into a charger, where it is exposed to unipolar positive ions. The charged aerosol then flows into the mobility analyzer, where particles with mobilities above a value determined by the potential on a cylindrical electrostatic precipitator are removed from the flow. Particles with lower mobilities pass through the field and are conveyed to Faraday cup, where the current they carry is measured with an electrometer. The concentration in different size intervals is determined by measurements at a series of electric field strengths. The sequential difference measurements of the EAA limit its usefulness in the measurement of size distributions from unsteady aerosol sources. Only a fraction

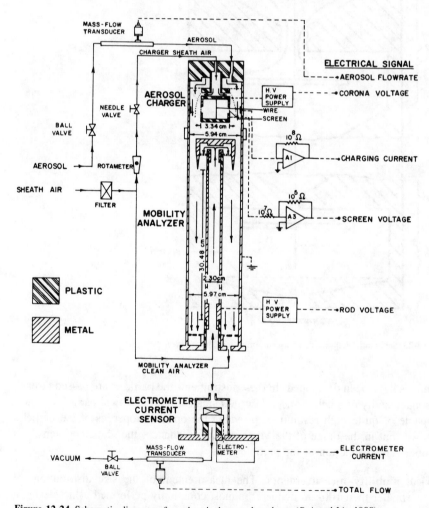

Figure 12-24 Schematic diagram of an electrical aerosol analyzer (*Pui and Liu 1988*).

of the particles smaller than about 50 nm are charged, but larger particles can carry multiple charges. This leads to different sizes of particles having the same mobilities, and severely complicates the data analysis.

The differential mobility classifier allows higher resolution using the mobility classification technique (Liu and Pui 1974). A small flow is extracted from holes in the collection rod of the classifier, so only particles with mobilities within a narrow range are removed for analysis. The resolution is further increased by using a bipolar diffusion charger, which reduces the charging efficiency but extends the range for which particles are at most singly charged to larger sizes. Gupta and McMurry (1989) have recently taken this approach one step further, extending the size range for singly charged particles up to 1 μm. The reduced charging efficiency of these methods reduces the current flow that must be measured if an electrometer is used for particle detection.

Electrometer techniques can still be used with differential mobility analyzers, but single-particle counting techniques are more common. To count extremely small particles, a condensation nuclei counter (CNC) must be used. In the CNC, a vapor is condensed on the small particles to grow them to a size that can be counted using light scattering. By making measurements at a number of voltages, high-resolution measurements of the particle size distribution can be made. Sampling times range from a minimum of about 10 min to more than 1 hour, depending on the resolution desired. Particles ranging in size from 3nm to 1 μm can be measured with various combinations of DMA (differential mobility analyzer) columns and detectors. The sampling time can be greatly reduced if, instead of making measurements at constant field strength, the electric field is scanned and particles are counted continuously (Wang and Flagan 1990). With this scanning electrical mobility spectrometer, the time required to obtain a high-resolution particle size distribution measurement can be reduced to 30 s or less.

12-5 PORE STRUCTURE MEASUREMENTS

The combustion of coal and char involves surface reactions. Combustion proceeds by a combination of reactions on the exterior surfaces of the char and within the pores. The traditional methods for probing the porous microstructure of the char are based on the penetration of a gas or liquid into the pores. The adsorption of gases onto the surface of the char is commonly used to measure the total surface area of the char, e.g., the adsorption of nitrogen in the Brunauer-Emmett-Teller (BET) method. By measuring the sorption to relatively high pressures, estimates of the pore size distribution can also be obtained. Mercury intrusion under high pressure is a commonly used method for the measurement of the distribution of pore volume with pore size, particularly at relatively large pore sizes. Estimates of the pore and surface area distribution can also be made using X-ray or neutron-scattering data. These measurements must, with a few possible exceptions, be made on samples of coal or char that have been removed from the combustion system. Although the measurements are not made directly in the combustion process, they are essential to the understanding of coal combustion and will therefore be briefly reviewed here.

Gas Adsorption

The quantity of gas adsorbed on a solid surface depends on the temperature of the system, the partial pressure of the particular adsorptive, and, of course, the nature of the solid. The extent of adsorption is a function of pressure of the adsorptive at a fixed temperature (an adsorption isotherm). With knowledge of the surface area covered by an adsorbed molecule, the surface area available for adsorption can be inferred. Ideally, sufficient time is allowed for the adsorptive to diffuse throughout the porous solid, so the total surface area can be estimated.

Brunauer et al. (1940) identified five basic types of isotherms. These are illustrated in Fig. 12-25. The isotherm type depends on both the nature of the gas (adsorbate) and that of the solid (adsorbent). For a typical nonpolar adsorbate like nitrogen, the isotherms depend on the nature of the solid. Microporous solids generally give a type I isotherm. These rise sharply at low relative pressures and reach a plateau. This is consistent with the hypothesis that no further adsorption takes place upon formation of a monolayer. Types II–V result from multilayer adsorption. Type II is found for multilayer adsorption on nonporous solids unless the adsorbent-adsorbate is very weak, in which case a type III results. Types IV and V are characteristic isotherms for porous solids. For types III and V, enhanced adsorption at relative pressures greater than 0.4 is attributed to capillary condensation of adsorbate in the small pores of the solid due to the reduced equilibrium vapor pressure of the liquid above an interface with a small

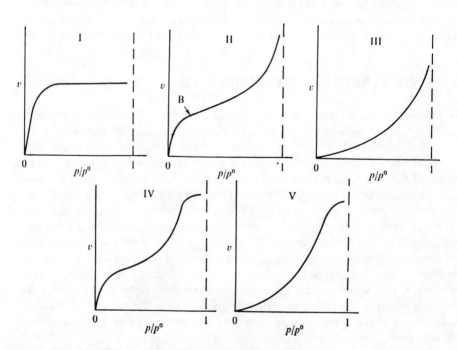

Figure 12-25 Classification of adsorption isotherms (*Brunauer et al. 1940*).

negative radius of curvature as illustrated in Fig. 12-26. This is the region of the isotherm where information about pore size distribution can be obtained.

The conventional (static) method of gas adsorption involves outgassing the sample under vacuum at a relatively high temperature, cooling it to the required temperature (usually the normal boiling point of the adsorbate) while it is isolated, filling a control volume with adsorbate to a controlled pressure, opening the sample volume to the control volume, and finally, recording the pressure after a specified lapse of time. Measurements are made at a number of pressures to generate the isotherm. The amount of gas adsorbed at each pressure level is simply the amount of gas introduced into the system less the gas remaining in the system after the pressure measurement is taken. While this method allows sufficient time for equilibration at each pressure point, it can be very tedious and time consuming. A variation of this technique is the discrete flow method. At first, the sample is out gassed by heating in a flow of helium. Thereafter, a mixture of nitrogen and helium (the relative amount determines the nitrogen partial pressure) is admitted over the sample, and the situation is allowed to equilibrate. Finally, the flow is switched back to helium, and the sample desorbs nitrogen. The amount of desorbed nitrogen in the presence of helium is detected by a thermal conductivity detector, since the thermal conductivities of helium and nitrogen are very different. The run is repeated at different partial pressures of nitrogen to complete the isotherm. While this method too is time consuming, it obviates the need for high-vacuum equipment. More recently, continuous pressure flow methods have been reported. In these, a steady supply of metered gas is admitted (via a mass flow controller or an orifice under choked flow conditions) to a system of known volume containing the outgassed sample. The system pressure is continuously monitored. The

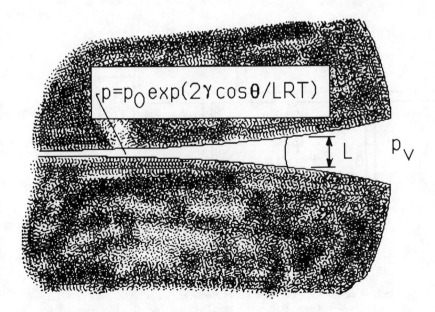

Figure 12-26 Capillary condensation in a porous matrix.

rate of gas introduction into the sample cell should be small enough to ensure quasi-equilibrium conditions between the gas and the sample. This method is much faster and if the gas introduction rate is slow enough, equilibration is not a major problem (Northrop, Flagan, and Gavalas 1987). A schematic apparatus for continuous flow gas adsorption is shown in Fig. 12-27.

Desorption isotherms often do not coincide with adsorption isotherms for porous materials. This hysteresis (Fig. 12-28) is attributed to different physical states of the gas-solid system that occur during the processes of evaporation and condensation. Different types of hysteresis loops are discussed by Gregg and Sing (1976). Hysteresis has been used to infer about pore shape (de Boer 1958); however, it is debatable as to how accurate this information is, insofar as actual pores are rarely, if ever, describable by regular geometrical shapes.

The dynamics of adsorption and the exact nature of solid surfaces are not well understood on the molecular level, so interpretation of experimental data is uncertain. The interpretation of data in the regime of capillary condensation is even more questionable. Simplifying assumptions have been made in the development of models to

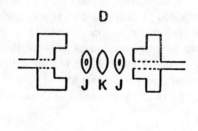

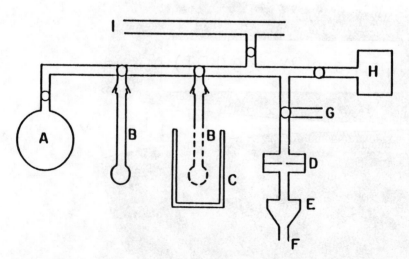

Figure 12-27 Schematic of the gas adsorption apparatus of Northrop, Flagan, and Gavalas (1987). Inset is the orifice mount. A, Calibration volume; B, sample holder; C, Dewar flask; D, orifice mount; E, filter; F, adsorptive line; G, helium line; H, pressure gage; I, vacuum line; J, teflon washer; and K, orifice plate.

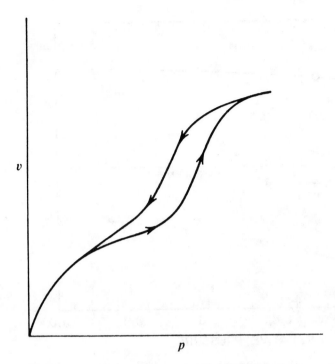

Figure 12-28 Typical adsorption hysteresis for a type IV isotherm.

elucidate the nature of the underlying porous structure of the solid. The most common problem with these models is that some ideal pore shape is assumed a priori. Pores are often assumed to be cylindrical or slit shaped, although actual pore geometries are generally much more complex. In spite of this shortcoming, some models provide a qualitative feel for particle morphology. The analysis of particle morphology in the microporous region is extremely difficult because of the presence of strong overlapping pore potentials from pore walls, about which little is known. The concept of rigid pore geometry is too limiting for a realistic description of the structure at molecular dimensions. Some of the more common methods for analyzing the pore structure in this regime are discussed in the next section. Details of experimental apparatus and procedure are given elsewhere.

Adsorption isotherms on an oxidized char are shown in Fig. 12-29 for various gases. The nitrogen and argon isotherms on the oxidized char are examples of the classic pseudo-type I isotherm. This shows that the char contains a large number of micropores. Once these micropores are filled, there is little subsequent adsorption. The pore volume of the char is 0.039 cm^3/g. The CO_2 isotherm at 196 K is also pseudo-type I. The large freon molecule interacts with the char in a fundamentally different manner, leading to an isotherm that does not fit the existing classification.

A number of different theories have been proposed to describe the adsorption of gases over solids and to explain the isotherms described above. The simplest is the

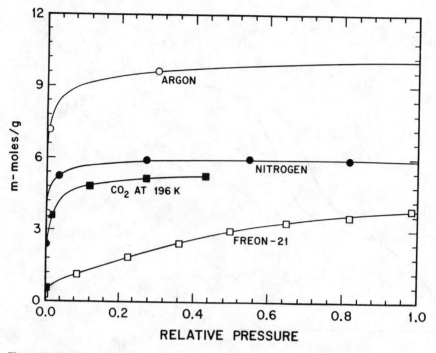

Figure 12-29 Isotherms of an oxidized char for different gases (*Sahu 1988*).

Langmuir theory, which describes monolayer adsorption only. The Langmuir equation is given by

$$v = \frac{v_m bp}{1 + bp} \qquad (12\text{-}53)$$

where v is the volume of gas adsorbed, v_m is the volume of gas required to cover the surface with a complete monolayer of molecules, p is the pressure of the gas, and b is a constant. At low and intermediate pressures, $v \simeq v_m bp$, so that the adsorbed volume increases with pressure. At higher pressures (when $bp \gg 1$), $v \simeq v_m$, indicating that the adsorption tends to the limiting value for monolayer coverage. This behavior describes the type I isotherm. For applicability of Eq. (12-53) to experimental data, it is generally written as

$$\frac{p}{v} = \frac{p}{v_m} + \frac{1}{v_m b} \qquad (12\text{-}54)$$

so that a plot of p/v versus p should be a straight line. The constants v_m and b can then be obtained from the plot.

The BET model extends the monolayer Langmuir model to multilayer adsorption. Although many models have been developed to explain physical adsorption on free surfaces (Brunauer, Emmett, and Teller 1938; Halsey 1948; Young and Crowell 1962; Sircar 1985), the BET theory remains the most widely used model for the estimation of surface areas from adsorption isotherms. It assumes that the surface is homogeneous and that the different layers of molecules do not interact. For adsorption on flat surfaces, the theory places no restriction on the number of layers that can be adsorbed. These assumptions are difficult to justify, yet the model is very useful in a qualitative sense and describes type II and III isotherms. Derivations of the BET equation are given by Gregg and Sing (1976). In a form that will fit experimental data, the BET equation is given by

$$\frac{x}{v(1 - x)} = \frac{1}{v_m C} + \frac{(C - 1)x}{v_m C} \tag{12-55}$$

where x is the relative pressure of the gas given by $x = p/p_0$, p_0 being the saturation pressure, and C is a constant described below. For type II isotherms the BET equation will generally fit data in the range of relative pressures between 0.05 and 0.35. The constant C and v_m can be obtained from a plot of $x/v(1 - x)$ versus x. Knowing v_m and the cross-sectional area of the adsorbate molecule (for nitrogen this is 16.2 Å^2), the surface area of the sample is readily calculated. The constant C in Eq. (12-55) is given as

$$C \simeq \exp \frac{E_1 - L}{RT} \tag{12-56}$$

where E_1 is the heat of adsorption of the first layer, L is the latent heat of condensation of the vapor to bulk liquid (assumed to be equal to the heat of adsorption of all layers after the first), R is the universal gas constant, and T is the temperature. Thus a large value of C implies that the adsorption in the first layer is strong relative to adsorption in the higher layers. This indicates that the first layer should be almost complete before other layers can form. In these situations a "knee" forms in the isotherm at low x values (typical type II isotherm). The change in isotherm type from II to III occurs at a C value of 2, when the point of inflection in the isotherm coincides with the origin. Although the BET equation does not describe capillary condensation and is therefore inapplicable for type IV and V isotherms, a modification of the BET equation for adsorption in pores (where due to pore walls there must be only a finite number of layers adsorbed) is given by the n layer BET equation. For x larger than 0.35, this equation results in a better fit of the experimental data. The n layer equation is given by

$$\frac{v}{v_m} = \frac{xC}{(1 - x)} \frac{[1 - (n + 1)x^n + nx^{n+1}]}{[1 + (C - 1)x - Cx^{n+1}]} \tag{12-57}$$

Equation (12-57) reduces to the Langmuir equation for $n = 1$ and to Eq. (12-55) for $n = \infty$.

Figure 12-30 shows the BET plots for the isotherms on the oxidized char given in Fig. 12-29. The upward deviation at higher relative pressures is due to the fact that there is little additional adsorption in this region. Once the micropores of the solid are filled with adsorbate, the solid appears virtually nonporous. The specific area was found to be around 400 m^2/g.

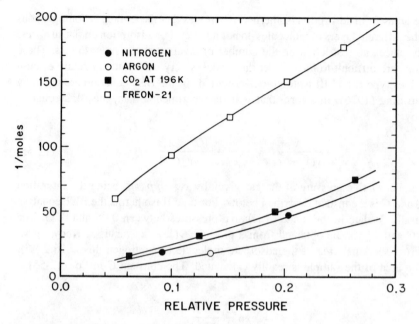

Figure 12-30 BET plots for oxidized char (*Sahu 1988*).

At high relative pressures, capillary condensation occurs in the pores of the particles leading to isotherms of type IV and V. Capillary condensation occurs in porous solids below the vapor pressure of the adsorbate due to the pressure difference that exists across the curved meniscus of the condensed vapor.

The relationship between the size of the capillary and pressure is given by the Kelvin equation:

$$\ln (p/p_0) \equiv 2\gamma\overline{V} \cos \theta/LRT \tag{12-58}$$

where γ is the surface tension, $\overline{V}$ is the molar volume, θ is the angle of contact, R is the gas constant, and T is the absolute temperature. L is a measure of the width of a pore; for a cylinder it is the radius, while for a slit it is half the distance between walls.

Wheeler (1946) considered a model for capillary condensation that included multilayer adsorption. He deduced a relation between specific pore volume $V(r)$ and the volume of nitrogen desorbed over a small segment on the desorption branch of the isotherm $v(r)$:

$$v(r) = \frac{[r - t(r)]^2}{r} V(r) + dt/dr \int_r^{r_0} \frac{2r' - t(r')}{r'^2} V(r')dr' \tag{12-59}$$

where r is the radius of the pore and t is the thickness of the adsorbed layer of gas. The first term on the right accounts for liquid evaporated from the pore cores, while the second term represents the amount of gas desorbed from free surfaces.

The thickness of the layer t is obtained from what is known as a t curve (Lippens and deBoer 1965). This is simply a plot of volume of adsorptive divided by BET surface area at monolayer coverage versus relative pressure for a nonporous substance. It has been pointed out that a t curve of a material with a similar C value must be used to obtain the proper t values (Lecloux and Pirard 1979).

In principle, both the Kelvin radius and the average thickness of adsorbate can be calculated for a given relative pressure. The appropriate values can then be used in the integral equation Eq. (12-59). However, Wheeler (1946) did not develop an iterative scheme to calculate $V(r)$. This was done later by several others, among them Pierce (1953), Dollimore and Heal (1964), and Cranston and Inkley (1957). More recently, Yan and Zhang (1986) developed an efficient means of calculating pore volume distribution based on Wheeler's model. According to their procedure, the volume of pores of mean radius r_i is given by

$$\Delta V_i = R_i \left(\Delta v_i - 2\Delta t_i \sum_{j=1}^{i-1} \frac{\Delta V_j}{\bar{r}_j} + 2\bar{t}_i \Delta t_i \sum_{j=1}^{i-1} \frac{\Delta V_j}{\bar{r}_j^2} \right)$$

$$i = 1, 2, \ldots, n$$

(12-60)

where i refers to the pore group and the overbar denotes average values. The Yan-Zhang inversion described above was applied to the nitrogen isotherm (in Fig. 12-29) to determine the pore volume distribution of the char (Fig. 12-31). The pore volume distributions are calculated instead of pore surface area distributions for the following reasons.

1. Fundamentally, the concept of surface area in porous (or microporous) solids is nebulous on close scrutiny. Surface areas are useful only as a comparison between different materials; therefore their intrinsic value has been questioned by many researchers (Debelak and Schrodt 1979).
2. Geometrical assumptions, often simplistic, must be made to derive surface area distributions from the experimental data.
3. Most inversions naturally lend themselves to pore volume distribution determinations. The volume distributions are presented as plots of $dV/d \log r$ versus $\log r$ to facilitate visual interpretation of the plots. When presented in this way, the contributions of different intervals can readily be seen by visual integration of the pore distributions. This presentation is preferred over cumulative distributions, since the results are not biased by pore sizes that are not probed by a particular measurement method.

The lower limit of applicability of the Kelvin equation Eq. (12-58) occurs at a relative pressure of about 0.4, which corresponds to a radius of 16 Å. Below this, the concept of bulk surface tension becomes harder to justify. Of course, smaller pores may be present in the solid. Thus it is necessary to consider analyses that deal with these micropores.

Dubinin (1960) originally developed a theory for micropore volume analysis of carbonaceous solids. The theory postulates the presence of pore potentials whose distribution is of the form

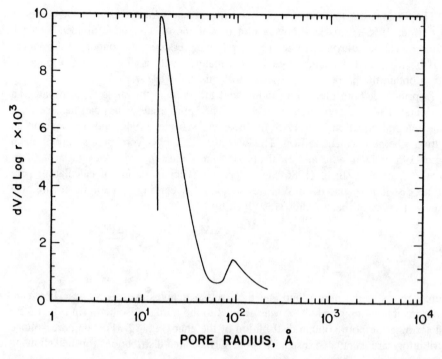

Figure 12-31 Pore volume distribution from capillary condensation for oxidized char (*Sahu 1988*).

$$N = N_0 \exp\left(-KE_0^2\right) \tag{12-61}$$

where N is the number of moles of gas adsorbed, N_0 the total number of moles at monolayer coverage, E_0 the energy of adsorption of the reference adsorbate, and K a constant characteristic of the system. Also, all adsorbates can be scaled to a single reference adsorbate:

$$N = f(E/\beta) \tag{12-62}$$

where $\beta = E/E_0$. The work required to isothermally compress a gas from p to p_0 is

$$E = RT \ln (p_0/p) \tag{12-63}$$

Simple substitution of Eqs. (12-62) and (12-63) into Eq. (12-61) gives

$$N = N_0 \exp\left[-K(RT)^2 \ln^2 (p_0/p)/\beta^2\right] \tag{12-64}$$

or, taking logarithms,

$$\log N = \log N_0 - 2.303K(RT/\beta)^2 \log^2(p_0/p) \tag{12-65}$$

A plot of $\log N$ versus $\log^2 (p_0/p)$ should give a straight line, from which the number of moles at monolayer coverage can be obtained.

Mercury Porosimetry

Mercury porosimetry is widely used to study the pore structure properties of porous solids (Orr 1969). Mercury, due to its nonwetting nature with most substances, can be forced under pressure to penetrate pores, openings, or voids in materials. For pores of regular geometry, the relationship between the pore radius penetrated at a given intrusion pressure is given by the Washburn equation:

$$r_p = \frac{2\gamma \cos \theta}{p} \tag{12-66}$$

where r_p is the pore radius, γ is the surface tension of mercury (normally taken to be 480 ergs/cm^2), θ is the angle of contact between mercury and the solid, and p is the applied pressure. The contact angle with coal/char is conventionally assumed to be 140°. This equation assumes that the pores are cylindrical in shape. On a conventional porosimeter, the volume of mercury penetrated is recorded as a function of pressure. From this, the pore volume and pore surface area distributions can be calculated. Moscou and Lub (1981) discuss the use of porosimetry to study porous materials. Data interpretation, including hysteresis and pore geometry, is discussed by Lowell and Shields (1979). Lowell and Shields (1981) also discuss the equivalence of mercury porosimetry and gas adsorption.

Figure 12-32 shows an example of the raw data derived from mercury porosimetry on a cenospheric bituminous coal char, with particles in the 90- to 104-μm size range. The abscissa is the volume in cubic centimeters, and the ordinate is the gage pressure in pounds per square inch. The intrusion and extrusion curves are both shown. The former is shown in two parts, the scale being expanded at lower pressures for better accuracy in that range. This cumulative curve (intrusion branch) was numerically differentiated to give the pore volume distribution as shown in Fig. 12-33. A dead volume correction was applied before the numerical differentiation.

The distribution for the unburned material shows that there is significant transition porosity (200–1000 Å radius) as well as macroporosity in this char. Pores larger than 1.76 μm in radius are penetrated at pressures below 60 psig and are assumed to be interparticle spaces rather than internal porosity. Of course, it is not possible to rule out the presence of cracks and voids of these sizes or even larger ones within particles. Indeed electron micrographs do show the presence of voids of this size in the particles. Thus at these low pressures it is impossible to resolve the controversy regarding the relative contributions of the interparticle and intraparticle voids to the total voidage, particularly in a material that has large void openings and cracks.

X Ray and Neutron Scattering

Some of the earliest physical characterizations of coal were done by Hirsch (1954) using X ray scattering to elucidate the molecular structure of coal. In this section, the use of X ray and neutron scattering to determine the pore structure of coals and chars will be discussed. The subject of coal molecular structure including the presence of different functional groups and the "molecular weight" of coal deserves more space than can be devoted to it in this review.

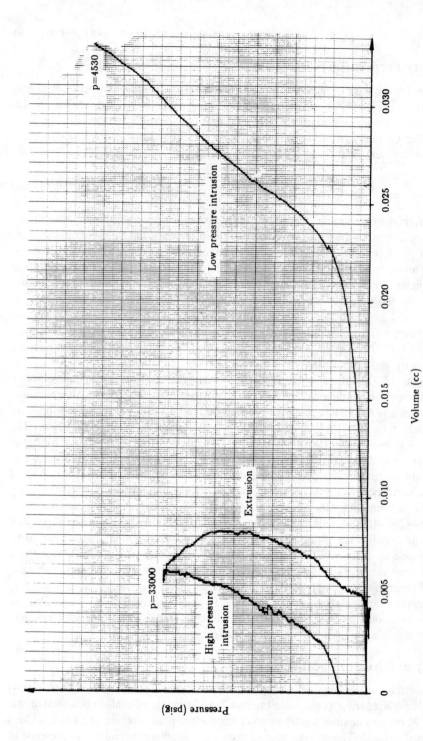

Figure 12-32 Raw data from a mercury porosimetry intrusion measurement (*Sahu 1988*).

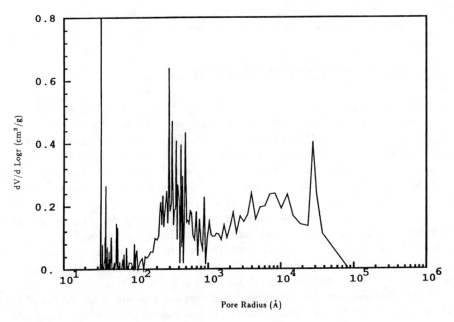

Figure 12-33 Pore volume distribution from mercury porosimetry (*Sahu 1988*).

While adsorption of gases in coal provides much of the currently known information regarding pore structure and porosity in coals, in some situations, using gas adsorption may not be appropriate. First, neither gas adsorption nor mercury porosimetry provides information on closed porosity that can contribute to a dramatic increase in the surface area early in the char oxidation phase of combustion. For some low-rank coals that can undergo severe shrinkage upon drying (usually required before gas adsorption), gas adsorption techniques may underestimate their porosity. Finally, in gas adsorption, chemisorption may be involved between the adsorbate and the adsorbent in some situations, complicating data interpretation.

SAXS obviates these problems. Mahajan (1984) discusses some of the applications of SAXS. Although there is some controversy in the literature regarding the ability of SAXS to measure micropore areas (Kalliat, Kwak, and Schmidt 1981; Spitzer and Ulicky 1976), areas measured by SAXS agree with those obtained from CO_2 adsorption. Setek, Snook, and Wagenfeld (1984) discuss the application of SAXS to determine the pore structures of a wet and a dry Australian brown coal. They discuss some of the drawbacks of the method and also the need to do both adsorption and SAXS for a more complete picture of the internal structure. Bale et al. (1984) have applied SAXS to determine the structure of low-rank American coals, and proposed an inversion method to determine the pore dimensions from scattering data for lignites. Hall and Williams (1985) measured surface areas and porosities of a wide range of materials using SANS. Although closely related to SAXS, Hall and Williams (1985) suggest that SANS has advantages over SAXS, relating to wider variations of wavelength and momentum transfer without absorption problems. Their paper also summarizes the related theoretical and data analyses needed in using SANS.

12-6 CARBON CONVERSION MEASUREMENTS

The single most important variable in characterizing the partially oxidized samples is carbon conversion or burnoff. Conversion is most commonly determined gravimetrically from mass measurements made before and after combustion. Assuming that there are no sampling losses, the extent of conversion is easily estimated. However, in most cases, sampling losses are unavoidable, so this method is relatively inaccurate. A more direct method of determination, although less practical, is to measure size and density independently and thereby infer the conversion.

Tracers can be employed to determine the carbon conversion. The important restriction on a tracer is that its mass remain constant during the combustion process. Based on this assumption, and knowing the fractional mass of the tracer in the initial and final materials, the loss of carbon can be easily estimated, provided carbon is assumed to be the only material that can be gasified.

Typical materials that have been used as tracers are total ash in the coal or char and various refractory oxides in coal or char. Unless temperatures are very high, ash vaporization is quite small (Flagan and Friedlander 1978), and therefore for most applications it is safe to assume that the mass of total ash remains constant in the sample. This is more true in the case of chars (that have been made at high temperatures and thereby already lost the low volatile fractions) than coals. In the case of coals, suitable oxides like alumina or titania have been used as tracers. The amount of ash in the sample is usually determined by completely oxidizing the samples. This is definitely a drawback of using ash as the tracer, since it is a destructive process. Also rapid heating rates have to be used in ashing to prevent graphitization of the carbon. This is necessary, since otherwise, the resulting graphitic structure is difficult to oxidize except at very high temperatures. Graphitic residues left with the ash could lead to erroneous carbon conversion values. Oxide contents are measured by atomic absorption or neutron scattering. These techniques are capable of great accuracy even with small sample sizes.

Carbon conversion can also be measured by determining the amounts of CO and CO_2 formed by burning a known sample of char. Usually gas chromatography and mass spectrometry are employed to determine the amounts of the carbon oxides. Details of the implementation of this method can be found in the work by Northrop (1988).

12-7 MODEL FUELS

The variability of coal properties hinders quantitative interpretation of coal or char combustion data, since in most cases, measurements involve averages over an ensemble of many fuel particles. This has led researchers to develop model materials to elucidate reaction mechanisms. The variability arises from the carbonaceous structure of coal, the included mineral matter, and processing technologies. There are several carbon-aceous structures known as macerals present in coal. The macerals are microscopically identifiable organic portions of the coal that have different origins, physical, and chemical features, and reactivity. Morley and Jones (1986) observed that the surface oxidation rate coefficients at 1350 K of different macerals varied by as much as a factor of 2.

The mix of minerals is representative of the minerology in the region where the coal was found. The fate of mineral matter is directly coupled to the complex chemistry of the mixed mineral systems that result when coal minerals melt and coalesce during combustion. Even within a relatively homogeneous coal sample from a single seam, the properties can vary significantly. The mineral content may also influence the char oxidation rate, either through catalytic reactions or physical obstruction of surface or pores.

As coal is ground to small sizes, the properties of the particles vary dramatically from one another. For example, Littlejohn (1965) found that the mineral content of pulverized coal increased substantially with decreasing particle size. This enrichment of the small coal particles results from the aerodynamic classification that takes place within power plant pulverizers. All this variability hinders quantitative interpretation of coal or char combustion data, since in most cases, measurements involve averages over an ensemble of many fuel particles.

One further complication is the highly variable particle shape of pulverized coal. Theoretical interpretation of coal combustion data is complicated by the complex shapes and variable sizes of coal particles. Theoretical descriptions of char particle oxidation have frequently modeled the char as a dense sphere of carbon, e.g., Libby and Blake (1979). Pulverized coal combustion models either treat the coal as uniformly sized spheres or as a collection of spheres with several discrete sizes.

Many studies of the fundamental mechanisms of char oxidation or the details of mineral transformations have employed surrogate materials that maintain the essential features of coal or coal char while eliminating some of the complicating features. Senior and Flagan (1984) developed synthetic chars for the study of ash vaporization. Amorphous or glassy carbons were generated with different pore structures and mineral contents. These materials were generated by high-temperature curing of a polymer of furfuryl alcohol. The mineral content of the chars was limited to quartz, eliminating the complex thermochemistry of the ash mixtures encountered in coal combustion. In drop-tube combustion experiments, as much as 32% of the silica vaporized and formed a submicron fume. Inferences were made regarding the volatilization mechanism. The mechanism appeared to differ from previous results on silica vaporization in coal combustion, but reasons for the differences were not resolved.

Levendis and Flagan (1987) used an acoustically excited capillary jet atomizer to produce mineral-free, spherical, synthetic char particles of uniform size such as those shown in Fig. 12-34. The original particles were microporous and initially had little surface area that was accessible to nitrogen adsorption ($2 \text{ m}^2 \text{ g}^{-1}$). After about one-third of the carbon had been consumed in drop-tube combustion experiments, the pore structure had changed dramatically, with the surface area increasing to $350 \text{ m}^2 \text{ g}^{-1}$. Similar methods had previously been used (Szekely and Faeth 1982; Szekely, Turns,

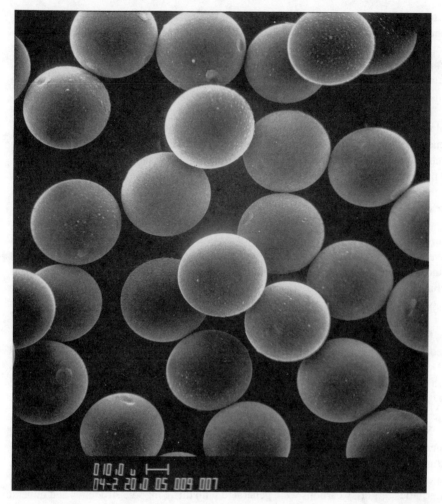

Figure 12-34 Electron micrograph of spherical glassy carbon particles (*Levendis and Flagan 1987*).

and Faeth 1984) for the production of fairly uniformly sized aggregates of carbon black.

Levendis and Flagan later developed procedures for producing uniformly sized, spherical, synthetic char particles with a wide range of porous microstructures through the addition of high boiling organics or dispersed carbon black to the original polymer solution (Levendis and Flagan 1989; Levendis, Flagan, and Gavalas 1989). With SAXS the large surface area seen after partial combustion was shown to be present in the initial synthetic char even though it was not available for nitrogen adsorption. X ray diffraction studies on this mineral-free char showed evidence of oxygen-catalyzed graphitization of the amorphous carbon.

Intrinsic oxidation rates of glassy carbon particles were determined from single-particle pyrometry studies at high temperatures and gravimetric measurements at lower temperatures. The rate exhibited a plateau at high temperature as previously observed for pyrolytic graphite (Nagle and Strickland-Constable 1963) and carbon black (Park and Appleton 1973).

Other synthetic materials have also been used as surrogates for coal char in the study of the fundamental mechanisms of char combustion. Floess, Longwell, and Sarofim (1988a) used chars made of sucrose and commercially available spherical carbons known as Spherocarb (Analabs, North Haven, Connecticut) for thermogravimetric studies of char oxidation kinetics. In studies of the catalytic enhancement of char oxidation by calcium, they observed a 2 order of magnitude enhancement (Floess, Longwell, and Sarofim 1988b), as later confirmed by Levendis et al. (1989b) using glassy carbons.

Helble and Sarofim (1989) used the Spherocarb and sucrose chars to probe the role of the porous microstructure on char fragmentation and the evolution of the residual ash particle size distribution in pulverized coal combustion. Macroporosity was found to control the degree of fragmentation. The microporous sucrose chars produced only one ash particle per char particle, while each macroporous Spherocarb generated approximately 75 ash particles larger than 1-μm diameter under identical combustion conditions.

12-8 NOMENCLATURE

a_T total specific surface area, Eq. (12-12) (m^2/kg)

A_a apparent frequency factor, Eq. (12-6) (m/s)

A_i intrinsic frequency factor, Eq. (12-6) (m/s)

A_E active surface area, Eq. (12-5) (m^2)

A_T total surface area, Eq. (12-5) (m^2)

b constant in Langmuir equation, Eq. (12-53)

c_p particle specific heat, Eq. (12-19) [J/(kg K)]

C slip correction factor, Eq. (12-51)

C constant in BET equation, Eq. (12-55)

C_1 first radiation constant, Eq. (12-27) [(W μm)/m^2]

C_2 second radiation constant, Eq. (12-27) (μm K)

C_g O_2 concentration in bulk gas, Eq. (12-7) (kg/m^3)

C_s O_2 concentration at surface, Eq. (12-3) (kg/m^3)

D fractal dimension, Eq. (12-26)

D_p particle diameter, Eq. (12-2) (m)

$\mathscr{D}$ bulk gas diffusivity, Eq. (12-8) (m^2/s)

$\mathscr{D}$ particle diffusivity, Eq. (12-42) (m^2/s)

$\mathscr{D}_e$ effective diffusivity, Eq. (12-14) (m^2/s)

$\mathscr{D}_{pore}$ pore diffusivity, Eq. (12-16) (m^2/s)

e_λ Graybody emittance, Eq. (12-28) [W/(μm m^2)]

$e_{\lambda b}$ blackbody emittance, Eq. (12-27) [W/(μm m^2)]

E_0 heat of adsorption, Eq. (12-61) (J/kg)

E_1	heat of adsorption of first layer, Eq. (12-56) (J/kg)
E_a	apparent activation energy, Eq. (12-6) (J/kg)
E_i	intrinsic activation energy, Eq. (12-6) (J/kg)
f	A_E/A_T, Eq. (12-5)
h_m	mass transfer coefficient, Eq. (12-7) (m/s)
ΔH	heat of reaction, Eq. (12-19) (J/kg)
I	nucleation rate, Eq. (12-25) [L/(m^3 s)]
I	intensity of scattered light, Eq. (12-42) [W/(m^2 sr)]
I_0	intensity of incident light, Eq. (12-44) [W/(m^2 sr)]
I_λ	intensity at detector, Eq. (12-29) [W/(m^2 sr)]
J_1	Bessel function, Eq. (12-44)
k_a	apparent rate coefficient, Eq. (12-3) (m/s)
k_i	intrinsic rate coefficient, Eq. (12-4) (m/s)
k_{abs}	absorption coefficient, Eq. (12-40)
K	coagulation coefficient, Eq. (12-25) (m^3/s)
K	scattered wave vector, Eq. (12-42) (L/μm)
K	pore potential constant, Eq. (12-61) (kg^2/J^2)
Kn	Knudsen number, Eq. (12-51)
l	length, Eq. (12-47) (m)
L	collector dimension, Eq. (12-20) (m)
L	latent heat of condensation, Eq. (12-56) (J/kg)
m	mass of carbon, Eq. (12-1) (kg)
m	true reaction order, Eq. (12-4)
m	particle mass, Eq. (12-26) (kg)
m	Imaginative part of refractive index, Eq. (12-40)
M	molecular weight, Eq. (12-17)
M	mass distribution function, Eq. (12-23)
M	aperture transmittance function, Eq. (12-36)
n	apparent reaction order, Eq. (12-3)
n	number of adsorbed layers, Eq. (12-57)
N	number size distribution, Eq. (12-21)
N	number of moles adsorbed, Eq. (12-61) (moles)
N_0	moles in a monolayer, Eq. (12-61) (moles)
p	gas pressure, Eq. (12-53) (Pa)
P	particle image function, Eq. (12-36)
P	line shape function, Eq. (12-42)
Q	momentum transfer, Eq. (12-47) (L/m)
Q_{vv}	total scattering coefficient, Eq. (12-41)
$\bar{r}_p$	average pore radius, Eq. (12-17) (m)
r_p	pore radius, Eq. (12-66) (m)
R	mass loss rate, Eq. (12-1) (L/s)
R	universal gas constant, Eq. (12-6) [J/(kg K)]
R_a	apparent reactivity, Eq. (12-2) [kg/(m^2 s)]
R_i	intrinsic reactivity, Eq. (12-4) [kg/(m^2 s)]

R_m	diffusion-limited reactivity, Eq. (12-10) [kg/(m^2 s)]
R_p	particle radius, Eq. (12-29) (m)
R_{12}	signal ratio, Eq. (12-32)
S_i	detector response function, Eq. (12-30)
St	Stokes number, Eq. (12-20)
t	time, Eq. (12-1) (s)
t	adsorbed layer thickness, Eq. (12-59) (m)
T	temperature, Eq. (12-27) (K)
T_p	particle temperature, Eq. (12-6) (K)
T_∞	ambient temperature, Eq. (12-19) (K)
Th	thermophoresis group, Eq. (12-50)
U	fluid velocity, Eq. (12-20) (m/s)
v	velocity, Eq. (12-25) (m/s)
v	volume of adsorbed gas, Eq. (12-53) (m^3)
v_m	monolayer volume, Eq. (12-53) (m^3)
v_T	migration velocity, Eq. (12-50) (m/s)
V	pore volume, Eq. (12-60) (m^3)
$\bar{V}$	molar volume, Eq. (12-58) (m^3/mole)
W_2	slit width, Eq. (12-37) (m)
x	relative pressure, Eq. (12-55)
α	convection coefficient, Eq. (12-19) [W/(m^2 K)]
α	constant in slip factor, Eq. (12-52)
β	constant in slip factor, Eq. (12-52)
γ	constant in slip factor, Eq. (12-52)
γ	surface tension, Eq. (12-58) (N/m)
ϵ	porosity of solid, Eq. (12-16)
ϵ_p	particle emissivity, Eq. (12-19)
ϵ_λ	surface spectral emissivity, Eq. (12-37)
η	effectiveness factor, Eq. (12-12)
θ	scattering angle, Eq. (12-43) (deg)
θ	angle of contact, Eq. (12-58) (deg)
κ	calibration factor, Eq. (12-33)
λ	wavelength, Eq. (12-27) (μm)
λ_0	incident wavelength, Eq. (12-43) (μm)
λ_1	wavelength of first band, Eq. (12-32) (μm)
λ_2	wavelength of second beam, Eq. (12-32) (μm)
μ	fluid viscosity, Eq. (12-20) [kg/(m s)]
μ	gas viscosity, Eq. (12-50) [kg/(m s)]
ρ_a	apparent particle density, Eq. (12-2) (kg/m^3)
σ	Stefan-Boltzmann constant, Eq. (12-19) [W/(m^2 K^4)]
τ	tortuosity factor, Eq. (12-16)
τ_{coag}	coagulation time, Eq. (12-49) (s)
χ	$= R_a/R_m$, Eq. (12-9)
χ	dimensionless particle size, Eq. (12-39)

ω	scattered light frequency, Eq. (12-42) (L/s)
ω_0	incident light frequency, Eq. (12-42) (L/s)
Ω	solid angle, Eq. (12-29) (sr)

REFERENCES

Ariessohn, P. C., Self, S. A., and Eustis, R. H. 1980. *Appl. Opt.* 19:3775.

Ayling, A. B., and Smith, I. W. 1972. *Combust. Flame* 18:173.

Bachalo, W. D. 1980. *Appl. Opt.* 19:363.

Bale, H., Carlson, M. L., Kalliat, M., Kwak, C. Y., and Schmidt, P. W. 1984. ACS Symp. Ser. 264:79.

Bar-Ziv, E., Jones, D. B., Spjut, R. E., Dudek, D. R., Sarofim, A. F., and Longwell, J. P. 1989. *Combust. Flame* 75:81.

Baxter, L. L., Fletcher, T. H., and Ottesen, D. K. 1988. *Energy Fuels* 2:423.

Bhatia, S. K., and Perlmutter, D. D. 1980. *AIChE J.* 26:379.

Binford, J. S., and Eyring, H. 1956. *J. Phys. Chem.* 60:486.

Biswas, P., and Flagan, R. C. 1988. *J. Aerosol Sci.* 19:113.

Blyholder, G., and Eyring, H. 1959. *J. Phys. Chem.* 63:1004.

Brunauer, S., Emmett, P. H., and Teller, E. 1938. *J. Am. Chem. Soc.* 60:309.

Brunauer, S., Demming, L. S., Demming, W. S., and Teller, E. 1940. *J. Am. Chem. Soc.* 62:1723.

Calo, J. M., and Perkins, M. T. 1987. *Carbon* 25:395.

Chowdhury, D. P., Sorenson, C. M., Taylor, T. W., Merklin, J. F., and Lester, T. W. 1984. *Appl. Opt.* 23:4149.

Clark, N. A., Lunacek, J. H., and Benedek, G. B. 1970. *Am. J. Phys.* 38:575.

Cranston, R., and Inkley, F. 1957. *Adv. Cat.* 9:143.

Dalzell, W. H., Williams, G. C., and Hottel, H. C. 1970. *Combust. Flame* 14:161.

Davies, E. J. 1987. *Surface and colloid science*, vol. 14, ed. E. Matijevic, p. 1. New York: Plenum.

Davison, R. L., Natusch, D. F. S., Wallace, J. R., Jr., and Evans, C. A., Jr. 1974. *Environ. Sci. Technol.* 8:1107–13.

Debelak, K. A., and Schrodt, J. T. 1979. *Fuel* 58:732.

deBoer, J. H. 1958. *The structure and properties of porous materials*. London: Butterworths.

De la Mora, J. F., Halpern, B. L., Kramer, M., Yamashita, A., and Schmitt, J. 1984. In *Aerosols*, ed. B. Y. H. Liu, D. Y. H. Pui, and H. Fissan, p. 109. New York: Elsevier.

Dixon, J. 1988. *J. Phys. E: Sci. Instrum.* 21:425.

Dobbins, R. A., and Mulholland, G. W. 1984. *Combust. Sci. Technol.* 40:175.

Dollimore, D., and Heal, G. R. 1964. *J. Appl. Chem.* 14:109.

Dubinin, M. M. 1960. *Chem. Rev.* 60:235.

Dudek, D. R., Longwell, J. P., and Sarofim, A. F. 1989. *Energy Fuels* 3:24.

Dunn-Rankin, D., and Kerstein, A. R. 1986. *Western states section fall meeting*. Paper WSS/CI 86-10. Pittsburgh, Pa.: The Combustion Institute.

Dutta, S., and Wen, C. Y. 1977. *Ind. Eng. Chem. Process Des. Dev.* 16:31.

Essenhigh, R. H. 1981. In *Chemistry of coal utilization*, 2nd suppl. vol., ed. M. A. Elliott. New York: John Wiley.

Essenhigh, R. H., Froberg, R., and Howard, J. B. 1965. *Ind. Eng. Chem.* 57:32.

Farmer, W. M. 1972. *Appl. Opt.* 11:2603.

Field, M. A. 1969. *Combust. Flame* 13:237.

Field, M. A., Gill, D. W., Morgan, B. B., and Hawksley, P. G. W. 1967. *Combustion of pulverized coal*. British Coal Utilization Research Association, Leatherhead, U.K.

Flagan, R. C., and Friedlander, S. K. 1978. In *Recent developments in aerosol science*, ed. D. T. Shaw, p. 25. New York: John Wiley.

Flagan, R. C., and Seinfeld, J. H. 1988. *Fundamentals of air pollution engineering*. New York: Prentice Hall.

Floess, J. K., Longwell, J. P., and Sarofim, A. F. 1988a. *Energy Fuels* 2:18.

Floess, J. K., Longwell, J. P., and Sarofim, A. F. 1988b. *Energy Fuels* 2:756.

Flower, W. L. 1983. *Combust. Sci. Technol.* 33:17.

Forrest, S. R., and Witten, T. A. 1979. *J. Phys. A* 12:L109.

Gavalas, G. R. 1980. *AIChE J.* 26:577.

Gavalas, G. R. 1981. *Combust. Sci. Technol.* 24:197.

Gaydon, A. G., and Wolfhard, H. G. 1970. *Flames: their structure radiation and temperature*, London: Chapman and Hall.

Goodwin, D. G., and Mitchner, M. 1989. *J. Thermophys. Heat Transfer* 3:53.

Gregg, S. J., and Sing, K. S. W. 1976. In *Surface and colloid science*, vol. 9, ed. E. Matijevic, p. 231. New York: Plenum.

Grosshandler, W. L. 1984. *Combust. Flame* 55:59.

Gupta, A., and McMurry, P. H. 1989. *Aerosol Sci. Technol.* 10:451.

Hall, P. G., and Williams, R. T. 1985. *J. Colloid Interface Sci.* 104:151.

Halsey, G. D. 1948. *J. Chem. Phys.* 16:931.

Hamor, R. J., Smith, I. W., and Tyler, R. J. 1973. *Combust. Flame* 21:153.

Hardesty, D. R., Pohl, J. H., and Stark, A. H. 1978. Sandia Laboratories Energy Report SAND 78-8234, p. 26.

Hawtin, P., and Gibson, J. A. 1966. *Carbon (Oxford)* 4:501.

Heintzenberg, J., and Welch, R. M. 1982. *Appl. Opt.* 21:822.

Helble, J. J., and Sarofim, A. F. 1989. *Combust. Flame* 76:183.

Hering, S. V., Flagan, R. C., and Friedlander, S. K. 1978. *Environ. Sci. Technol.* 12:667.

Hinds, W., and Reist, P. C. 1973a. *Aerosol Sci. Technol.* 3:501.

Hinds, W., and Reist, P. C. 1973b. *Aerosol Sci. Technol.* 3:515.

Hippo, E. J., and Walker, P. L., Jr. 1975. *Fuel* 54:245.

Hirleman, E. D., and Wittig, S. 1976. *Sixteenth symp. (international) on combustion.* Pittsburgh, Pa.: The Combustion Institute.

Hirsch, P. B. 1954. *Proc. Roy. Soc. London* A226:143.

Hoff, H. A., Rothman, S. J., and Mundy, J. N. 1985. *J. Phys. E: Sci. Instrum.* 18:197.

Holve, D. J. 1980. *Combust. Sci. Technol.* 44:269.

Holve, D. J., and Davis, G. W. 1985. *Appl. Opt.* 24:998.

Holve, D. J., Tichenor, D., Wang, J. C. F., and Hardesty, D. R. 1981. *Opt. Eng.* 20:529.

Hottel, H. C., Williams, G. C., and Wu, P. C. 1966. *Prepar. Am. Chem. Soc. Div. Fuel Chem.* 10(3):58.

Howard, J. B., and Essenhigh, R. H. 1967. *Ind. Eng. Chem. Proc. Des. Dev.* 6:74.

Hurd, A. J., and Flower, W. J. 1988. *J. Colloid Interface Sci.* 122:178.

Jenkins, R. G., and Piotrowski, A. 1987. *Am. Chem. Soc. Div. Fuel Chem.* 32(4):147.

Jenkins, R. G., Nandi, S. P., and Walker, P. L., Jr. 1973. *Fuel* 52:288.

Jorgensen, F. R. A., and Zuiderwyk, M. J. 1985. *J. Phys. E: Sci. Instrum.* 18:486.

Kalliat, M., Kwak, C. Y., and Schmidt, P. W. 1981. *ACS Symp. Ser.* 169:3.

Kerstein, A. R., and Edwards, B. F. 1987. *Chem. Eng. Sci.* 42:1629.

Kerstein, A. R., and Niksa, S. 1984. *Twentieth symposium (international) on combustion*, p. 941. Pittsburgh, Pa.: The Combustion Institute.

Knill, K. J., Chambers, A. K., Parkash, S., Ungarian, D. E., and Zacharkiw, R. 1986. *Canadian and western states section spring meeting.* Pittsburgh, Pa.: The Combustion Institute.

Kobayashi, H., Howard, J. B., and Sarofim, A. F. 1976. *Sixteenth symposium (international) on combustion*, p. 411. Pittsburgh, Pa.: The Combustion Institute.

Lai, F. S., Friedlander, S. K., Pich, J., and Hidy, G. M. 1972. *J. Colloid Interface Sci.* 39:395.

Laurendeau, N. M. 1978. *Progr. Energy Combust. Sci.* 4:221.

Lecloux, A., and Pirard, J. P. 1979. *J. Colloid Interface Sci.* 70:265.

Lester, T. W., Seeker, W. R., Merklin, J. F. 1981. *Eighteenth symposium (international) on combustion*, p. 1257. Pittsburgh, Pa.: The Combustion Institute.

Levendis, Y. A., and Flagan, R. C. 1987. *Combust. Sci. Technol.* 53:117.

Levendis, Y. A., and Flagan, R. C. 1989. *Carbon* 27:265.

Levendis, Y. A., Flagan, R. C., and Gavalas, G. R. 1989. *Combust. Flame* 76:221.

Levendis, Y. A., Sahu, R., Flagan, R. C., and Gavalas, G. R. 1989a. *Fuel* 68:849.

Levendis, Y. A., Nam, S. W., Lowenberg, M., Flagan, R. C., and Gavalas, G. R. 1989*b*. *Energy Fuels* 3:28.

Libby, P. A., and Blake, T. R. 1979. *Combust. Flame* 36:139.

Lightman, P., and Street, P. J. 1968. *Fuel* 47:7.

Lippens, B. C., and deBoer, J. H. 1965. *J. Cat.* 4:319.

Littlejohn, R. F. 1965. *J. Inst. Fuel* 38:59.

Liu, B. Y. H., and Pui, D. Y. H. 1974. *J. Colloid Interface Sci.* 47:155.

Lowell, S., and Shields, J. E. 1979. *Powder surface area and porosity.* London: Chapman and Hall.

Lowell, S., and Shields, J. E. 1981. *Powder Technol.* 29:225.

Lowenberg, M., Bellan, J., and Gavalas, G. R. 1987. *Chem. Eng. Sci.* 58:89.

Lundgren, D. A., and Rangaraj, C. N. 1982. *Environ. Progr.* 1:79.

Macek, A., and Bulik, C. 1984. *Twentieth symposium (international) on combustion*, p. 1223. Pittsburgh, Pa.: The Combustion Institute.

Mahajan, O. P. 1984. *Powder Technol.* 40:1.

Mahajan, O. P., Yarzab, R., and Walker, P. L., Jr. 1978. *Fuel* 57:643.

Markowski, G. R., Ensor, D. S., Hooper, R. G., and Carr, R. C. 1980. *Environ. Sci. Technol.* 14:1400.

Marple, V. A., Liu, B. Y. H., and Kuhlmey, G. A. 1981. *J. Aerosol Sci.* 12:333.

Marsh, H., O'Hair, E., and Wynne-Jones, W. F K. 1965. *Trans. Faraday Soc.* 61:274.

McLean, W. J., Hardesty, D. R., and Pohl, J. H. 1981. *Eighteenth symposium (international) on combustion*, p. 1239. Pittsburgh, Pa.: The Combustion Institute.

Meakin, P. 1984. *Phys. Rev. A* 29:997.

Mehta, B. N., and Aris, R. 1971. *Chem. Eng. Sci.* 26:1699.

Mie, G. 1908. *Ann. Phys.* 25:377.

Mims, C. A., Neville, M., Quann, R. J., House, K., and Sarofim, A. F. 1980. *AIChE Symp. Ser.* 76(201).

Mitchell, R. E. 1987. *Combust. Sci. Technol.* 53:165.

Mitchell, R. E., and Madsen, O. H. 1986. *Twenty-first symposium (international) on combustion*, p. 173. Pittsburgh, Pa.: The Combustion Institute.

Mitchell, R. E., and McLean, W. J. 1982. *Nineteenth symposium (international) on combustion*, p. 1113. Pittsburgh, Pa.: The Combustion Institute.

Morgan, P. A., Robertson, S. D., and Unsworth, J. F. 1987. *Fuel* 66:210.

Morley, C., and Jones, R. B. 1986. *Twenty-first symposium (international) on combustion*, p. 239. Pittsburgh, Pa.: The Combustion Institute.

Moscou, L., and Lub, S. 1981. *Powder Technol.* 29:45.

Mountain, R. D., Mulholland, G. W., and Baum, H. 1986. *J. Colloid Interface Sci.* 114:67.

Mulcahy, M. F. R., and Smith, I. W. 1969. *Rev. Pure Appl. Chem.* 19:81.

Nagle, J., and Strickland-Constable, R. F. 1963. *Proceedings, fifth conference on carbon (1962)*, vol. 1, p. 154. New York: Pergamon.

Niksa, S., Mitchell, R. E., Hencken, K. R., and Tichenor, D. A. 1984. *Combust. Flame* 60:183.

Northrop, P. S. 1988. Ph.D. thesis, California Institute of Technology.

Northrop, P. S., Flagan, R. C., and Gavalas, G. R. 1987. *Langmuir* 3:300.

Orr, C., Jr. 1969. *Powder Technol.* 3:117.

Park, C., and Appleton, J. P. 1973. *Combust. Flame* 20:369.

Pershing, D. W., and Wendt, J. O. L. 1976. *Sixteenth symposium (international) on combustion*, p. 389. Pittsburgh, Pa.: The Combustion Institute.

Pettit, D. R., and Peterson, T. W. 1983. *Aerosol Sci. Technol.* 2:351.

Pettit, D. R., and Peterson, T. W. 1984. *Aerosol Sci. Technol.* 3:305.

Pierce, C. 1953. *J. Phys. Chem.* 57:149.

Pui, D. Y. H., and Liu, B. Y. H. 1988. *Phys. Scr.* 37:252.

Raask, E. 1985. *Progr. Energy Combust. Sci.* 11:97.

Radovic, L. R., Walker, P. L., Jr., and Jenkins, R. G. 1983. *Am. Chem. Soc. Div. Fuel Chem.* 28(1):1.

Saffman, P., and Turner, J. 1956. *J. Fluid Mech.* 1:16.

Sahimi, M., and Tsotsis, T. T. 1987. *Chem. Eng. Sci.* 43:113.

Sahu, R. 1988. On the combustion of bituminous coal chars, Ph.D. thesis, California Institute of Technology.

Sahu, R., Flagan, R. C., and Gavalas, G. R. 1989. *Combust. Flame* 77:337.

Sahu, R., Levendis, Y. A., Flagan, R. C., and Gavalas, G. R. 1988*a. Fuel* 67:275.

Sahu, R., Northrop, P. S., Flagan, R. C., and Gavalas, G. R. 1988*b. Combust. Sci. Technol.* 60:215.

Sandmann, C. W., Jr., and Zygourakis, K. 1986. *Chem. Eng. Sci.* 41:733.

Satterfield, C. N. 1970. *Mass transfer in heterogeneous catalysis*, p. 56. Cambridge, Mass.: MIT Press.

Schaefer, D. W. 1987. In *Scattering, deformation, and fracture in polymers*. Materials Research Society Symposia Proceedings, vol. 79, ed. G. D. Wignall, B. Crist, T. P. Russel, and E. L. Thomas, p. 47. Pittsburgh, Pa.: Materials Research Society.

Schaefer, D. W. 1988. *MRS Bull.* 13:22.

Schmitt-Ott, A. 1988. *Aerosol Sci.* 19:553.

Seeker, W. R., Samuelsen, G. S., Heap, M. P., and Trolinger, J. D. 1981. *Eighteenth symposium (international) on combustion*, p. 1213. Pittsburgh, Pa.: The Combustion Institute.

Senior, C. L., and Flagan, R. C. 1984. *Twentieth symposium (international) on combustion*, p. 921. Pittsburgh, Pa.: The Combustion Institute.

Setek, M., Snook, I. K., and Wagenfeld, H. K. 1984. *ACS Symp. Ser.* 264:95.

Simons, G. A., and Finson, M. L. 1979. *Combust. Sci. Technol.* 19:217.

Sircar, S. 1985. *Adsorption Sci. Technol.* 2:23.

Smith, I. W. 1971*a. Combust. Flame* 17:303.

Smith, I. W. 1971*b. Combust. Flame* 17:421.

Smith, I. W. 1978. *Fuel* 57:409.

Smith, I. W. 1982. *Nineteenth symposium (international) on combustion*, p. 1045. Pittsburgh, Pa.: The Combustion Institute.

Smith, I. W., and Tyler, R. J. 1974. *Combust. Sci. Technol.* 9:87.

Smoot, L. D., and Smith, P. J. 1985. *Coal combustion and gasification.* New York: Plenum.

Solomon, P. R., Carangelo, R. M., Best, P. E., Markham, J. R., and Hamblen, D. G. 1986. *Twenty-first symposium (international) on combustion*, p. 437. Pittsburgh, Pa.: The Combustion Institute.

Spitzer, Z., and Ulicky, L. 1976. *Fuel* 55:212.

Spjut, R. E., Bar-Ziv, E., Sarofim, A. F., and Longwell, J. P. 1986. *Rev. Sci. Instrum.* 57:1604.

Strange, J. F., and Walker, P. L., Jr. 1976. *Carbon (Oxford)* 14:345.

Swithenbank, J., Beer, J. M., Taylor, D. S., Abbott, D., and McCreath, G. C. 1976. Paper 76-69 presented at the Fourteenth Aerospace Sciences Meeting, Washington, D.C.

Sy, O., and Calo, J. M. 1983. *Am. Chem. Soc. Div. Fuel Chem.* 28(1):6.

Szekely, G. A., Jr., and Faeth, G. M. 1982. *Nineteenth symposium (international) on combustion*, p. 1077. Pittsburgh, Pa.: The Combustion Institute.

Szekely, G. A., Jr., Turns, S. R., and Faeth, G. M. 1984. *Combust. Flame* 58:31.

Talbot, L., Cheng, R. K., Schefer, R. W., and Willis, D. R. 1980. *J. Fluid Mech.* 101:737.

Taylor, D. D., and Flagan, R. C. 1981. *ACS Symp. Ser.* 167:157.

Tichenor, D. A., and Wang, J. C. F. 1982. *Western states section spring meeting*. Paper WSS/CI 82-26. Pittsburgh, Pa.: The Combustion Institute.

Tichenor, D. A., Mitchell, R. E., Hencken, K. R., and Niksa, S. 1984. *Twentieth symposium (international) on combustion*, p. 1213. Pittsburgh, Pa.: The Combustion Institute.

Timothy, L. D., Sarofim, A. F., and Beer, J. M. 1982. *Nineteenth symposium (international) on combustion*, p. 1123. Pittsburgh, Pa.: The Combustion Institute.

Trolinger, J. D. 1976. *Opt. Eng.* 14:383.

Tseng, H. P., and Edgar, T. F. 1985. *Fuel* 64:373.

Tyler, R.J., Wouterlood, H. J., and Mulcahy, M. F. R. 1976. *Carbon (Oxford)* 14:271.

Walker, P. L., Jr., Foresti, R. J., and Wright, C. C. 1953. *Ind. Eng. Chem.* 45:1703.

Wang, S. C., and Flagan, R. C. 1990. *Aerosol Sci. Technol.* in press.

Wells, W. F., Kramer, S. K., Smoot, L. D., and Blackham, A. U. 1984. *Twentieth symposium (international) on combustion*, p. 1539. Pittsburgh, Pa.: The Combustion Institute,

Wheeler, A. 1946. *Catalyst symposium*. Gibson Island AAAS Conference.

Whitby, K. T., and Clark, W. E. 1966. *Tellus* 18:573.

Witten, T. A., and Sander, L. M. 1981. *Phys. Rev. Lett.* 47:1400.

Wojtowicz, M., Calo, J. M., and Suuberg, E. M. 1985. *AIChE annual meeting*. Paper 30f.

Yan, J., and Zhang, Q. 1986. *Particle Characterization* 3:20.

Young, B. C., and Smith, I. W. 1981. *Eighteenth symposium (international) on combustion*, p. 1249. Pittsburgh, Pa.: The Combustion Institute.

Young, D. M., and Crowell, A. D. 1962. *Physical adsorption of gases*. London: Butterworths.

AUTHOR INDEX

SUBJECT INDEX